Models and Games

This gentle introduction to logic and model theory is based on a systematic use of three important games in logic: the Semantic Game, the Ehrenfeucht–Fraïssé Game, and the Model Existence Game. The third game has not been isolated in the literature before, but it underlies the concepts of Beth tableaux and consistency properties.

Jouko Väänänen shows that these games are closely related and, in turn, govern the three interrelated concepts of logic: truth, elementary equivalence, and proof. All three methods are developed not only for first-order logic, but also for infinitary logic and generalized quantifiers. Along the way, the author also proves completeness theorems for many logics, including the cofinality quantifier logic of Shelah, a fully compact extension of first-order logic.

With over 500 exercises, this book is ideal for graduate courses, covering the basic material as well as more advanced applications.

Jouko Väänänen is a Professor of Mathematics at the University of Helsinki, and a Professor of Mathematical Logic and Foundations of Mathematics at the University of Amsterdam.

T0350078

Models and Games

JOUKO VÄÄNÄNEN
University of Helsinki
University of Amsterdam

CAMBRIDGE
UNIVERSITY PRESS

CAMBRIDGE UNIVERSITY PRESS

Cambridge, New York, Melbourne, Madrid, Cape Town,
Singapore, São Paulo, Delhi, Mexico City

Cambridge University Press
The Edinburgh Building, Cambridge CB2 8RU, UK

Published in the United States of America by Cambridge University Press, New York

www.cambridge.org
Information on this title: www.cambridge.org/9780521518123

First published 2011

A catalogue record for this publication is available from the British Library

ISBN 978-0-521-51812-3 Hardback

To Juliette

Contents

Preface

When I was a beginning mathematics student a friend gave me a set of lecture notes for a course on infinitary logic given by Ronald Jensen. On the first page was the definition of a partial isomorphism: a set of partial mappings between two structures with the back-and-forth property. I became immediately interested and now – 37 years later – I have written a book on this very concept.

This book can be used as a text for a course in model theory with a game- and set-theoretic bent.

I am indebted to the students who have given numerous comments and corrections during the courses I have given on the material of this book both in Amsterdam and in Helsinki. I am also indebted to members of the Helsinki Logic Group, especially Tapani Hyttinen and Juha Oikkonen, for discussions, criticisms, and new ideas over the years on Ehrenfeucht–Fraïssé Games in uncountable structures. I am grateful to Fan Yang for reading and commenting on parts of the manuscript.

I am extremely grateful to my wife Juliette Kennedy for encouraging me to finish this book, for reading and commenting on the manuscript pointing out necessary corrections, and for supporting me in every possible way during the writing process.

The preparation of this book has been supported by grant 40734 of the Academy of Finland and by the EUROCORES LogICCC LINT programme. I am grateful to the Institute for Advanced Study, the Mittag-Leffler Institute, and the Philosophy Department of Princeton University for providing hospitality during the preparation of this book.

1

Introduction

A recurrent theme in this book is the concept of a game. There are essentially three kinds of games in logic. One is the Semantic Game, also called the Evaluation Game, where the *truth* of a given sentence in a given model is at issue. Another is the Model Existence Game, where the *consistency* in the sense of having a model, or equivalently in the sense of impossibility to derive a contradiction, is at issue. Finally there is the Ehrenfeucht–Fraïssé Game, where *separation* of a model from another by finding a property that is true in one given model but false in another is the goal. The three games are closely linked to each other and one can even say they are essentially variants of just one basic game. This basic game arises from our understanding of the quantifiers. The purpose of this book is to make this strategic aspect of logic perfectly transparent and to show that it underlies not only first-order logic but infinitary logic and logic with generalized quantifiers alike.

We call the close link between the three games the *Strategic Balance of Logic* (Figure 1.1). This balance is perfectly commutative, in the sense that winning strategies can be transferred from one game to another. This mere fact is testimony to the close connection between logic and games, or, thinking semantically, between games and models. This connection arises from the nature of quantifiers. Introducing infinite disjunctions and conjunctions does not upset the balance, barring some set-theoretic issues that may surface. In the last chapter of this book we consider generalized quantifiers and show that the Strategic Balance of Logic persists even in the presence of generalized quantifiers.

The purpose of this book is to present the Strategic Balance of Logic in all its glory.

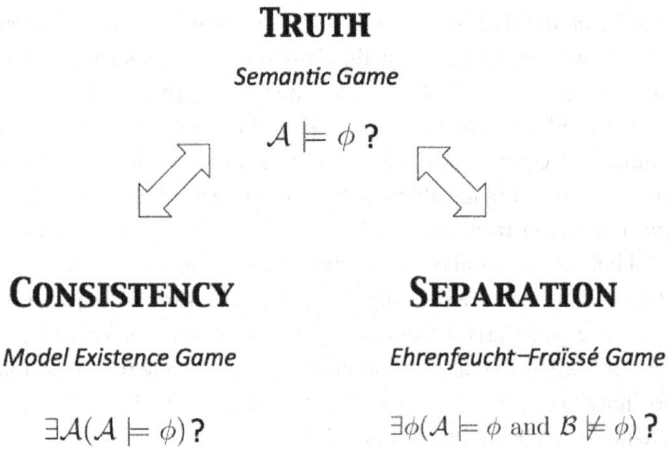

TRUTH

Semantic Game

$$\mathcal{A} \models \phi \, ?$$

CONSISTENCY

Model Existence Game

$$\exists \mathcal{A} (\mathcal{A} \models \phi) \, ?$$

SEPARATION

Ehrenfeucht–Fraïssé Game

$$\exists \phi (\mathcal{A} \models \phi \text{ and } \mathcal{B} \not\models \phi) \, ?$$

Figure 1.1   The Strategic Balance of Logic.

2

Preliminaries and Notation

We use some elementary set theory in this book, mainly basic properties of countable and uncountable sets. We will occasionally use the concept of countable ordinal when we index some uncountable sets. There are many excellent books on elementary set theory. (See Section 2.7.) We give below a simplified account of some basic concepts, the barest outline necessary for this book.

We denote the set $\{0, 1, 2, \ldots\}$ of all natural numbers by $\mathbb{N}$, the set of rational numbers by $\mathbb{Q}$, and the set of all real numbers by $\mathbb{R}$. The power-set operation is written

$$\mathcal{P}(A) = \{B : B \subseteq A\}.$$

We use $A \setminus B$ to denote the set-theoretical difference of the sets A and B. If f is a function, $f''X$ is the set $\{f(x) : x \in X\}$ and $f^{-1}(X)$ is the set $\{x \in \mathrm{dom}(f) : f(x) \in X\}$. *Composition* of two functions f and g is denoted $g \circ f$ and defined by $(g \circ f)(x) = g(f(x))$. We often write fa for $f(a)$. The notation id_A is used for the *identity function* $A \to A$ which maps every element of A to itself, i.e. $id_A(a) = a$ for $a \in A$.

2.1 Finite Sequences

The concept of a finite (ordered) sequence

$$s = (a_0, \ldots, a_{n-1})$$

of elements of a given set A plays an important role in this book. Examples of finite sequences of elements of $\mathbb{N}$ are

$$(8, 3, 9, 67, 200, 0)$$

$$(8, 8, 8)$$

$$(24).$$

We can identify the sequence $s = (a_0, \ldots, a_{n-1})$ with the function

$$s' : \{0, \ldots, n-1\} \to A,$$

where

$$s'(i) = a_i.$$

The main property of finite sequences is: $(a_0, \ldots, a_{n-1}) = (b_0, \ldots, b_{m-1})$ if and only if $n = m$ and $a_i = b_i$ for all $i < n$. The number n is called the *length* of the sequence $s = (a_0, \ldots, a_{n-1})$ and is denoted $\text{len}(s)$. A special case is the case $\text{len}(s) = 0$. Then s is called the empty sequence. There is exactly one empty sequence and it is denoted by $\emptyset$.

The Cartesian product of two sets A and B is written

$$A \times B = \{(a, b) : a \in A, b \in B\}.$$

More generally

$$A_0 \times \ldots \times A_{n-1} = \{(a_0, \ldots, a_{n-1}) : a_i \in A_i \text{ for all } i < n\}.$$

$$A^n = A \times \ldots \times A \ (n \text{ times}).$$

According to this definition, $A^1 \neq A$. The former consists of sequences of length 1 of elements of A. Note that $A^0 = \{\emptyset\}$.

Finite Sets

A set A is *finite* if it is of the form $\{a_0, \ldots, a_{n-1}\}$ for some natural number n. This means that the set A has at most n elements. If A has exactly n elements we write $|A| = n$ and call $|A|$ the cardinality of A. A set which is not finite is *infinite*. Finite sets form a so-called *ideal*, which means that:

1. $\emptyset$ is finite.
2. If A and B are finite, then so is $A \cup B$.
3. If A is finite and $B \subseteq A$, then also B is finite.

Further useful properties of finite sets are:

1. If A and B are finite, then so is $A \times B$.
2. If A is finite, then so is $\mathcal{P}(A)$.

The *Axiom of Choice* says that for every set A of non-empty sets there is a function f such that $f(a) \in a$ for all $a \in A$. We shall use the Axiom of Choice freely without specifically mentioning it. It needs some practice in set theory to see how the axiom is used. Often an intuitively appealing argument involves a hidden use of it.

Lemma 2.1 *A set A is finite if and only if every injective $f : A \to A$ is a bijection.*

Proof Suppose A is finite and $f : A \to B$ is an injection with $B \subset A$ and $a \in A \setminus B$. Let $a_0 = a$ and $a_{n+1} = f(a_n)$. It is easy to see that $a_n \neq a_m$ whenever $n < m$, so we contradict the finiteness of A. On the other hand, if A is infinite, we can (by using the Axiom of Choice) pick a sequence b_n, $n \in \mathbb{N}$, of distinct elements from A. Then the function g which maps each b_n to b_{n+1} and is the identity elsewhere is an injective mapping from A into A but not a bijection. $\qquad\qquad\square$

The set of all n-element subsets $\{a_0, \dots, a_{n-1}\}$ of A is denoted by $[A]^n$.

2.2 Equipollence

Sets A and B are *equipollent*

$$A \sim B$$

if there is a bijection $f : A \to B$. Then $f^{-1} : B \to A$ is a bijection and

$$B \sim A$$

follows. The composition of two bijections is a bijection, whence

$$A \sim B \sim C \implies A \sim C.$$

Thus $\sim$ divides sets into equivalence classes. Each equivalence class has a canonical representative (a cardinal number, see the Subsection "Cardinals" below) which is called the *cardinality* of (each of) the sets in the class. The cardinality of A is denoted by $|A|$ and accordingly $A \sim B$ is often written

$$|A| = |B|.$$

One of the basic properties of equipollence is that if

$$A \sim C, B \sim D \text{ and } A \cap B = C \cap D = \emptyset,$$

then

$$A \cup B \sim C \cup D.$$

Indeed, if $f : A \to C$ is a bijection and $g : B \to D$ is a bijection, then $f \cup g : A \cup B \to C \cup D$ is a bijection. If the assumption

$$A \cap B = C \cap D = \emptyset$$

is dropped, the conclusion fails, of course, as we can have $A \cap B = \emptyset$ and $C = D$. It is also interesting to note that even if $A \cap B = C \cap D = \emptyset$, the assumption $A \cup B \sim C \cup D$ does not imply $B \sim D$ even if $A \sim C$ is assumed: Let $A = \mathbb{N}, B = \emptyset, C = \{2n : n \in \mathbb{N}\}$, and $D = \{2n + 1 : n \in \mathbb{N}\}$. However, for finite sets this holds: if $A \cup B$ is finite,

$$A \cup B \sim C \cup D, \ A \sim C, \ A \cap B = C \cap D = \emptyset$$

then

$$B \sim D.$$

We can interpret this as follows: the cancellation law holds for finite numbers but does not hold for cardinal numbers of infinite sets.

There are many interesting and non-trivial properties of equipollence that we cannot enter into here. For example the Schröder–Bernstein Theorem: If $A \sim B$ and $B \subseteq C \subseteq A$, then $A \sim C$. Here are some interesting consequences of the Axiom of Choice:

- For all A and B there is C such that $A \sim C \subseteq B$ or $B \sim C \subseteq A$.
- For all infinite A we have $A \sim A \times A$.

It is proved in set theory by means of the Axiom of Choice that $|A| \leq |B|$ holds in the above sense if and only if the cardinality $|A|$ of the set A is at most the cardinality $|B|$ of the set B. Thus the notation $|A| \leq |B|$ is very appropriate.

2.3 Countable sets

A set A which is empty or of the form $\{a_0, a_1, \ldots\}$, i.e. $\{a_n : n \in \mathbb{N}\}$, is called *countable*. A set which is not countable is called *uncountable*. The countable sets form an ideal just as the finite sets do. We now prove two important results about countability. Both are due to Georg Cantor:

Theorem 2.2 *If A and B are countable, then so is $A \times B$.*

Proof If either set is empty, the Cartesian product is empty. So let us assume the sets are both non-empty. Suppose $A = \{a_0, a_1, \ldots\}$ and $B = \{b_0, b_1, \ldots\}$. Let

$$c_n = \begin{cases} (a_i, b_j), & \text{if } n = 2^i 3^j \\ (a_0, b_0), & \text{otherwise.} \end{cases}$$

Now $A \times B = \{c_n : n \in \mathbb{N}\}$, whence $A \times B$ is countable. $\qquad \square$

Theorem 2.3 *The union of a countable family of countable sets is countable.*

Proof The empty sets do not contribute anything to the union, so let us assume all the sets are non-empty. Suppose A_n is countable for each $n \in \mathbb{N}$, say, $A_n = \{a_m^n : m \in \mathbb{N}\}$ (we use here the Axiom of Choice to choose an enumeration for each A_n). Let $B = \bigcup_n A_n$. We want to represent B in the form $\{b_n : n \in \mathbb{N}\}$. If n is given, we consider two cases: If n is $2^i 3^j$ for some i and j, we let $b_n = a_j^i$. Otherwise we let $b_n = a_0^0$. $\qquad \square$

Theorem 2.4 *The power-set of an infinite set is uncountable.*

Proof Suppose A is infinite and $\mathcal{P}(A) = \{b_n : n \in \mathbb{N}\}$. Since A is infinite, we can choose distinct elements $\{a_n : n \in \mathbb{N}\}$ from A. (This uses the Axiom of Choice. For an argument which avoids the Axiom of Choice see Exercise 2.14.) Let

$$B = \{a_n : a_n \notin b_n\}.$$

Since $B \subseteq A$, there is some n such that $B = b_n$. Is a_n an element of B or not? If it is, then $a_n \notin b_n$ which is a contradiction. So it is not. But then $a_n \in b_n = B$, again a contradiction. $\qquad \square$

2.4 Ordinals

The ordinal numbers introduced by Cantor are a marvelous general theory of measuring the *potentially infinite*. They are intimately related to inductive definitions and occur therefore widely in logic. It is easiest to understand ordinals in the context of games, although this was not Cantor's way. Suppose we have a game with two players **I** and **II**. It does not matter what the game is, but it could be something like chess. If **II** can force a win in n moves we say that the game has *rank* n. Suppose then **II** cannot force a win in n moves for any n, but after she has seen the first move of **I**, she can fix a number n and say that she can force a win in n moves. This situation is clearly different from being able to say in advance what n is. So we invent a symbol ω for the rank of this game. In a clear sense ω is greater than each n but there does not seem to be any possible rank between all the finite numbers n and ω. We can think of ω as an infinite number. However, there is nothing metaphysical about the infiniteness of ω. It just has infinitely many predecessors. We can think of ω as a tree T_ω with a root and a separate branch of length n for each n above the root as in the tree on the left in Figure 2.1.

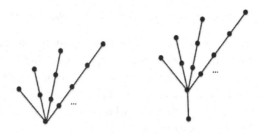

Figure 2.1 T_ω and $T_{\omega+1}$.

Suppose then **II** is not able to declare after the first move how many moves she needs to beat **II**, but she knows how to play her first move in such a way that after **I** has played his second move, she can declare that she can win in n moves. We say that the game has rank $\omega + 1$ and agree that this is greater than ω but there is no rank between them. We can think of $\omega + 1$ as the tree which has a root and then above the root the tree T_ω, as in the tree on the right in Figure 2.1. We can go on like this and define the ranks $\omega + n$ for all n.

Suppose now the rank of the game is not any of the above ranks $\omega + n$, but still **II** can make an interesting declaration: she says that after the first move of **I** she can declare a number m so that after m moves she declares another number n and then in n moves she can force a win. We would say that the rank of the game is $\omega + \omega$. We can continue in this way defining ranks of games that are always finite but potentially infinite. These ranks are what set theorists call ordinals.

We do not give an exact definition of the concept of an ordinal, because it would take us too far afield and there are excellent textbooks on the topic. Let us just note that the key properties of ordinals and their total order $<$ are:

1. Natural numbers are ordinals.
2. For every ordinal α there is an immediate successor $\alpha + 1$.
3. Every non-empty set of ordinals has a smallest element.
4. Every non-empty set of ordinals has a supremum (i.e. a smallest upper bound).

The supremum of the set $\{0, 1, 2, 3, \ldots\}$ of ordinals is denoted by ω. An ordinal is said to be *countable* if it has only countably many predecessors, otherwise *uncountable*. The supremum of all countable ordinals is denoted by ω_1. Here is a picture of the ordinal number "line":

$$0 < 1 < 2 < \ldots < \omega < \omega + 1 < \ldots < \alpha < \alpha + 1 < \ldots < \omega_1 < \ldots$$

Ordinals that have a last element, i.e. are of the form $\alpha + 1$, are called *successor* ordinals; the rest are *limit* ordinals, like ω and $\omega + \omega$.

Ordinals are often used to index elements of uncountable sets. For example, $\{a_\alpha : \alpha < \beta\}$ denotes a set whose elements have been indexed by the ordinal β, called the *length* of the sequence. The set of all such sequences of length β of elements of a given set A is denoted by A^β. The set of all sequences of length $< \beta$ of elements of a given set A is denoted by $A^{<\beta}$.

2.5 Cardinals

Historically cardinals (or more exactly cardinal numbers) are just representatives of equivalence classes of equipollence. Thus there is a cardinal number for countable sets, denoted $\aleph_0$, a cardinal number for the set of all reals, denoted $\mathfrak{c}$, and so on. There is some question as to what exactly are these cardinal numbers. The Axiom of Choice offers an easy answer, which is the prevailing one, as it says that every set can be well-ordered. Then we can let the cardinal number of a set be the order-type of the smallest well-order equipollent with the set. Equivalently, the cardinal number of a set is the smallest ordinal equipollent with the set. If we leave aside the Axiom of Choice, some sets need not have have a cardinal number. However, as is customary in current set theory, let us indeed assume the Axiom of Choice. Then every set has a cardinal number and the cardinal numbers are ordinals, hence well-ordered. The αth infinite cardinal number is denoted $\aleph_\alpha$. Thus $\aleph_1$ is the next in order of magnitude from $\aleph_0$. The famous *Continuum Hypothesis* is the statement that $\aleph_1 = \mathfrak{c}$.

For every set A there exists (by the Axiom of Choice) an ordinal α such that the elements of A can be listed as $\{a_\beta : \beta < \alpha\}$. The smallest such α is called the *cardinal number*, or *cardinality*, of A and denoted by $|A|$. Thus certain ordinals are cardinal numbers of sets. Such ordinals are called *cardinals*. They are considered as canonical representatives of each equivalence class of equipollent sets. For example, all finite numbers are cardinals, as are ω and ω_1. The smallest cardinal such that the smaller infinite cardinals can be enumerated in increasing order as κ_β, $\beta < \alpha$, is denoted ω_α, or alternatively $\aleph_\alpha$. If

$\kappa = \aleph_\alpha$, then $\aleph_{\alpha+1}$ is denoted κ^+ and is called a *successor cardinal*. Cardinals that are not successor cardinals are called *limit cardinals*.

Arithmetic operations $\kappa + \lambda$, $\kappa \cdot \lambda$, κ^λ for cardinals are defined as follows:

$$\kappa + \lambda = |\kappa \cup \lambda|, \quad \kappa \cdot \lambda = |\kappa \times \lambda|.$$

Moreover, exponentiation κ^λ of cardinal numbers is defined as the cardinality of the set κ^λ of sequences of elements of κ of length λ. A certain amount of knowledge about the arithmetic of cardinal numbers in necessary in this book, especially in the later chapters, and Chapters 8 and 9 in particular.

The *cofinality* of an ordinal α is the smallest ordinal β for which there is a function $f : \beta \to \alpha$ such that (1) $\xi < \zeta < \beta$ implies $f(\xi) < f(\zeta)$, and (2) for all $\xi < \alpha$ there is some $\zeta < \beta$ such that $\xi < f(\zeta)$. We use $\mathrm{cf}(\alpha)$ to denote the cofinality of α. A cardinal κ is said to be *regular* if $\mathrm{cf}(\kappa) = \kappa$, and *singular* if $\mathrm{cf}(\kappa) < \kappa$. Successor cardinals are always regular. The smallest singular cardinal is $\aleph_\omega$.

The *Continuum Hypothesis* (CH) is the hypothesis $|\mathcal{P}(\mathbb{N})| = \aleph_1$. Neither it nor its negation can be derived from the usual *Zermelo–Fraenkel axioms* of set theory and therefore it (or its negation), like many other similar hypotheses, has to be explicitly mentioned as an assumption, when it is used.

2.6 Axiom of Choice

We have already mentioned the Axiom of Choice. There are so many equivalent formulations of this axiom that books have been written about it. The most notable formulation is the Well-Ordering Principle: every set is equipollent with an ordinal. The Axiom of Choice is sometimes debated because it brings arbitrariness or abstractness into mathematics, often with examples that can be justifiably called pathological, like the Banach–Tarski Paradox: The unit sphere in three-dimensional space can be split into five pieces so that if the pieces are rigidly moved and rotated they form two spheres, each of the original size. The trick is that the splitting exists only in the abstract world of mathematics and can never actually materialize in the physical world. Conclusion: infinite abstract objects do not obey the rules we are used to among finite concrete objects. This is like the situation with sub-atomic elementary particles, where counter-intuitive phenomena, such as entanglement, occur.

Because of the abstractness brought about by the Axiom of Choice it has received criticism and some authors always mention explicitly if they use it in their work. The main problem in working *without* the Axiom of Choice is that there is no clear alternative and just leaving it out leaves many areas of mathematics, like measure theory, without proper foundation. In this book we assume the Axiom of Choice. This seems essential at least in Chapters 8 and 9 where uncountable models are considered. On the other hand, the game-theoretic approach to logic that we have adopted in this book heavily rests on the Axiom of Choice. We think of the truth of $\forall x \exists y R(x, y)$ as the existence of a function f which for any given a in the domain of discourse picks an element $b = f(a)$ such that $R(a, b)$. We go on to call such a function a *winning strategy* of player **II** in the appropriate game, where player **I** first picks a and then player **II** picks b and **II** wins if $R(a, b)$ holds. If we allow completely arbitrary binary predicates as our R there is no way to derive the existence of f without the Axiom of Choice. Indeed, the existence of f is one equivalent formulation of the Axiom of Choice.

2.7 Historical Remarks and References

The distinction between countable and uncountable sets is due to Georg Cantor. There are many elementary books providing an introduction to set theory, for example Devlin (1993), Enderton (1977), and Rotman and Kneebone (1966). A textbook covering a wide spectrum of modern set theory is Jech (1997).

Exercises

2.1 List the elements of $\{0, 1, 2\} \times \{0, 1\}$.

2.2 Show that the Cartesian product of two finite sets is finite.

2.3 Show that the power set of a finite set is finite.

2.4 Show that the following sets are ideals:

$$\{A \subseteq \mathbb{N} : 4 \notin A\}$$
$$\{A \subseteq \mathbb{N} : 2^n \notin A \text{ for all sufficiently large } n\}.$$

2.5 Show that if $A \sim B$ and $C \sim D$, then $A \times C \sim B \times D \sim C \times A$.

2.6 Show that $A \sim B$ implies $\mathcal{P}(A) \sim \mathcal{P}(B)$.

2.7 Assume $A \sim B \cup D$. Find $C \subseteq A$ such that $C \sim B$.

2.8 Assume $A \sim B$ and $C \sim D$. Does $A \setminus C \sim B \setminus D$ follow?

2.9 Assume $A \times B \sim C$. Show that there is $D \subseteq C$ such that $A \sim D$ or $B \sim D$.

2.10 Assume $\mathcal{P}(A) \sim C$. Show that there is $D \subseteq C$ such that $A \sim D$.

2.11 Let $^A B$ be the set of all functions $f : A \to B$. Show that if $A \sim B$ and $C \sim D$, then $^A C \sim {}^B D$.

2.12 Show that $^A(B \times C) \sim {}^A B \times {}^A C$.

2.13 Show that if $A \cap B = \emptyset$, then $^{(A \cup B)}C \sim {}^A C \times {}^B C$. What if $A \cap B \neq \emptyset$?

2.14 Prove $A \not\sim \mathcal{P}(A)$ for all sets A. (Hint: Assume $f : A \to \mathcal{P}(A)$ is onto and consider $a = \{x \in A : x \notin f(x)\}$.)

2.15 Show that if $\{1, \ldots, n\} \sim \{1, \ldots, m\}$ then $n = m$.

2.16 Show that for any A and B there is an injection $f : A \to B$ or an injection $f : B \to A$. (Hint: You have to invoke Zorn's Lemma.[1] Let C be the set of partial injections $A \to B$. Clearly C is closed under unions of $\subseteq$-chains. By Zorn's Lemma C has a maximal element.)

2.17 Show that a subset of a countable set is countable.

2.18 Show that the set of finite sequences of elements of a countable set is countable.

2.19 Show that the set of polynomials with rational coefficients is countable.

2.20 Show that above every ordinal there is a limit ordinal.

2.21 Show that the equations

$$\begin{cases} \alpha + 0 = \alpha \\ \alpha + (\beta + 1) = (\alpha + \beta) + 1 \\ \alpha + \beta = \sup_{\gamma < \beta}(\alpha + \gamma) \text{ for limit } \beta \end{cases}$$

define uniquely a binary function $\alpha + \beta$, called *ordinal addition*. This is an example of a definition by the so called *transfinite recursion*. Show that $(\alpha + \beta) + \gamma = \alpha + (\beta + \gamma)$.

2.22 Show that the equations

$$\begin{cases} \alpha \cdot 0 = 0 \\ \alpha \cdot (\beta + 1) = \alpha \cdot \beta + \alpha \\ \alpha \cdot \beta = \sup_{\gamma < \beta}(\alpha \cdot \gamma) \text{ for limit } \beta \end{cases}$$

define uniquely a binary function $\alpha \cdot \beta$, called *ordinal multiplication*, and also denoted $\alpha\beta$. Show that $(\alpha\beta)\gamma = \alpha(\beta\gamma)$ and $\alpha(\beta + \gamma) = \alpha\beta + \alpha\gamma$.

2.23 Show that every ordinal has a unique representation as $\alpha + n$, where α is a limit ordinal and $n \in \mathbb{N}$.

2.24 Show that for any ordinals $\alpha < \beta$ there is a unique γ such that $\alpha + \gamma = \beta$.

2.25 Show that for any ordinals $\alpha < \beta$ there are unique γ and δ such that $\beta = \alpha\gamma + \delta$ and $\delta < \alpha$.

---

[1] Zorn's Lemma, a consequence of the Axiom of Choice, says that if a partial order has the property that every chain has an upper bound, then the partial order has a maximal element.

2.26 Show that the equations

$$\begin{cases} \alpha^0 = 1 \\ \alpha^{\beta+1} = \alpha^\beta \alpha \\ \alpha^\beta = \sup_{\gamma < \beta}(\alpha^\gamma) \text{ for limit } \beta \end{cases}$$

define uniquely a binary function α^β, called *ordinal exponentiation*. Show that $\alpha^{\beta+\gamma} = \alpha^\beta \alpha^\gamma$ and $(\alpha^\beta)^\gamma = \alpha^{\beta\gamma}$.

2.27 Show that every ordinal can be uniquely expressed in the form $\omega^{\alpha_0} \cdot n_0 + \omega^{\alpha_1} \cdot n_1 + \cdots + \omega^{\alpha_{k-1}} \cdot n_{k-1} + n_k$, where $\alpha_0 > \alpha_1 > \cdots > \alpha_{k-1}$ are ordinals and $n_0, \ldots, n_k > 0, n_k \geq 0$ are natural numbers. (This is called the Cantor Normal Form).

2.28 Show that the supremum of a countable non-empty set of countable ordinal numbers is countable. Show that ω_1 is an uncountable ordinal number.

2.29 Show that a set A is uncountable if and only if there is an injection $f : \omega_1 \to A$.

2.30 Suppose A_α, $\alpha < \omega_1$, are sets such that $\alpha < \beta$ implies $A_\alpha \subsetneq A_\beta$. Show that the set $\bigcup_{\alpha < \omega_1} A_\alpha$ is uncountable.

2.31 Show that a set has cardinality $\leq \aleph_1$ if and only if it is the union of an increasing sequence of countable sets.

2.32 Show that if κ is an infinite cardinal then $\kappa \times \kappa \sim \kappa$. (Hint: Suppose κ is the smallest infinite κ for which $\kappa \times \kappa \not\sim \kappa$. By Theorem 2.2, $\kappa > \omega$. Now construct in a canonical way a well-order of $\kappa \times \kappa$. Use the assumption $\forall \lambda < \kappa (\lambda \times \lambda \sim \lambda)$ to show that all initial segments of your order have cardinality $< \kappa$.)

2.33 Suppose κ is an infinite cardinal. Suppose also that $A = \bigcup_{i \in I} A_i$, where $|I| \leq \kappa$, and $|A_i| \leq \kappa$ for all $i \in I$. Show that $|A| \leq \kappa$.

2.34 Show that $\aleph_{\alpha+1}$ is a regular cardinal for each ordinal α.

3

Games

3.1 Introduction

In this first part we march through the mathematical details of zero-sum two-person games of perfect information in order to be well prepared for the introduction of the three games of the Strategic Balance of Logic (see Figure 1.1) in the subsequent parts of the book. Games are useful as intuitive guides in proofs and constructions but it is also important to know how to make the intuitive arguments and concepts mathematically exact.

3.2 Two-Person Games of Perfect Information

Two-person games of perfect information are like chess: two players set their wits against each other with no role for chance. One wins and the other loses. Everything is out in the open, and the winner wins simply by having a better strategy than the loser.

A Preliminary Example: Nim

In the game of Nim, if it is simplified to the extreme, there are two players **I** and **II** and a pile of six identical tokens. During each round of the game player **I** first removes one or two tokens from the top of the pile and then player **II** does the same, if any tokens are left. Obviously there can be at most three rounds. The player who removes the last token wins and the other one loses.

The game of Figure 3.1 is an example of a zero-sum two-person game of perfect information. It is zero-sum because the victory of one player is the loss of the other. It is of perfect information because both players know what the other player has played. A moment's reflection reveals that player **II** has a way

Figure 3.1 The game of Nim.

Play	Winner
111111	**II**
11112	**I**
11121	**I**
11211	**I**
1122	**II**
12111	**I**
1212	**II**
1221	**II**
21111	**I**
2112	**II**
2121	**II**
2211	**II**
222	**I**

Figure 3.2 Plays of Nim.

of playing which guarantees that she[1] wins: During the first round she takes away one token if player **I** takes away two tokens, and two tokens if player **I** takes away one token. Then we are left with three tokens. During the second round she does the same: she takes away the last token if player **I** takes away two tokens, and the last two tokens if player **I** takes away one token. We say that player **II** has a winning strategy in this game.

If we denote the move of a player by a symbol – 1 or 2 – we can form a list of all sequences of ones and twos that represent a play of the game. (See Figure 3.2.)

The set of finite sequences displayed in Figure 3.2 has the structure of a tree, as Figure 3.3 demonstrates. The tree reveals easily the winning strategy of player **II**. Whatever player **I** plays during the first round, player **II** has an option which leaves her in such a position (node 12 or 21 in the tree) that whether the opponent continues with 1 or 2, she has a winning move (1212, 1221, 2112, or 2121).

We can express the existence of a winning strategy for player **II** in the above game by means of first-order logic as follows: Let us consider a vo-

---

[1] We adopt the practice of referring to the first player by "he" and the second player by "she".

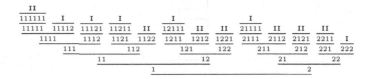

Figure 3.3

cabulary $L = \{W\}$, where W is a four-place predicate symbol. Let $\mathcal{M}$ be an L-structure[2] with $M = \{1, 2\}$ and

$$W^{\mathcal{M}} = \{(a_0, b_0, a_1, b_1) \in M^4 : a_0 + b_0 + a_1 + b_1 = 6\}.$$

Now we have just proved

$$\mathcal{M} \models \forall x_0 \exists y_0 \forall x_1 \exists y_1 W(x_0, y_0, x_1, y_1). \tag{3.1}$$

Conversely, if $\mathcal{M}$ is an arbitrary L-structure, condition (3.1) defines *some* game, maybe not a very interesting one but a game nonetheless: Player **I** picks an element $a_0 \in M$, then player **II** picks an element $b_0 \in M$. Then the same is repeated: player **I** picks an element $a_1 \in M$, then player **II** picks an element $b_1 \in M$. After this player **II** is declared the winner if $(a_0, b_0, a_1, b_1) \in W^{\mathcal{M}}$, and otherwise player **I** is the winner. By varying the structure $\mathcal{M}$ we can model in this way various two-person two-round games of perfect information. This gives a first hint of the connection between games and logic.

Games – a more general formulation

Above we saw an example of a two-person game of perfect information. This concept is fundamental in this book. In general, the simplest formulation of such a game is as follows (see Figure 3.4): There are two players[3] **I** and **II**, a domain A, and a natural number n representing the length of the game. Player **I** starts the game by choosing some element $x_0 \in A$. Then player **II** chooses $y_0 \in A$. After x_i and y_i have been played, and $i + 1 < n$, player **I** chooses $x_{i+1} \in A$ and then player **II** chooses $y_{i+1} \in A$. After n rounds the game ends. To decide who wins we fix beforehand a set $W \subseteq A^{2n}$ of sequences

$$(x_0, y_0, \ldots, x_{n-1}, y_{n-1}) \tag{3.2}$$

---

[2] For the definition of an L-structure see Definition 5.1.

[3] There are various names in the literature for player **I** and **II**, such as player **I** and player **II**, spoiler and duplicator, Nature and myself, or Abelard and Eloise.

I	II
x_0	
	y_0
x_1	
	y_1
$\vdots$	
	$\vdots$
x_{n-1}	
	y_{n-1}

Figure 3.4 A game.

and declare that player **II** wins the game if the sequence formed during the game is in W; otherwise player **I** wins. We denote this game by $\mathcal{G}_n(A, W)$. For example, if $W = \emptyset$, player **II** cannot possibly win, and if $W = A^{2n}$, player **I** cannot possibly win. If W is a set of sequences $(x_0, y_0, \ldots, x_{n-1}, y_{n-1})$ where $x_0 = x_1$ and if moreover A has at least two elements, then **II** could not possibly win, as she cannot prevent player **I** from playing x_0 and x_1 differently. On the other hand, W could be the set of all sequences (3.2) such that $y_0 = y_1$. Then **II** can always win because all she has to do during the game is make sure that she chooses y_0 and y_1 to be the same element.

If player **II** has a way of playing that guarantees a sure win, i.e. the opponent **I** loses whatever moves he makes, we say that player **II** has a winning strategy in the game. Likewise, if player **I** has a way of playing that guarantees a sure win, i.e. player **II** loses whatever moves she makes, we say that player **I** has a winning strategy in the game. To make intuitive concepts, such as "way of playing" more exact in the next chapter we define the basic concepts of game theory in a purely mathematical way.

Example 3.1 The game of Nim presented in the previous chapter is in the present notation $\mathcal{G}_3(\{1, 2\}, W)$, where

$$W = \left\{ (a_0, b_0, a_1, b_1, a_2, b_2) \in \{1, 2\}^6 : \sum_{i=0}^{n} (a_i + b_i) = 6 \text{ for some } n \leq 2 \right\}.$$

We allow three rounds as theoretically the players could play three rounds even if player **II** can force a win in two rounds.

Example 3.2 Consider the following game on a set A of integers:

I	II
a	
	b

where **II** is declared the winner if $a+b \in A$. So here $A \subseteq \mathbb{Z}$ and $W = \{(a,b) : a+b \in A\}$. The game proceeds as follows: Player **I** starts by choosing an element $a \in A$. Then player **II** chooses an element $b \in A$. The game ends here and we check who won. If $a + b \in A$, player **II** won, otherwise player **I** won. Does player **II** have a winning strategy? Clearly this depends on whether every element of A is the difference of two elements of A, and on nothing else. In logical formalism, let $L = \{W\}$, where W is a two-place predicate symbol. Consider an L-structure M, where $|M| = A \subseteq \mathbb{Z}$ and

$$W^M = \{(a_0, b_0) \in A^2 : a_0 + b_0 \in A\}.$$

Now player **II** has a winning strategy in the game described if and only if

$$M \models \forall x_0 \exists y_0 W(x_0, y_0).$$

If A is the set of all integers, or the set of all even integers, or any set containing 0, then player **II** has a winning strategy. If A is a singleton $\neq \{0\}$, or the set of integers between 100 and 200, then player **I** has a winning strategy. If player **II** does not have a winning strategy, then there is $a \in A$ such that $a+b \notin A$ for all $b \in A$. Now player **I** has a winning strategy: he plays a. Whatever element b player **II** chooses, she loses the game since $a+b \notin A$. Thus one of the players has a winning strategy, whatever A is.

Example 3.3 Suppose f is a mapping $f : \mathbb{R} \to \mathbb{R}$ and a and b are reals. Let $A = \mathbb{R}$ and W the set of (ϵ, δ, x, y), where:

If $\epsilon > 0$ then $\delta > 0$ and if moreover $0 < |x - a| < \delta$ then $|f(x) - b| < \epsilon$.

Consider the game

I	II
ϵ	
	δ
x	
	y

where **II** is declared the winner if $(\epsilon, \delta, x, y) \in W$. Note that the second move of player **II** is a "dummy" move, i.e. a move which has no relevance. Its only role is to complete the second round. Clearly, **II** has a winning strategy if and only if $\lim_{x \to a} f(x) = b$. Using logical formalism we could write this as follows: Let $L = \{<, F, A, c, d\}$, where $<$ is a binary predicate symbol, F is a unary function symbol, A is a binary function symbol, and c and d are constant symbols. Consider the L-structure $\mathcal{M}$, where $M = \mathbb{R}$, $a <^{\mathcal{M}} b$ iff $a < b$, $F^{\mathcal{M}}(x) = f(x)$, $A^{\mathcal{M}}(x, y) = |x - y|$, $c^{\mathcal{M}} = a$ and $d^{\mathcal{M}} = b$. Then player **II** has a winning strategy in the above game if and only if $\mathcal{M}$ satisfies

$$\forall x_0(0 < x_0 \to \exists y_0(0 < y_0 \wedge \forall x_1(A(x_1, c) < y_0 \to A(F(x_1), d) < x_0))).$$

Example 3.4 Let A be the set of cities of Europe. Let G be a collection of German cities and F a collection of French cities. Consider the following game on airline connections from cities in $G \cup F$ to Scandinavia: During the first round of the game player **I** chooses a city $f \in F$ in France and then player **II** chooses a city $g \in G$ in Germany. During the second round of the game player **I** chooses a city s_0 in Scandinavia. Finally player **II** chooses a city s_1 also in Scandinavia. The game ends and player **II** wins if she was able to choose city s_1 in such a way that it is in the same country as s_0 and if there is a direct commercial flight from city f to city s_0, then there is also a direct commercial flight from city g to city s_1. Thus this game is $\mathcal{G}_2(A, W)$

I	II
f	
	g
s_0	
	s_1

where W is the set of sequences (f, g, s_0, s_1) such that

1. $f \notin F$ or
2. $g \in G$, s_0 and s_1 are in the same country, and if there is a commercial airline connection from f to s_0 then there is also a commercial airline connection from g to s_1.

To find out under what conditions player **II** has a winning strategy, let us define an auxiliary concept. The *type* $\tau(c)$ of a city c is the set of Scandinavian countries to which there is a commercial airline connection from c. It is clear that player **II** has to choose city g in such a way that $\tau(f) \subseteq \tau(g)$. Let the *type* $\tau(C)$ of a set C of cities be the set of types of its elements. Player **II** has a winning strategy if and only if every element of $\tau(F)$ is contained in an element of $\tau(G)$. This is trivially satisfied if $\tau(F)$ consists of merely the set $\emptyset$, or if $\tau(G)$ contains the set of all Scandinavian countries as an element.

The simplest game is so simple it is hard to even recognize as a game:

Example 3.5 The following game has no moves:

$$\text{I} \quad \text{II}$$

If $W = \{\emptyset\}$, player **II** is the winner. If $W = \emptyset$, player **I** is the winner. So this is a game with 0 rounds. In practice one of the players would find these games unfair as he or she loses without even having a chance to make a move. It is like being invited to play a game of chess starting in a position where you are already in check-mate.

3.3 The Mathematical Concept of Game

Let A be an arbitrary set and n a natural number. Let $W \subseteq A^{2n}$. We redefine the game

$$\mathcal{G}_n(A, W)$$

in a purely mathematical way. Let us fix two players **I** and **II**. A *play* of one of the players is any sequence $\bar{x} = (x_0, \ldots, x_{n-1})$ of elements of A. A sequence

$$(\bar{x}; \bar{y}) = (x_0, y_0, \ldots, x_{n-1}, y_{n-1}),$$

of elements of A is called a *play* (of $\mathcal{G}_n(A, W)$). So we have defined the concept of play without any reference to playing the game as an act. The play $(\bar{x}; \bar{y})$ is a *win for* player **II** if

$$(x_0, y_0, \ldots, x_{n-1}, y_{n-1}) \in W$$

and otherwise a *win for* player **I**.

Example 3.6 Let us consider the game of chess in this mathematical framework. We modify the game so that the number of rounds is for simplicity exactly n and Black wins a draw, i.e. if neither player has check-mated the other player during those up to n rounds. If a check-mate is reached the rest of the n-round game is of course irrelevant and we can think that the game is finished with "dummy" moves. Let A be the set of all possible positions, i.e. configurations of the pieces on the board. A play $\bar{x}$ of **I** (White) is the sequence of positions where White has just moved. A play $\bar{y}$ of **II** is the sequence of positions where Black has just moved. We let W be the set of plays $(\bar{x}; \bar{y})$, where either White has not obeyed the rules, or Black has obeyed the rules and White has not check-mated Black. With the said modifications, chess is just the game $\mathcal{G}_n(A, W)$ with White playing as player **I** and Black playing as player **II**.

A *strategy* of player **I** in the game $\mathcal{G}_n(A, W)$ is a sequence

$$\sigma = (\sigma_0, \ldots, \sigma_{n-1})$$

of functions $\sigma_i : A^i \to A$. We say that player **I** has *used the strategy σ in the play* $(\bar{x}; \bar{y})$ if for all $0 < i < n$:

$$x_i = \sigma_i(y_0, \ldots, y_{i-1})$$

and

$$x_0 = \sigma_0.$$

The strategy σ of player **I** is a *winning strategy*, if every play where **I** has used σ is a win for player **I**. Note that the strategy depends only on the opponent's moves. It is tacitly assumed that when the function σ_{i+1} is used to determine x_{i+1}, the previous functions $\sigma_0, \ldots, \sigma_i$ were used to determine the previous moves $x_0, \ldots, x_n$. Thus a strategy σ is a winning strategy because of the concerted effect of all the functions $\sigma_0, \ldots, \sigma_{n-1}$.

A *strategy* of player **II** in the game $\mathcal{G}_n(A, W)$ is a sequence

$$\tau = (\tau_0, \ldots, \tau_{n-1})$$

of functions $\tau_i : A^{i+1} \to A$. We say that player **II** has *used the strategy τ in the play* $(\bar{x}; \bar{y})$ if for all $i < n$:

$$y_i = \tau_i(x_0, \ldots, x_i).$$

The strategy τ of player **II** is a *winning strategy*, if every play where player **II** has used τ is a win for player **II**. A player who has a winning strategy in $\mathcal{G}_n(A, W)$ is said to *win the game $\mathcal{G}_n(A, W)$*.

3.4 Game Positions

A *position* of the game $\mathcal{G}_n(A, W)$ is any initial segment

$$p = (x_0, y_0, \ldots, x_{i-1}, y_{i-1})$$

of a play $(\bar{x}; \bar{y})$, where $i \leq n$. Positions have a natural ordering: a position p' extends a position p, if p is an initial segment of p'. Of course, this extension-relation is a partial ordering[4] of the set of all positions, that is, if p' extends p and p'' extends p', then p'' extends p, and if p and p' extend each other, then $p = p'$. The empty sequence $\emptyset$ is the smallest element, and the plays $(\bar{x}; \bar{y})$ are

---

[4] See Example 5.7 for the definition of partial order. Indeed this is a tree-ordering. See Example 5.8 for the definition of tree-ordering.

maximal elements of this partial ordering. A common problem of games is that
the set of all positions is huge.

A *strategy of player* **I** *in position* $p = (x_0, y_0, \ldots, x_{i-1}, y_{i-1})$ in the game
$\mathcal{G}_n(A, W)$ is a sequence

$$\sigma = (\sigma_0, \ldots, \sigma_{n-1-i})$$

of functions $\sigma_j : A^j \to A$. We say that player **I** has *used strategy* σ *after
position* p *in the play* $(\bar{x}; \bar{y})$, if $(\bar{x}; \bar{y})$ extends p and for all j with $i < j < n$
we have

$$x_j = \sigma_{j-i}(y_i, \ldots, y_{j-1})$$

and

$$x_i = \sigma_0.$$

The strategy σ of player **I** in position p is a *winning strategy in position* p, if
every play extending p where player **I** has used σ after position p is a win for
player **I**.

A *strategy of player* **II** *in position* p in the game $\mathcal{G}_n(A, W)$ is a sequence

$$\tau = (\tau_0, \ldots, \tau_{n-1-i})$$

of functions $\tau_j : A^{j+1} \to A$. We say that player **II** has *used strategy* τ *after
position* p *in the play* $(\bar{x}; \bar{y})$ if $(\bar{x}; \bar{y})$ extends p and for all j with $i \leq j < n$ we
have

$$y_j = \tau_{j-i}(x_i, \ldots, x_j).$$

The strategy τ of player **II** in position p is a *winning strategy in position* p, if
every play extending p where player **II** has used τ after p is a win for player
II.

The following important lemma shows that if player **II** has a chance in the
beginning, i.e. player **I** does not already have a winning strategy, she has a
chance all the way.

Lemma 3.7 (Survival Lemma) *Suppose A is a set, n is a natural num-
ber, $W \subseteq A^{2n}$ and $p = (x_0, y_0, \ldots, x_{i-1}, y_{i-1})$ is a position in the game
$\mathcal{G}_n(A, W)$, with $i < n$. Suppose furthermore that player **I** does not have a
winning strategy in position p. Then for every $x_i \in A$ there is $y_i \in A$ such that
player **I** does not have a winning strategy in position $p' = (x_0, y_0, \ldots, x_i, y_i)$.*

Proof The proof is by contradiction. The intuition is clear: if player **I** had a
smart move x_i so that he has a strategy for winning whatever the response y_i of
player **II** is, then we could argue that, contrary to the hypothesis, player **I** had

a winning strategy already in position p, as he wins whatever **II** moves. Let us now make this idea more exact. Suppose there were an $x_i \in A$ such that for all $y_i \in A$ player **I** has a winning strategy σ^{y_i} in position $p' = (x_0, y_0, \ldots, x_i, y_i)$. We define a strategy $\sigma = (\sigma_0, \ldots, \sigma_{n-1-i})$ of player **I** in position p as follows: $\sigma_0(\emptyset) = x_i$ and

$$\sigma_{j-i}(y_i, \ldots, y_{j-i}) = \sigma^{y_i}(y_{i+1}, \ldots, y_{j-i}).$$

This is a winning strategy of **I** in position p, contrary to our assumption that none exists. □

The following concept is of fundamental importance in game theory and in applications to logic, in particular:

Definition 3.8 A game is called *determined* if one of the players has a winning strategy. Otherwise the game is *non-determined*.

Virtually all games that one comes across in logic are determined. The following theorem is the crucial fact behind this phenomenon:

Theorem 3.9 (Zermelo) *If A is any set, n is a natural number, and $W \subseteq A^{2n}$, then the game $\mathcal{G}_n(A, W)$ is determined.*

Proof Suppose player **I** has no winning strategy. Then player **II** has a winning strategy based on repeated use of Lemma 3.7. Player **II** notes that in the beginning of the game, that is, in position $\emptyset$, player **I** does not have a winning strategy. Then by the Survival Lemma 3.7 she can, whatever player **I** moves, find a move such that afterwards player **I** still does not have a winning strategy. In short, the strategy of player **II** is to prevent player **I** from having a winning strategy. After n rounds the game ends and player **I** still does not have a winning strategy. That means player **I** has lost and player **II** has won. Let us now make this more precise: We define a strategy

$$\tau = (\tau_0, \ldots, \tau_{n-1})$$

of player **II** in the game $\mathcal{G}_n(A, W)$ as follows: Let a be some arbitrary element of A. By Lemma 3.7 we have for each position $p = (x_0, y_0, \ldots, x_{i-1}, y_{i-1})$ in the game $\mathcal{G}_n(A, W)$ such that player **I** does not have a winning strategy in position p and each $x_i \in A$ some $y_i \in A$ such that player **I** does not have a winning strategy in position $p' = (x_0, y_0, \ldots, x_i, y_i)$. Let us denote this y_i by

$$y_i = f(p, x_i).$$

If $p = (x_0, y_0, \ldots, x_{i-1}, y_{i-1})$ is a position in which player **I** *does* have a winning strategy, we let $f(p, x_i) = a$. We have defined a function f defined on positions p and elements $x_i \in A$. Let $\tau_0(x_0) = f(\emptyset, x_0)$. Assuming $\tau_0, \ldots, \tau_{i-1}$ have been defined already, let

$$\tau_i(x_0, \ldots, x_i) = f(p, x_i),$$

where

$$p = (x_0, y_0, \ldots, x_{i-1}, y_{i-1})$$

and

$$y_0 = \tau_0(x_0)$$
$$y_{i-1} = \tau_{i-1}(x_0, \ldots, x_{i-1}).$$

It is easy to see that in every play in which player **II** uses this strategy, every position p is such that player **I** *does not* have a winning strategy in position p. It is also easy to see that this is a winning strategy of player **II**. $\quad\square$

3.5 Infinite Games

The concept of a game is by no means limited to games with just finitely many rounds. Imagine a chess board which extends the usual board left and right without end. Then the chess game could go on for infinitely many rounds without the same configuration of pieces coming up twice. A simple infinite game is one in which two players pick natural numbers each choosing a bigger number, if he or she can, than the opponent. There is no end to this game, since there are infinitely many natural numbers. A third kind of infinite game is the following:

Example 3.10 Suppose A is a set of real numbers on the unit interval. We describe a game we denote by $G(A)$. During the game the players decide the decimal expansion of a real number $r = 0.d_0 d_1 \ldots$ on the interval $[0, 1]$. Player **I** decides the even digits d_{2n} and player **II** the odd digits d_{2n+1}. Player **II** wins if $r \in A$. If A is countable, say $A = \{b_n : n \in \mathbb{N}\}$, player **I** has a winning strategy: during round n he chooses the digit d_{2n} so that $r \neq b_n$. If the complement of A is countable, player **II** wins with the same strategy. What if A and its complement are uncountable? This is a well-known and much studied hard question. (See e.g. Jech (1997).)

I	II
x_0	
	y_0
x_1	
	y_1
$\vdots$	$\vdots$

Figure 3.5 An infinite game.

If A is any set, we use $A^{\mathbb{N}}$ to denote infinite sequences

$$(x_0, x_1, \ldots)$$

of elements of A. We can think of such sequences as limits of an increasing sequence

$$(x_0), (x_0, x_1), (x_0, x_1, x_2), \ldots$$

of finite sequences.

Let A be an arbitrary set. Let $W \subseteq A^{\mathbb{N}}$. We define the game

$$\mathcal{G}_\omega(A, W)$$

as follows (see Figure 3.5): An infinite sequence

$$(\bar{x}; \bar{y}) = (x_0, y_0, x_1, y_1, \ldots),$$

of elements of A is called a *play* (of $\mathcal{G}_\omega(A, W)$). A *play* of one of the players is likewise any infinite sequence $\bar{x} = (x_0, x_1, \ldots)$ of elements of A. The play $(\bar{x}; \bar{y})$ is a *win for* player **II** if

$$(x_0, y_0, x_1, y_1, \ldots) \in W$$

and otherwise a *win for* player **I** .

A *strategy of* player **I** in the game $\mathcal{G}_\omega(A, W)$ is an infinite sequence

$$\sigma = (\sigma_0, \sigma_1, \ldots)$$

of functions $\sigma_i : A^i \to A$. We say that player **I** has *used the strategy σ in the play* $(\bar{x}; \bar{y})$ if for all $i \in \mathbb{N}$:

$$x_i = \sigma_i(y_0, \ldots, y_{i-1})$$

and

$$x_0 = \sigma_0.$$

The strategy σ of player **I** is a *winning strategy*, if every play where **I** has used σ is a win for player **I**.

A *strategy of* player **II** in the game $\mathcal{G}_\omega(A, W)$ is an infinite sequence

$$\tau = (\tau_0, \tau_1, \dots)$$

of functions $\tau_i : A^{i+1} \to A$. We say that player **II** has *used the strategy τ in the play* $(\bar{x}; \bar{y})$ if for all $i < n$:

$$y_i = \tau_i(x_0, \dots, x_i).$$

The strategy τ of player **II** is a *winning strategy*, if every play where player **II** has used τ is a win for player **II**. A player is said to *win the game* $\mathcal{G}_\omega(A, W)$ if he or she has a winning strategy in it.

A *position* of the infinite game $\mathcal{G}_\omega(A, W)$ is any initial segment

$$p = (x_0, y_0, \dots, x_{i-1}, y_{i-1})$$

of a play $(\bar{x}; \bar{y})$. We say that player **I** has *used strategy* $\sigma = (\sigma_0, \sigma_1, \dots)$ *after position p in the play* $(\bar{x}; \bar{y})$, if $(\bar{x}; \bar{y})$ extends p and for all j with $i < j$ we have $x_j = \sigma_{j-i}(y_i, \dots, y_{j-1})$ and $x_i = \sigma_0$. The strategy σ of player **I** is a *winning strategy in position p*, if every play extending p where player **I** has used σ after position p is a win for player **I**. We say that player **II** has *used strategy* $\tau = (\tau_0, \tau_1, \dots)$ *after position p in the play* $(\bar{x}; \bar{y})$ if for all j with $i \le j$ we have $y_j = \tau_{j-i}(x_i, \dots, x_j)$. The strategy τ of player **II** is a *winning strategy in position p*, if every play extending p where player **II** has used τ after p is a win for player **II**.

An important example of a class of infinite games is the class of *open* or *closed* games of length ω. A subset W of $A^{\mathbb{N}}$ is *open*,[5] if

$$(x_0, y_0, x_1, y_1, \dots) \in W$$

implies the existence of $n \in \mathbb{N}$ such that

$$(x_0, y_0, \dots, x_{n-1}, y_{n-1}, x'_n, y'_n, x'_{n+1}, y'_{n+1}, \dots) \in W$$

for all $x'_n, y'_n, x'_{n+1}, y'_{n+1}, \dots \in A$. Respectively, W is *closed* if $A^{\mathbb{N}} \setminus W$ is open. Finally, W is *clopen* if it is both open and closed. We call a game $\mathcal{G}_\omega(A, W)$ closed (or open or clopen) if the set W is. We are mainly concerned in this book with closed games. A typical strategy of player **II** in a closed game is to "hang in there", as she knows that if player **I** ends up winning the play $p = (x_0, y_0, \dots)$, that is, $p \notin W$, there is some n such that player **I** won the game already in position $(x_0, y_0, \dots, x_{n-1}, y_{n-1})$.

---

[5] The collection of open subsets of $A^{\mathbb{N}}$ is a topology, hence the name.

We can think of infinite games as *limits* of finite games as follows: Any finite game $G_n(A, W)$ can be made infinite by disregarding the moves after the usual n moves. The resulting infinite game is clopen (see Exercise 3.31). On the other hand, if $G_\omega(A, W)$ is an infinite game and $n \in \mathbb{N}$ we can form an n-round game by simply considering only the first n rounds of $G_\omega(A, W)$ and declaring a play of n rounds a win for player **II** if *any* infinite play extending it is in W. Unless W is open or closed, there may be very little connection between the resulting finite games and the original infinite game (see however Exercise 3.32).

Lemma 3.11 (Infinite Survival Lemma) *Suppose A is a set, $W \subseteq A^\mathbb{N}$, and $p = (x_0, y_0, \ldots, x_{i-1}, y_{i-1})$ is a position in the game $G_\omega(A, W)$, with $i \in \mathbb{N}$. Suppose furthermore that player **I** does not have a winning strategy in position p. Then for every $x_i \in A$ there is $y_i \in A$ such that player **I** does not have a winning strategy in position $p' = (x_0, y_0, \ldots, x_i, y_i)$.*

Proof The proof is by contradiction. Suppose there were an $x_i \in A$ such that for all $y_i \in A$ player **I** has a winning strategy σ^{y_i} in position $p' = (x_0, y_0, \ldots, x_i, y_i)$. We define a strategy $\sigma = (\sigma_0, \sigma_1, \ldots)$ of player **I** in position p as follows: $\sigma_0(\emptyset) = x_i$ and for $j > i$,

$$\sigma_{j-i}(y_i, \ldots, y_{j-1}) = \sigma^{y_i}(y_{i+1}, \ldots, y_{j-i}).$$

This is a winning strategy of player **I** in position p, contrary to assumption. $\square$

Theorem 3.12 (Gale–Stewart) *If A is any set and $W \subseteq A^\mathbb{N}$ is open or closed, then the game $G_\omega(A, W)$ is determined.*

Proof Suppose first W is closed and player **I** has no winning strategy. We define a strategy

$$\tau = (\tau_0, \tau_1, \ldots)$$

of player **II** in the game $G_\omega(A, W)$ as follows: Let a be some arbitrary element of A. By Lemma 3.11 we have for each position $p = (x_0, y_0, \ldots, x_{i-1}, y_{i-1})$ in the game $G_\omega(A, W)$ such that player **I** does not have a winning strategy in position p, and each $x_i \in A$, some $y_i \in A$ such that player **I** does not have a winning strategy in position $p' = (x_0, y_0, \ldots, x_i, y_i)$. Let us denote this y_i by

$$y_i = f(p, x_i).$$

If $p = (x_0, y_0, \ldots, x_{i-1}, y_{i-1})$ is a position in which player **I** *does* have a winning strategy, we let $f(p, x_i) = a$. We have defined a function f defined on positions p and elements $x_i \in A$. Let $\tau_0(x_0) = f(\emptyset, x_0)$. Assuming $\tau_0, \ldots, \tau_{i-1}$ have been defined already, let $\tau_i(x_0, \ldots, x_i) = f(p, x_i)$, where

$p = (x_0, y_0, \ldots, x_{i-1}, y_{i-1})$ and $y_0 = \tau_0(x_0), y_{i-1} = \tau_{i-1}(x_0, \ldots, x_{i-1})$. It is easy to see that in every play in which player **II** uses this strategy, every position p is such that player **I** does not have a winning strategy in position p. It is also easy to see that this is a winning strategy of player **II**.

The proof is similar if W is open. It follows that $G_\omega(A, W)$ is determined. $\square$

Theorem 3.12 can been vastly generalized, see e.g. (Jech, 1997, Chapter 33). The *Axiom of Determinacy* says that the game $G_\omega(A, W)$ is determined for all sets A and W. However, this axiom contradicts the Axiom of Choice. By using the Axiom of Choice one can show that there are sets A of real numbers such that the game $G(A)$ is not determined (see Exercise 3.37).

3.6 Historical Remarks and References

The mathematical theory of games was started by von Neumann and Morgenstern (1944). For the early history of two-person zero-sum games of perfect information, see Schwalbe and Walker (2001). See Mycielski (1992) for a more recent survey on games of perfect information. Theorem 3.12 goes back to Gale and Stewart (1953).

Exercises

3.1   Consider the following game: Player **I** picks a natural number n. Then player **II** picks a natural number m. If $2^m = n$, then **II** wins, otherwise **I** wins. Express this game in the form $G_1(A, W)$.

3.2   Consider the following game: Player **I** picks a natural number n. Then player **II** picks two natural numbers m and k. If $m \cdot k = n$, then **II** wins, otherwise **I** wins. Express this game in the form $G_2(A, W)$.

3.3   Consider $G_3(A, W)$, where $A = \{0, 1, 2\}$ and
   1. $W = \{(x_0, y_0, x_1, y_1, x_2, y_2) \in A^3 : x_0 = y_2\}$.
   2. $W = \{(x_0, y_0, x_1, y_1, x_2, y_2) \in A^3 : y_0 \neq x_2 \text{ or } y_2 \neq x_0\}$.
   3. $W = \{(x_0, y_0, x_1, y_1, x_2, y_2) \in A^3 : x_0 \neq y_2 \text{ and } x_1 \neq y_2 \text{ and } x_2 \neq y_2\}$.

   Who has a winning strategy?

3.4   Suppose $f : \mathbb{R} \to \mathbb{R}$ is a mapping. Express the condition that f is uniformly continuous as a game and as the truth of a first-order sentence in a suitable structure.

3.5 This is a variation of Example 3.4. Let A be the set of cities of Europe. Let G be a collection of German cities and F a collection of French cities. During the first round of the game player **I** chooses a city $f \in F$ in France and then player **II** chooses a city $g \in G$ in Germany. During the second round of the game player **I** chooses a city s_0 in Scandinavia. Finally player **II** chooses a city s_1 also in Scandinavia. The game ends and player **II** wins if she was able to choose city s_1 in such a way that it is in the same country as s_0 and there is a direct commercial flight from city g to city s_0 if and only if there is also a direct commercial flight from city f to city s_1. Represent this game as $\mathcal{G}_2(A, W)$ for a suitable W and give a necessary and sufficient condition for player **II** to have a winning strategy.

3.6 This is a variation of Example 3.4. Let A be the set of cities of Europe. Let G be a collection of German cities and F a collection of French cities. During the first round of the game player **I** chooses a city $x \in A$ and then player **II** chooses a city $y \in A$. During the second round of the game player **I** chooses a city s_0 in Scandinavia. Finally player **II** chooses a city s_1 also in Scandinavia. The game ends and player **II** wins if

1. City s_1 is in the same country as s_0.
2. If $x \in F$, then $y \in G$.
3. If $x \in G$, then $y \in F$.
4. If there is a a direct commercial flight from city y to city s_0 then there is also a direct commercial flight from city x to city s_1.

Represent this game as $\mathcal{G}_2(A, W)$ for a suitable W and give a necessary and sufficient condition for player **II** to have a winning strategy.

3.7 Modify the game of Example 3.4 in such a way that player **II** has a winning strategy if and only if $\tau(F) = \tau(G)$.

3.8 Examine the game determined by condition (3.1) when $M = \mathbb{N}$ and

1. $W^{\mathcal{M}} = \{(a_0, b_0, a_1, b_1) \in M^4 : a_0 + b_0 + a_1 + b_1 = 6\}$.
2. $W^{\mathcal{M}} = \{(a_0, b_0, a_1, b_1) \in M^4 : a_0 + b_0 + a_1 + b_1 > 6\}$.
3. $W^{\mathcal{M}} = \{(a_0, b_0, a_1, b_1) \in M^4 : a_0 + b_0 + a_1 + b_1 < 6\}$.

Who has a winning strategy?

3.9 Examine the game determined by condition (3.1) when

1. $M = \mathbb{N}$
2. $M = \mathbb{Z}$

and $W^{\mathcal{M}} = \{(a_0, b_0, a_1, b_1) \in M^4 : a_0 + b_0 = a_1 + b_1\}$. Who has a winning strategy?

3.10 Examine the game determined by condition (3.1) $M = \mathbb{N}$ and $W^{\mathcal{M}} = \{(a_0, b_0, a_1, b_1) \in M^4 : a_0 < b_0$ and either a_1 does not divide b_0 or $b_1 = a_1 = 1$ or $b_1 = a_1 = b_0\}$. Who has a winning strategy?

3.11 Suppose X is a set of positions of the game $G_n(A, W)$ such that

  1. $\emptyset \in X$.
  2. For all $i < n$, all $(x_0, y_0, \dots, x_{i-1}, y_{i-1}) \in X$, and all $x_i \in A$ there is $y_i \in A$ such that $(x_0, y_0, \dots, x_i, y_i) \in X$.
  3. If $p = (x_0, y_0, \dots, x_{n-1}, y_{n-1}) \in X$, then $p \in W$.

  Show that player **II** has a winning strategy in the game $G_n(A, W)$. Give such a set for the game of Example 3.1.

3.12 Suppose that player **II** has a winning strategy in the game $G_n(A, W)$. Show that there is a set X of positions of the game $G_n(A, W)$ satisfying conditions 1–3 of the previous exercise.

3.13 Suppose X is a set of positions of the game $G_n(A, W)$ such that

  1. $\emptyset \in X$.
  2. For all $i < n$, all $(x_0, y_0, \dots, x_{i-1}, y_{i-1}) \in X$ there is $x_i \in A$ such that for all $y_i \in A$ we have $(x_0, y_0, \dots, x_i, y_i) \in X$.
  3. If $p = (x_0, y_0, \dots, x_{n-1}, y_{n-1}) \in X$, then $p \notin W$.

  Show that player **I** has a winning strategy in the game $G_n(A, W)$. Give such a set for the game of Example 3.1 when we start with seven tokens.

3.14 Suppose that player **I** has a winning strategy in the game $G_n(A, W)$. Show that there is a set X of positions of the game $G_n(A, W)$ satisfying conditions 1–3 of the previous exercise.

3.15 Suppose A is finite. Describe an algorithm which searches for a winning strategy for a player in $\mathcal{G}_n(A, W)$, provided the player has one.

3.16 Finish the proof of Lemma 3.7 by showing that the strategy described in the proof is indeed a winning strategy of player **I**.

3.17 Finish the proof of Theorem 3.9 by showing that the strategy described in the proof is indeed a winning strategy of player **II**.

3.18 Consider $\mathcal{G}_2(A, W)$, where $A = \{0, 1\}$ and

  1. $W = \{(x_0, y_0, x_1, y_1) \in A^2 : x_0 = y_1\}$.
  2. $W = \{(x_0, y_0, x_1, y_1) \in A^2 : y_0 \neq x_1 \text{ or } y_1 \neq x_0\}$.
  3. $W = \{(x_0, y_0, x_1, y_1) \in A^2 : x_0 \neq y_1 \text{ and } x_1 \neq y_1\}$.

  In each case give a winning strategy for one of the players.

3.19 Suppose σ is a strategy of player **I** and τ a strategy of player **II** in $\mathcal{G}_n(A, W)$. Show that there is exactly one play $(\bar{x}; \bar{y})$ of $\mathcal{G}_n(A, W)$ such that player **I** has used σ and player **II** has used τ in it.

3.20 Show that at most one player can have a winning strategy in $\mathcal{G}_n(A, W)$.

3.21 Give the winning strategy of player **II** in Nim (Example 3.1) in the form $\tau = (\tau_0, \tau_1)$.

3.22 Consider the game of Example 3.3 when $f(x) = 2x + 3$, $a \in \mathbb{R}$, and $b = 2a + 3$. Give some winning strategy of player **II**.

3.23 Consider the game of Example 3.3 when $f(x) = 2x + 3$, $a = 1$ and $b = 4$. Give some winning strategy of player **I**.

3.24 Consider the game of Example 3.3 when $f(x) = x^2$, $a \in \mathbb{R}$, and $b = a^2$. Give some winning strategy of player **II**.

3.25 A more general version of Nim has m tokens rather than six. Decide who has a winning strategy for each m and give the winning strategy.

3.26 Suppose we have two games $\mathcal{G}_n(A, W)$ and $\mathcal{G}_n(A', W')$, where $A \cap A' = \emptyset$. Let $A'' = A \cup A'$ and let W'' be the set of sequences

$$(x_0, y_0, \ldots, x_{2n-1}, y_{2n-1}),$$

which satisfy the following condition:

$$(x_0, y_0, x_2, y_2 \ldots, x_{2n-2}, y_{n-2}) \in W$$

and

$$(x_1, y_1, x_3, y_3, \ldots, x_{2n-1}, y_{2n-1}) \in W'.$$

Show that:

1. If player **I** has a winning strategy in $\mathcal{G}_n(A, W)$ *or* in $\mathcal{G}_n(A', W')$, then he has one in $\mathcal{G}_{2n}(A'', W'')$.
2. If player **II** has a winning strategy in $\mathcal{G}_n(A, W)$ *and* in $\mathcal{G}_n(A', W')$, then she has one in $\mathcal{G}_{2n}(A'', W'')$.

3.27 Suppose we have two games $\mathcal{G}_n(A, W)$ and $\mathcal{G}_n(A', W')$. Let $A'' = A \times A'$ and let W'' be the set of sequences

$$(((x_0, x_0'), (y_0, y_0')), \ldots, ((x_{n-1}, x_{n-1}'), (y_{n-1}, y_{n-1}'))),$$

where

$$(x_0, y_0, \ldots, x_{n-1}, y_{n-1}) \in W$$

and

$$(x_0', y_0', \ldots, x_{n-1}', y_{n-1}') \in W'.$$

Show that:

1. If player **I** has a winning strategy in $\mathcal{G}_n(A, W)$ *or* in $\mathcal{G}_n(A', W')$, then he has one in $\mathcal{G}_n(A'', W'')$.
2. If player **II** has a winning strategy in $\mathcal{G}_n(A, W)$ *and* in $\mathcal{G}_n(A', W')$, then she has one in $\mathcal{G}_n(A'', W'')$.

3.28 This exercise demonstrates why we can let a winning strategy depend on the opponent's moves only. Let us consider the game $\mathcal{G}_n(A, W)$.

1. Let

$$\sigma = (\sigma_0, \ldots, \sigma_{n-1})$$

be a sequence of functions $\sigma_i : A^{2i} \to A$. Suppose every play $(\bar{x}; \bar{y})$, where for all $i < n$

$$x_i = \sigma_i(x_0, y_0, \ldots, x_{i-1}, y_{i-1})$$

is a win for player **I**. Show that player **I** has a winning strategy in $\mathcal{G}_n(A, W)$.

2. Let

$$\tau = (\tau_0, \ldots, \tau_{n-1})$$

be a sequence of functions $\tau_i : A^{2i+1} \to A$. Suppose every play $(\bar{x}; \bar{y})$, where for all $i < n$

$$y_i = \tau_i(x_0, y_0, \ldots, x_{i-1}, y_{i-1}, x_i),$$

is a win for player **II**. Show that player **II** has a winning strategy in $\mathcal{G}_n(A, W)$.

3.29 Suppose $|A| = m$. How many plays of the game $\mathcal{G}_n(A, W)$ are there?

3.30 Suppose $|A| = m$. How many strategies does each of the players have in $\mathcal{G}_n(A, W)$?

3.31 Suppose $G_n(A, W)$ is a given finite game. Let W' consist of all infinite sequences $(x_0, y_0, x_1, y_1, \ldots)$, where $(x_0, y_0, \ldots, x_{n-1}, y_{n-1}) \in W$. Show that $G_\omega(A, W')$ is clopen, and a player has a winning strategy in the game $G_\omega(A, W')$ if and only he or she has one in $G_n(A, W)$.

3.32 Suppose $G_\omega(A, W)$ is a given infinite game and $n \in \mathbb{N}$. Let W_n^0 consist of all sequences $(x_0, y_0, \ldots, x_{n-1}, y_{n-1})$ for which there is no infinite sequence $(x_0, y_0, x_1, y_1, \ldots) \in W$ extending $(x_0, y_0, \ldots, x_{n-1}, y_{n-1})$. Let W_n^1 be the set consisting of sequences $(x_0, y_0, \ldots, x_{n-1}, y_{n-1})$ all extensions $(x_0, y_0, x_1, y_1, \ldots)$ of which are in W. Let W_n^2 consist of all sequences $(x_0, y_0, \ldots, x_{n-1}, y_{n-1})$ for which there is some infinite sequence $(x_0, y_0, x_1, y_1, \ldots) \in W$ extending $(x_0, y_0, \ldots, x_{n-1}, y_{n-1})$. Prove the following:

1. If player **I** has a winning strategy in $G_n(A, W_n^0)$ for some $n \in \mathbb{N}$, then he has one in $G_\omega(A, W)$.

2. If player **II** has a winning strategy in $G_n(A, W_n^1)$ for some $n \in \mathbb{N}$, then she has one in $G_\omega(A, W)$.

3. If player **I** has a winning strategy in $G_\omega(A, W)$, then he has one in $G_n(A, W_n^1)$ for each $n \in \mathbb{N}$.

4. If player **II** has a winning strategy in $G_\omega(A, W)$, then she has one in $G_n(A, W_n^2)$ for each $n \in \mathbb{N}$.

3.33 Suppose $W \subseteq A^\mathbb{N}$ is closed and X is a set of positions of the game $G_\omega(A, W)$ such that

1. $\emptyset \in X$.

2. For all $i \in \mathbb{N}$ and all $(x_0, y_0, \ldots, x_{i-1}, y_{i-1}) \in X$ and all $x_i \in A$ there is $y_i \in A$ such that $(x_0, y_0, \ldots, x_i, y_i) \in X$.

3. If $(x_0, y_0, \ldots, x_{i-1}, y_{i-1}) \in X$, then $(x_0, y_0, x_1, y_1, \ldots) \in W$ for some $(x_0, y_0, x_1, y_1, \ldots)$ extending $(x_0, y_0, \ldots, x_{i-1}, y_{i-1})$.

Show that player **II** has a winning strategy in the game $G_\omega(A, W)$. Give such a set for the game of Example 3.10 when $A = \mathbb{R}$ and W is the set of reals on the unit interval whose decimal expansion has every fourth digit after the decimal point zero. (Note that W is a closed set.)

3.34 Suppose that player **II** has a winning strategy in the game $G_\omega(A, W)$. Show that there is a set X of positions of the game $G_\omega(A, W)$ satisfying conditions 1–3 of the previous exercise.

3.35 Suppose $W \subseteq A^\mathbb{N}$ is open and X is a set of positions of the game $G_\omega(A, W)$ such that

1. $\emptyset \in X$.

2. For all $i \in \mathbb{N}$ and all $(x_0, y_0, \ldots, x_{i-1}, y_{i-1}) \in X$ there is $x_i \in A$ such that for all $y_i \in A$ we have $(x_0, y_0, \ldots, x_i, y_i) \in X$.

3. If $(x_0, y_0, \ldots, x_{n-1}, y_{n-1}) \in X$, then $(x_0, y_0, x_1, y_1, \ldots) \notin W$ for some $(x_0, y_0, x_1, y_1, \ldots)$ extending $(x_0, y_0, \ldots, x_{i-1}, y_{i-1})$

Show that player **I** has a winning strategy in the game $G_\omega(A, W)$. Give such a set for the game of Example 3.10 when $A = \mathbb{R}$ and W is the set of reals on the unit interval whose ternary expansion has a digit 1. (W is the complement of the so-called *Cantor ternary set*.)

3.36 Suppose that player **I** has a winning strategy in the game $G_\omega(A, W)$. Show that there is a set X of positions of the game $G_\omega(A, W)$ satisfying conditions 1–3 of the previous exercise.

3.37 Show that there is a set $A \subseteq \mathbb{R}$ such that the game $G(A)$ is not determined. (Hint: This exercise presupposes knowledge of set theory. List all reals on the unit interval as $\{r_\alpha : \alpha < 2^{\aleph_0}\}$. List all strategies of player **I** as $\{\sigma^\alpha : \alpha < 2^{\aleph_0}\}$. List all strategies of player **II** as $\{\tau^\alpha : \alpha < 2^{\aleph_0}\}$. Define A by transfinite induction taking care that no σ^α can be a winning strategy of player **I**, and no τ^α can be a winning strategy of player **II**.)

3.38 Suppose player **II** has a winning strategy in $G(A)$. Show that A contains a non-empty *perfect* subset, i.e. a non-empty subset B such that B is closed and if $0.d_0 d_1 \ldots \in B$ and $n \in \mathbb{N}$, then there is $0.c_0 c_1 \ldots \neq 0.d_0 d_1 \ldots$ in B such that $c_i = d_i$ for $i < n$.

4

Graphs

4.1 Introduction

Graphs are among the simplest binary structures and yet they still present formidable challenges to combinatorists, computer scientists, and logicians alike. We introduce the basic concepts of logic, such as the Semantic Game, the Model Existence Game, and the Ehrenfeucht–Fraïssé Game, first in the context of graphs, and then in the next chapter for arbitrary structures.

4.2 First-Order Language of Graphs

A *graph* is a pair $\mathcal{G} = (V, E)$ where V is a set of elements called *vertices* and E is an anti-reflexive symmetric binary relation on V called the *edge* relation. We write $V = V_{\mathcal{G}}$ and $E = E_{\mathcal{G}}$ when it is necessary to distinguish which graph is in question. The *first-order language of graphs* is built from equations $x = y$ and edge statements xEy, both called *atomic* formulas, by means of the propositional operations $\neg$, $\wedge$, $\vee$, and the quantifiers $\exists x$ and $\forall x$, where x ranges over vertices. (An exact definition of the term "first-order language" will be given later.) *Assignments* give values to variables. Thus an assignment in $\mathcal{G}$ is a function s the domain of which is a finite set of variables and the values of which are elements of G. We write

$$\mathcal{G} \models_s \varphi$$

if the formula φ is true in the graph $\mathcal{G}$ when the free variables occurring in φ are interpreted according to s, i.e. a variable x is interpreted as $s(x) \in V$. The assignment $s[a/x]$ is defined as follows:

$$s[a/x](y) = \left\{ \begin{array}{ll} a & \text{if } y = x \\ s(y) & \text{if } y \neq x. \end{array} \right.$$

With this concept the truth condition of quantified formulas can be given easily:

$$\mathcal{G} \models_s \exists x \varphi \iff \text{there is a vertex } v \text{ such that } \mathcal{G} \models_{s[v/x]} \varphi.$$

We can think of the truth of a sentence in a graph as the existence of a winning strategy of player **II** in the following *Semantic Game*: In the beginning player **II** holds a pair (φ, s) consisting of a formula φ and an assignment s of the free variables of φ.

1. If φ is atomic, and s satisfies it in $\mathcal{G}$, then the player who holds (φ, s) wins the game, otherwise the other player wins.
2. If $\varphi = \neg\psi$, then the player who holds (φ, s), gives (ψ, s) to the other player.
3. If $\varphi = \psi \wedge \theta$, then the player who holds (φ, s), switches to hold (ψ, s) or (θ, s), and the other player decides which.
4. If $\varphi = \psi \vee \theta$, then the player who holds (φ, s), switches to hold (ψ, s) or (θ, s), and can himself or herself decide which.
5. If $\varphi = \forall x\psi$, then the player who holds (φ, s), switches to hold $(\psi, s[v/x])$ for some v, and the other player decides for which.
6. If $\varphi = \exists x\psi$, then the player who holds (φ, s), switches to hold $(\psi, s[v/x])$ for some v, and can himself or herself decide for which.

Now $\mathcal{G} \models_s \varphi$ if and only if player **II** has a winning strategy in the above game, starting with (φ, s). Why? If $\mathcal{G} \models_s \varphi$, then the winning strategy of player **II** is to play so that if she holds (φ', s'), then $\mathcal{G} \models_{s'} \varphi'$, and if player **I** holds (φ', s'), then $\mathcal{G} \not\models_{s'} \varphi'$. Conversely, we use induction on φ to prove for all s that if player **II** has a winning strategy starting with (φ, s), then $\mathcal{G} \models_s \varphi$, and if **II** has a winning strategy starting with $(\neg\varphi, s)$, then $\mathcal{G} \not\models_s \varphi$. Jaakko Hintikka has advocated this approach to first-order logic, and recently also a variant in which the winning strategy of player **II** may be required to depend only on some specific moves of player **I** , rather than on all moves.

Example 4.1 A typical property of a vertex of a graph that is first-order (i.e. can be written in the first-order language of graphs) is the *degree* of a vertex (but only if the degree is finite). The degree of a vertex v is the number of vertices that are connected by a (single) edge to v. Thus it is the number of "neighbors" of v. This number can be infinite. If the degree is zero the vertex is called *isolated*. A graph where every vertex is isolated is called an *empty* graph (even though the graph is non-empty in the sense that there are vertices). We call a vertex *anti-isolated* if it has an edge to every other vertex. A graph where every vertex is anti-isolated is called a *complete* graph. Consider the formulas:

$$\mathrm{NI}_n(x_1, \ldots, x_n) \equiv \neg x_1 = x_2 \wedge \ldots \wedge \neg x_1 = x_n \wedge \ldots \neg x_{n-1} = x_n$$

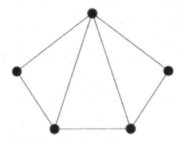

Figure 4.1 A graph.

$$DEG_{\geq n}(x) \equiv \exists y_1 \ldots \exists y_n (NI_n(y_1, \ldots, y_n) \wedge xEy_1 \wedge \ldots \wedge xEy_n)$$
$$DEG_n(x) \equiv DEG_{\geq n}(x) \wedge \neg DEG_{\geq n+1}(x).$$

Clearly, $NI_n(v_1, \ldots, v_n)$ is satisfied by vertices $x_1, \ldots, x_n$ of a graph if and only if these vertices are all different from each other. Equally clear is that $DEG_{\geq n}(v)$ is satisfied by a vertex v of a graph if and only if the degree of v is at least n. Therefore, $DEG_n(v)$ is satisfied by a vertex v of a graph if and only if the degree of v is exactly n.

It is important to note here that all quantifiers range over just vertices of the graph. So the truth or falsity of $DEG_n(v)$ for a particular vertex v can be checked by merely going through the vertices of the graph several times.

Definition 4.2 The *quantifier rank*, denoted by $QR(\varphi)$, of a formula φ is defined as follows: $QR(x = y) = QR(xEy) = 0$, $QR(\neg\varphi) = QR(\varphi)$, $QR(\varphi \wedge \psi) = QR(\varphi \vee \psi) = \max\{QR(\varphi), QR(\psi)\}$, $QR(\exists x\varphi) = QR(\forall x\varphi) = QR(\varphi) + 1$.

Example 4.3 $NI_n(x_1, \ldots, x_n)$ has quantifier rank 0. $DEG_{\geq n}(x)$ has quantifier rank n, and while $DEG_n(x)$ has $n + 1$.

A formula that has quantifier rank n may very well be equivalent to a formula of lower rank. For example, $DEG_n(x) \wedge \neg DEG_n(x)$ is equivalent to a formula of quantifier rank 1 (or 0 if we have constants or a symbol for falsity). On the other hand, we will develop in the next section a method for proving that, for example, $DEG_n(x)$ is *not* logically equivalent to a formula of lower quantifier rank.

Definition 4.4 The *number of variables* of a formula φ is defined to be the total number of distinct variables occurring (one or several times) in the formula.

Example 4.5 $\exists x \forall y \forall z ((\neg y E x \land \neg z E x \land \neg (y = z)) \rightarrow \exists x (y E x \land z E x))$
is true in a graph if and only if the graph has a vertex such that outside its
neighbors any two distinct vertices have a common neighbor. The number of
variables in this formula is 3. Note that x is used in two different roles.

As it turns out, the number of variables of a formula is an important property
of the formula. For example it has been proved that if the number of variables
of a consistent sentence is 2, the sentence has a finite model (Mortimer (1975)),
something which is certainly not true of sentences with three variables.

4.3 The Ehrenfeucht–Fraïssé Game on Graphs

The Ehrenfeucht–Fraïssé Game is played on two graphs. Player **I** tries to
demonstrate a difference between the graphs, while player **II** tries to defend
the claim that there is no difference. The challenge to player **I** is that even
if there really is a difference between the graphs, he has to demonstrate this
during a game which may have a very limited number of rounds. During each
round player **I** may exhibit one vertex only (there are versions of this game
where player **I** can exhibit more vertices during each round). Typical properties of graphs that player **I** will keep an eye on are "local" properties such
as

- Isolated vertices.
- Cycles, e.g. triangles.
- Complete subgraphs.

Suppose $\mathcal{G} = (V, E)$ and $\mathcal{G}' = (V', E')$ are graphs. They could be the
two graphs of Figure 4.2. Let us fix a natural number n. We shall now define
the game $\mathrm{EF}_n(\mathcal{G}, \mathcal{G}')$, which is called the Ehrenfeucht–Fraïssé Game on $\mathcal{G}$
and $\mathcal{G}'$. The number n is the number of rounds of the game. During the game
player **I** picks n vertices $x_0, \ldots, x_{n-1}$ from the graphs. At the same time,
alternating with player **I**, player **II** also picks n vertices $y_0, \ldots, y_{n-1}$ from
the two graphs. The position in the game is as in Figure 4.3. Note that player
I has played his first two elements in $\mathcal{G}$, the third in $\mathcal{G}'$, and the last in $\mathcal{G}$.
Respectively, player **II** has played the first two moves in $\mathcal{G}'$, the third $\mathcal{G}$, and
the last in $\mathcal{G}'$. Player **II** plays always in the other model than where player **I**
has played, as her role is to try to imitate the moves of player **I**. If $n = 4$,
the game $\mathrm{EF}_n(\mathcal{G}, \mathcal{G}')$ ends here, and we can check who won. It appears from
Figure 4.3 that there is an edge between two chosen vertices in $\mathcal{G}$ if and only
if the there is an edge between the corresponding edges in $\mathcal{G}'$. This means that

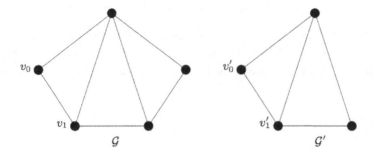

Figure 4.2 A play of $\mathrm{EF}_2(\mathcal{G}, \mathcal{G}')$.

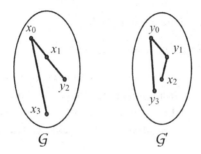

Figure 4.3 $\mathrm{EF}_4(\mathcal{G}, \mathcal{G}')$ underway.

player **II** has won. She has been able to imitate the edge structure of the two graphs to the extent that player **I** has challenged her.

In the mathematical definition of the game $\mathrm{EF}_n(\mathcal{G}, \mathcal{G}')$ below we use special notation to facilitate reference to the chosen elements. In our special notation we use v_i to denote an element picked by either player in graph $\mathcal{G}$. Respectively, we use v_i' to denote an element picked by either player in graph $\mathcal{G}'$. This makes it easier to write down the winning condition for player **II**. It is simply the condition that there is an edge between v_i and v_j if and only there is an edge between v_i' and v_j'. A different kind of simplification is the assumption below that the two graphs have no vertices together, that is, the graphs are disjoint. This is a completely harmless assumption as we can always swap, e.g. $\mathcal{G}'$ with an isomorphic copy which satisfies the disjointness condition. The advantage of this convention is that when a player chooses an element we know immediately whether we should interpret it as a move in model $\mathcal{G}$ or in $\mathcal{G}'$.

We now give the definition of the Ehrenfeucht–Fraïssé Game $EF_n(\mathcal{G}, \mathcal{G}')$:

Definition 4.6 Suppose $\mathcal{G} = (V, E)$ and $\mathcal{G}' = (V', E')$ are graphs with $V \cap V' = \emptyset$, and $n \in \mathbb{N}$. The *Ehrenfeucht–Fraïssé Game* $EF_n(\mathcal{G}, \mathcal{G}')$ is the game $\mathcal{G}_n(V \cup V', W)$, where W is the set of $(x_0, y_0, \ldots, x_{n-1}, y_{n-1})$ such that:

(G1) For all $i < n$: $x_i \in V \iff y_i \in V'$.
(G2) If we denote

$$v_i = \begin{cases} x_i & \text{if } x_i \in V \\ y_i & \text{if } y_i \in V \end{cases} \quad \text{and} \quad v_i' = \begin{cases} x_i & \text{if } x_i \in V' \\ y_i & \text{if } y_i \in V', \end{cases}$$

then

(G2.1) $v_i = v_j$ if and only if $v_i' = v_j'$.
(G2.2) $v_i E v_j$ if and only if $v_i' E' v_j'$.

We call v_i and v_i' above *corresponding* vertices.

The game $EF_2(\mathcal{G}, \mathcal{G}')$ could proceed as follows (see Figure 4.2): Player **I** chooses a vertex $v_0 = x_0$ from $\mathcal{G}$, and then Player **II** chooses a vertex $v_0' = y_0$ from $\mathcal{G}'$. After this player **I** chooses a vertex $v_1 = x_1$ from $\mathcal{G}$, and then Player **II** chooses a vertex $v_1' = y_1$ from $\mathcal{G}'$. The game ends and player **II** is the winner if

(1) $y_1 = y_0 \iff x_1 = x_0$
(2) $y_1 E' y_0 \iff x_1 E x_0$

as is indeed the case in Figure 4.2.

We can immediately observe: If one graph has just one vertex while the other has more, player **I** has an obvious winning strategy. Namely, he lets x_0 and x_1 be two different vertices in the *other* graph. Then **II** cannot force the final play (x_0, y_0, x_1, y_1) to satisfy condition (G2.1). If one graph has an isolated vertex v, while the other has none, player **I** can let $x_0 = v$ and after player **II** has chosen y_0 (which cannot be isolated), player **I** chooses x_1 to be connected by an edge to y_0. Player **II** is at a loss as to which to choose as she cannot choose y_1 to be connected to v, the latter being an isolated vertex. Thus player **I** wins. The same happens if one graph has an anti-isolated vertex but the other does not. On the other hand, if neither graph has isolated or anti-isolated vertices, player **II** has the following winning strategy in $EF_2(\mathcal{G}, \mathcal{G}')$: She chooses y_0 in an arbitrary way. For y_1 she chooses a vertex which has an edge to x_0 or y_0 if and only if x_1 has an edge to, respectively, y_0 or x_0. Since x_0 and y_0 are neither isolated nor anti-isolated, player **II** can make the choice. We have proved:

Figure 4.4 Player **I** wins.

Figure 4.5 Player **II** wins.

Proposition 4.7 *Suppose $\mathcal{G} = (V, E)$ and $\mathcal{G}' = (V', E')$ are graphs with $V \cap V' = \emptyset$. Player **II** has a winning strategy in $\mathrm{EF}_2(\mathcal{G}, \mathcal{G}')$ if and only if the following conditions hold:*

(1) $\mathcal{G}$ has an isolated vertex if and only if $\mathcal{G}'$ has an isolated vertex.
(2) $\mathcal{G}$ has an anti-isolated vertex if and only if $\mathcal{G}'$ has an anti-isolated vertex.
(3) $\mathcal{G}$ has a vertex which is neither isolated nor anti-isolated if and only if $\mathcal{G}'$ does.

Example 4.8   Player **I** has a winning strategy in $\mathrm{EF}_2(\mathcal{G}, \mathcal{G}')$ if $\mathcal{G}$ and $\mathcal{G}'$ are the two graphs of Figure 4.4. Player **II** has a winning strategy in $\mathrm{EF}_2(\mathcal{G}, \mathcal{G}')$ if $\mathcal{G}$ and $\mathcal{G}'$ are the two graphs of Figure 4.5.

Example 4.9   Player **I** has a winning strategy in $\mathrm{EF}_2(\mathcal{G}, \mathcal{G}')$ if $\mathcal{G}$ and $\mathcal{G}'$ are the two graphs of Figure 4.6. Player **II** has a winning strategy in $\mathrm{EF}_2(\mathcal{G}, \mathcal{G}')$ if $\mathcal{G}$ and $\mathcal{G}'$ are the two graphs of Figure 4.7.

Example 4.10   Suppose $\mathcal{G}$ and $\mathcal{G}'$ are empty graphs, both with at least n vertices. Then player **II** has a winning strategy in $\mathrm{EF}_n(\mathcal{G}, \mathcal{G}')$. She need not worry about condition (G2.2) at all. To satisfy condition (G2.1) she just needs to know that there are enough elements to choose from in both graphs. During n rounds it is necessary to pick at most n elements, so as both graphs have at

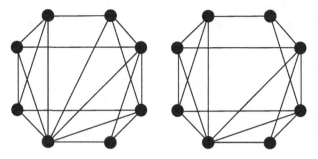

Figure 4.6 Player **I** wins.

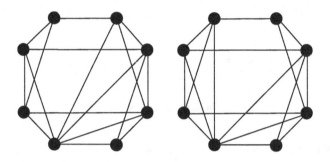

Figure 4.7 Player **II** wins.

least that many elements, player **II** can always win. The situation is exactly the same in complete graphs: If $\mathcal{G}$ and $\mathcal{G}'$ are complete graphs with at least n vertices, then player **II** has a winning strategy in $\text{EF}_n(\mathcal{G}, \mathcal{G}')$. So in a game on graphs with fewer rounds than vertices player **I** cannot distinguish whether the number of vertices is, e.g. even or not.

Example 4.11 Suppose $\mathcal{G}$ and $\mathcal{G}'$ are graphs, both with at least $2n$ vertices, and both with every vertex of degree exactly 1. Then player **II** has a winning strategy in $\text{EF}_n(\mathcal{G}, \mathcal{G}')$. All she has to make sure of is that if player **I** chooses a vertex with an edge to a previously chosen vertex, she does the same. So in a game on graphs with fewer rounds than the number of edges, player **I** cannot distinguish whether the number of edges is, e.g. even or not.

Example 4.12 Suppose $\mathcal{G}$ and $\mathcal{G}'$ are graphs, both with at least n vertices, and both with every vertex of degree exactly 1 except one vertex (the "center") which has an edge to every other element. These graphs look like stars, with one center vertex and all edges emerging from this center. Player **II** has a winning strategy in $\mathrm{EF}_n(\mathcal{G}, \mathcal{G}')$. All she has to make sure of is that if player **I** chooses the center of one graph, she also chooses the center of the other graph. So in a game on graphs with fewer rounds than the number of edges, player **I** cannot distinguish whether the degree of some vertex is, e.g. even, odd, or infinite.

Example 4.13 Suppose $\mathcal{G}$ and $\mathcal{G}'$ are finite graphs, both with at least 2^{n+1} vertices, and both with every vertex of degree exactly 2. These graphs look like collections of cycles. Let us further assume that $\mathcal{G}$ is just one cycle, while $\mathcal{G}'$ consists of two cycles of size 2^n. Player **II** has a winning strategy in $\mathrm{EF}_n(\mathcal{G}, \mathcal{G}')$. Her strategy is the following. Suppose player **I** has just played a vertex v_i in $\mathcal{G}$. Let v_k and v_l be the closest neighbors of v_i played so far. We define the *distance* of two vertices of a graph to be the number of edges on the shortest path connecting the vertices, and ∞ if no such path exists. Player **II** chooses her vertex v_i' so that if the distance d from v_i to v_k (or v_l) is at most 2^{n-i}, then the distance from v_i' to v_k' (or v_l') is exactly d. On the other hand, if the distance from v_i to v_k (or v_l) is $> 2^{n-i}$, then the distance from v_i' to v_k' (or v_l') is also $> 2^{n-i}$. Is this possible to do? Yes, if player **II** is systematic about it. She has to play so that when round m of the game starts, for any two vertices v_i and v_j played in $\mathcal{G}$ so far, and the corresponding vertices v_i' and v_j' in $\mathcal{G}'$, the following holds: If the distance d from v_i to v_k (or v_l) is at most 2^{n-i}, then the distance from v_i' to v_k' (or v_l') is exactly d, but if the distance from v_i to v_k (or v_l) is $> 2^{n-i}$, then the distance from v_i' to v_k' (or v_l') is also $> 2^{n-i}$. Starting with our graphs of size at least 2^{n+1} it is possible for player **II** to maintain this strategy. So in a game on graphs with fewer rounds than the one half of the number of vertices, player **I** cannot distinguish whether the graph is connected or not, i.e. whether any two distinct vertices can be connected by a finite *chain* $(a_0 E a_1 \ldots a_{n-1} E a_n)$ of edges.

4.4 Ehrenfeucht–Fraïssé Games and Elementary Equivalence

In this section we prove a theorem which is the motivation for introducing the Ehrenfeucht–Fraïssé Game in the first place. This theorem relates the game with logic. We prove that if player **II** has a winning strategy in the game $\mathrm{EF}_n(\mathcal{G}, \mathcal{G}')$, then the graphs $\mathcal{G}$ and $\mathcal{G}'$ are so similar that the same sentences of quantifier rank at most n are true in them. In other words, the graphs cannot be distinguished by a first-order sentence without using quantifier rank $> n$. This is a very useful theorem. We then prove the converse, too.

A *position* in an Ehrenfeucht–Fraïssé Game $EF_n(\mathcal{G}, \mathcal{G}')$ is defined as for general games. However, because of the special character of the Ehrenfeucht–Fraïssé Game we can define a more useful concept of position. Suppose $p = (x_0, y_0, \ldots, x_{i-1}, y_{i-1})$ is a position of the game $EF_n(\mathcal{G}, \mathcal{G}')$. We call the corresponding set of pairs

$$f_p = \{(v_0, v_0'), \ldots, (v_{i-1}, v_{i-1}')\}$$

the *map* of the position p, or a *position map* of the game $EF_n(\mathcal{G}, \mathcal{G}')$. Note that unless player **II** has already lost the game, f_p is a function with $\mathrm{dom}(f_p) \subseteq V_\mathcal{G}$ and $\mathrm{rng}(f_p) \subseteq V_{\mathcal{G}'}$. Such functions are called *partial mappings* $V_\mathcal{G} \to V_{\mathcal{G}'}$.

If s is an assignment of $\mathcal{G}$ and f is a partial mapping $V_\mathcal{G} \to V_{\mathcal{G}'}$ such that $\mathrm{rng}(s) \subseteq \mathrm{dom}(f)$, then the equation

$$(f \circ s)(x) = f(s(x))$$

defines an assignment $f \circ s$ of $\mathcal{G}'$.

Now we are ready to prove the main theorem about Ehrenfeucht–Fraïssé Games on graphs and logic:

Theorem 4.14 *Suppose $\mathcal{G}$ and $\mathcal{G}'$ are two graphs. If player **II** has a winning strategy in $EF_n(\mathcal{G}, \mathcal{G}')$, then the graphs $\mathcal{G}$ and $\mathcal{G}'$ satisfy the same sentences of quantifier rank $\leq n$.*

Proof Before giving a detailed proof it is perhaps helpful to consider a special case. So let us suppose player **II** has a winning strategy τ in $EF_3(\mathcal{G}, \mathcal{G}')$ and we have the simple sentence $\exists z \forall x \exists y (xEy \wedge zEy)$. Note that this sentence says that there is a special "popular" vertex a such that every vertex has a neighbor which is a neighbor of a. We show that if $\mathcal{G}$ satisfies $\exists z \forall x \exists y (xEy \wedge zEy)$ then so does $\mathcal{G}'$. For this, let $\mathcal{G}$ indeed satisfy $\exists z \forall x \exists y (xEy \wedge zEy)$. Thus $\mathcal{G}$ has a "popular" vertex a:

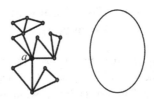

To prove that $\mathcal{G}'$ satisfies $\exists z \forall x \exists y (xEy \wedge zEy)$ we let player **I** start the game $EF_3(\mathcal{G}, \mathcal{G}')$ by playing $x_0 = a$. Player **II** responds according to her winning strategy τ with an element a' of $\mathcal{G}$:

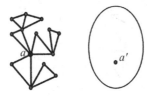

We need to show that a' is a "popular" vertex in $\mathcal{G}'$. So let us take an element b' from $\mathcal{G}'$ and try to find it a neighbor that is also a neighbor of a':

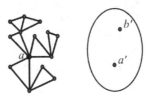

The trick is that we continue the game $\mathrm{EF}_3(\mathcal{G}, \mathcal{G}')$ by letting player **I** play $x_1 = b'$ in $\mathcal{G}'$. Player **II** responds according to her winning strategy τ with an element b of $\mathcal{G}$:

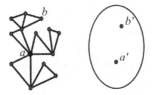

Since a is "popular" in $\mathcal{G}$, the vertex b has a neighbor c which is also a neighbor of a:

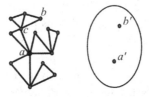

We finish the game $\mathrm{EF}_3(\mathcal{G}, \mathcal{G}')$ by letting player **I** play $x_2 = c$ in $\mathcal{G}$. Player **II** responds according to her winning strategy τ with an element c' of $\mathcal{G}'$:

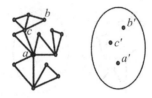

Since τ is a winning strategy, and c is a neighbor of both a and b, c' is a neighbor of both a' and b':

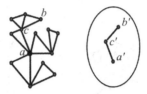

To prove the theorem we actually prove a more general claim from which the theorem follows: If player **II** has a winning strategy in $\mathrm{EF}_n(\mathcal{G}, \mathcal{G}')$ in position $p = (x_0, y_0, \ldots, x_{n-k-1}, y_{n-k-1})$, then $\mathcal{G} \models_s \varphi$ if and only if $\mathcal{G}' \models_{f_p \circ s} \varphi$ for all formulas φ of quantifier rank $\leq k$ and all assignments s of $\mathcal{G}$ such that $\mathrm{rng}(s) \subseteq \mathrm{dom}(f_p)$.

The proof is by induction on k. For $k = 0$ there is nothing to prove. Let us assume the claim for k as an induction hypothesis and prove it for $k + 1$. Let $\tau = (\tau_0, \ldots, \tau_k)$ be a winning strategy of player **II** in $\mathrm{EF}_n(\mathcal{G}, \mathcal{G}')$ in position $p = (x_0, y_0, \ldots, x_{n-k-2}, y_{n-k-2})$. Suppose φ is a formula of quantifier rank $\leq k + 1$ and s is an assignment of $\mathcal{G}$ such that $\mathrm{rng}(s) \subseteq \mathrm{dom}(f_p)$. We show: $\mathcal{G} \models_s \varphi$ if and only if $\mathcal{G}' \models_{f_p \circ s} \varphi$.

Case 1 φ is atomic. The claim follows from the fact that player **II** has not lost yet.

Case 2 $\varphi = \exists x \psi$, where ψ is of quantifier rank $\leq k$. The induction hypothesis applies to ψ. We assume that $\mathcal{G} \models_s \exists x \psi$ and show that $\mathcal{G}' \models_{f_p \circ s} \exists x \psi$. Let therefore $a \in V_\mathcal{G}$ be such that $\mathcal{G} \models_{s[a/x]} \psi$. Our goal is to find $b \in V_{\mathcal{G}'}$ such that $\mathcal{G}' \models_{(f_p \circ s)[b/x]} \psi$. Let us play $\mathrm{EF}_n(\mathcal{G}, \mathcal{G}')$ in position p so that player **I** first chooses $x_{n-k-1} = a \in V_\mathcal{G}$ and then player **II** uses her winning strategy τ to pick $y_{n-k-1} = b \in V_{\mathcal{G}'}$. We have a new position $p' = (x_0, y_0, \ldots, x_{n-k-1}, y_{n-k-1})$ and player **II** still has a winning strategy in the position p'. By the Induction Hypothesis, $\mathcal{G} \models_{s[a/x]} \psi$ implies $\mathcal{G}' \models_{(f_{p'}) \circ (s[a/x])} \psi$. Now comes a crucial but simple observation: $(f_{p'}) \circ (s[a/x]) = (f_p \circ s)[b/x]$. Thus $\mathcal{G}' \models_{(f_p \circ s)[b/x]} \psi$ as was wanted.

Similarly we could start from $\mathcal{G}' \models_{f_p \circ s} \exists x \psi$ and derive $\mathcal{G} \models_s \exists x \psi$.

Case 3 φ is one of $\neg\psi$, $\psi \wedge \theta$, $\psi \vee \theta$ of $\forall x\psi$. Now the claim follows from cases 1 and 2. □

Corollary *There is no sentence φ of the first-order language of graphs such that the following are equivalent for all graphs $\mathcal{G}$:*

(1) $\mathcal{G}$ has an even number of vertices.
(2) $\mathcal{G} \models \varphi$.

We can replace condition (1) by any of

(1.1) $\mathcal{G}$ has an even number of edges.
(1.2) $\mathcal{G}$ has a vertex of infinite degree.
(1.3) $\mathcal{G}$ is connected.

Proof Suppose such a φ exists and it has quantifier rank n. In Example 4.10 we observed that if $\mathcal{G}$ is an empty graph of size n and $\mathcal{G}'$ an empty graph of size $n + 1$, then player **II** wins the game $\mathrm{EF}_n(\mathcal{G}, \mathcal{G}')$. By Theorem 4.14 we have $\mathcal{G} \models \varphi$ if and only if $\mathcal{G}' \models \varphi$. This contradicts the choice of φ. In the remaining cases we appeal to Example 4.11, Example 4.12, and Example 4.13. □

In the next proposition we estimate roughly the number of formulas of a given quantifier rank. We count this number only up to logical equivalence, for otherwise there would always be infinitely many, as we can repeat the same formula: φ, $\varphi \wedge \varphi$, $\varphi \wedge \varphi \wedge \varphi$, etc.

Proposition 4.15 *For every n and for every set $\{x_1, \ldots, x_n\}$ of variables, there are only finitely many logically non-equivalent formulas of the first-order language of graphs of quantifier rank $< n$ with the free variables $\{x_1, \ldots, x_n\}$.*

Proof We use induction on n. For $n = 0$ the number of formulas is also 0. Assume then that there are only finitely many logically non-equivalent formulas of the first-order language of graphs of quantifier rank $< n$ with the free variables $\{x_1, \ldots, x_n\}$. We prove the same for $n + 1$. We have formulas of the form $\exists x\varphi$ and $\forall x\varphi$, where the quantifier rank of φ is $< n$. They are only finitely many, up to logical equivalence – say k many. Then we have Boolean combinations of such formulas. A truth table for k propositional symbols has 2^k rows. Each row can have a value of 0 or 1. Thus there are at most 2^{2^k} such truth tables. Therefore there can be at most 2^{2^k} logically non-equivalent formulas of quantifier rank $< n + 1$. □

Theorem 4.16 *Suppose $\mathcal{G}$ and $\mathcal{G}'$ are two graphs. If $\mathcal{G}$ and $\mathcal{G}'$ satisfy the same sentences of quantifier rank $\leq n$, then player **II** has a winning strategy in $\mathrm{EF}_n(\mathcal{G}, \mathcal{G}')$.*

Proof Assume $\mathcal{G}$ and $\mathcal{G}'$ satisfy the same sentences of quantifier rank $\leq n$. Player **II** then has the following winning strategy in $\mathrm{EF}_n(\mathcal{G}, \mathcal{G}')$: She makes sure that after she has played, the position $p = (x_0, y_0, \ldots, x_{i-1}, y_{i-1})$ satisfies $\mathcal{G} \models_s \varphi$ if and only if $\mathcal{G}' \models_{f_p \circ s} \varphi$ for all formulas φ of quantifier rank $\leq n - i$ with free variables $z_0, \ldots, z_{i-1}$ and all assignments s of $\mathcal{G}$ such that $s(z_j) = v_j$ for $j = 0, \ldots, i - 1$. This condition holds in the beginning of the game by assumption. Let us suppose we are in a position $p = (x_0, y_0, \ldots, x_{i-1}, y_{i-1})$ and player **II** has so far been able to follow her strategy. Now player **I** plays $x_i = a$. We assume $x_i \in V_{\mathcal{G}}$, as the two cases are symmetrical. Let s be an assignment of $\mathcal{G}$ such that $s(z_j) = v_j$ for $j = 0, \ldots, i-1$. Let Φ be the conjunction of all formulas φ of quantifier rank $< n - i - 1$ with free variables in the set $\{z_0, \ldots, z_{i-1}, z_i\}$ such that $\mathcal{G} \models_{s[a/z_i]} \varphi$. Now $\mathcal{G} \models_s \exists z_i \Phi$. Since the quantifier rank of Φ is $\leq n - i$, by the induction hypothesis $\mathcal{G}' \models_{f_p \circ s} \exists z_i \Phi$. Let $y_i = b \in V_{\mathcal{G}'}$ such that $\mathcal{G}' \models_{(f_p \circ s)[b/z_i]} \Phi$. Let $p' = (x_0, y_0, \ldots, x_i, y_i)$. Clearly, $\mathcal{G} \models_s \varphi$ if and only if $\mathcal{G}' \models_{f_{p'} \circ s} \varphi$ for all φ of quantifier rank $\leq n - i - 1$. So player **II** has been able to maintain her strategy in the new position p'. $\square$

Corollary (Ehrenfeucht, Fraïssé) *Suppose $\mathcal{G}$ and $\mathcal{G}'$ are two graphs. The following are equivalent:*

(1) $\mathcal{G}$ and $\mathcal{G}'$ satisfy the same sentences of quantifier rank $\leq n$.
*(2) Player **II** has a winning strategy in $\mathrm{EF}_n(\mathcal{G}, \mathcal{G}')$.*

This corollary has numerous applications. We can immediately conclude that for each n there are only finitely many graphs such that player **I** has a winning strategy in $\mathrm{EF}_n(\mathcal{G}, \mathcal{G}')$ for any distinct $\mathcal{G}$ and $\mathcal{G}'$ of them. Why? Because there are, up to logical equivalence, only finitely many sets S of sentences of quantifier rank $\leq n$. If two graphs $\mathcal{G}$ and $\mathcal{G}'$ satisfy the same such set S, then by the above Corollary player **II** has a winning strategy in $\mathrm{EF}_n(\mathcal{G}, \mathcal{G}')$.

4.5 Historical Remarks and References

The Ehrenfeucht–Fraïssé Game was introduced in Ehrenfeucht (1957) and Ehrenfeucht (1960/1961). The idea of Extension Axioms (see Exercise 4.16) was introduced in Gaifman (1964) and used effectively in Fagin (1976) and later, e.g. in Kolaitis and Vardi (1992). Distributive Normal Forms (See Exercise 4.19) were introduced in Hintikka (1953). There is a rich literature on Ehrenfeucht–Fraïssé Games on graphs. See Grädel et al. (2007) for a recent handbook. See Spencer (2001) for applications of Ehrenfeucht–Fraïssé Games to random graphs.

n	Has a winning strategy
1	**II**
2	
3	
4	**I**

Figure 4.8 Who wins?

Exercises

4.1 A *path* in a graph $\mathcal{G}$ is a finite sequence $v_0, \ldots, v_n$ such that for every $i = 0, \ldots, n - 1$ there is an edge from v_i to v_{i+1}. The number n is the *length* of the path. The path is said to start from v_0 and end in v_n (or vice versa). Write for each $n > 0$ a first-order formula $\text{PTH}_n(x, y)$ such that an assignment s satisfies the formula in a graph if and only if there is a path of length n starting from $s(x)$ and ending in $s(y)$.

4.2 Solve the previous problem so that the number of variables of the formula is three irrespective of n.

4.3 Show that if $\mathcal{G} \models_s \varphi$, then player **II** has the following winning strategy in the Semantic Game starting with (her holding) (φ, s): she plays so that if she holds (φ', s'), then $\mathcal{G} \models_{s'} \varphi'$, and if player **I** holds (φ', s'), then $\mathcal{G} \not\models_{s'} \varphi'$.

4.4 Use induction on φ to prove for all s that if player **II** has a winning strategy in the Semantic Game starting with (φ, s), then $\mathcal{G} \models_s \varphi$, and if **II** has a winning strategy starting with $(\neg\varphi, s)$, then $\mathcal{G} \not\models_s \varphi$.

4.5 Suppose $\mathcal{G}$ and $\mathcal{G}'$ are chains, i.e. connected graphs in which every vertex has degree 2 except two (the endpoints) which have degree 1. Suppose $\mathcal{G}$ has five vertices and $\mathcal{G}'$ has six. Complete the table of Figure 4.8 which indicates for $1 \leq n \leq 6$ which of the two players has a winning strategy in $\text{EF}_n(\mathcal{G}, \mathcal{G}')$.

4.6 Solve Exercise 1 in the case that $\mathcal{G}$ has six vertices and $\mathcal{G}'$ has seven.

4.7 Show that if player **II** has a winning strategy in $\text{EF}_n(\mathcal{G}, \mathcal{G}')$ and also in $\text{EF}_n(\mathcal{G}', \mathcal{G}'')$, then she has one in $\text{EF}_n(\mathcal{G}, \mathcal{G}'')$.

4.8 Show that if player **I** has a winning strategy in $\text{EF}_n(\mathcal{G}, \mathcal{G}')$ and player **II** has a winning strategy in $\text{EF}_n(\mathcal{G}', \mathcal{G}'')$, then player **I** has a winning strategy in $\text{EF}_n(\mathcal{G}, \mathcal{G}'')$.

4.9 Let us consider graphs $\mathcal{G}$ whose set of vertices is $\{1, \ldots, n\}$. There are $n(n-1)/2$ possible edges and therefore $2^{n(n-1)/2}$ possible graphs with these vertices. How many such graphs $\mathcal{G}$ are there such that player **I** has a winning strategy in $\mathrm{EF}_2(\mathcal{G}, \mathcal{G}')$ for any two distinct $\mathcal{G}$ and $\mathcal{G}'$ of them?

4.10 Show that the property "$V_{\mathcal{G}}$ is finite" of a graph $\mathcal{G}$ is not expressible in the first-order language of graphs, i.e. there is no sentence φ of the first-order language of graphs such that the following are equivalent for all graphs $\mathcal{G}$:

    (1) $V_{\mathcal{G}}$ is finite.

    (2) $\mathcal{G} \models \varphi$.

4.11 Show that the property "$\mathcal{G}$ has a Hamiltonian path" (i.e. a sequence of connected edges which pass through every vertex exactly once) of a graph $\mathcal{G}$ is not expressible in the first-order language of graphs.

4.12 Show that the property "$\mathcal{G}$ has an infinite clique" (i.e. a complete subgraph, or still in other words, a subset any two vertices of which are connected by an edge) of a graph $\mathcal{G}$ is not expressible in the first-order language of graphs.

4.13 Show that the property "$\mathcal{G}$ has a cycle path" (i.e. a sequence of connected edges which starts and ends with the same vertex) of a graph $\mathcal{G}$ is not expressible in the first-order language of graphs.

4.14 Show that for every $n > 3$ there are at least 2^{n-1} graphs with at most $2n$ vertices such that no two of them satisfy exactly the same sentences of quantifier rank $< n + 1$.

4.15 Let n be a natural number. Define an equivalence relation among all graphs as follows: $\mathcal{G} \sim \mathcal{G}'$ if $\mathcal{G}$ and $\mathcal{G}'$ satisfy the same sentences of quantifier rank $\leq n$. Show that $\sim$ has but finitely many equivalence classes and each equivalence class C is definable[1] by a sentence of quantifier rank $\leq n$ (i.e. there is a sentence of quantifier rank $\leq n$ such that for all graphs $\mathcal{G}$ we have: $\mathcal{G} \in C$ if and only if $\mathcal{G} \models \varphi$).

4.16 A graph $\mathcal{G}$ is said to satisfy the *extension axiom* E_n if for any two disjoint sets W and U of vertices, with $W \cup U$ of size $< n$, there is a vertex, not in $W \cup U$, which is connected by an edge to every element of W but to none of U. Show that if $\mathcal{G}$ and $\mathcal{G}'$ both satisfy the extension axiom E_n, then player **II** has a winning strategy in the game $\mathrm{EF}_n(\mathcal{G}, \mathcal{G}')$. (Note: if we take a set of m vertices and toss a coin to decide about each pair of different vertices whether we put and edge between them or not, then with a probability which tends to 1 as m tends to infinity, we get a graph which satisfies the extension axioms E_n.)

4.17 The k-*Pebble Game* on two graphs $\mathcal{G}$ and $\mathcal{G}'$ is defined as follows. There are k pairs of pebbles. Each pair is numbered with numbers $1, \ldots, k$.

[1] The concepts "definable" and "expressible" are synonymous.

The two pebbles in a pair are said to correspond to each other. During the game player **I** takes a pebble and places it on a vertex of $\mathcal{G}$ (or $\mathcal{G}'$). Then player **II** places the corresponding pebble on a vertex of $\mathcal{G}'$ (or $\mathcal{G}$). This is repeated ad infinitum. When player **I** takes a pebble, he can take one that has not been played yet or one that has been played already. In the end player **II** wins if there is an edge in $\mathcal{G}$ between two pebbled vertices if and only if there is an edge in $\mathcal{G}'$ between the corresponding vertices. Show that if the graphs $\mathcal{G}$ and $\mathcal{G}'$ both satisfy the extension axiom E_k of Exercise 4.16, then player **II** has a winning strategy in the k-Pebble Game.

4.18 Suppose $\mathcal{G}$ is a graph. Let $\sigma^0_{\mathcal{G},a_0,\dots,a_{m-1}}(x_0,\dots,x_{m-1})$ be the conjunction of

1. $x_i = x_j$, if $a_i = a_j$.
2. $\neg x_i = x_j$, if $a_i \neq a_j$.
3. $x_i E x_j$, if $a_i E_{\mathcal{G}} a_j$.
4. $\neg x_i E x_j$, if not $a_i E_{\mathcal{G}} a_j$.

Let

$$\sigma^{n+1}_{\mathcal{G},a_0,\dots,a_{m-1}}(x_0,\dots,x_{m-1}) = \forall x_m \bigvee_{a_m \in V_{\mathcal{G}}} \sigma^n_{\mathcal{G},a_0,\dots,a_m}(x_0,\dots,x_m)$$
$$\wedge \bigwedge_{a_m \in V_{\mathcal{G}}} \exists x_m \sigma^n_{\mathcal{G},a_0,\dots,a_m}(x_0,\dots,x_m).$$

Note that even if the graph $\mathcal{G}$ may be infinite, the disjunction $\bigvee_{a_m \in V_{\mathcal{G}}}$ and the conjunction $\bigwedge_{a_m \in V_{\mathcal{G}}}$ above is over a finite set. Prove that the following are equivalent for all graphs $\mathcal{G}$ and $\mathcal{G}'$:

1. $\mathcal{G}' \models \sigma^n_{\mathcal{G}}$.
2. Player **II** has a winning strategy in $\mathrm{EF}_n(\mathcal{G},\mathcal{G}')$.

4.19 Suppose $\mathcal{G}$ is a graph. Let $\sigma^0_{\mathcal{G},a_0,\dots,a_{m-1}}(x_0,\dots,x_{m-1})$ be defined as in the previous problem. Let $S^0_m(x_0,\dots,x_{m-1})$ be the set of all sentences $\sigma^0_{\mathcal{G},a_0,\dots,a_{m-1}}(x_0,\dots,x_{m-1})$ where $\mathcal{G}$ runs through all graphs and $a_0,\dots,a_{m-1}$ through all sequences of elements of each $\mathcal{G}$. Note that this is a finite set. Suppose $S^n_{m+1}(x_0,\dots,x_m) = \{\sigma^n_i(x_0,\dots,x_m) : i \in I\}$ has been defined and is finite. Let for $J \subseteq I$:

$$\sigma^{n+1}_J(x_0,\dots,x_{m-1}) = \forall x_m \bigvee_{i \in J} \sigma^n_i(x_0,\dots,x_m) \wedge \bigwedge_{i \in J} \exists x_m \sigma^n_i(x_0,\dots,x_m)$$

and

$$S^{n+1}_m = \{\sigma^{n+1}_J(x_0,\dots,x_{m-1}) : J \subseteq I\}.$$

Prove that the following are equivalent for all formulas φ with free variables $x_0, \ldots, x_{m-1}$:

1. φ is equivalent to a formula of quantifier rank $\leq n$.
2. There are unique $\psi_1, \ldots, \psi_n$ in S_m^n such that $\models \varphi \leftrightarrow \psi_1 \vee \cdots \vee \psi_n$.

This is called the Distributive Normal Form of φ.

5

Models

5.1 Introduction

The concept of a model (or structure) is one of the most fundamental in logic. In brief, while the meaning of logical symbols $\wedge, \vee, \exists, \ldots$ is always fixed, models give meaning to non-logical symbols such as constant, predicate, and function symbols. When we have agreed about the meaning of the logical and non-logical symbols of logic, we can then define the meaning of arbitrary formulas.

Depending on context and preference, models appear in logic in two roles. They can serve the auxiliary role of clarifying logical derivation. For example, one quick way to tell what it means for φ to be a logical consequence of ψ is to say that in every model where ψ is true also φ is true. It is then an almost trivial matter to understand why for example $\forall x \exists y \varphi$ is a logical consequence of $\exists y \forall x \varphi$ but $\forall y \exists x \varphi$ is in general not.

Alternatively models can be the prime objects of investigation and it is the logical derivation that is in an auxiliary role of throwing light on properties of models. This is manifestly demonstrated by the Completeness Theorem which says that any set T of first-order sentences has a model unless a contradiction can be logically derived from T, which entails that the two alternative perspectives of models are really equivalent. Since derivations are finite, this implies the important Compactness Theorem: If a set of first-order sentences is such that each of its finite subsets has a model it itself has a model. The Compactness Theorem has led to an abundance of non-isomorphic models of first-order theories, and constitutes the origin of the whole subject of Model Theory. In this chapter models are indeed the prime objects of investigation and we introduce auxiliary concepts such as the Ehrenfeucht–Fraïssé Game that help us understand models.

We use the words "model" and "structure" as synonyms. We have a slight preference for the word "structure" in a context where absolute generality prevails and the structures are not assumed to satisfy any particular axioms. Respectively, our preference is to call a structure that satisfies some given axioms a model, so a structure satisfying a theory is called a model of the theory.

5.2 Basic Concepts

A *vocabulary* is any set L of predicate symbols $P, Q, R, \ldots$, function symbols $f, g, h, \ldots$, and constant symbols $c, d, e, \ldots$. Each vocabulary has an *arity-function*

$$\#_L : L \to \mathbb{N}$$

which tells the arity of each symbol. Thus if $P \in L$, then P is a $\#_L(P)$-ary predicate symbol. If $f \in L$, then f is a $\#_L(f)$-ary function symbol. Finally, $\#_L(c)$ is assumed to be 0 for constants $c \in L$. Predicate or function symbols of arity 1 are called *unary* or *monadic*, and those of arity 2 are called *binary*. A vocabulary is called unary (or binary) if it contains only unary (respectively, binary) symbols. A vocabulary is called *relational* if it contains no function or constant symbols.

Definition 5.1 An *L-structure* (or *L-model*) is a pair $\mathcal{M} = (M, \text{Val}_{\mathcal{M}})$, where M is a non-empty set called the *universe (or the domain)* of $\mathcal{M}$, and $\text{Val}_{\mathcal{M}}$ is a function defined on L with the following properties:

1. If $R \in L$ is a relation symbol and $\#_L(R) = n$, then $\text{Val}_{\mathcal{M}}(R) \subseteq M^n$.
2. If $f \in L$ is a function symbol and $\#_L(f) = n$, then $\text{Val}_{\mathcal{M}}(f) : M^n \to M$.
3. If $c \in L$ is a constant symbol, then $\text{Val}_{\mathcal{M}}(c) \in M$.

We use $\text{Str}(L)$ to denote the class of all L-structures.

We usually shorten $\text{Val}_{\mathcal{M}}(R)$ to $R^{\mathcal{M}}$, $\text{Val}_{\mathcal{M}}(f)$ to $f^{\mathcal{M}}$, and $\text{Val}_{\mathcal{M}}(c)$ to $c^{\mathcal{M}}$. If no confusion arises, we use the notation

$$\mathcal{M} = (M, R_1^{\mathcal{M}}, \ldots, R_n^{\mathcal{M}}, f_1^{\mathcal{M}}, \ldots, f_m^{\mathcal{M}}, c_1^{\mathcal{M}}, \ldots, c_k^{\mathcal{M}})$$

for an L-structure $\mathcal{M}$, where $L = \{R_1, \ldots, R_n, f_1, \ldots, f_m, c_1 \ldots, c_k\}$.

Example 5.2 Graphs are L-structures for the relational vocabulary $L = \{E\}$, where E is a predicate symbol with $\#_L(E) = 2$. Groups are L-structures for $L = \{\circ\}$, where $\circ$ is a binary function symbol. Fields are L-structures for $L = \{+, \cdot, 0, 1\}$, where $+, \cdot$ are binary function symbols and $0, 1$ are constant symbols. Ordered sets (i.e. linear orders) are L-structures for the relational vocabulary $L = \{<\}$, where $<$ is a binary predicate symbol. If $L = \emptyset$, an L-structure (M) is a structure with just the universe and no structure in it.

If $\mathcal{M}$ is a structure and π maps M bijectively onto another set M', we can use π to copy the relations, functions, and constants of $\mathcal{M}$ on M'. In this way we get a perfect copy $\mathcal{M}'$ of $\mathcal{M}$ which differs from $\mathcal{M}$ only in the respect that the underlying elements are different. We then say that $\mathcal{M}'$ is an isomorphic copy of $\mathcal{M}$. For all practical purposes we consider the structures $\mathcal{M}$ and $\mathcal{M}'$ as one and the same structure. However, they are not the same structure, just isomorphic. This may sound as if isomorphism was a rather trivial matter, but this is not true. In many cases it is a highly non-trivial enterprise to investigate whether two structures are isomorphic or not. In the realm of finite structures the question of deciding whether two given structures are isomorphic or not is a famous case of a complexity question which is between P (polynomial time) and NP (non-deterministic polynomial time) and about which we do not know whether it is NP-complete. In the light of present knowledge it is conceivable that this question is strictly between P and NP.

Definition 5.3 L-structures $\mathcal{M}$ and $\mathcal{M}'$ are *isomorphic* if there is a bijection

$$\pi : M \to M'$$

such that

1. For all $a_1, \ldots, a_{\#_L(R)} \in M$:

$$(a_1, \ldots, a_{\#_L(R)}) \in R^{\mathcal{M}} \iff (\pi(a_1), \ldots, \pi(a_{\#_L(R)})) \in R^{\mathcal{M}'}.$$

2. For all $a_1, \ldots, a_{\#_L(f)} \in M$:

$$f^{\mathcal{M}'}(\pi(a_1), \ldots, \pi(a_{\#_L(f)})) = \pi(f^{\mathcal{M}}(a_1, \ldots, a_{\#_L(f)})).$$

3. For all $c \in L$: $\pi(c^{\mathcal{M}}) = c^{\mathcal{M}'}$.

In this case we say that π is an *isomorphism* $\mathcal{M} \to \mathcal{M}'$, denoted

$$\pi : \mathcal{M} \cong \mathcal{M}'.$$

If also $\mathcal{M} = \mathcal{M}'$, we say that π is an *automorphism* of $\mathcal{M}$.

Example 5.4 *Unary* (or *monadic*) structures, i.e. L-structures for unary L, are particularly simple and easy to deal with. Figure 5.1 depicts a unary structure. Suppose L consists of unary predicate symbols $R_1, \ldots, R_n$ and $\mathcal{A}$ is an L-structure. If $X \subseteq A$ and $d \in \{0, 1\}$, let $X^d = X$ if $d = 0$ and $X^d = A \setminus X$

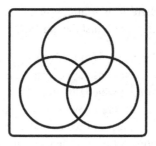

Figure 5.1 A unary structure.

otherwise. Suppose $\epsilon : \{1, \ldots, n\} \to \{0, 1\}$. The ϵ-*constituent* of $\mathcal{A}$ is the set

$$C_\epsilon(\mathcal{A}) = \bigcap_{i=1}^{n} (R_i^{\mathcal{A}})^{\epsilon(i)}.$$

A priori, the 2^n sets $C_\epsilon(\mathcal{A})$ can each have any cardinality whatsoever. It is the nature of unary structures that the constituents are totally independent of each other. If $\mathcal{A} \cong \mathcal{B}$, then

$$|C_\epsilon(\mathcal{A})| = |C_\epsilon(\mathcal{B})| \tag{5.1}$$

for every ϵ. Conversely, if two L-structures $\mathcal{A}$ and $\mathcal{B}$ satisfy Equation (5.1) for every ϵ, then $\mathcal{A} \cong \mathcal{B}$ (see Exercise 5.6). We can say that the function $\epsilon \mapsto |C_\epsilon(\mathcal{A})|$ characterizes completely (i.e. up to isomorphism) the unary structure $\mathcal{A}$. There is nothing more we can say about $\mathcal{A}$ but this function.

Example 5.5 *Equivalence relations*, i.e. L-structures $\mathcal{M}$ for $L = \{\sim\}$ such that $\sim^{\mathcal{M}}$ is a symmetric ($x \sim y \Rightarrow y \sim x$), transitive ($x \sim y \sim z \Rightarrow x \sim z$), and reflexive ($x \sim x$) relation on M can be characterized almost as easily as unary structures. Let for every cardinal number $\kappa \leq |M|$ the number of equivalence classes of $\sim^{\mathcal{M}}$ of cardinality κ be denoted by $EC_\kappa(\mathcal{M})$. If $\mathcal{A} \cong \mathcal{B}$, then

$$EC_\kappa(\mathcal{A}) = EC_\kappa(\mathcal{B}) \tag{5.2}$$

for every $\kappa \leq |A|$. Conversely, if two L-structures $\mathcal{A}$ and $\mathcal{B}$ satisfy Equation (5.2) for every $\kappa \leq |A \cup B|$, then $\mathcal{A} \cong \mathcal{B}$ (see Exercise 5.12). We can say that the function $\kappa \mapsto EC_\kappa(\mathcal{A})$ characterizes completely (i.e. up to isomorphism) the equivalence relation $\mathcal{A}$. There is nothing more we can say about $\mathcal{A}$ but this function. For equivalence relations on a finite universe of size n this function is a function $f : \{1, \ldots, n\} \to \{0, \ldots, n\}$ such that

$$\sum_{i=1}^{n} i f(i) = n.$$

The so-called Hardy–Ramanujan asymptotic formula says that the number of equivalence relations on a fixed set of n elements is asymptotically

$$\frac{1}{4\sqrt{3n}} e^{\pi \sqrt{2/3}\sqrt{n}}.$$

So this is also an asymptotic upper bound for the number of non-isomorphic equivalence relations on a universe of n elements.

Example 5.6 The ordered sets (i.e. linear orders)

$$\mathcal{M} = \left(\left(-\tfrac{1}{2}, \tfrac{1}{2}\right), <\right)$$
$$\mathcal{M}' = (\mathbb{R}, <)$$

are isomorphic, as the mapping $x \mapsto \tan(\frac{x}{\pi})$ demonstrates. The ordered sets $\mathcal{M}'$ and

$$\mathcal{M}'' = \left(\left[-\tfrac{1}{2}, \tfrac{1}{2}\right], <\right)$$

are not isomorphic as is easily seen by considering what would be mapped onto the endpoints $-\tfrac{1}{2}$ and $\tfrac{1}{2}$. These were easy examples, but it is in general very difficult to tell in simple terms when two linear orders are isomorphic. For finite linear orders the answer is trivial, but already countable linear orders present great problems. There are special cases that are easier to deal with, for example the case of countable dense linear orders. A linear order is *dense* if it has at least two elements and between any two distinct elements there is always a third element. Examples of such are

$$\begin{aligned}
&(\mathbb{Q}, <) \\
&(\{r \in \mathbb{R} : 0 \le r \le 1\}, <) \\
&(\{q \in \mathbb{Q} : 0 \le q < 1\}, <) \\
&(\{q \in \mathbb{Q} : 0 < q \le 1\}, <).
\end{aligned} \tag{5.3}$$

For a linear order $\mathcal{M}$, let $SG(\mathcal{M}) = (d_0, d_1)$, where d_0 is 1 if $\mathcal{M}$ has a smallest element and otherwise 0, and d_1 is 1 if $\mathcal{M}$ has a largest element and otherwise 0. All four possibilities of $SG(\mathcal{M})$ occur in the list (5.3). If $\mathcal{M}$ and $\mathcal{N}$ are countable dense linear orders, then $\mathcal{M} \cong \mathcal{N}$ if and only if $SG(\mathcal{M}) = SG(\mathcal{N})$ (see Exercise 5.14 and Figure 5.2). Thus for countable dense linear orders SG reveals everything. This is not at all true of non-dense linear orders or uncountable dense linear orders (see Exercise 5.14). A *well-order* is a linear order in

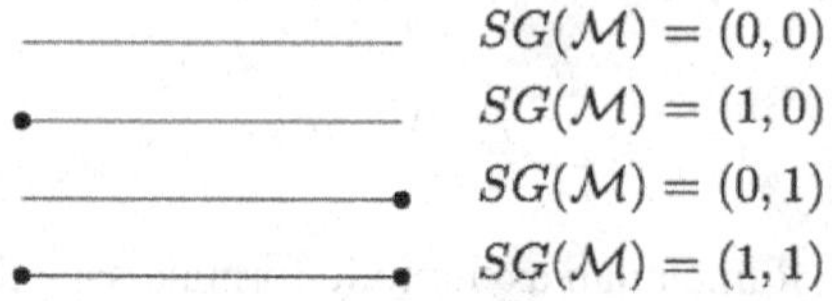

$$SG(\mathcal{M}) = (0,0)$$
$$SG(\mathcal{M}) = (1,0)$$
$$SG(\mathcal{M}) = (0,1)$$
$$SG(\mathcal{M}) = (1,1)$$

Figure 5.2 The four types of countable dense linear orders.

which every non-empty subset has a smallest element (or equivalently, there are no infinite descending sequences). Any well-order $\mathcal{M}$ can be assigned an invariant, or *ordinal* as follows: If $x \in M$, let

$$o_{\mathcal{M}}(x) = \sup\{o_{\mathcal{M}}(y) + 1 : y <^{\mathcal{M}} x\},$$

$$o(\mathcal{M}) = \sup\{o_{\mathcal{M}}(x) : x \in M\}.$$

Now two well-orders $\mathcal{M}$ and $\mathcal{N}$ are isomorphic if and only if $o(\mathcal{M}) = o(\mathcal{N})$ (see Exercise 5.15). The *sum* $\mathcal{M} + \mathcal{N}$ of two linear orders $\mathcal{M}$ and $\mathcal{N}$ is the linear order obtained by putting the linear orders one after the other, first $\mathcal{M}$ and then $\mathcal{N}$. The more general sum $\Sigma_{i \in \mathcal{N}} \mathcal{M}_i$, where $\mathcal{N}$ and $\mathcal{M}_i$, $i \in N$, are linear orders, is defined as follows. It is a linear order obtained by replacing in $\mathcal{N}$ each element i by a copy of $\mathcal{M}_i$ Thus it is the set $\{(i, a) : i \in N, a \in M_i\}$ ordered by $(i, a) < (j, b)$ iff either $i <_{\mathcal{N}} j$ or $i = j$ and $a <_{\mathcal{M}_i} b$.

Example 5.7 A *partially ordered set* is an L-structure $\mathcal{M} = (M, \leq^{\mathcal{M}})$ for the vocabulary $L = \{\leq\}$, where $\leq^{\mathcal{M}}$ is assumed to be reflexive, transitive, and anti-symmetric ($x \leq y \leq x \Rightarrow x = y$). We shorten ($x \leq^{\mathcal{M}} y$ & $x \neq y$) to $x <^{\mathcal{M}} y$. A *predecessor* of x is any $y \in M$ with $y <^{\mathcal{M}} x$. An *immediate predecessor* of x is any y such that $y <^{\mathcal{M}} x$ and there is no z with $y <^{\mathcal{M}} z <^{\mathcal{M}} x$. A *successor* of x is any $y \in M$ with $x <^{\mathcal{M}} y$. An *immediate successor* of x is any y such that $x <^{\mathcal{M}} y$ and there is no z with $x <^{\mathcal{M}} z <^{\mathcal{M}} y$. Prime examples of partially ordered sets are the structures $\mathcal{M} = (M, \subseteq)$, where M is a collection of subsets of a fixed set A. Here an immediate successor of an element x is any set $x \cup \{a\}$, where $a \in A$. If M happens to be closed under intersection and union (i.e. $x, y \in M \Rightarrow x \cap y, x \cup y \in M$), then $\mathcal{M}$ is a *lattice*, which means that for any two elements x and y of M there is in $\mathcal{M}$ the smallest upper bound $x \vee y$ (the "join") and the largest lower bound $x \wedge y$ (the "meet"). If a lattice satisfies the conditions $x \wedge (y \vee z) = (x \wedge y) \vee (x \wedge z)$ and $x \vee (y \wedge z) = (x \vee y) \wedge (x \vee z)$, as $(M, \subseteq)$ always does, then it is called a *distributive lattice*. If M is even closed under complements and

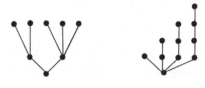

Figure 5.3 Two trees.

contains $\emptyset$, then $\mathcal{M}$ is a Boolean algebra, i.e. a distributive lattice with zero, 0, in which every element x has a complement $-x$ defined by the equations $x \wedge -x = 0$ and $x \vee -x = 1(= -0)$. The most well-known Boolean algebras are the power-set Boolean algebras $(\mathcal{P}(A), \subseteq)$. For finite Boolean algebras these are the only examples (up to isomorphism). An infinite Boolean algebra need not be isomorphic to a power-set Boolean algebra. An example of this is the Boolean algebra of finite or co-finite subsets of an infinite set. A subset F of a Boolean algebra $\mathcal{M}$ is a *filter* if $(x \in F$ and $x \leq^{\mathcal{M}} y) \Rightarrow y \in F$ and $(x \in F$ and $y \in F) \Rightarrow x \wedge y \in F$. A filter F is an *ultrafilter* if $x \in F$ or $-x \in F$ for all $x \in M$. An ultrafilter F is *non-trivial* if $0 \notin F$. An easy example of an ultrafilter is $\{x \in M : a \leq_{\mathcal{M}} x\}$ where a is a fixed element of M. Such an F is called *a principal* ultrafilter and a the *generator* of F. For every $x \in M \setminus \{0\}$, M infinite, there is a non-trivial ultrafilter containing x.[1]

Example 5.8 A *tree* is a partially ordered set $\mathcal{M}$ such that the set $\{x \in M : x <^{\mathcal{M}} t\}$ of predecessors of any $t \in M$ is well-ordered by $\leq^{\mathcal{M}}$ and there is a unique smallest element in $\mathcal{M}$, called the *root* of the tree. Thus for any $t <^{\mathcal{M}} s$ in $\mathcal{M}$ there is an immediate successor r of t such that $t <^{\mathcal{M}} r \leq^{\mathcal{M}} s$. Figure 5.3 depicts two trees. Let us denote the set of immediate successors of t by $\mathrm{ImSuc}_{\mathcal{M}}(t)$. The *height* $\mathrm{ht}(t)$ of an element t of a tree $\mathcal{M}$ is defined as follows: $\mathrm{ht}(t) = 0$ if t is the root, and otherwise

$$\mathrm{ht}(t) = \sup\{\mathrm{ht}(s) + 1 : s <^{\mathcal{M}} t\}.$$

Thus the height of an element of a tree is the order type of the set of predecessors of the element. The height $\mathrm{ht}(\mathcal{M})$ of the tree $\mathcal{M}$ is now $\sup\{\mathrm{ht}(t) : t \in \mathcal{M}\}$. The α-th *level* of a tree T, T_α, is the set of elements of T of height α. Good examples of trees are trees of sequences of elements of a fixed set: Let A be a set and $n \in \mathbb{N}$. Let us denote the set of all sequences $(a_0, \dots, a_{i-1}), i \leq n$, of elements of A by $A^{<n}$. For $i = 0$ the sequence $(a_0, \dots, a_{i-1})$ is the empty

---

[1] The proof of this uses the Axiom of Choice. See e.g. Jech (1997).

sequence. We can partially order this set by

$$(a_0, \ldots, a_{i-1}) \leq (a'_0, \ldots, a'_{j-1}) \tag{5.4}$$

if $i \leq j$ and $a_k = a'_k$ for $k < i$. The structure $\mathcal{S}(A, n) = (A^{<n}, \leq)$ is a tree of height n in which the height of an element $(a_0, \ldots, a_{i-1})$ is always i. Let us denote the set of all finite sequences $(a_0, \ldots, a_{i-1})$ of any finite length $i \in \mathbb{N}$ of elements of A by $A^{<\omega}$. We can partially order this set by the relation (5.4). The structure $\mathcal{S}(A, \omega) = (A^{<\omega}, \leq)$ is a tree of height ω in which the height of an element $(a_0, \ldots, a_{i-1})$ is again i. The relation (5.4) defines a partial order on any subset of $A^{<n}$ (of $A^{<\omega}$ too). In this way we identify a lot of new trees. A *chain* of a tree is a subset of M which is linearly ordered by $\leq^{\mathcal{M}}$. A *branch* of a tree is a chain which would not be a chain if even one element were added to it (i.e. a branch is a maximal chain). Every branch in the tree $\mathcal{S}(A, n)$ looks like:

$$\emptyset \leq (a_0) \leq (a_0, a_1) \leq \ldots \leq (a_0, \ldots, a_{n-1}).$$

Branches of the tree $\mathcal{S}(A, \omega)$ are infinite:

$$\emptyset \leq (a_0) \leq (a_0, a_1) \leq \ldots \leq (a_0, \ldots, a_{n-1}) \leq \ldots.$$

A tree is *well-founded* if it has no infinite branches. Examples of well-founded trees are, first of all, trees of finite height, such as $\mathcal{S}(A, n)$. An example of a well-founded tree of infinite height is the subtree $\mathcal{M}_0$ of $\mathcal{S}(\mathbb{N}, \omega)$ consisting of sequences $(a_0, \ldots, a_{n-1})$ such that

$$a_0 > a_1 > \ldots > a_{n-1}. \tag{5.5}$$

For example,

$$\emptyset \leq (10) \leq (10, 3) \leq (10, 3, 1) \leq (10, 3, 1, 0).$$

The *rank* $\mathrm{rk}_{\mathcal{M}}(t)$ of an element t of a well-founded tree $\mathcal{M}$ is defined as follows: $\mathrm{rk}(t) = 0$ if t is a maximal element (i.e. no element t' of the tree satisfies $t < t'$), and otherwise

$$\mathrm{rk}_{\mathcal{M}}(t) = \sup\{\mathrm{rk}_{\mathcal{M}}(s) + 1 : s \in \mathrm{ImSuc}_{\mathcal{M}}(t)\}.$$

The rank $\mathrm{rk}_{\mathcal{M}}$ of a well-founded tree $\mathcal{M}$ is the rank of the root of the tree. The rank of an element $(a_0, \ldots, a_{n-1})$ of the above tree (defined by (5.5)) $\mathcal{M}_0$ is a_{n-1}, and the rank of the root $\emptyset$ of this tree is the ordinal ω. For elements t of a well-founded tree $\mathcal{M}$ we can define a function $\mathrm{stp}_{\mathcal{M}, t}$, which we call the *successor type* of t, as follows:

$$\mathrm{dom}(\mathrm{stp}_{\mathcal{M}, t}) = \{\mathrm{stp}_{\mathcal{M}, s} : s \in \mathrm{ImSuc}_{\mathcal{M}}(t)\},$$

$$\mathrm{stp}_{\mathcal{M},t}(\mathrm{stp}_{\mathcal{M},s}) = |\{s' \in \mathrm{ImSuc}_{\mathcal{M}}(t) : \mathrm{stp}_{\mathcal{M},s} = \mathrm{stp}_{\mathcal{M},s'}\}|.$$

The successor type has all the successor types of immediate successors of t as its domain and the number of immediate successors of each successor type $\mathrm{stp}_{\mathcal{M},s}$ as its value. The successor type $\mathrm{stp}_{\mathcal{M}}$ of the tree $\mathcal{M}$ is the successor type of its root. It is not hard to prove that two well-founded trees $\mathcal{M}$ and $\mathcal{N}$ are isomorphic if and only if $\mathrm{stp}_{\mathcal{M}} = \mathrm{stp}_{\mathcal{N}}$ (see Exercise 5.16). An *Aronszajn tree* is a tree of height ω_1 in which every level is countable and there are no uncountable branches. An *antichain* of a tree is a subset A of M which satisfies neither $x \leq^{\mathcal{M}} y$ nor $y \leq^{\mathcal{M}} x$ for any $x \neq y$ in A. A tree is *special* if it is the union of countably many antichains, and otherwise *non-special*. Any well-founded tree is special since every level of any tree is an antichain. For an example of a non-special tree, see Example 9.56. For examples of Aronszajn trees, see Jech (1997).

Example 5.9   A *successor structure* is an L-structure $\mathcal{M} = (M, S^{\mathcal{M}}, 0^{\mathcal{M}})$ for the vocabulary $L = \{S, 0\}$, where $\#_L(S) = 1$ and $\#_L(0) = 0$, satisfying the three properties: $S^{\mathcal{M}}(x) = S^{\mathcal{M}}(y) \Rightarrow x = y$, $S^{\mathcal{M}}(x) \neq 0^{\mathcal{M}}$, and $x \neq 0^{\mathcal{M}} \Rightarrow \exists y (S^{\mathcal{M}}(y) = x)$. The canonical successor structure is $(\mathbb{N}, f_+, 0)$, where $f_+(n) = n+1$. Another example is obtained by taking the disjoint union $\mathbb{N} \cup \mathbb{Z}'$ of $\mathbb{N}$ and a copy of $\mathbb{Z}$, disjoint from $\mathbb{N}$. In the structure $(\mathbb{N} \cup \mathbb{Z}', f_+, 0)$ we let 0 be the zero of $\mathbb{N}$ and $f_+(x)$ be defined as the canonical successor in both $\mathbb{N}$ and $\mathbb{Z}'$. A slightly different successor structure is the union of $\mathbb{N}$ and a set of cycles. On $\mathbb{N}$ the successor function is defined canonically and on the cycles the successor function is defined also canonically by deciding an orientation. Figure 5.4 is a picture of a successor structure. If $\mathcal{M} = (M, f, 0)$ is a successor structure and $a \in M$, we can define the forward and backward *iterations* of f as follows:

$$f^0(a) = a$$
$$f^{n+1}(a) = f(f^n(a))$$
$$f^{-n-1}(f(a)) = f^{-n}(a)$$
$$f^{-n-1}(0) = 0.$$

The *component* of $a \in M$ is the set

$$\mathrm{Cmp}_{\mathcal{M}}(a) = \{f^z(a) : z \in \mathbb{Z}\}.$$

It is easy to see that every successor structure contains an isomorphic copy of $(\mathbb{N}, f_+, 0)$. We call it the *standard component*. The remaining components are either n-cycles for some n, and we call them *n-cycle components* (or just cycle-components), or isomorphic copies of $(\mathbb{Z}, f_+)$, which we call *$\mathbb{Z}$-components*.

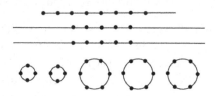

Figure 5.4 A successor structure.

If $\mathcal{M}$ is a successor structure, let $Cmp_{\mathcal{M}}$ be the set of components of $\mathcal{M}$ and

$$CC_n(\mathcal{M}) = |\{C \in Cmp_{\mathcal{M}} : C \text{ is an } n\text{-cycle component}\}|,$$

$$CC_\infty(\mathcal{M}) = |\{C \in Cmp_{\mathcal{M}} : C \text{ is a } \mathbb{Z}\text{-component}\}|.$$

Two successor structures $\mathcal{M}$ and $\mathcal{N}$ are isomorphic if and only if $CC_a(\mathcal{M}) = CC_a(\mathcal{N})$ for all $a \in \mathbb{N} \cup \{\infty\}$.

5.3 Substructures

The concept of a substructure is in principle a very simple one, especially for relational vocabularies. There are however subtleties which deserve special attention when function symbols are involved.

Definition 5.10   An L-structure $\mathcal{M}$ is a *substructure* of another L-structure $\mathcal{M}'$, in symbols $\mathcal{M} \subseteq \mathcal{M}'$, if:

1. $M \subseteq M'$.
2. $R^{\mathcal{M}} = R^{\mathcal{M}'} \cap M^n$ if $R \in L$ is an n-ary predicate symbol.
3. $f^{\mathcal{M}} = f^{\mathcal{M}'} \upharpoonright M^n$ if $f \in L$ is an n-ary function symbol.
4. $c^{\mathcal{M}} = c^{\mathcal{M}'}$ if $c \in L$ is a constant symbol.

Substructures are particularly easy to understand in the case that L is relational. Then any subset M of an L-structure $\mathcal{M}'$ determines a substructure $\mathcal{M}$ the universe of which is M. If L is not relational we have to worry about the question whether M is closed under the functions $f^{\mathcal{M}'}$, $f \in L$, and whether the interpretations $c^{\mathcal{M}'}$ of constant symbols $c \in L$ are in M. For example, if $L = \{f\}$ where f is a unary function symbol, then any substructure of an L-structure which contains an element a has to contain also $f^{\mathcal{M}'}(a), f^{\mathcal{M}'}(f^{\mathcal{M}'}(a))$, etc. A substructure of a group need not be a subgroup even when it is closed under the group operation. For example, $(\mathbb{N}, +)$ is a substructure of $(\mathbb{Z}, +)$ but it is not a group. A substructure of a linear order is again a linear order. Similarly, a substructure of a partial order is again a partial order. A substructure of a tree is a tree if it has a smallest element.

Lemma 5.11 *Suppose L is a vocabulary, $\mathcal{M}$ an L-structure, and $X \subseteq M$. Suppose furthermore that either L contains constant symbols or $X \neq \emptyset$. There is a unique L-structure $\mathcal{N}$ such that:*

1. *$\mathcal{N} \subseteq \mathcal{M}$.*
2. *$X \subseteq N$.*
3. *If $\mathcal{N}' \subseteq \mathcal{M}$ and $X \subseteq N'$, then $\mathcal{N} \subseteq \mathcal{N}'$.*

Proof Let $X_0 = X \cup \{c^{\mathcal{M}} : c \in L\}$ and inductively

$$X_{n+1} = X_n \cup \{f^{\mathcal{M}}(a_1, \ldots, a_{\#_L(f)}) : a_1, \ldots, a_{\#_L(f)} \in X_n, f \in L\}.$$

It is easy to see that the set $N = \bigcup_{n \in \mathbb{N}} X_n$ is the universe of the unique structure $\mathcal{N}$ claimed to exist in the lemma. □

We call the unique structure $\mathcal{N}$ of Lemma 5.11 the substructure of $\mathcal{M}$ *generated* by X and denote it by $[X]_{\mathcal{M}}$. The following lemma is used repeatedly in the sequel.

Lemma 5.12 *Suppose L is a vocabulary. Suppose $\mathcal{M}$ and $\mathcal{N}$ are L-structures and $\pi : M \to N$ is a partial mapping. There is at most one isomorphism $\pi^* : [\mathrm{dom}(\pi)]_{\mathcal{M}} \to [\mathrm{rng}(\pi)]_{\mathcal{N}}$ extending π.*

5.4 Back-and-Forth Sets

One of the main themes of this book is the question: Given two structures $\mathcal{M}$ and $\mathcal{N}$, how do we measure how close they are to being isomorphic? They may be non-isomorphic for a totally obvious reason, e.g. two graphs one of which has a triangle while the other does not. They may also be non-isomorphic for an extremely subtle reason which involves the use of the Axiom of Choice (see e.g. Lemma 9.9). One of the basic tools in trying to answer this question is the concept of partial isomorphism.

Definition 5.13 Suppose L is a vocabulary and $\mathcal{M}, \mathcal{M}'$ are L-structures. A partial mapping $\pi : M \to M'$ is a *partial isomorphism* $\mathcal{M} \to \mathcal{M}'$ if there is an isomorphism $\pi^* : [\mathrm{dom}(\pi)]_{\mathcal{M}} \to [\mathrm{rng}(\pi)]_{\mathcal{M}'}$ extending π. We use $\mathrm{Part}(\mathcal{M}, \mathcal{M}')$ to denote the set of partial isomorphisms $\mathcal{M} \to \mathcal{M}'$. If $\mathcal{M} = \mathcal{M}'$ we call π a *partial automorphism*.

Note that the extension π^* referred to in Definition 5.13 is by Lemma 5.12 necessarily unique.

The main topic of this section, the back-and-forth sets, are very useful weaker versions of isomorphisms. To get a picture of this, suppose $f : \mathcal{A} \cong \mathcal{B}$. Then $f \in \mathrm{Part}(\mathcal{A}, \mathcal{B})$ and we can go back and forth between $\mathcal{A}$ and $\mathcal{B}$ with f in the following sense:

$$\forall a \in A \exists b \in B(f(a) = b) \tag{5.6}$$

$$\forall b \in B \exists a \in A(f(a) = b). \tag{5.7}$$

We now generalize this to a situation where we do not quite have an isomorphism but only a set P which reflects the back and forth conditions (5.8) and (5.9) of an isomorphism.

Definition 5.14 Suppose $\mathcal{A}$ and $\mathcal{B}$ are L-structures. A *back-and-forth set* for $\mathcal{A}$ and $\mathcal{B}$ is any non-empty set $P \subseteq \mathrm{Part}(\mathcal{A}, \mathcal{B})$ such that

$$\forall f \in P \forall a \in A \exists g \in P(f \subseteq g \text{ and } a \in \mathrm{dom}(g)) \tag{5.8}$$

$$\forall f \in P \forall b \in B \exists g \in P(f \subseteq g \text{ and } b \in \mathrm{rng}(g)). \tag{5.9}$$

The structures $\mathcal{A}$ and $\mathcal{B}$ are said to be *partially isomorphic*, in symbols $\mathcal{A} \simeq_p \mathcal{B}$, if there is a back-and-forth set for them.

Lemma 5.15 *The relation $\simeq_p$ is an equivalence relation on $\mathrm{Str}(L)$.*

Proof The relation $\simeq_p$ is reflexive, because $\{id_\mathcal{A}\}$ is a back-and-forth set for $\mathcal{A}$ and $\mathcal{B}$. If P is a back-and-forth set for $\mathcal{A}$ and $\mathcal{B}$, then $\{f^{-1} : f \in P\}$ is a back-and-forth set for $\mathcal{B}$ and $\mathcal{A}$. Finally, if P_1 is a back-and-forth set for $\mathcal{A}$ and $\mathcal{B}$ and P_2 is a back-and-forth set for $\mathcal{B}$ and $\mathcal{C}$, then $\{f_2 \circ f_1 : f_1 \in P_1, f_2 \in P_2\}$ is a back-and-forth set for $\mathcal{A}$ and $\mathcal{C}$, where we stipulate $\mathrm{dom}(f_2 \circ f_1) = f_1^{-1}(\mathrm{dom}(f_2))$. $\qquad\square$

Proposition 5.16 *If $\mathcal{A} \simeq_p \mathcal{B}$, where $\mathcal{A}$ and $\mathcal{B}$ are countable, then $\mathcal{A} \cong \mathcal{B}$.*

Proof Let us enumerate A as $(a_n : n < \omega)$ and B as $(b_n : n < \omega)$. Let P be a back-and-forth set for $\mathcal{A}$ and $\mathcal{B}$. Since $P \neq \emptyset$, there is some $f_0 \in P$. We define a sequence $(f_n : n < \omega)$ of elements of P as follows: Suppose $f_n \in P$ is defined. If n is even, say $n = 2m$, let $y \in B$ and $f_{n+1} \in P$ such that $f_n \cup \{(a_m, y)\} \subseteq f_{n+1}$. If n is odd, say $n = 2m + 1$, let $x \in A$ and $f_{n+1} \in P$ such that $f_n \cup \{(x, b_m)\} \subseteq f_{n+1}$. Finally, let

$$f = \bigcup_{n=0}^{\infty} f_n.$$

Clearly, $f : \mathcal{A} \cong \mathcal{B}$. $\qquad\square$

This proposition is not true for uncountable structures. Indeed, let $L = \emptyset$ and let $\mathcal{A}$ and $\mathcal{B}$ be any infinite L-structures. Then there is a back-and-forth set for $\mathcal{A}$ and $\mathcal{B}$ (Exercise 5.28). Thus $\mathcal{A} \simeq_p \mathcal{B}$. But $\mathcal{A} \not\cong \mathcal{B}$ if, for example, $A = \mathbb{Q}$ and $B = \mathbb{R}$. The failure of Proposition 5.16 to generalize is a major topic in the sequel.

Proposition 5.17 *Suppose $\mathcal{A}$ and $\mathcal{B}$ are dense linear orders without endpoints. Then $\mathcal{A} \simeq_p \mathcal{B}$.*

Proof Let $P = \{f \in \mathrm{Part}(\mathcal{A}, \mathcal{B}) : \mathrm{dom}(f) \text{ is finite}\}$. It turns out that this straightforward choice works. Clearly, $P \neq \emptyset$. Suppose then $f \in P$ and $a \in A$. Let us enumerate f as $\{(a_1, b_1), \ldots, (a_n, b_n)\}$ where $a_1 < \ldots < a_n$. Since f is a partial isomorphism, also $b_1 < \ldots < b_n$. Now we consider different cases. If $a < a_1$, we choose $b < b_1$ and then $f \cup \{(a, b)\} \in P$. If $a_i < a < a_{i+1}$, we choose $b \in B$ so that $b_i < b < b_{i+1}$ and then $f \cup \{(a, b)\} \in P$. If $a_n < a$, we choose $b > b_n$ and again $f \cup \{(a, b)\} \in P$. Finally, if $a = a_i$, we let $b = b_i$ and then $f \cup \{(a, b)\} = f \in P$. We have proved (5.8). Condition (5.9) is proved similarly. $\qquad\square$

Putting Proposition 5.16 and Proposition 5.17 together yields the famous result of Cantor (1895): countable dense linear orders without endpoints are isomorphic. See Exercise 6.29 for a more general result.

5.5 The Ehrenfeucht–Fraïssé Game

In Section 4.3 we introduced the Ehrenfeucht–Fraïssé Game played on two graphs. This game was used to measure to what extent two graphs have similar properties, especially properties expressible in the first-order language of graphs limited to a fixed quantifier rank. In this section we extend this game to the context of arbitrary structures, not just graphs.

Let us recall the basic idea behind the Ehrenfeucht–Fraïssé Game. Suppose $\mathcal{A}$ and $\mathcal{B}$ are L-structures for some relational L. We imagine a situation in which two mathematicians argue about whether $\mathcal{A}$ and $\mathcal{B}$ are isomorphic or not. The mathematician that we denote by **II** claims that they are isomorphic, while the other mathematician whom we call **I** claims the models have an intrinsic structural difference and they cannot possibly be isomorphic.

The matter would be quickly resolved if **II** was required to show the claimed isomorphism. But the rules of the game are different. The rules are such that **II** is required to show only small pieces of the claimed isomorphism.

More exactly, **I** asks what is the image of an element a_1 of A that he chooses

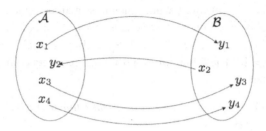

Figure 5.5 The Ehrenfeucht–Fraïssé Game.

at will. Then **II** is required to respond with some element b_1 of B so that

$$\{(a_1, b_1)\} \in \mathrm{Part}(\mathcal{A}, \mathcal{B}). \tag{5.10}$$

Alternatively, **I** might have chosen an element b_1 of B and then **II** would have been required to produce an element a_1 of A such that (5.10) holds. The one-element mapping $\{(a_1, b_1)\}$ is called the *position* in the game after the first move.

Now the game goes on. Again **I** asks what is the image of an element a_2 of A (or alternatively he can ask what is the pre-image of an element b_2 of B). Then **II** produces an element b_2 of B (or in the alternative case an element a_2 of A). In either case the choice of **II** has to satisfy

$$\{(a_1, b_1), (a_2, b_2)\} \in \mathrm{Part}(\mathcal{A}, \mathcal{B}). \tag{5.11}$$

Again, $\{(a_1, b_1), (a_2, b_2)\}$ is called the position after the second move.

We continue until the position

$$\{(a_1, b_1), \ldots, (a_n, b_n)\} \in \mathrm{Part}(\mathcal{A}, \mathcal{B})$$

after the nth move has been produced. If **II** has been able to play all the moves according to the rules she is declared the winner. Let us call this game $\mathrm{EF}_n(\mathcal{A}, \mathcal{B})$. Figure 5.5 pictures the situation after four moves. If **II** can win repeatedly whatever moves **I** plays, we say that **II** has a *winning strategy*.

Example 5.18   Suppose $\mathcal{A}$ and $\mathcal{B}$ are two L-structures and $L = \emptyset$. Thus the structures $\mathcal{A}$ and $\mathcal{B}$ consist merely of a universe with no structure on it. In this singular case any one-to-one mapping is a partial isomorphism. The only thing player **II** has to worry about, say in (5.11), is that $a_1 = a_2$ if and only if $b_1 = b_2$. Thus **II** has a winning strategy in $\mathrm{EF}_n(\mathcal{A}, \mathcal{B})$ if A and B both have at least n elements. So **II** can have a winning strategy even if A and B have different cardinality and there could be no isomorphism between them for the trivial reason that there is no bijection. The intuition here is that by playing a finite number of elements, or even $\aleph_0$ many, it is not possible to get hold of the cardinality of the universe if it is infinite.

Example 5.19 Let $\mathcal{A}$ be a linear order of length 3 and $\mathcal{B}$ a linear order of length 4. How many moves does **I** need to beat **II**? Suppose $A = \{a_1, a_2, a_3\}$ in increasing order and $B = \{b_1, b_2, b_3, b_4\}$ in increasing order. Clearly, if **I** plays at any point the smallest element, also **II** has to play the smallest element or face defeat on the next move. Also, if **I** plays at any point the smallest but one element, also **II** has to play the smallest but one element or face defeat in two moves. Now in $\mathcal{A}$ the smallest but one element is the same as the largest but one element, while in $\mathcal{B}$ they are different. So if **I** starts with a_2, **II** has to play b_2 or b_3, or else she loses in one move. Suppose she plays b_2. Now **I** plays b_3 and **II** has no good moves left. To obey the rules, she must play a_3. That is how long she can play, for now when **I** plays b_4, **II** cannot make a legal move anymore. In fact **II** has a winning strategy in $\text{EF}_2(\mathcal{A}, \mathcal{B})$ but **I** has a winning strategy in $\text{EF}_3(\mathcal{A}, \mathcal{B})$.

We now proceed to a more exact definition of the Ehrenfeucht–Fraïssé Game.

Definition 5.20 Suppose L is a vocabulary and $\mathcal{M}, \mathcal{M}'$ are L-structures such that $M \cap M' = \emptyset$. The *Ehrenfeucht–Fraïssé Game* $\text{EF}_n(\mathcal{M}, \mathcal{M}')$ is the game $\mathcal{G}_n(M \cup M', W_n(\mathcal{M}, \mathcal{M}'))$, where $W_n(\mathcal{M}, \mathcal{M}') \subseteq (M \cup M')^{2n}$ is the set of $p = (x_0, y_0, \ldots, x_{n-1}, y_{n-1})$ such that:

(G1) For all $i < n$: $x_i \in M \iff y_i \in M'$.
(G2) If we denote

$$v_i = \left\{ \begin{array}{ll} x_i & \text{if } x_i \in M \\ y_i & \text{if } y_i \in M \end{array} \right. \quad v_i' = \left\{ \begin{array}{ll} x_i & \text{if } x_i \in M' \\ y_i & \text{if } y_i \in M', \end{array} \right.$$

then

$$f_p = \{(v_0, v_0'), \ldots, (v_{n-1}, v_{n-1}')\}$$

is a partial isomorphism $\mathcal{M} \to \mathcal{M}'$.

We call v_i and v_i' *corresponding* elements. The infinite game $\text{EF}_\omega(\mathcal{M}, \mathcal{M}')$ is defined quite similarly, that is, it is the game $\mathcal{G}_\omega(M \cup M', W_\omega(\mathcal{M}, \mathcal{M}'))$, where $W_\omega(\mathcal{M}, \mathcal{M}')$ is the set of $p = (x_0, y_0, x_1, y_1, \ldots)$ such that for all $n \in \mathbb{N}$ we have $(x_0, y_0, \ldots, x_{n-1}, y_{n-1}) \in W_n(\mathcal{M}, \mathcal{M}')$.

Note that the game EF_ω is a closed game.

Proposition 5.21 *Suppose L is a vocabulary and $\mathcal{A}$ and $\mathcal{B}$ are L-structures. The following are equivalent:*

1. $\mathcal{A} \simeq_p \mathcal{B}$.
2. **II** *has a winning strategy in* $\mathrm{EF}_\omega(\mathcal{A}, \mathcal{B})$.

Proof Assume $A \cap B = \emptyset$. Let P be first a back-and-forth set for $\mathcal{A}$ and $\mathcal{B}$. We define a winning strategy $\tau = (\tau_i : i < \omega)$ for **II**. Since $P \neq \emptyset$ we can fix an element f of P. Condition (5.8) tells us that if $a_1 \in A$, then there are $b_1 \in B$ and g such that

$$f \cup \{(a_1, b_1)\} \subseteq g \in P. \tag{5.12}$$

Let $\tau_0(a_1)$ be one such b_1. Likewise, if $b_1 \in B$, then there are $a_1 \in A$ such that (5.12) holds and we can let $\tau_0(b_1)$ be some such a_1. We have defined $\tau_0(c_1)$ whatever c_1 is. To define $\tau_1(c_1, c_2)$, let us assume **I** played $c_1 = a_1 \in A$. Thus (5.12) holds with $b_1 = \tau_0(a_1)$. If $c_2 = a_2 \in A$ we can use (5.8) again to find $b_2 = \tau_1(a_1, a_2) \in B$ and h such that

$$f \cup \{(a_1, b_1), (a_2, b_2)\} \subseteq h \in P.$$

The pattern should now be clear. The back-and-forth set P guides **II** to always find a valid move. Let us then write the proof in more detail: Suppose we have defined τ_i for $i < j$ and we want to define τ_j. Suppose player **I** has played $x_0, \ldots, x_{j-1}$ and player **II** has followed τ_i during round $i < j$. During the inductive construction of τ_i we took care to define also a partial isomorphism $f_i \in P$ such that $\{v_0, \ldots, v_{i-1}\} \subseteq \mathrm{dom}(f_{i-1})$. Now player **I** plays x_j. By assumption there is $f_j \in P$ extending f_{j-1} such that if $x_j \in A$, then $x_j \in \mathrm{dom}(f_j)$ and if $x_j \in B$, then $x_j \in \mathrm{rng}(f_j)$. We let $\tau_j(x_0, \ldots, x_j) = f_j(x_j)$ if $x_j \in A$ and $\tau_j(x_0, \ldots, x_j) = f_j^{-1}(x_j)$ otherwise. This ends the construction of τ_j. This is a winning strategy because every f_p extends to a partial isomorphism $\mathcal{M} \to \mathcal{N}$.

For the converse, suppose $\tau = (\tau_n : n < \omega)$ is a winning strategy of **II**. Let Q consist of all plays of $\mathrm{EF}_\omega(\mathcal{A}, \mathcal{B})$ in which player **II** has used τ. Let P consist of all possible f_p where p is a position in the game $\mathrm{EF}_\omega(\mathcal{A}, \mathcal{B})$ with an extension in Q. It is clear that P is non-void and has the properties (5.8) and (5.9). $\qquad\square$

To prove partial isomorphism of two structures we now have two alternative methods:

1. Construct a back-and-forth set.
2. Show that player **II** has a winning strategy in EF_ω.

By Proposition 5.21 these methods are equivalent. In practice one uses the game as a guide to intuition and then for a formal proof one usually uses a back-and-forth set.

5.6 Back-and-Forth Sequences

Back-and-forth sets and winning strategies of player **II** in the Ehrenfeucht–Fraïssé Game EF_ω correspond to each other. There is a more refined concept, called a back-and-forth sequence, which corresponds to a winning strategy of player **II** in the finite game EF_n.

Definition 5.22 A *back-and-forth sequence* $(P_i : i \leq n)$ is defined by the conditions

$$\emptyset \neq P_n \subseteq \ldots \subseteq P_0 \subseteq \text{Part}(\mathcal{A}, \mathcal{B}). \tag{5.13}$$

$$\forall f \in P_{i+1} \forall a \in A \exists b \in B \exists g \in P_i(f \cup \{(a,b)\} \subseteq g) \text{ for } i < n. \tag{5.14}$$

$$\forall f \in P_{i+1} \forall b \in B \exists a \in A \exists g \in P_i(f \cup \{(a,b)\} \subseteq g) \text{ for } i < n. \tag{5.15}$$

If P is a back-and-forth set, we can get back-and-forth sequences $(P_i : i \leq n)$ of any length by choosing $P_i = P$ for all $i \leq n$. But the converse is not true: the sets P_i need by no means be themselves back-and-forth sets. Indeed, pairs of countable models may have long back-and-forth sequences without having any back-and-forth sets. Let us write

$$\mathcal{A} \simeq_p^n \mathcal{B}$$

if there is a back-and-forth sequence of length n for $\mathcal{A}$ and $\mathcal{B}$.

Lemma 5.23 *The relation* $\simeq_p^n$ *is an equivalence relation on* $Str(L)$.

Proof Exactly as Lemma 5.15. $\qquad\qquad\square$

Example 5.24 We use $(\mathbb{N} + \mathbb{N}, <)$ to denote the linear order obtained by putting two copies of $(\mathbb{N}, <)$ one after the other. (The ordinal of this order is $\omega + \omega$.) Now $(\mathbb{N}, <) \simeq_p^2 (\mathbb{N} + \mathbb{N}, <)$, for we may take

$$P_2 = \{\emptyset\}.$$
$$P_1 = \{\{(a,b)\} : 0 < a \in \mathbb{N}, \ 0 < b \in \mathbb{N} + \mathbb{N}\} \cup \{(0,0)\} \cup P_2.$$
$$P_0 = \{\{(a_0, b_0), (a_1, b_1)\} : a_0 < a_1 \in \mathbb{N}, \ b_0 < b_1 \in \mathbb{N} + \mathbb{N}\} \cup P_1.$$

Note that $(\mathbb{N}, <) \not\simeq_p^3 (\mathbb{N} + \mathbb{N}, <)$.

Proposition 5.25 *Suppose $\mathcal{A}$ and $\mathcal{B}$ are discrete linear orders (i.e. every element with a successor has an immediate successor and every element with a predecessor has an immediate predecessor) with no endpoints, and $n \in \mathbb{N}$. Then $\mathcal{A} \simeq_p^n \mathcal{B}$.*

Proof   Let P_i consist of $f \in \text{Part}(\mathcal{A}, \mathcal{B})$ with the following property: $f = \{(a_0, b_0), \ldots, (a_{n-i-1}, b_{n-i-1})\}$ where

$$a_0 \leq \ldots \leq a_{n-i-1},$$
$$b_0 \leq \ldots \leq b_{n-i-1},$$

and for all $0 \leq j < n - i - 1$ if $|(a_j, a_{j+1})| < 2^i$ or $|(b_j, b_{j+1})| < 2^i$, then $|(a_j, a_{j+1})| = |(b_j, b_{j+1})|$. $\square$

Example 5.26 $(\mathbb{Z}, <) \simeq_p^n (\mathbb{Z} + \mathbb{Z}, <)$ for all $n \in \mathbb{N}$, but note that $(\mathbb{Z}, <) \not\simeq_p (\mathbb{Z} + \mathbb{Z}, <)$.

Proposition 5.27   *Suppose L is a vocabulary and $\mathcal{A}$ and $\mathcal{B}$ are L-structures. The following are equivalent:*

1. $\mathcal{A} \simeq_p^n \mathcal{B}$.
2. **II** *has a winning strategy in* $\text{EF}_n(\mathcal{A}, \mathcal{B})$.

Proof   Let us assume $A \cap B = \emptyset$. Let $(P_i : i \leq n)$ be a back-and-forth sequence for $\mathcal{A}$ and $\mathcal{B}$. We define a winning strategy $\tau = (\tau_i : i \leq n)$ for **II**. Since $P_n \neq \emptyset$ we can fix an element f of P_n. Condition (5.14) tells us that if $a_1 \in A$, then there are $b_1 \in B$ and g such that

$$f \cup \{(a_1, b_1)\} \subseteq g \in P_{n-1}. \tag{5.16}$$

Let $\tau_0(a_1)$ be one such b_1. Likewise, if $b_1 \in B$, then there are $a_1 \in A$ such that (5.16) holds and we can let $\tau_0(b_1)$ be some such a_1. We have defined $\tau_0(c_1)$ whatever c_1 is. To define $\tau_1(c_1, c_2)$, let us assume **I** played $c_1 = a_1 \in A$. Thus (5.16) holds with $b_1 = \tau_0(a_1)$. If $c_2 = a_2 \in A$ we can use (5.13) again to find $b_2 = \tau_1(a_1, a_2) \in B$ and h such that

$$f \cup \{(a_1, b_1), (a_2, b_2)\} \subseteq h \in P_{n-2}.$$

The pattern should be clear now. As before, the back-and-forth sequence guides **II** to always find a valid move. Let us then write the proof in more detail: Suppose we have defined τ_i for $i < j$ and we want to define τ_j. Suppose player **I** has played $x_0, \ldots, x_{j-1}$ and player **II** has followed τ_i during round $i < j$. During the inductive construction of τ_i we took care to define also a partial isomorphism $f_i \in P_{n-i}$ such that $\{v_0, \ldots, v_{i-1}\} \subseteq \text{dom}(f_i)$. Now player **I** plays x_j. By assumption there is $f_j \in P_{n-j}$ extending f_{j-1} such that if $x_j \in A$, then $x_j \in \text{dom}(f_j)$ and if $x_j \in B$, then $x_j \in \text{rng}(f_j)$. We let $\tau_j(x_0, \ldots, x_j) = f_j(x_j)$ if $x_j \in A$ and $\tau_j(x_0, \ldots, x_j) = f_j^{-1}(x_j)$ otherwise. This ends the construction of τ_j. This is a winning strategy because every f_p extends to a partial isomorphism $\mathcal{M} \to \mathcal{N}$.

For the converse, suppose $\tau = (\tau_i : i \leq n)$ is a winning strategy of **II**. Let Q consist of all plays of $\mathrm{EF}_n(\mathcal{A}, \mathcal{B})$ in which player **II** has used τ. Let P_{n-i} consist of all possible f_p where $p = (x_0, y_0, \ldots, x_{i-1}, y_{i-1})$ is a position in the game $\mathrm{EF}_n(\mathcal{A}, \mathcal{B})$ with an extension in Q. It is clear that $(P_i : i \leq n)$ has the properties (5.13) and (5.14). Note that:

$$P_n = \{\emptyset\}$$

$$P_{n-1} = \{(x_0, \tau_0(x_0)) : x_0 \in A \cup B\}$$

$$P_{n-2} = \{(x_0, \tau_0(x_0), x_1, \tau_1(x_0, x_1)) : x_0, x_1 \in A \cup B\}$$

$$P_0 = \{(x_0, \tau_0(x_0), \ldots, x_{n-1}, \tau_{n-1}(x_0, \ldots, x_{n-1})) : x_0, \ldots, x_{n-1} \in A \cup B\}.$$

$\square$

5.7 Historical Remarks and References

Back-and-forth sets are due to Fraïssé (1955). The Ehrenfeucht–Fraïssé Game was introduced in Ehrenfeucht (1957) and Ehrenfeucht (1960/1961). Back-and-forth sequences were introduced in Karp (1965). Exercise 5.40 is from Ellentuck (1976). Exercise 5.40 is from Ellentuck (1976). Exercise 5.54 is from Barwise (1975). Exercise 5.71 is from Rosenstein (1982).

Exercises

5.1 Show that isomorphism of structures is an equivalence relation in the sense that it is reflexive, symmetric, and transitive.

5.2 Suppose L is a finite vocabulary, $\mathcal{B}$ is a countable L-model, and $\{b_n : n < \omega\}$ is an enumeration of the domain B of $\mathcal{B}$. Suppose $\mathcal{A}$ is a countable L-model. Show that the following are equivalent:

(1) $\mathcal{A} \cong \mathcal{B}$.

(2) There is an enumeration $\{a_n : n < \omega\}$ of the domain of $\mathcal{A}$ so that for all atomic L-formulas $\theta(x_0, \ldots, x_n)$ and all $n < \omega$ we have

$$\mathcal{A} \models \theta(a_0, \ldots, a_n) \iff \mathcal{B} \models \theta(b_0, \ldots, b_n).$$

5.3   Suppose L is a vocabulary and $\mathcal{M}$ is an L-structure. Show that the set $\mathrm{Aut}(\mathcal{M})$ of automorphisms of $\mathcal{M}$ forms a group under the operation of composition of functions.

5.4   Give an example of $\mathcal{M}$ such that $\mathrm{Aut}(\mathcal{M})$ (see the previous exercise) is:

1. The trivial one-element group.
2. A non-trivial abelian group (e.g. the additive group of the integers).
3. A non-abelian group (e.g. the symmetric group S_3).

5.5   How many automorphisms do the following structures have.

1. A linear order of n elements.
2. $(\mathbb{N}, <)$.
3. $(\mathbb{Z}, <)$.
4. $(\mathbb{Q}, <)$.

5.6   Show that if $\mathcal{A}$ and $\mathcal{B}$ are unary structures, then $\mathcal{A} \cong \mathcal{B}$ if and only if for all $\epsilon : \{1, \ldots, n\} \to \{0, 1\}$ we have $|C_\epsilon(\mathcal{A})| = |C_\epsilon(\mathcal{B})|$. Easier version: Show that if $\mathcal{A}$ and $\mathcal{B}$ are unary structures with a finite universe of size n, then $\mathcal{A} \cong \mathcal{B}$ if and only if for all $\epsilon : \{1, \ldots, n\} \to \{0, 1\}$ we have $|C_\epsilon(\mathcal{A})| = |C_\epsilon(\mathcal{B})|$.

5.7   Suppose $\mathcal{M}$ is a unary structure in which every ϵ-constituent has exactly three elements. How many elements does $\mathcal{M}$ have? How many automorphisms does $\mathcal{M}$ have?

5.8 $L = \{P_1, \ldots, P_m\}$, where each P_i is unary. Show that the number of non-isomorphic L-structures on the universe $\{1, \ldots, n\}$ is $\binom{n+2^m-1}{2^m-1}$.

5.9   Describe the group of automorphisms of a finite unary structure.

5.10   Suppose $\mathcal{M}$ is an equivalence relation with a finite universe such that $EC_n(\mathcal{M}) = 2$ for each $n = 1, \ldots, 5$ and $EC_n(\mathcal{M}) = 0$ for other n. How many elements are there in the universe of $\mathcal{M}$? How many automorphisms does $\mathcal{M}$ have?

5.11   Show that for any $m \in \mathbb{N}$ there is $m^* \in \mathbb{N}$ such that if $n \geq m^*$ then there are more than n^m non-isomorphic equivalence relations on the universe $\{1, \ldots, n\}$. Conclude that for any $m \in \mathbb{N}$ there is $m^* \in \mathbb{N}$ such that if $n \geq m^*$ then there are more non-isomorphic equivalence relations on the universe $\{1, \ldots, n\}$ than non-isomorphic $\{P_1, \ldots, P_m\}$-structures, where each P_i is unary.

5.12 Show that if $\mathcal{A}$ and $\mathcal{B}$ are equivalence relations, then $\mathcal{A} \cong \mathcal{B}$ if and only if for all $\kappa \leq |A \cup B|$ we have $EC_\kappa(\mathcal{A}) = EC_\kappa(\mathcal{B})$. Easier version: Show that if $\mathcal{A}$ and $\mathcal{B}$ are equivalence relations with a finite universe of size n, then $\mathcal{A} \cong \mathcal{B}$ if and only if for all $m \leq n$ we have $EC_m(\mathcal{A}) = EC_m(\mathcal{B})$.

5.13 Describe the group of automorphisms of a finite equivalence relation.

5.14 Show that if $\mathcal{M}$ and $\mathcal{N}$ are countable dense linear orders, then $\mathcal{M} \cong \mathcal{N}$ if and only if $SG(\mathcal{M}) = SG(\mathcal{N})$. Demonstrate that this is not true for non-dense countable linear orders or for uncountable dense linear orders.

5.15 Show that two well-orders $\mathcal{M}$ and $\mathcal{N}$ are isomorphic if and only if $o(\mathcal{M}) = o(\mathcal{N})$.

5.16 Prove that two well-founded trees $\mathcal{M}$ and $\mathcal{N}$ are isomorphic if and only if $\mathrm{stp}_{\mathcal{M}} = \mathrm{stp}_{\mathcal{N}}$.

5.17 Prove that two successor structures $\mathcal{M}$ and $\mathcal{N}$ are isomorphic if and only if $CC_a(\mathcal{M}) = CC_a(\mathcal{N})$ for all $a \in \mathbb{N} \cup \{\infty\}$. Easier version: Prove that two successor structures $\mathcal{M}$ and $\mathcal{N}$ both of which have only finitely many components are isomorphic if and only if $CC_a(\mathcal{M}) = CC_a(\mathcal{N})$ for all $a \in \mathbb{N} \cup \{\infty\}$.

5.18 Show that any uncountable collection of countable non-isomorphic successor structures has to contain a successor structure with infinitely many cycle components.

5.19 Describe the group of automorphisms of a successor structure with n $\mathbb{Z}$-components and m_i i-cycle components for $i = 1, \ldots, k$.

5.20 Give an example of an infinite structure $\mathcal{M}$ with no substructures $\mathcal{N} \neq \mathcal{M}$.

5.21 Consider $\mathcal{M} = (\mathbb{Z}, +)$. What is $[X]_{\mathcal{M}}$, if X is

1. $\{0\}$,
2. $\{1\}$,
3. $\{2, -2\}$.

5.22 Consider $\mathcal{M} = (\mathbb{Z}, +, -)$. What is $[X]_{\mathcal{M}}$, if X is $\{13, 17\}$?

5.23 Suppose $\mathcal{M}$ is a successor structure consisting of the standard component and two five-cycles. Show that there are exactly four possibilities for the set $[X]_{\mathcal{M}}$.

5.24 Show that the universe of $[X]_{\mathcal{M}}$ is the intersection of all universes of substructures $\mathcal{N}$ of $\mathcal{M}$ such that $X \subseteq N$.

5.25 Prove Lemma 5.12.

5.26 Show that every Boolean algebra $\mathcal{M}$ is isomorphic to a substructure of $(\mathcal{P}(A), \subseteq)$, where A is the set of all ultrafilters of $\mathcal{M}$. (This is the so-called *Stone's Representation Theorem*.)

5.27 Show that every tree every element of which has height $< \omega$ is isomorphic to a substructure of the tree $(A^{<\omega}, \leq)$ for some set A.

5.28 Suppose $L = \emptyset$. Show that any two infinite L-structures are partially isomorphic.

5.29 Suppose $L = \{P_1, \ldots, P_n\}$ is a *unary* vocabulary. Suppose we have two L-structures $\mathcal{M}$ and $\mathcal{N}$ satisfying the following condition: For all $\epsilon : \{1, \ldots, n\} \to \{0, 1\}$ and all $m \in \mathbb{N}$ it holds that

$$|C_\epsilon(\mathcal{M})| = m \iff |C_\epsilon(\mathcal{N})| = m.$$

Show that this is a necessary and sufficient condition for the two structures to be partially isomorphic.

5.30 Suppose that two equivalence relations $\mathcal{M}$ and $\mathcal{N}$ satisfy the following conditions for all $n, m < \omega$:

1. $EC_n(\mathcal{M}) = m \iff EC_n(\mathcal{N}) = m$.
2. If one has exactly m infinite classes, then so does the other. In symbols:

$$\sum_{\aleph_0 \leq \kappa \leq |M|} EC_\kappa(\mathcal{M}) = m \iff \sum_{\aleph_0 \leq \kappa \leq |N|} EC_\kappa(\mathcal{N}) = m.$$

Show that these are a necessary and sufficient condition for the two structures to be partially isomorphic.

5.31 For elements t of a well-founded tree $\mathcal{M}$ we can define

$$\mathrm{dom}(\mathrm{stp}'_{\mathcal{M},t}) = \{\mathrm{stp}'_{\mathcal{M},s} : s \in \mathrm{ImSuc}(t)\}$$

$$\mathrm{stp}'_{\mathcal{M},t}(\mathrm{stp}'_{\mathcal{M},s}) = \min(\aleph_0, |\{s' \in \mathrm{ImSuc}(t) : \mathrm{stp}'_{\mathcal{M},s} = \mathrm{stp}'_{\mathcal{M},s'}\}|).$$

Suppose $\mathcal{M}$ and $\mathcal{N}$ are well-founded trees such that $\mathrm{stp}'_{\mathcal{M}} = \mathrm{stp}'_{\mathcal{N}}$. Show that $\mathcal{M}$ and $\mathcal{N}$ are partially isomorphic. Give an example of two well-founded partially isomorphic trees that are not isomorphic.

5.32 Suppose that $\mathcal{M}$ and $\mathcal{N}$ are successor structures, $f \in \mathrm{Part}(\mathcal{M}, \mathcal{N})$. Show:

1. f maps elements of the standard component of $\mathcal{M}$ to elements of the standard component of $\mathcal{N}$.
2. f maps elements of a cycle component of $\mathcal{M}$ of size n to elements of a cycle component of $\mathcal{N}$ of size n.
3. f maps elements of a $\mathbb{Z}$-component of $\mathcal{M}$ to elements of a $\mathbb{Z}$-component of $\mathcal{N}$.

5.33 Suppose that two successor structures $\mathcal{M}$ and $\mathcal{N}$ satisfy the following conditions for all $n, m < \omega$:

1. $CC_n(\mathcal{M}) = m \iff CC_n(\mathcal{N}) = m$.
2. $CC_\infty(\mathcal{M}) = m \iff CC_\infty(\mathcal{N}) = m$.

Show that the successor structures are partially isomorphic.

5.34 Show that $\text{Part}(\mathcal{M}, \mathcal{N})$ is closed under unions of chains, i.e. if $f_0 \subseteq f_1 \subseteq f_2 \subseteq \ldots$ are in $\text{Part}(\mathcal{M}, \mathcal{N})$, then so is $\bigcup_{n=0}^{\infty} f_n$.

5.35 Suppose $(\mathbb{R}, <, f) \simeq_p (\mathbb{R}, <, g))$, where $f : \mathbb{R} \to \mathbb{R}$ is continuous. Show that g is also continuous.

5.36 If $(M, d), d : M \times M \to \mathbb{R}$, is a metric space, we can think of (M, d) as a an L-structure $\mathcal{M} = (M, d, \mathbb{R}, <_\mathbb{R})$, where L contains a binary function symbol, a unary predicate symbol, and a binary relation symbol. Show that there are a separable metric space $\mathcal{M} = (M, d, \mathbb{R}, <_\mathbb{R})$ and a non-separable metric space $\mathcal{M}' = (M', d', \mathbb{R}, <_\mathbb{R})$ such that $\mathcal{M} \simeq_p \mathcal{M}'$.

5.37 Show that there is a complete separable metric space (Polish space) $\mathcal{M} = (M, d, \mathbb{R}, <_\mathbb{R})$ and a non-complete separable metric space $\mathcal{M}' = (M', d', \mathbb{R}, <_\mathbb{R})$ such that $\mathcal{M} \simeq_p \mathcal{M}'$.

5.38 Suppose $\mathcal{A}$ and $\mathcal{B}$ are structures of the same relational vocabulary L and $A \cap B = \emptyset$. The *disjoint sum* of $\mathcal{A}$ and $\mathcal{B}$ is the L-structure

$$(A \cup B, (R^{\mathcal{A}} \cup R^{\mathcal{B}})_{R \in L}).$$

Show that partial isomorphism is preserved by disjoint sums of models.

5.39 Suppose $\mathcal{A}$ and $\mathcal{B}$ are structures of the same vocabulary L. The *direct product* of $\mathcal{A}$ and $\mathcal{B}$ is the L-structure

$$(A \times B, (R^{\mathcal{A}} \times R^{\mathcal{B}})_{R \in L},$$

$$(((a_0, b_0) \ldots, (a_n, b_n)) \mapsto (f^{\mathcal{A}}(a_0, \ldots, a_n), f^{\mathcal{B}}(b_0, \ldots, b_n)))_{f \in L},$$

$$((c^{\mathcal{A}}, c^{\mathcal{B}}))_{c \in L}).$$

Show that partial isomorphism is preserved by direct products of models.

5.40 Show that if two structures are partially isomorphic, then they are *potentially isomorphic*,[2] i.e. there is a forcing extension in which they are isomorphic. Conversely, show that if two structures are potentially isomorphic, then they are partially isomorphic.

5.41 Consider $\text{EF}_2(\mathcal{M}, \mathcal{N})$, where $\mathcal{M} = (\mathbb{R} \times \{0\}, f)$, $f(x, 0) = (x^2, 0)$ and $\mathcal{N} = (\mathbb{R} \times \{1\}, g)$, $g(x, 1) = (x^3, 1)$. Player **I** can win even without looking at the moves of **II**. How?

5.42 Consider $\text{EF}_\omega(\mathcal{M}, \mathcal{N})$, where $\mathcal{M} = (\mathbb{R} \times \{0\}, f)$, $f(x, 0) = (x^3, 0)$ and $\mathcal{N} = (\mathbb{R} \times \{1\}, g)$, $g(x, 1) = (x^5, 1)$. After a few moves player **I** resigns. Can you explain why?

---

[2] Some authors use the term potential isomorphism for partial isomorphism.

5.43 Consider $EF_2(\mathcal{M}, \mathcal{N})$, where $\mathcal{M} = (\mathbb{Z}, \{(a, b) : a - b = 10\})$ and $\mathcal{N} = (\mathbb{Q}, \{(a, b) : a - b = 2/3\})$. Suppose we are in position $(-8, -1/4)$ (i.e. $x_0 = -8$ and $y_0 = -1/4$). Then **I** plays $x_1 = 11/12$. What would be a good move for **II**?

5.44 Consider $EF_\omega(\mathcal{M}, \mathcal{N})$, where $\mathcal{M}$ and $\mathcal{N}$ are as in the previous exercise. Player **I** resigns before the game even starts. Can you explain why?

5.45 Suppose M and N are disjoint sets with 10 elements each. Let $c \in M$ and $d \in N$. Who has a winning strategy in $EF_\omega(\mathcal{M}, \mathcal{N})$ in the following cases:

1. $\mathcal{M} = (M, \{(a, b, c) : a = b\}), \mathcal{N} = (N, \{(a, b, d) : a = b\})$,
2. $\mathcal{M} = (M, \{(a, b, e) : a = b\}), \mathcal{N} = (N, \{(a, b, e) : b = e\})$.

5.46 Who has a winning strategy in $EF_\omega(\mathcal{M}, \mathcal{N})$ in the following cases:

1. $\mathcal{M} = (\mathbb{Q}, <, 1855), \mathcal{N} = (\mathbb{R}, <, 1854)$,
2. $\mathcal{M} = (\mathbb{N}, <, 1855), \mathcal{N} = (\mathbb{N}, <, 1854)$.

5.47 Show that $(\mathcal{P}(X), \subseteq) \simeq_p (\mathcal{P}(Y), \subseteq)$, if X and Y are disjoint infinite sets. (Hint: Consider the set of finite partial isomorphisms of the form $\{(A_0, B_0), \ldots, (A_{i-1}, B_{i-1})\}$, such that $(X, A_0, \ldots, A_{i-1})$ and $(Y, B_0, \ldots, B_{i-1})$ are partially isomorphic, and then use Exercise 5.29 of Section 5.4.)

5.48 Show that player **II** has a winning strategy in the game $EF_\omega(\mathcal{M}, \mathcal{N})$ for any two atomless (i.e. if $0 < x$ then there is y with $0 < y < x$) Boolean algebras $\mathcal{M}$ and $\mathcal{N}$.

5.49 Show that player **I** has a winning strategy in $EF_2((\mathbb{Q}, +, -), (\mathbb{R}, +, -))$.

5.50 Consider $EF_\omega((\mathbb{R}, +, -), (\mathbb{R} \times \mathbb{R}, +, -))$, where addition and substraction in $\mathbb{R} \times \mathbb{R}$ are defined componentwise. Show that player **II** has a winning strategy.

5.51 Show that partially isomorphic linear orders are isomorphic, if one is a well-order.

5.52 Show that infinite partially isomorphic structures have countably infinite isomorphic substructures.

5.53 Show that if one of two partially isomorphic trees is well-founded, then both are and the trees have the same rank. (Hint: For the second claim, prove first that if $\mathcal{M}$ is a well-founded tree, $t \in M$ and $\alpha < \mathrm{rk}_\mathcal{M}(t)$, then there is $t' \in M$ such that $\alpha = \mathrm{rk}_\mathcal{M}(t')$ and $t <^\mathcal{M} t'$.)

5.54 Suppose T is an axiomatization of set theory, at least as strong as the Kripke–Platek set theory KP (see Barwise (1975)). We say that a formula $\varphi(x_1, \ldots, x_n)$ of the language of set theory is *absolute* relative to T if for all transitive models M and N of T and for all $a_1, \ldots, a_n \in M$ we have

$$M \models \varphi(a_1, \ldots, a_n) \iff M' \models \varphi(a_1, \ldots, a_n).$$

Show that "x is a vocabulary, y and z are x-structures, and $y \simeq_p z$" can be defined with a formula $\varphi(x, y, z)$ which is absolute relative to T.

5.55 Suppose $\mathcal{A}$ is a linear order of length three and $\mathcal{B}$ a linear order of length four. Give a back-and-forth sequence of length two for $\mathcal{A}$ and $\mathcal{B}$.

5.56 Suppose $\mathcal{A}$ is a cycle of four vertices and $\mathcal{B}$ a cycle of five vertices. Give a back-and-forth sequence of length two for $\mathcal{A}$ and $\mathcal{B}$.

5.57 Suppose $\mathcal{A}$ is an equivalence relation of four classes each of size 3 and $\mathcal{B}$ an equivalence relation of three classes each of size 4. Give a back-and-forth sequence of length three for $\mathcal{A}$ and $\mathcal{B}$.

5.58 Suppose $\mathcal{A}$ is an equivalence relation of four classes each of size 2 and $\mathcal{B}$ an equivalence relation of three classes each of size 2. Give a back-and-forth sequence of length three for $\mathcal{A}$ and $\mathcal{B}$.

5.59 Suppose $\mathcal{A}$ is an equivalence relation of four classes each of size 2 plus one class of size 3, and $\mathcal{B}$ an equivalence relation of three classes each of size 2 plus one class of size 4. Give a back-and-forth sequence of length three for $\mathcal{A}$ and $\mathcal{B}$.

5.60 Suppose $\mathcal{A}$ and $\mathcal{B}$ are successor structures, both consisting of the standard component plus some cycle components. Suppose $\mathcal{A}$ has three five-cycles and $\mathcal{B}$ has four five-cycles. Give a back-and-forth sequence of length three for $\mathcal{A}$ and $\mathcal{B}$.

5.61 Show that $(7, <) \simeq_p^3 (8, <)$.

5.62 Show that $(\mathbb{Z}, <) \not\simeq_p^3 (\mathbb{Q}, <)$.

5.63 Show that $(\mathbb{N}, <) \not\simeq_p^3 (\mathbb{N} + \mathbb{N}, <)$.

5.64 Show that $(\mathbb{Z}, <) \not\simeq_p (\mathbb{Z} + \mathbb{Z}, <)$.

5.65 Show that $(\mathbb{N} + \mathbb{N}, <) \simeq_p^3 (\mathbb{N} + \mathbb{N} + \mathbb{N}, <)$

5.66 Finish the proof of Proposition 5.25.

5.67 Prove the claim of Example 5.26.

5.68 Let the game $\mathrm{EF}_\omega^*(\mathcal{A}, \mathcal{B})$ be like the game $\mathrm{EF}_\omega(\mathcal{A}, \mathcal{B})$ except that **I** has to play $x_{2n} \in A$ and $x_{2n+1} \in B$ for all $n \in \mathbb{N}$. Show that player **II** has a winning strategy in $\mathrm{EF}_\omega^*(\mathcal{A}, \mathcal{B})$ if and only if she has a winning strategy in $\mathrm{EF}_\omega(\mathcal{A}, \mathcal{B})$.

5.69 Suppose $B = \{b_n : n \in \mathbb{N}\}$. Let the game $\mathrm{EF}_\omega^{**}(\mathcal{A}, \mathcal{B})$ be like the game $\mathrm{EF}_\omega(\mathcal{A}, \mathcal{B})$ except that **I** has to play $x_{2n} \in A$ and $x_{2n+1} = b_n$ for all $n \in \mathbb{N}$. Show that player **II** has a winning strategy in $\mathrm{EF}_\omega^{**}(\mathcal{A}, \mathcal{B})$ if and only if she has a winning strategy in $\mathrm{EF}_\omega(\mathcal{A}, \mathcal{B})$.

5.70 Suppose $\mathcal{A}_0 = (A_0, <_0)$ and $\mathcal{A}_1 = (A_1, <_1)$ are linearly ordered sets. Show that if player **II** has a winning strategy both in $\mathrm{EF}_n(\mathcal{A}_0, \mathcal{B}_0)$ and in $\mathrm{EF}_n(\mathcal{A}_1, \mathcal{B}_1)$, then she has one in $\mathrm{EF}_n(\mathcal{A}_0 + \mathcal{A}_1, \mathcal{B}_0 + \mathcal{B}_1)$.

5.71 If $\mathcal{A} = (A, <)$ is a linearly ordered set and $a \in A$, then $\mathcal{A}^{<a}$ is the substructure of $\mathcal{A}$ generated by the set $\{x \in A : x < a\}$. Thus $\mathcal{A}^{<a}$ is the initial segment of $\mathcal{A}$ determined by a. Likewise, $\mathcal{A}^{>a}$ is the substructure of $\mathcal{A}$ generated by the set $\{x \in A : x > a\}$. Thus $\mathcal{A}^{>a}$ is the final segment of $\mathcal{A}$ determined by a. Show that if $\mathcal{A}$ and $\mathcal{B}$ are ordered sets, then player **II** has a winning strategy in $\mathrm{EF}_{n+1}(\mathcal{A}, \mathcal{B})$ if and only if

1. For every $a \in A$ there is $b \in B$ such that player **II** has a winning strategy in $\mathrm{EF}_n(\mathcal{A}^{<a}, \mathcal{B}^{<b})$ and in $\mathrm{EF}_n(\mathcal{A}^{>a}, \mathcal{B}^{>b})$.
2. For every $b \in B$ there is $a \in A$ such that player **II** has a winning strategy in $\mathrm{EF}_n(\mathcal{A}^{<a}, \mathcal{B}^{<b})$ and in $\mathrm{EF}_n(\mathcal{A}^{>a}, \mathcal{B}^{>b})$.

5.72 Suppose $n > 0$. Show that player **II** has a winning strategy in $\mathrm{EF}_n(\mathcal{A}, \mathcal{B})$, where $\mathcal{A}$ and $\mathcal{B}$ are linear orders with at least $2^n - 1$ elements.

5.73 Suppose $n > 0$. Show that player **I** has a winning strategy in $\mathrm{EF}_n(\mathcal{A}, \mathcal{B})$, where $\mathcal{A}$ and $\mathcal{B}$ are linear orders such that $\mathcal{A}$ has at least $2^n - 1$ elements and $\mathcal{B}$ has fewer than $2^n - 1$ elements.

5.74 Show that player **II** has a winning strategy in $\mathrm{EF}_n((\mathbb{N}, <), (\mathbb{N} + \mathbb{Z}, <))$ for every $n \in \mathbb{N}$.

5.75 An ordered set is scattered if it contains no substructure isomorphic to $(\mathbb{Q}, <)$. Show that if $\mathcal{M} \simeq_p \mathcal{N}$, where $\mathcal{N}$ is scattered, then $\mathcal{M}$ is scattered.

5.76 Suppose $\mathcal{T}$ is the tree of finite increasing sequences of rationals, and $\mathcal{T}'$ is the tree of finite increasing sequences of reals. Prove $\mathcal{T} \simeq_p \mathcal{T}'$.

5.77 Suppose $\mathcal{T}$ is the tree of finite sequences of rationals, and $\mathcal{T}'$ is the tree of finite sequences of reals. Prove $\mathcal{T} \simeq_p \mathcal{T}'$.

5.78 Suppose $\mathcal{T}$ is the tree of increasing sequences of length $\leq n$ of rationals, and $\mathcal{T}'$ is the tree of increasing sequences of length $\leq n$ of reals. Prove $\mathcal{T} \simeq_p \mathcal{T}'$.

5.79 Suppose $\mathcal{T}$ is the tree of sequences of length $\leq n$ of rationals, and $\mathcal{T}'$ is the tree of sequences of length $\leq n$ of reals. Prove $\mathcal{T} \simeq_p \mathcal{T}'$.

5.80 Suppose $\mathcal{T}$ is the tree of sequences of length $\leq n$ of elements of the set $\{1, \ldots, m\}$, and $\mathcal{T}'$ is the tree of sequences of length $\leq n$ of elements of $\{1, \ldots, m+1\}$. Prove $\mathcal{T} \simeq_p^m \mathcal{T}'$.

6

First-Order Logic

6.1 Introduction

We have already discussed the *first-order language of graphs*. We now define the basic concepts of a more general first-order language, denoted FO, one which applies to any vocabulary, not just the vocabulary of graphs. First-order logic fits the Strategic Balance of Logic better than any other logic. It is arguably the most important of all logics. It has enough power to express interesting and important concept and facts, and still it is weak and flexible enough to permit powerful constructions as demonstrated, e.g. by the Model Existence Theorem below.

6.2 Basic Concepts

Suppose L is a vocabulary. The *logical symbols* of the first-order language (or logic) of the vocabulary L are $\approx, \neg, \wedge, \vee, \forall, \exists, (,), x_0, x_1, \ldots$. *Terms* are defined as follows: Constant symbols $c \in L$ are L-terms. Variables $x_0, x_1, \ldots$ are L-terms. If $f \in L$, $\#(f) = n$, and $t_1, \ldots, t_n$ are L-terms, then so is $ft_1 \ldots t_n$. *L-equations* are of the form $\approx tt'$ where t and t' are L-terms. *L-atomic formulas* are either L-equations or of the form $Rt_1 \ldots t_n$, where $R \in L$, $\#(R) = n$ and $t_1, \ldots, t_n$ are L-terms. A *basic formula* is an atomic formula or the negation of an atomic formula. *L-formulas* are of the form

$$\approx tt'$$
$$Rt_1 \ldots t_n$$
$$\neg \varphi$$
$$(\varphi \wedge \psi), (\varphi \vee \psi)$$
$$\forall x_n \varphi, \exists x_n \varphi$$

where $t, t', t_1, \ldots, t_n$ are L-terms, $R \in L$ with $\#(R) = n$, and φ and ψ are L-formulas.

Definition 6.1 An *assignment* for a set M is any function s with $\mathrm{dom}(s)$ a set of variables and $\mathrm{rng}(s) \subseteq M$. The *value* $t^{\mathcal{M}}(s)$ of an L-term t in $\mathcal{M}$ under the assignment s is defined as follows: $c^{\mathcal{M}}(s) = \mathrm{Val}_{\mathcal{M}}(c)$, $x_n^{\mathcal{M}}(s) = s(x_n)$ and $(ft_1 \ldots t_n)^{\mathcal{M}}(s) = \mathrm{Val}_{\mathcal{M}}(f)(t_1^{\mathcal{M}}(s), \ldots, t_n^{\mathcal{M}}(s))$. The *truth* of L-formulas in $\mathcal{M}$ under s is defined as follows:

$$
\begin{aligned}
\mathcal{M} \models_s Rt_1 \ldots t_n &\quad \text{iff} \quad (t_1^{\mathcal{M}}(s), \ldots, t_n^{\mathcal{M}}(s)) \in \mathrm{Val}_{\mathcal{M}}(R) \\
\mathcal{M} \models_s {\approx} t_1 t_2 &\quad \text{iff} \quad t_1^{\mathcal{M}}(s) = t_2^{\mathcal{M}}(s) \\
\mathcal{M} \models_s \neg\varphi &\quad \text{iff} \quad \mathcal{M} \not\models_s \varphi \\
\mathcal{M} \models_s (\varphi \wedge \psi) &\quad \text{iff} \quad \mathcal{M} \models_s \varphi \text{ and } \mathcal{M} \models_s \psi \\
\mathcal{M} \models_s (\varphi \vee \psi) &\quad \text{iff} \quad \mathcal{M} \models_s \varphi \text{ or } \mathcal{M} \models_s \psi \\
\mathcal{M} \models_s \forall x_n \varphi &\quad \text{iff} \quad \mathcal{M} \models_{s[a/x_n]} \varphi \text{ for all } a \in M \\
\mathcal{M} \models_s \exists x_n \varphi &\quad \text{iff} \quad \mathcal{M} \models_{s[a/x_n]} \varphi \text{ for some } a \in M,
\end{aligned}
$$

$$
\text{where } s[a/x_n](y) = \begin{cases} a & \text{if } y = x_n \\ s(y) & \text{otherwise.} \end{cases}
$$

We assume the reader is familiar with such basic concepts as free variable, sentence, substitution of terms for variables, etc. A standard property of first-order (or any other) logic is that $\mathcal{M} \models_s \varphi$ depends only on $\mathcal{M}$ and the values of s on the variables that are free in φ. A *sentence* is a formula φ without free variables. Then $\mathcal{M} \models \varphi$ means $\mathcal{M} \models_\emptyset \varphi$. In this case we say that φ is *true* in $\mathcal{M}$.

Convention: If φ is an L-formula with the free variables $x_1, \ldots, x_n$, we indicate this by writing φ as $\varphi(x_1, \ldots, x_n)$. If $\mathcal{M}$ is an L-structure and s is an assignment for M such that $\mathcal{M} \models_s \varphi$, we write $\mathcal{M} \models \varphi(a_1, \ldots, a_n)$, where $a_i = s(x_i)$ for $i = 1, \ldots, n$.

Definition 6.2 The *quantifier rank* of a formula φ, denoted $\mathrm{QR}(\varphi)$, is defined as follows: $\mathrm{QR}({\approx}tt') = \mathrm{QR}(Rt_1 \ldots t_n) = 0$, $\mathrm{QR}(\neg\varphi) = \mathrm{QR}(\varphi)$, $\mathrm{QR}((\varphi \wedge \psi)) = \mathrm{QR}((\varphi \vee \psi)) = \max\{\mathrm{QR}(\varphi), \mathrm{QR}(\psi)\}$, $\mathrm{QR}(\exists x \varphi) = \mathrm{QR}(\forall x \varphi) = \mathrm{QR}(\varphi) + 1$. A formula φ is *quantifier free* if $\mathrm{QR}(\varphi) = 0$.

The quantifier rank is a measure of the longest sequence of "nested" quantifiers. In the first three of the following formulas the quantifiers $\forall x_n$ and $\exists x_n$ are nested but in the last unnested:

$$\forall x_0(P(x_0) \vee \exists x_1 R(x_0, x_1)) \tag{6.1}$$

$$\exists x_0(P(x_0) \wedge \forall x_1 R(x_0, x_1)) \tag{6.2}$$

$$\forall x_0(P(x_0) \vee \exists x_1 Q(x_1)) \tag{6.3}$$

$$(\forall x_0 P(x_0) \vee \exists x_1 Q(x_1)). \tag{6.4}$$

Note that formula (6.3) of quantifier rank 2 is logically equivalent to the formula (6.4) which has quantifier rank 1. So the nesting can sometimes be eliminated. In formulas (6.1) and (6.2) nesting cannot be so eliminated.

Proposition 6.3 *Suppose L is a finite vocabulary without function symbols. For every n and for every set $\{x_1, \ldots, x_n\}$ of variables, there are only finitely many logically non-equivalent first-order L-formulas of quantifier rank $< n$ with the free variables $\{x_1, \ldots, x_n\}$.*

Proof The proof is exactly like that of Proposition 4.15. □

Note that Proposition 6.3 is not true for infinite vocabularies, as there would be infinitely many logically non-equivalent atomic formulas, and also not true for vocabularies with function symbols, as there would be infinitely many logically non-equivalent equations obtained by iterating the function symbols.

6.3 Characterizing Elementary Equivalence

We now show that the concept of a back-and-forth sequence provides an alternative characterization of elementary equivalence

$$\mathcal{A} \equiv \mathcal{B} \quad \text{i.e.} \quad \forall \varphi \in FO(\mathcal{A} \models \varphi \iff \mathcal{B} \models \varphi).$$

This is the original motivation for the concepts of a back-and-forth set, back-and-forth sequence, and Ehrenfeucht–Fraïssé Game. To this end, let

$$\mathcal{A} \equiv_n \mathcal{B}$$

mean that $\mathcal{A}$ and $\mathcal{B}$ satisfy the same sentences of FO of quantifier rank $\leq n$.

We now prove an important leg of the Strategic Balance of Logic, namely the marriage of truth and separation:

Proposition 6.4 *Suppose L is an arbitrary vocabulary. Suppose $\mathcal{A}$ and $\mathcal{B}$ are L-structures and $n \in \mathbb{N}$. Consider the conditions:*

(i) $\mathcal{A} \equiv_n \mathcal{B}$.
(ii) $\mathcal{A}{\upharpoonright}_{L'} \simeq_p^n \mathcal{B}{\upharpoonright}_{L'}$ *for all finite $L' \subseteq L$.*

We have always $(ii) \to (i)$ and if L has no function symbols, then $(ii) \leftrightarrow (i)$.

Proof (ii)→(i). If $\mathcal{A} \not\equiv_n \mathcal{B}$, then there is a sentence φ of quantifier rank $\leq n$ such that $\mathcal{A} \models \varphi$ and $\mathcal{B} \not\models \varphi$. Since φ has only finitely many symbols, there is a finite $L' \subseteq L$ such that $\mathcal{A}\restriction_{L'} \not\equiv_n \mathcal{B}\restriction_{L'}$. Suppose $(P_i : i \leq n)$ is a back-and-forth sequence for $\mathcal{A}\restriction_{L'}$ and $\mathcal{B}\restriction_{L'}$. We use induction on $i \leq n$ to prove the following

Claim   If $f \in P_i$ and $a_1, \ldots, a_k \in \mathrm{dom}(f)$, then

$$(\mathcal{A}\restriction_{L'}, a_1, \ldots, a_k) \equiv_i (\mathcal{B}\restriction_{L'}, fa_1, \ldots, fa_k).$$

If $i = 0$, the claim follows from $P_0 \subseteq \mathrm{Part}(\mathcal{A}\restriction_{L'}, \mathcal{B}\restriction_{L'})$. Suppose then $f \in P_{i+1}$ and $a_1, \ldots, a_k \in \mathrm{dom}(f)$. Let $\varphi(x_0, x_1, \ldots, x_k)$ be an L'-formula of FO of quantifier rank $\leq i$ such that

$$\mathcal{A}\restriction_{L'} \models \exists x_0 \varphi(x_0, a_1, \ldots, a_k).$$

Let $a \in A$ so that $\mathcal{A}\restriction_{L'} \models \varphi(a, a_1, \ldots, a_k)$ and $g \in P_i$ such that $a \in \mathrm{dom}(g)$ and $f \subseteq g$. By the induction hypothesis, $\mathcal{B}\restriction_{L'} \models \varphi(ga, ga_1, \ldots, ga_k)$. Hence

$$\mathcal{B}\restriction_{L'} \models \exists x_0 \varphi(x_0, fa_1, \ldots, fa_k).$$

The claim is proved. Putting $i = n$ and using the assumption $P_n \neq \emptyset$, gives a contradiction with $\mathcal{A}\restriction_{L'} \not\equiv_n \mathcal{B}\restriction_{L'}$.

(i)$\rightarrow$ (ii). Assume L has no function symbols. Fix $L' \subseteq L$ finite. Let P_i consist of $f : A \to B$ such that $\mathrm{dom}(f) = \{a_0, \ldots, a_{n-i-1}\}$ and

$$(\mathcal{A}\restriction_{L'}, a_0, \ldots, a_{n-i-1}) \equiv_i (\mathcal{B}\restriction_{L'}, fa_0, \ldots, fa_{n-i-1}).$$

We show that $(P_i : i \leq n)$ is a back-and-forth sequence for $\mathcal{A}\restriction_{L'}$ and $\mathcal{B}\restriction_{L'}$. By (i), $\emptyset \in P_n$ so $P_n \neq \emptyset$. Suppose $f \in P_i, i > 0$, as above, and $a \in A$. By Proposition 6.3 there are only finitely many pairwise non-equivalent L'-formulas of quantifier rank $i - 1$ of the form $\varphi(x, x_0, \ldots, x_{n-i-1})$ in FO. Let them be $\varphi_j(x, x_0, \ldots, x_{n-i-1}), j \in J$. Let

$$J_0 = \{j \in J : \mathcal{A}\restriction_{L'} \models \varphi_j(a, a_0, \ldots, a_{n-i-1})\}.$$

Let

$$\psi(x, x_0, \ldots, x_{n-i-1}) = \bigwedge_{j \in J_0} \varphi_j(x, x_0, \ldots, x_{n-i-1}) \wedge$$

$$\bigwedge_{j \in J \setminus J_0} \neg \varphi_j(x, x_0, \ldots, x_{n-i-1}).$$

Now $\mathcal{A}\restriction_{L'} \models \exists x \psi(x, a_0, \ldots, a_{n-i-1})$, so as we have assumed $f \in P_i$, we have $\mathcal{B}\restriction_{L'} \models \exists x \psi(x, fa_0, \ldots, fa_{n-i-1})$. Thus there is some $b \in B$ with $\mathcal{B}\restriction_{L'} \models \psi(b, fa_0, \ldots, fa_{n-i-1})$. Now $f \cup \{(a, b)\} \in P_{i-1}$. The other condition (5.15) is proved similarly. $\square$

The above proposition is the standard method for proving models elementary equivalent in FO. For example, Proposition 6.4 and Example 5.26 together give $(Z, <) \equiv (Z + Z, <)$. The exercises give more examples of partially isomorphic pairs – and hence elementary equivalent – structures. The restriction on function symbols can be circumvented by first using quantifiers to eliminate nesting of function symbols and then replacing the unnested equations $f(x_1, \ldots, x_{n-1}) = x_n$ by new predicate symbols $R(x_1, \ldots, x_n)$.

Let $\mathrm{Str}(L)$ denote the class of all L-structures. We can draw the following important conclusion from Proposition 6.4 (see Figure 6.1):

Corollary *Suppose L is a vocabulary without function symbols. Then for all $n \in \mathbb{N}$ the equivalence relation*

$$\mathcal{A} \equiv_n \mathcal{B}$$

divides $\mathrm{Str}(L)$ into finitely many equivalence classes C_i^n, $i = 1, \ldots, m_n$, such that for each C_i^n there is a sentence φ_i^n of FO with the properties:

1. *For all L-structures $\mathcal{A}$: $\mathcal{A} \in C_i^n \iff \mathcal{A} \models \varphi_i^n$.*
2. *If φ is an L-sentence of quantifier rank $\leq n$, then there are $i_1, \ldots, i_k$ such that $\models \varphi \leftrightarrow (\varphi_{i_1}^n \vee \ldots \vee \varphi_{i_k}^n)$.*

Proof Let φ_i^n be the conjunction of all the finitely many L-sentences of quantifier rank $\leq n$ that are true in some (every) model in C_i^n (to make the conjunction finite we do not repeat logically equivalent formulas). For the second claim, let $\varphi_{i_1}^n, \ldots, \varphi_{i_k}^n$ be the finite set of all L-sentences of quantifier rank $\leq n$ that are consistent with φ. If now $\mathcal{A} \models \varphi$, and $\mathcal{A} \in C_i^n$, then $\mathcal{A} \models \varphi_i^n$. On the other hand, if $\mathcal{A} \models \varphi_i^n$ and there is $\mathcal{B} \models \varphi_i^n$ such that $\mathcal{B} \models \varphi$, then $\mathcal{A} \equiv_n \mathcal{B}$, whence $\mathcal{A} \models \varphi$. $\square$

We can actually read from the proof of Proposition 6.4 a more accurate description for the sentences φ_i. This leads to the theory of so-called *Scott formulas* (see Section 7.4).

Theorem 6.5 *Suppose K is a class of L-structures. Then the following are equivalent (see Figure 6.2):*

1. *K is FO-definable, i.e. there is an L-sentence φ of FO such that for all L-structures $\mathcal{M}$ we have $\mathcal{M} \in K \iff \mathcal{M} \models \varphi$.*
2. *There is $n \in \mathbb{N}$ such that K is closed under $\simeq_p^n$.*

As in the case of graphs, Theorem 6.5 can be used to demonstrate that certain properties of models are not definable in FO:

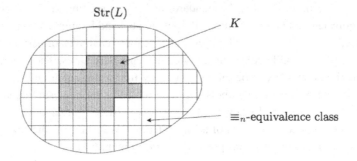

Figure 6.1 First-order definable model class K.

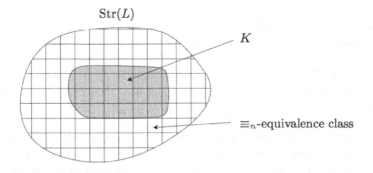

Figure 6.2 Not first-order definable model class K.

Example 6.6 Let $L = \emptyset$. The following properties of L-structures $\mathcal{M}$ are not expressible in FO:

1. M is infinite.
2. M is finite and even.

In both cases it is easy to find, for each $n \in \mathbb{N}$, two models $\mathcal{M}_n$ and $\mathcal{N}_n$ such that $\mathcal{M}_n \simeq_p^n \mathcal{N}_n$, $\mathcal{M}$ has the property, but $\mathcal{N}$ does not.

Example 6.7 Let $L = \{P\}$ be a unary vocabulary. The following properties of L-structures (M, A) are not expressible in FO:

1. $|A| = |M|$.
2. $|A| = |M \setminus A|$.

3. $|A| \leq |M \setminus A|$.

This is demonstrated by the models $(\mathbb{N}, \{1, \ldots, n\})$, $(\mathbb{N}, \mathbb{N} \setminus \{1, \ldots, n\})$, and $(\{1, \ldots, 2n\}, \{1, \ldots, n\})$.

Example 6.8 Let $L = \{<\}$ be a binary vocabulary. The following properties of L-structures $\mathcal{M} = (M, <)$ are not expressible in FO:

1. $\mathcal{M} \cong (\mathbb{Z}, <)$.
2. All closed intervals of $\mathcal{M}$ are finite.
3. Every bounded subset of $\mathcal{M}$ has a supremum.

This is demonstrated in the first two cases by the models $\mathcal{M}_n = (\mathbb{Z}, <)$ and $\mathcal{N}_n = (\mathbb{Z} + \mathbb{Z}, <)$ (see Example 5.26), and in the third case by the partially isomorphic models: $\mathcal{M} = (\mathbb{R}, <)$ and $\mathcal{N} = (\mathbb{R} \setminus \{0\}, <)$.

6.4 The Löwenheim–Skolem Theorem

In this section we show that if a first-order sentence φ is true in a structure $\mathcal{M}$, it is true in a countable substructure of $\mathcal{M}$, and even more, there are countable substructures of $\mathcal{M}$ in a sense "everywhere" satisfying φ. To make this statement precise we introduce a new game from Kueker (1977) called the Cub Game.

Definition 6.9 Suppose A is an arbitrary set. $\mathcal{P}_\omega(A)$ is defined as the set of all countable subsets of A.

The set $\mathcal{P}_\omega(A)$ is an auxiliary concept useful for the general investigation of countable substructures of a model with universe A. One should note that if A is infinite, the set $\mathcal{P}_\omega(A)$ is uncountable.[1] For example, $|\mathcal{P}_\omega(\mathbb{N})| = |\mathbb{R}|$. The set $\mathcal{P}_\omega(A)$ is closed under intersections and countable unions but not necessarily under complements, so it is a (distributive) lattice under the partial order $\subseteq$, but not a Boolean algebra. The sets in $\mathcal{P}_\omega(A)$ cover the set A entirely, but so do many proper subsets of $\mathcal{P}_\omega(A)$ such as the set of all singletons in $\mathcal{P}_\omega(A)$ and the set of all finite sets in $\mathcal{P}_\omega(A)$.

Definition 6.10 Suppose A is an arbitrary set and $\mathcal{C}$ a subset of $\mathcal{P}_\omega(A)$. The *Cub Game of* $\mathcal{C}$ is the game $G_{cub}(\mathcal{C}) = G_\omega(A, W)$, where W consists of sequences $(a_1, a_2, \ldots)$ with the property that $\{a_1, a_2, \ldots\} \in \mathcal{C}$.

---

[1] Its cardinality is $|A|^\omega$.

I	II
a_0	
	a_1
a_2	
	a_3
$\vdots$	$\vdots$

Figure 6.3 The game $G_{\text{cub}}(\mathcal{C})$.

In other words, during the game $G_{\text{cub}}(\mathcal{C})$ the players pick elements of the set A, player **I** being the one who starts. After all the infinitely many moves a set $X = \{a_1, a_2, \ldots\}$ has been formed. Player **II** tries to make sure that $X \in \mathcal{C}$ while player **I** tries to prevent this. If $\mathcal{C} = \emptyset$, player **II** has no chance. On the other hand, if $\mathcal{C} = \mathcal{P}_\omega(A)$, player **I** has no chance. When $\emptyset \neq \mathcal{C} \neq \mathcal{P}_\omega(A)$, there is a challenge for both players.

Example 6.11   Suppose $B \in \mathcal{P}_\omega(A)$ and $\mathcal{C} = \{X \in \mathcal{P}_\omega(A) : B \subseteq X\}$. Then player **II** has a winning strategy in $G_{\text{cub}}(\mathcal{C})$. Respectively, player **I** has a winning strategy in $G_{\text{cub}}(\mathcal{P}_\omega(A) \setminus \mathcal{C})$

Lemma 6.12   *Suppose $\mathcal{F}$ is a countable set of functions $f : A^{n_f} \to A$ and*

$$\mathcal{C} = \{X \in \mathcal{P}_\omega(A) : X \text{ is closed under each } f \in \mathcal{F}\}.$$

*Then player **II** has a winning strategy in the game $G_{\text{cub}}(\mathcal{C})$.*

Proof   We use the notation of Figure 6.3 for $G_{\text{cub}}(\mathcal{C})$, The strategy of player **II** is to make sure that the images of the elements a_m under the functions in $\mathcal{F}$ are eventually played. She cannot control player **I**'s moves, so she has to do it herself. On the other hand, she has nothing else to do in the game. Let $\mathcal{F} = \{f_i : i \in \mathbb{N}\}$. Let $b \in A$. If

$$m = \prod_{i=0}^{k} p_i^{m_i+1},$$

where $p_0, p_1, \ldots$ is the sequence of consecutive primes, and k is the arity of f_{m_0}, then player **II** plays

$$a_{2m+1} = f_{m_0}(a_{m_1}, \ldots, a_{m_k}).$$

Otherwise **II** plays $a_{2m+1} = b$. After all $a_0, a_1, \ldots$ have been played, the set $X = \{a_0, a_1, \ldots\}$ is closed under each f_i. Why? Suppose $f_{m_0} \in \mathcal{F}$ is k-ary

I	II
a_0^0	
	b_0^0
a_1^0	
	b_1^0

$$\vdots \quad \vdots$$

Figure 6.4 The game $G_{\text{cub}}(\bigcap_{n \in \mathbb{N}} \mathcal{C}_n)$.

and $a_{m_1}, \ldots, a_{m_k} \in X$. Let

$$m = \prod_{i=0}^{k} p_i^{m_i + 1}.$$

Then $a_{2m+1} = f_{m_0}(a_{m_1}, \ldots, a_{m_k})$. Therefore $X \in \mathcal{C}$. For example, if $f_2 \in \mathcal{F}$ is binary, then

$$a_{2 \cdot 2^3 \cdot 3^6 \cdot 5^7 + 1} = f_2(a_5, a_6).$$

$\square$

In a countable vocabulary there are only countably many function symbols. On the other hand, the functions are the main concern in checking whether a subset of a structure is the universe of a substructure. This leads to the following application of Lemma 6.12:

Proposition 6.13 *Suppose L is a countable vocabulary and $\mathcal{M}$ is an L-structure. Let $\mathcal{C}$ be the set of domains of countable submodels of $\mathcal{M}$. Then player II has a winning strategy in $G_{\text{cub}}(\mathcal{C})$.*

Intuitively this means that the countable submodels of $\mathcal{M}$ extend everywhere in $\mathcal{M}$. We will improve this observation considerably below.

Let $\pi : \mathbb{N} \times \mathbb{N} \to \mathbb{N}$ be the bijection $\pi(x, y) = \frac{1}{2}((x + y)^2 + 3x + y)$ with the inverses ρ and σ such that $\rho(\pi(x, y)) = x$ and $\sigma(\pi(x, y)) = y$.

Lemma 6.14 *Suppose player II has a winning strategy in $G_{\text{cub}}(\mathcal{C}_n)$, where $\mathcal{C}_n \subseteq \mathcal{P}_\omega(A)$, for each $n \in \mathbb{N}$. Then she has one in $G_{\text{cub}}(\bigcap_{n \in \mathbb{N}} \mathcal{C}_n)$.*

Proof We use the notation of Figure 6.4 for $G_{\text{cub}}(\bigcap_{n=1}^{\infty} \mathcal{C}_n)$, and the notation of Figure 6.5 for $G_{\text{cub}}(\mathcal{C}_n)$. The idea is that while we play $G_{\text{cub}}(\bigcap_{n \in \mathbb{N}} \mathcal{C}_n)$, player II is playing the infinitely many games $G_{\text{cub}}(\mathcal{C}_n)$, using there her winning strategy. The strategy of player II is to choose

$$b_{\pi(n,k)}^0 = b_k^{n+1},$$

I	II
a_0^n	
	b_0^n
a_1^n	
	b_1^n
$\vdots$	$\vdots$

Figure 6.5 The game $G_{\text{cub}}(\mathcal{C}_n)$.

I	II
a_0	
	b_0
a_1	
	b_1
$\vdots$	$\vdots$

Figure 6.6 The game $G_{\text{cub}}(\triangle_{a \in A} \mathcal{C}_a)$.

where b_k^{n+1} is obtained from the the Cub Game of $\mathcal{C}_{n+1}$, where player **I** plays

$$a_{2j}^{n+1} = a_j^0, a_{2j+1}^{n+1} = b_j^0.$$

$\square$

Lemma 6.15 *Suppose player **II** has a winning strategy in $G_{\text{cub}}(\mathcal{C}_a)$, where $\mathcal{C}_a \subseteq \mathcal{P}_\omega(A)$ for each $a \in A$. Then she has one in the Cub Game of the diagonal intersection $\triangle_{a \in A} \mathcal{C}_a = \{X \in \mathcal{P}_\omega(A) : \forall a \in X(X \in \mathcal{C}_a)\}$.*

Proof We use the notation of Figure 6.6 for $G_{\text{cub}}(\triangle_{a \in A} \mathcal{C}_a)$, the notation of Figure 6.7 for $G_{\text{cub}}(\mathcal{C}_{a_i})$, and the notation of Figure 6.8 for $G_{\text{cub}}(\mathcal{C}_{b_i})$. The idea is that while we play $G_{\text{cub}}(\triangle_{a \in A} \mathcal{C}_a)$, player **II** is playing the induced games

I	II
x_0^i	
	y_0^i
x_1^i	
	y_1^i
$\vdots$	$\vdots$

Figure 6.7 The game $G_{\text{cub}}(\mathcal{C}_{a_i})$.

I	II
u_0^i	
	v_0^i
u_1^i	
	v_1^i
$\vdots$	$\vdots$

Figure 6.8 The game $G_{\mathrm{cub}}(\mathcal{C}_{b_i})$.

I	II
a_0	
	b_0
a_1	
	b_1
$\vdots$	$\vdots$

Figure 6.9 The game $G_{\mathrm{cub}}(\bigtriangledown_{a \in A}\mathcal{C}_a)$.

$G_{\mathrm{cub}}(\mathcal{C}_{a_i})$ and $G_{\mathrm{cub}}(\mathcal{C}_{b_i})$, using there her winning strategy. The strategy of player **II** is to choose

$$b_{2\pi(n,k)} = y_k^n, b_{2\pi(n,k)+1} = v_k^n,$$

where b_k^{n+1} is obtained from $G_{\mathrm{cub}}(\mathcal{C}_{a_i})$, where player **I** plays

$$x_{2j}^{i+1} = a_j, x_{2j+1}^{i+1} = b_j,$$

and from $G_{\mathrm{cub}}(\mathcal{C}_{b_i})$, where player **I** plays

$$u_{2j}^{i+1} = a_j, u_{2j+1}^{i+1} = b_j.$$

$\square$

Lemma 6.16 *Suppose player **II** has a winning strategy in $G_{\mathrm{cub}}(\mathcal{C}_a)$, where $\mathcal{C}_a \subseteq \mathcal{P}_\omega(A)$, for some $a \in A$. Then she has one in the Cub Game of the diagonal union $\bigtriangledown_{a \in A}\mathcal{C}_a = \{X \in \mathcal{P}_\omega(A) : \exists a \in X(X \in \mathcal{C}_a)\}$.*

Proof We use the notation of Figure 6.9 for $G_{\mathrm{cub}}(\bigtriangledown_{a \in A}\mathcal{C}_a)$, and the notation of Figure 6.10 for $G_{\mathrm{cub}}(\mathcal{C}_a)$.

The idea is that while we play $G_{\mathrm{cub}}(\bigtriangledown_{a \in A}\mathcal{C}_a)$, player **II** is playing the game $G_{\mathrm{cub}}(\mathcal{C}_a)$ using there her winning strategy. The strategy of player **II** is to choose

$$b_0 = a, b_{n+1} = y_n,$$

I	II
x_0	
	y_0
x_1	
	y_1
$\vdots$	$\vdots$

Figure 6.10 The game $G_{\text{cub}}(\mathcal{C}_a)$.

where y_n is obtained from $G_{\text{cub}}(\mathcal{C}_a)$, where player **I** plays

$$x_0 = a, x_{i+1} = a_i.$$

$\square$

The following new concept gives an alternative characterization of the Cub Game:

Definition 6.17   A subset $\mathcal{C}$ of $\mathcal{P}_\omega(A)$ is *unbounded* if for every $X \in \mathcal{P}_\omega(A)$ there is $X' \in \mathcal{C}$ with $X \subseteq X'$. A subset $\mathcal{C}$ of $\mathcal{P}_\omega(A)$ is *closed* if the union of any increasing sequence $X_0 \subseteq X_1 \subseteq \ldots$ of elements of $\mathcal{C}$ is again an element of $\mathcal{C}$. A subset $\mathcal{C}$ of $\mathcal{P}_\omega(A)$ is *cub* if it is closed and unbounded.

A cub set of countable subsets of A covers A completely and permits the taking of unions of increasing sequences of sets.

Lemma 6.18   *Suppose $\mathcal{F}$ is a countable set of functions $f : A^{n_f} \to A$. Then the set*

$$\mathcal{C} = \{X \subseteq A : X \text{ is closed under each } f \in \mathcal{F}\}$$

is a cub set in $\mathcal{P}_\omega(A)$.

Proof   Let us first prove that $\mathcal{C}$ is unbounded. Suppose $B \in \mathcal{P}_\omega(A)$. Let

$$B^0 = B,$$
$$B^{n+1} = B^n \cup \{f(a_1, \ldots, a_{n_f}) : a_1, \ldots, a_{n_f} \in B^n\},$$
$$B^* = \bigcup_{n \in \mathbb{N}} B^n.$$

As a countable union of countable sets, B^* is countable. Since clearly $B^* \in \mathcal{C}$, we have proved the unboundedness of $\mathcal{C}$. To prove that $\mathcal{C}$ is closed, let $X_0 \subseteq X_1 \subseteq \ldots$ be elements of $\mathcal{C}$ and $X = \bigcup_{n \in \mathbb{N}} X_n$. If $f \in \mathcal{F}$ and $a_1, \ldots, a_{n_f} \in X$, then there is $n \in \mathbb{N}$ such that $a_1, \ldots, a_{n_f} \in X_n$. Since $X_n \in \mathcal{C}$, $f(a_1, \ldots, a_{n_f}) \in X_n \subseteq X$. Thus $\mathcal{C}$ is indeed closed. $\square$

Now we can prove a characterization of the Cub Game in terms of cub sets:

Proposition 6.19 *Suppose A is an arbitrary set and $\mathcal{C} \subseteq \mathcal{P}(A)$. Player* **II** *has a winning strategy in $G_{cub}(\mathcal{C})$ if and only if $\mathcal{C}$ contains a cub set.*

Proof Suppose first player **II** has a winning strategy $(\tau_0, \tau_1, \ldots)$ in $G_{cub}(\mathcal{C})$. Let $\mathcal{D}$ be the family of subsets of A that are closed under each τ_n, $n \in \mathbb{N}$. By Lemma 6.18 the set $\mathcal{D}$ is a cub set. To prove that $\mathcal{D} \subseteq \mathcal{C}$, let $X \in \mathcal{D}$. Let $X = \{a_0, a_1, \ldots\}$. Suppose player **I** plays $G_{cub}(\mathcal{C})$ by playing the elements $a_0, a_1, \ldots$ one at a time. If player **II** uses her strategy $(\tau_0, \tau_1, \ldots)$, her responses are all in X, the set X being closed under the functions τ_n. Thus at the end of the game we have the set X and since player **II** wins, $X \in \mathcal{C}$.

For the converse, suppose $\mathcal{C}$ contains a cub set $\mathcal{D}$. We need to show that player **II** has a winning strategy in $G_{cub}(\mathcal{C})$. She plays as follows: Suppose $a_0, b_0, \ldots, a_{n-1}, b_{n-1}, a_n$ have been played so far. Player **II** has as a part of her strategy produced elements $X_0 \subseteq \ldots \subseteq X_{n-1}$ of $\mathcal{D}$ such that $a_i \in X_i$ for each $i \leq n$. Let

$$X_i = \{x_0^i, x_1^i, \ldots\}.$$

The choice of player **II** for her next move is now

$$b_n = x_{\sigma(n)}^{\rho(n)}.$$

In the end, player **II** has listed all sets X_n, as after all, $x_j^i = b_{\pi(i,j)}$. Thus the set X that the players produce has to contain each set X_n, $n \in \mathbb{N}$. On the other hand, the players only play elements of A which are members of some of the sets X_n. Thus $X = \bigcup_{n \in \mathbb{N}} X_n$. Since $\mathcal{D}$ is closed, $X \in \mathcal{D} \subseteq \mathcal{C}$. $\square$

If player **I** does not have a winning strategy in $G_{cub}(\mathcal{C})$, we call $\mathcal{C}$ a *stationary* subset of $\mathcal{P}_\omega(A)$. It is a non-trivial task to construct stationary sets which are not stationary for the trivial reason that they contain a cub (see Exercise 6.46).

Endowed with the powerful methods of the Cub Game and the cub sets, we can now return to the original problem of this section: how to find countable submodels satisfying a given sentence? We attack this problem by associating every first-order sentence φ with a family $\mathcal{C}_\varphi$ of countable sets and showing that this set necessarily contains a cub set. Let us say that a formula of first order logic is in *negation normal form*, NNF in symbols, if it has negation symbols in front of atomic formulas only. Well-known equivalences show that every first-order formula is logically equivalent to a formula in NNF.

Definition 6.20 Suppose L is a vocabulary and $\mathcal{M}$ an L-structure. Suppose φ is a first-order formula in NNF and s is an assignment for the set M the domain of which includes the free variables of φ. We define the set $\mathcal{C}_{\varphi,s}$ of countable subsets of M as follows: If φ is atomic, $C_{\varphi,s}$ contains as an element the domain A of a countable submodel $\mathcal{A}$ of $\mathcal{M}$ such that $\mathrm{rng}(s) \subseteq A$ and:

- If φ is $\approx tt'$, then $t^{\mathcal{A}}(s) = t'^{\mathcal{A}}(s)$.
- If φ is $\neg \approx tt'$, then $t^{\mathcal{A}}(s) \neq t'^{\mathcal{A}}(s)$.
- If φ is $Rt_1 \ldots t_n$, then $(t_1^{\mathcal{A}}(s), \ldots, t_n^{\mathcal{A}}(s)) \in R^{\mathcal{A}}$.
- If φ is $\neg Rt_1 \ldots t_n$, then $(t_1^{\mathcal{A}}(s), \ldots, t_n^{\mathcal{A}}(s)) \notin R^{\mathcal{A}}$.

For non-basic φ we define

- $C_{\varphi \wedge \psi, s} = C_{\varphi,s} \cap C_{\psi,s}$.
- $C_{\varphi \vee \psi, s} = C_{\varphi,s} \cup C_{\psi,s}$.
- $C_{\exists x \varphi, s} = \triangledown_{a \in M} C_{\varphi, s(a/x)}$.
- $C_{\forall x \varphi, s} = \triangle_{a \in M} C_{\varphi, s[a/x]}$.

If φ is a sentence, we denote $C_{\varphi,s}$ by C_φ. If φ is not in NNF, we define $C_{\varphi,s}$ and C_φ by first translating φ into a logically equivalent NNF formula.

The sets C_φ were defined with the following fact in mind:

Proposition 6.21   *Suppose $\mathcal{A}$ is an L-structure such that $A \in C_{\varphi,s}$. Then $\mathcal{A} \models_s \varphi$.*

Proof   This is trivial for atomic φ. The induction step is clear for $\varphi \wedge \psi$ and $\varphi \vee \psi$. Suppose $A \in C_{\exists x \varphi, s}$. Thus $A \in \triangledown_{a \in M} C_{\varphi, s[a/x]}$. By the definition of diagonal union $A \in C_{\varphi, s[a/x]}$ for some $a \in A$. By the induction hypothesis, $\mathcal{A} \models_{s[a/x]} \varphi$ for some $a \in A$. Thus $\mathcal{A} \models_s \exists x \varphi$. Finally, suppose $A \in C_{\forall x \varphi, s}$. Thus $A \in \triangle_{a \in M} C_{\varphi, s[a/x]}$. By the definition of diagonal intersection $A \in C_{\varphi, s[a/x]}$ for all $a \in A$. By the induction hypothesis, $\mathcal{A} \models_{s[a/x]} \varphi$ for all $a \in A$. Thus $\mathcal{A} \models_s \forall x \varphi$.   □

Proposition 6.22   *Suppose L is countable and $\mathcal{M}$ an L-structure such that $\mathcal{M} \models \varphi$. Then player II has a winning strategy in $G_{cub}(C_\varphi)$.*

Proof   We use induction on φ to prove that if $\mathcal{M} \models_s \varphi$, then II has a winning strategy in $G_{cub}(C_\varphi)$. For atomic formulas the claim follows from Proposition 6.13. The induction step is clear for $\varphi \vee \psi$. The induction step for $\varphi \wedge \psi$ follows from Lemma 6.14. The induction step for $\forall x \varphi$ and $\exists x \varphi$ follows from Lemma 6.16. Finally, the induction step for $\forall x \varphi$ follows from Lemma 6.15.   □

Theorem 6.23 (Löwenheim–Skolem Theorem)   *Suppose L is a countable vocabulary and T is a set of L-sentences. If $\mathcal{M}$ is a model of T, then player II has a winning strategy in*

$$G_{cub}(\{X \in \mathcal{P}_\omega(M) : [X]_{\mathcal{M}} \models T\}).$$

In particular, for every countable $X \subseteq M$ there is a countable submodel $\mathcal{N}$ of $\mathcal{M}$ such that $X \subseteq N$ and $\mathcal{N} \models T$.

Proof Let $T = \{\varphi_0, \varphi_1, \ldots\}$. By Proposition 6.22 player **II** has a winning strategy in $G_{\mathrm{cub}}(\mathcal{C}_{\varphi_n})$. By Lemma 6.14, player **II** has a winning strategy in $G_{\mathrm{cub}}(\bigcap_{n=0}^{\infty} \mathcal{C}_{\varphi_n})$. If $X \in \bigcap_{n=0}^{\infty} \mathcal{C}_{\varphi_n}$, then $[X]_{\mathcal{M}} \models T$. $\qquad\square$

6.5 The Semantic Game

The truth of a first-order sentence in a structure can be defined by means of a simple game called the Semantic Game. We examine this game in detail and give some applications of it.

Definition 6.24 Suppose L is a vocabulary, $\mathcal{M}$ is an L-structure, φ^* is an L-formula, and s^* is an assignment for M. The game $\mathrm{SG}^{\mathrm{sym}}(\mathcal{M}, \varphi^*)$ is defined as follows. In the beginning player **II** holds (φ^*, s^*). The rules of the game are as follows:

1. If φ is atomic, and s satisfies it in $\mathcal{M}$, then the player who holds (φ, s) wins the game, otherwise the other player wins.
2. If $\varphi = \neg\psi$, then the player who holds (φ, s), gives (ψ, s) to the other player.
3. If $\varphi = \psi \wedge \theta$, then the player who holds (φ, s), switches to hold (ψ, s) or (θ, s), and the other player decides which.
4. If $\varphi = \psi \vee \theta$, then the player who holds (φ, s), switches to hold (ψ, s) or (θ, s), and can himself or herself decide which.
5. If $\varphi = \forall x\psi$, then the player who holds (φ, s), switches to hold $(\psi, s[a/x])$ for some a, and the other player decides for which.
6. If $\varphi = \exists x\psi$, then the player who holds (φ, s), switches to hold $(\psi, s[a/x])$ for some a, and can himself or herself decide for which.

As was pointed out in Section 4.2, $\mathcal{M} \models_s \varphi$ if and only if player **II** has a winning strategy in the above game, starting with (φ, s). Why? If $\mathcal{M} \models_s \varphi$, then the winning strategy of player **II** is to play so that if she holds (φ', s'), then $\mathcal{M} \models_{s'} \varphi'$, and if player **I** holds (φ', s'), then $\mathcal{M} \not\models_{s'} \varphi'$.

For practical purposes it is useful to consider a simpler game which presupposes that the formula is in negation normal form. In this game, as in the Ehrenfeucht–Fraïssé Game, player **I** assumes the role of a doubter and player **II** the role of confirmer. This makes the game easier to use than the full game $\mathrm{SG}^{\mathrm{sym}}(\mathcal{M}, \varphi)$.

I	II
x_0	
	y_0
x_1	
	y_1
$\vdots$	$\vdots$

Figure 6.11 The game $G_\omega(W)$.

x_n	y_n	Explanation	Rule
$(\varphi, \emptyset)$		I enquires about $\varphi \in T$.	
	$(\varphi, \emptyset)$	II confirms.	Axiom rule
(φ_i, s)		I tests a played $(\varphi_0 \wedge \varphi_1, s)$ by choosing $i \in \{0, 1\}$.	
	(φ_i, s)	II confirms.	$\wedge$-rule
$(\varphi_0 \vee \varphi_1, s)$		I enquires about a played disjunction.	
	(φ_i, s)	II makes a choice of $i \in \{0, 1\}$.	$\vee$-rule
$(\varphi, s[a/x])$		I tests a played $(\forall x \varphi, s)$ by choosing $a \in M$.	
	$(\varphi, s[a/x])$	II confirms.	$\forall$-rule
$(\exists x \varphi, s)$		I enquires about a played existential statement.	
	$(\varphi, s[a/x])$	II makes a choice of $a \in M$.	$\exists$-rule

Figure 6.12 The game $SG(\mathcal{M}, T)$.

Definition 6.25 The *Semantic Game* $SG(\mathcal{M}, T)$ of the set T of L-sentences in NNF is the game (see Figure 6.11) $G_\omega(W)$, where W consists of sequences $(x_0, y_0, x_1, y_1, \ldots)$ where player **II** has followed the rules of Figure 6.12 and if player **II** plays the pair (φ, s), where φ is a basic formula, then $\mathcal{M} \models_s \varphi$.

In the game $SG(\mathcal{M}, T)$ player **II** claims that every sentence of T is true in $\mathcal{M}$. Player **I** doubts this and challenges player **II**. He may doubt whether a certain $\varphi \in T$ is true in $\mathcal{M}$, so he plays $x_0 = (\varphi, \emptyset)$. In this round, as in some other rounds too, player **II** just confirms and plays the same pair as player **I**. This may seem odd and unnecessary, but it is for book-keeping purposes only. Player **I** in a sense gathers a finite set of formulas confirmed by player **II** and tries to end up with a basic formula which cannot be true.

Theorem 6.26 *Suppose L is a vocabulary, T is a set of L-sentences, and $\mathcal{M}$ is an L-structure. Then the following are equivalent:*

1. $\mathcal{M} \models T$.
*2. Player **II** has a winning strategy in SG($\mathcal{M}, T$).*

Proof Suppose $\mathcal{M} \models T$. The winning strategy of player **II** in SG($\mathcal{M}, T$) is to maintain the condition $\mathcal{M} \models_{s_i} \psi_i$ for all $y_i = (\psi_i, s_i)$, $i \in \mathbb{N}$, played by her. It is easy to see that this is possible. On the other hand, suppose $\mathcal{M} \not\models T$, say $\mathcal{M} \not\models \varphi$, where $\varphi \in T$. The winning strategy of player **I** in SG($\mathcal{M}, T$) is to start with $x_0 = (\varphi, \emptyset)$, and then maintain the condition $\mathcal{M} \not\models_{s_i} \psi_i$ for all $y_i = (\psi_i, s_i)$, $i \in \mathbb{N}$, played by **II**:

1. If $y_i = (\psi_i, s_i)$, where ψ_i is basic, then player **I** has won the game, because $\mathcal{M} \not\models_{s_i} \psi_i$.
2. If $y_i = (\psi_i, s_i)$, where $\psi_i = \theta_0 \wedge \theta_1$, then player **I** can use the assumption $\mathcal{M} \not\models_{s_i} \psi_i$ to find $k < 2$ such that $\mathcal{M} \not\models_{s_i} \theta_k$. Then he plays $x_{i+1} = (\theta_k, s_i)$.
3. If $y_i = (\psi_i, s_i)$, where $\psi_i = \theta_0 \vee \theta_1$, then player **I** knows from the assumption $\mathcal{M} \not\models_{s_i} \psi_i$ that whether **II** plays (θ_k, s_i) for $k = 0$ or $k = 1$, the condition $\mathcal{M} \not\models_{s_i} \theta_k$ still holds. So player **I** can play $x_{i+1} = (\psi_i, s_i)$ and keep his winning criterion in force.
4. If $y_i = (\psi_i, s_i)$, where $\psi_i = \forall x\varphi$, then player **I** can use the assumption $\mathcal{M} \not\models_{s_i} \psi_i$ to find $a \in M$ such that $\mathcal{M} \not\models_{s_i[a/x]} \varphi$. Then he plays $x_{i+1} = (\varphi, s_i[a/x])$.
5. If $y_i = (\psi_i, s_i)$, where $\psi_i = \exists x\varphi$, then player **I** knows from the assumption $\mathcal{M} \not\models_{s_i} \psi_i$ that whatever $(\varphi, s_i[a/x])$ player **II** chooses to play, the condition $\mathcal{M} \not\models_{s_i[a/x]} \varphi$ still holds. So player **I** can play $(\exists x\varphi, s_i)$ and keep his winning criterion in force.

$\square$

Example 6.27 Let $L = \{f\}$ and $\mathcal{M} = (\mathbb{N}, f^{\mathcal{M}})$, where $f(n) = n + 1$. Let

$$\varphi = \forall x \exists y {\approx} fxy.$$

Clearly, $\mathcal{M} \models \varphi$. Thus player **II** has, by Theorem 6.26, a winning strategy in the game SG($\mathcal{M}, \{\varphi\}$). Figure 6.13 shows how the game might proceed. On

I	II	Rule
$(\forall x \exists y \approx fxy, \emptyset)$		
	$(\forall x \exists y \approx fxy, \emptyset)$	Axiom rule
$(\exists y \approx fxy, \{(x, 25)\})$		
	$(\exists y \approx fxy, \{(x, 25)\})$	$\forall$-rule
$(\exists y \approx fxy, \{(x, 25)\})$		
	$(\approx fxy, \{(x, 25), (y, 26)\})$	$\exists$-rule
$\vdots$	$\vdots$	

Figure 6.13 Player **II** has a winning strategy in SG($\mathcal{M}, \{\varphi\}$).

I	II	Rule
$(\forall x \exists y \approx fyx, \emptyset)$		
	$(\forall x \exists y \approx fyx, \emptyset)$	Axiom rule
$(\exists y \approx fyx, \{(x, 0)\})$		
	$(\exists y \approx fyx, \{(x, 0)\})$	$\forall$-rule
$(\exists y \approx fyx, \{(x, 0)\})$		
	$(\approx fyx, \{(x, 0), (y, 2)\})$	$\exists$-rule
	(**II** has no good move)	

Figure 6.14 Player **I** wins the game SG($\mathcal{M}, \{\psi\}$).

the other hand, suppose

$$\psi = \forall x \exists y \approx fyx.$$

Clearly, $\mathcal{M} \not\models \varphi$. Thus player **I** has, by Theorem 6.26 and Theorem 3.12, a winning strategy in the game SG($\mathcal{M}, \{\varphi\}$). Figure 6.14 shows how the game might proceed.

Example 6.28 Let $\mathcal{M}$ be the graph of Figure 6.15. and

$$\varphi = \forall x (\exists y \neg xEy \land \exists yxEy).$$

Clearly, $\mathcal{M} \models \varphi$. Thus player **II** has, by Theorem 6.26, a winning strategy in the game SG($\mathcal{M}, \{\varphi\}$). Figure 6.16 shows how the game might proceed. On the other hand, suppose

$$\psi = \exists x (\forall y \neg xEy \lor \forall yxEy).$$

Clearly, $\mathcal{M} \not\models \varphi$. Thus player **I** has, by Theorem 6.26 and Theorem 3.12, a winning strategy in the game SG($\mathcal{M}, \{\varphi\}$). Figure 6.17 shows how the game might proceed.

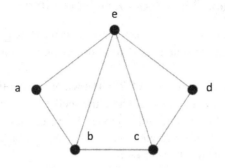

Figure 6.15 The graph $\mathcal{M}$.

I	II	Rule
$(\forall x(\exists y \neg xEy \wedge \exists yxEy), \emptyset)$	$(\forall x(\exists y \neg xEy \wedge \exists yxEy), \emptyset)$	Axiom rule
$(\exists y \neg xEy \wedge \exists yxEy, \{(x,d)\})$	$(\exists y \neg xEy \wedge \exists yxEy, \{(x,d)\})$	$\forall$-rule
$(\exists yxEy, \{(x,d)\})$	$(\exists yxEy, \{(x,d)\})$	$\wedge$-rule
$(\exists yxEy, \{(x,d)\})$	$(xEy, \{(x,d),(y,c)\})$	$\exists$-rule
$\vdots$	$\vdots$	

Figure 6.16 Player **II** has a winning strategy in $\mathrm{SG}(\mathcal{M}, \{\varphi\})$.

I	II	Rule
$(\exists x(\forall y \neg xEy \vee \forall yxEy), \emptyset)$	$(\exists x(\forall y \neg xEy \vee \forall yxEy), \emptyset)$	Axiom rule
$(\exists x(\forall y \neg xEy \vee \forall yxEy), \emptyset)$	$(\forall y \neg xEy \vee \forall yxEy), \{(x,a)\})$	$\exists$-rule
$(\forall y \neg xEy \vee \forall yxEy, \{(x,a)\})$	$(\forall y \neg xEy, \{(x,a)\})$	$\vee$-rule
$(\neg xEy, \{(x,a),(y,d)\})$	$(\neg xEy, \{(x,a),(y,d)\})$	$\forall$-rule

Figure 6.17 Player **I** wins the game $\mathrm{SG}(\mathcal{M}, \{\psi\})$.

6.6 The Model Existence Game

In this section we learn a new game associated with trying to construct a model for a sentence or a set of sentences. This is of fundamental importance in the sequel.

Let us first recall the game $SG(\mathcal{M}, T)$: The winning condition for **II** in the game $SG(\mathcal{M}, T)$ is the only place where the model $\mathcal{M}$ (rather than the set M) appears. If we do not start with a model $\mathcal{M}$ we can replace the winning condition with a slightly weaker one and get a very useful criterion for the existence of *some* $\mathcal{M}$ such that $\mathcal{M} \models T$:

Definition 6.29   The *Model Existence Game* $MEG(T, L)$ of the set T of L-sentences in NNF is defined as follows. Let C be a countably infinite set of new constant symbols. $MEG(T, L)$ is the game $G_\omega(W)$ (see Figure 6.11), where W consists of sequences $(x_0, y_0, x_1, y_1, \ldots)$ where player **II** has followed the rules of Figure 6.18 and for no atomic $L \cup C$-sentence φ both φ and $\neg\varphi$ are in $\{y_0, y_1, \ldots\}$.

The idea of the game $MEG(T, L)$ is that player **I** does not doubt the truth of T (as there is no model around) but rather the mere consistency of T. So he picks those $\varphi \in T$ that he thinks constitute a contradiction and offers them to player **II** for confirmation. Then he runs through the subformulas of these sentences as if there was a model around in which they cannot all be true. He wins if he has made player **II** play contradictory basic sentences. It turns out it did not matter that we had no model around, as two contradictory sentences cannot hold in any model anyway.

Definition 6.30   Let L be a vocabulary with at least one constant symbol. A *Hintikka set (for first-order logic)* is a set H of L-sentences in NNF such that:

1. $\approx tt \in H$ for every constant L-term t.
2. If $\varphi(x)$ is basic, $\varphi(c) \in H$ and $\approx tc \in H$, then $\varphi(t) \in H$.
3. If $\varphi \wedge \psi \in H$, then $\varphi \in H$ and $\psi \in H$.
4. If $\varphi \vee \psi \in H$, then $\varphi \in H$ or $\psi \in H$.
5. If $\forall x \varphi(x) \in H$, then $\varphi(c) \in H$ for all $c \in L$
6. If $\exists x \varphi(x) \in H$, then $\varphi(c) \in H$ for some $c \in L$.
7. For every constant L-term t there is $c \in L$ such that $\approx ct \in H$.
8. There is no atomic sentence φ such that $\varphi \in H$ and $\neg\varphi \in H$.

Lemma 6.31   *Suppose L is a vocabulary and T is a set of L-sentences. If T has a model, then T can be extended to a Hintikka set.*

x_n	y_n	Explanation
φ		**I** enquires about $\varphi \in T$.
	φ	**II** confirms.
$\approx tt$		**I** enquires about an equation.
	$\approx tt$	**II** confirms.
$\varphi(t')$		**I** chooses played $\varphi(t)$ and $\approx tt'$ with φ basic and enquires about substituting t' for t in φ.
	$\varphi(t')$	**II** confirms.
φ_i		**I** tests a played $\varphi_0 \wedge \varphi_1$ by choosing $i \in \{0, 1\}$.
	φ_i	**II** confirms.
$\varphi_0 \vee \varphi_1$		**I** enquires about a played disjunction.
	φ_i	**II** makes a choice of $i \in \{0, 1\}$
$\varphi(c)$		**I** tests a played $\forall x \varphi(x)$ by choosing $c \in C$.
	$\varphi(c)$	**II** confirms.
$\exists x \varphi(x)$		**I** enquires about a played existential statement.
	$\varphi(c)$	**II** makes a choice of $c \in C$
t		**I** enquires about a constant $L \cup C$-term t.
	$\approx ct$	**II** makes a choice of $c \in C$

Figure 6.18 The game $\mathrm{MEG}(T, L)$.

Proof Let us assume $\mathcal{M} \models T$. Let $L' \supseteq L$ such that L' has a constant symbol $c_a \notin L$ for each $a \in M$. Let $\mathcal{M}^*$ be an expansion of $\mathcal{M}$ obtained by interpreting c_a by a for each $a \in M$. Let H be the set of all L'-sentences true in $\mathcal{M}$. It is easy to verify that H is a Hintikka set.

$\square$

Lemma 6.32 *Suppose L is a countable vocabulary and T is a set of L-sentences. If player II has a winning strategy in $\text{MEG}(T, L)$, then the set T can be extended to a Hintikka set in a countable vocabulary extending L by constant symbols.*

Proof Suppose player **II** has a winning strategy in $\text{MEG}(T, L)$. We first run through one carefully planned play of $\text{MEG}(T, L)$. This will give rise to a model $\mathcal{M}$. Then we play again, this time providing a proof that $\mathcal{M} \models T$. To this end, let Trm be the set of all constant $L \cup C$-terms. Let

$$T = \{\varphi_n : n \in \mathbb{N}\},$$
$$C = \{c_n : n \in \mathbb{N}\},$$
$$Trm = \{t_n : n \in \mathbb{N}\}.$$

Let $(x_0, y_0, x_1, y_1, \ldots)$ be a play in which player **II** has used her winning strategy and player **I** has maintained the following conditions:

1. If $n = 3^i$, then $x_n = \varphi_i$.
2. If $n = 2 \cdot 3^i$, then x_n is $\approx c_i c_i$.
3. If $n = 4 \cdot 3^i \cdot 5^j \cdot 7^k \cdot 11^l$, y_i is $\approx t_j t_k$, and y_l is $\varphi(t_j)$, then x_n is $\varphi(t_k)$.
4. If $n = 8 \cdot 3^i \cdot 5^j$, y_i is $\theta_0 \wedge \theta_1$, and $j < 2$, then x_n is θ_j.
5. If $n = 16 \cdot 3^i$, and y_i is $\theta_0 \vee \theta_1$, then x_n is $\theta_0 \vee \theta_1$.
6. If $n = 32 \cdot 3^i \cdot 5^j$, y_i is $\forall x \varphi(x)$, then x_n is $\varphi(c_j)$.
7. If $n = 64 \cdot 3^i$, and y_i is $\exists x \varphi(x)$, then x_n is $\exists x \varphi(x)$.
8. If $n = 128 \cdot 3^i$, then x_n is t_i.

The idea of these conditions is that player **I** challenges player **II** in a maximal way. To guarantee this he makes a plan. The plan is, for example, that on round 3^i he always plays φ_i from the set T. Thus in an infinite game every element of T will be played. Also the plan involves the rule that if player **II** happens to play a conjunction $\theta_0 \wedge \theta_1$ on round i, then player **I** will necessarily play θ_0 on round $8 \cdot 3^i$ and θ_1 on round $8 \cdot 3^i \cdot 5$, etc. It is all just book-keeping – making sure that all possibilities will be scanned. This strategy of **I** is called the *enumeration strategy*. It is now routine to show that $H = \{y_0, y_1, \ldots\}$ is a Hintikka set. $\square$

Lemma 6.33 *Every Hintikka set has a model in which every element is the interpretation of a constant symbol.*

Proof Let $c \sim c'$ if $\approx c'c \in H$. The relation $\sim$ is an equivalence relation on C (see Exercise 6.77). Let us define an $L \cup C$-structure $\mathcal{M}$ as follows.

We let $M = \{[c] : c \in C\}$. For $c \in C$ we let $c^{\mathcal{M}} = [c]$. If $f \in L$ and $\#(f) = n$ we let $f^{\mathcal{M}}([c_{i_1}], \ldots, [c_{i_n}]) = [c]$ for some (any – see Exercise 6.78) $c \in C$ such that $\approx cfc_{i_1} \ldots c_{i_n} \in H$. For any constant term t there is a $c \in C$ such that $\approx ct \in H$. It is easy to see that $t^{\mathcal{M}} = [c]$. For the atomic sentence $\varphi = Rt_1 \ldots t_n$ we let $\mathcal{M} \models \varphi$ if and only if φ is in H. An easy induction on φ shows that if $\varphi(x_1, \ldots, x_n)$ is an L-formula and $\varphi(d_1, \ldots, d_n) \in H$ for some $d_1 \ldots, d_n$, then $\mathcal{M} \models \varphi(d_1, \ldots, d_n)$ (see Exercise 6.79). In particular, $\mathcal{M} \models T$. $\qquad \square$

Lemma 6.34 *Suppose L is a countable vocabulary and T is a set of L-sentences. If T can be extended to a Hintikka set in a countable vocabulary extending L, then player* **II** *has a winning strategy in* $\mathrm{MEG}(T, L)$

Proof Suppose L^* is a countable vocabulary extending L such that some Hintikka set H in the vocabulary L^* extends T. Let $C = \{c_n : n \in \mathbb{N}\}$ be a new countable set of constant symbols to be used in $\mathrm{MEG}(T, L)$. Suppose $D = \{t_n : n \in \mathbb{N}\}$ is the set of constant terms of the vocabulary L^*. The winning strategy of player **II** in $\mathrm{MEG}(T, L)$ is to maintain the condition that if y_i is $\varphi(c_1, \ldots, c_n)$, then $\varphi(t_1, \ldots, t_n) \in H$. $\qquad \square$

We can now prove the basic element of the Strategic Balance of Logic, namely the following equivalence between the Semantic Game and the Model Existence Game:

Theorem 6.35 (Model Existence Theorem) *Suppose L is a countable vocabulary and T is a set of L-sentences. The following are equivalent:*

1. *There is an L-structure $\mathcal{M}$ such that $\mathcal{M} \models T$.*
2. *Player* **II** *has a winning strategy in* $\mathrm{MEG}(T, L)$.

Proof If there is an L-structure $\mathcal{M}$ such that $\mathcal{M} \models T$, then by Lemma 6.31 there is a Hintikka set $H \supseteq T$. Then by Lemma 6.34 player **II** has a winning strategy in $\mathrm{MEG}(T, L)$. Suppose conversely that player **II** has a winning strategy in $\mathrm{MEG}(T, L)$. By Lemma 6.32 there is a Hintikka set $H \supseteq T$. Finally, this implies by Lemma 6.33 that T has a model. $\qquad \square$

Corollary *Suppose L is a countable vocabulary, T a set of L-sentences and φ an L-sentence. Then the following conditions are equivalent:*

1. $T \models \varphi$.
2. *Player* **I** *has a winning strategy in* $\mathrm{MEG}(T \cup \{\neg\varphi\}, L)$.

Proof By Theorem 3.12 the game $\mathrm{MEG}(T \cup \{\neg\varphi\}, L)$ is determined. So by Theorem 6.35, condition 2 is equivalent to $T \cup \{\neg\varphi\}$ not having a model, which is exactly what condition 1 says. $\qquad \square$

Condition 1 of the above Corollary is equivalent to φ having a *formal proof* from T. (See Enderton (2001), or any standard textbook in logic for a definition of formal proof.) We can think of a winning strategy of player **I** in MEG($T \cup \{\neg\varphi\}, L$) as a *semantic proof*. In the literature this concept occurs under the names *semantic tree* or *Beth tableaux*.

6.7 Applications

The Model Existence Theorem is extremely useful in logic. Our first application – The Compactness Theorem – is a kind of model existence theorem itself and very useful throughout model theory.

Theorem 6.36 (Compactness Theorem)   *Suppose L is a countable vocabulary and T is a set of L-sentences such that every finite subset of T has a model. Then T has a model.*

Proof   Let C be a countably infinite set of new constant symbols as needed in MEG(T, L). The winning strategy of player **II** in MEG(T, L) is the following. Suppose

$$(x_0, y_0, \ldots, x_{n-1}, y_{n-1})$$

has been played up to now, and then player **I** plays x_n. Player **II** has made sure that $T \cup \{y_0, \ldots, y_{n-1}\}$ is *finitely consistent*, i.e. each of its finite subsets has a model. Now she makes such a move y_n that $T \cup \{y_0, \ldots, y_n\}$ is still finitely consistent. Suppose this is the case and player **I** asks a confirmation for φ, where $\varphi \in T$. Now $T \cup \{y_0, \ldots, y_{n-1}, \varphi\}$ is finitely consistent as it is the same set as $T \cup \{y_0, \ldots, y_{n-1}\}$. Suppose then player **I** asks a confirmation for θ_0, where $\theta_0 \wedge \theta_1 = y_i$ for some $i < n$. If $T_0 \cup \{y_0, \ldots, y_{n-1}, \theta_0\}$ has no model, where T_0 is a finite subset of T, then surely $T_0 \cup \{y_0, \ldots, y_{n-1}\}$ has no models either, a contradiction. Suppose then player **I** asks for a decision about $\theta_0 \vee \theta_1$, where $\theta_0 \vee \theta_1 = y_i$ for some $i < n$. If $T_0 \cup \{y_0, \ldots, y_{n-1}, \theta_0\}$ has no models, where T_0 is a finite subset of T, and also $T_1 \cup \{y_0, \ldots, y_{n-1}, \theta_1\}$ has no models, where T_1 is another finite subset of T, then $T_0 \cup T_1 \cup \{y_0, \ldots, y_{n-1}\}$ has no models, a contradiction. Suppose then player **I** asks for a confirmation for $\varphi(c)$, where $\forall x \varphi(x) = y_i$ for some $i < n$ and $c \in C$. If $T_0 \cup \{y_0, \ldots, y_{n-1}, \varphi(c)\}$ has no models, where T_0 is a finite subset of T, then $T_0 \cup \{y_0, \ldots, y_{n-1}\}$ has no models either, a contradiction. Suppose then player **I** asks a decision about $\exists x \varphi(x)$, where $\exists x \varphi(x) = y_i$ for some $i < n$. Let $c \in C$ so that c does not occur in $\{y_0, \ldots, y_{n-1}\}$. We claim that $T \cup \{y_0, \ldots, y_{n-1}, \varphi(c)\}$ is finitely consistent. Suppose the contrary. Then there is a finite conjunction ψ of sentences in T such that

$$\{y_0, \ldots, y_{n-1}, \psi\} \models \neg\varphi(c).$$

Hence

$$\{y_0, \ldots, y_{n-1}, \psi\} \models \forall x \neg\varphi(x).$$

But this contradicts the fact that $\{y_0, \ldots, y_{n-1}, \psi\}$ has a model in which $\exists x\varphi(x)$ is true. Finally, if t is a constant term, it follows as above that there is a constant $c \in C$ such that $T \cup \{y_0, \ldots, y_{n-1}, \approx ct\}$ is finitely consistent. $\qquad\square$

It is a consequence of the Compactness Theorem that a theory in a countable vocabulary is consistent in the sense that every finite subset has a model if and only if it is consistent in the sense that T itself has a model. Therefore the word "consistent" is used in both meanings.

As an application of the Compactness Theorem consider the vocabulary $L = \{+, \cdot, 0, 1\}$ of number theory. An example of an L-structure is the so-called *standard model of number theory* $\mathcal{N} = (\mathbb{N}, +, \cdot, 0, 1)$. L-structures may be elementary equivalent to $\mathcal{N}$ and still be *non-standard* in the sense that they are not isomorphic to $\mathcal{N}$. Let c be a new constant symbol. It is easy to see that the theory

$$\{\varphi : \mathcal{N} \models \varphi\} \cup \{1 < c, +11 < c, ++111 < c, \ldots\}$$

is finitely consistent. By the Compactness Theorem it has a model $\mathcal{M}$. Clearly $\mathcal{M} \equiv \mathcal{N}$ and $\mathcal{M} \not\cong \mathcal{N}$.

Example 6.37 Suppose T is a theory in a countable vocabulary L, and T has for each $n > 0$ a model $\mathcal{M}_n$ such that $(M_n, E^{\mathcal{M}_n})$ is a graph with a cycle of length $\geq n$. We show that T has a model $\mathcal{N}$ such that $(N, E^{\mathcal{N}})$ is a graph with an infinite cycle (i.e. an infinite connected subgraph in which every node has degree 2). To this end, let $c_z, z \in \mathbb{Z}$, be new constant symbols. Let T' be the theory

$$T \cup \{c_z E c_{z+1} : z \in \mathbb{Z}\}.$$

Any finite subset of T' mentions only finitely constants c_z, so it can be satisfied in the model $\mathcal{M}_n$ for a sufficiently large n. By the Compactness Theorem T' has a model $\mathcal{M}$. Now $\mathcal{M} \upharpoonright L \models T$ and the elements $c_z^{\mathcal{M}}$, $z \in \mathbb{Z}$, constitute an infinite cycle in $\mathcal{M}$.

As another application of the Model Existence Game we prove the so-called

Omitting Types Theorem. Suppose we have a countable vocabulary L and T is a set of L-sentences. A *type* of T is a countable sequence

$$\psi_0(x), \psi_1(x), \ldots$$

of formulas with just one free variable x (therefore we write $\psi_i(x)$) such that for all n:

$$T \cup \{\exists x(\psi_0(x) \wedge \ldots \wedge \psi_n(x))\}$$

is consistent. A good example is

$$\mathrm{tp}_{\mathcal{M}}(a) = \{\psi(x) : \mathcal{M} \models \psi(a)\},$$

where $\mathcal{M} \models T$ and $a \in M$. This is called *the type of a* in $\mathcal{M}$.

A model $\mathcal{M}$ of T *realizes* a type p if some element a of M satisfies $\mathcal{M} \models \psi(a)$ for all $\psi(x) \in p$. A model $\mathcal{M}$ of T *omits* the type p if no element of M satisfies p. For example, the standard model $(\mathbb{N}, +, \cdot, 0, 1, <)$ of number theory realizes the type

$$\{x < 1, +xx < 1, ++\, xxx < 1, \ldots\}$$

but omits the type

$$\{1 < x, +11 < x, ++\, 111 < x, \ldots\}.$$

A type p of T is *principal* if there is a formula $\theta(x)$ such that

- $T \vdash \theta(x) \to \psi(x)$ for all $\psi(x) \in p$.
- $T \cup \{\exists x \theta(x)\}$ is consistent.

It is clear that a principal type cannot always be omitted, e.g. if T is complete. But every other type can:

Theorem 6.38 (Omitting Types Theorem)   *If L is a countable vocabulary, T a consistent set of L-sentences, and p a non-principal type of T, then T has a countable model which omits p.*

Proof Let $p = \{\varphi_n(x) : n \in \mathbb{N}\}$. Since T is consistent, player **II** has a winning strategy in $\mathrm{MEG}(T, L)$. Moreover, the proof of the Model Existence Theorem tells us that if the game $\mathrm{MEG}(T, L)$ is played, **II** playing her winning strategy and **I** playing his enumeration strategy, a model of T is generated in which every element is an interpretation of one of the new constants c_n. We now show that player **II** actually has a winning strategy satisfying the following extra condition: during round $n = 256 \cdot 3^i$ of the game player **I** is allowed to ask player **II** to provide a natural number $f(i)$ and play the sentence $\neg\varphi_{f(i)}(c_i)$. The winning strategy of **II** is to make sure that after round n the set

$$T \cup \{y_i : i \le n\}$$

is (finitely) consistent. If she can play this strategy against the enumeration strategy of player **I**, a model of T is generated in which every element is the interpretation of a constant c_n and thus p is omitted, as the model constructed in the proof of the Model Existence Theorem now satisfies also $\neg\varphi_{f(n)}(c_n)$ for each n.

Suppose **II** has been able to keep $T \cup \{y_i : i < n\}$ consistent and now $n = 256 \cdot 3^n$. If **II** can find no suitable $f(n)$, then for all $m \in \mathbb{N}$

$$T \cup \{y_i : i < n\} \models \varphi_m(c_n).$$

Let us write the sentences $\{y_i : i < n\}$ as

$$\{\theta_i(c_{j_1}, \ldots, c_{j_k}, c_n) : i < n\}.$$

Thus

$$T \models (\exists x_{j_1} \ldots \exists x_{j_k} \bigwedge_{i<n} \theta_i(x_{j_1}, \ldots, x_{j_k}, x)) \to \varphi_m(x)$$

for all $m \in \mathbb{N}$. Note that since **II** has been following her strategy up to now, the set

$$T \cup \{\exists x \exists x_{j_1} \ldots \exists x_{j_k} \bigwedge_{i<2n} \theta_i(x_{j_1}, \ldots, x_{j_k}, x)\}$$

is consistent, contrary to the non-principality of p. $\qquad\square$

As an application of the Omitting Types Theorem consider a language L which extends the language $L_0 = \{+, \cdot, 0, 1\}$ of number theory by a unary predicate P. Some examples of L-structures are

$$\mathcal{N}_A = (\mathbb{N}, +, \cdot, 0, 1, A), \tag{6.5}$$

where $A \subseteq \mathbb{N}$ can be, for example,

- $\emptyset$.
- $\mathbb{N}$.
- the set of all prime numbers.
- the set of all even numbers.

L-structures may also be *non-standard* in the sense that they are not isomorphic to any structure of the form (6.5). In fact for every $A \subseteq \mathcal{N}$ there is

$$\mathcal{M} = (M, +^{\mathcal{M}}, \cdot^{\mathcal{M}}, 0^{\mathcal{M}}, 1^{\mathcal{M}}, P^{\mathcal{M}}) \equiv (\mathbb{N}, +, \cdot, 0, 1, A) \qquad (6.6)$$

such that M has "infinitely large" elements. This is an easy consequence of the Compactness Theorem.

Conversely, we may ask whether for every non-standard

$$(M, +^M, \cdot^M, 0^M, 1^M) \equiv (\mathbb{N}, +, \cdot, 0, 1)$$

there is $A \subseteq \mathbb{N}$ such that

$$(M, +^M, \cdot^M, 0^M, 1^M, P^M) \equiv (\mathbb{N}, +, \cdot, 0, 1, A).$$

Suppose $\mathcal{M}$ is an L-structure with the following properties:

- If $\mathcal{M} \vDash \exists x \varphi(x)$, then there is $n \in \mathbb{N}$ such that $\mathcal{M} \vDash \varphi(\underline{n})$, where $\underline{0} = 0$ and $\underline{n+1} = \underline{n} + 1$.
- $\mathcal{M} \restriction L_0 \equiv (\mathbb{N}, +, \cdot, 0, 1)$.

Claim There is $A \subseteq \mathbb{N}$ such that

$$(\mathbb{N}, +, \cdot, 0, 1, A) \equiv \mathcal{M}.$$

Proof Let $T = \{\varphi : \mathcal{M} \vDash \varphi\}$. Let p be the type

$$\neg \approx \underline{1}x, \ \neg \approx \underline{2}x, \ \neg \approx \underline{3}x, \ \ldots$$

The type p is non-principal, for if $T \cup \{\exists x \varphi(x)\}$ is consistent, then $\mathcal{M} \vDash \exists x \varphi(x)$ whence $\mathcal{M} \vDash \varphi(\underline{n})$ for some $n \in \mathbb{N}$, so it is not possible that

$$T \vdash \varphi(x) \to \neg \approx \underline{m}x$$

for all $m \in \mathbb{N}$ (e.g. for $m = n$). By the Omitting Types Theorem there is a model $\mathcal{M}'$ of T which omits p. Thus every element of M' is equal to some $\underline{n}^{\mathcal{M}'}$. Let

$$f : M' \to \mathbb{N}$$

be defined by

$$f(n) = \underline{n}^{\mathcal{M}'}.$$

Since $\mathcal{M} \restriction L_0 \equiv (\mathbb{N}, +, \cdot, 0, 1)$, this function f is a bijection. Let

$$A = \{n \in \mathbb{N} : \mathcal{M}' \vDash P(\underline{n})\}.$$

Then

$$f : (\mathbb{N}, +, \cdot, 0, 1, A) \cong \mathcal{M}'.$$

Thus

$$(\mathbb{N}, +, \cdot, 0, 1, A) \equiv \mathcal{M}' \equiv \mathcal{M}.$$

$\square$

In general, the significance of the Omitting Types Theorem is the fact that it can be used – as above – to get "standard" models.

6.8 Interpolation

The Craig Interpolation Theorem says the following: Suppose $\models \varphi \to \psi$, where φ is an L_1-sentence and ψ is an L_2-sentence. Then there is an $L_1 \cap L_2$-sentence θ such that $\models \varphi \to \theta$ and $\models \theta \to \psi$. Here is an example:

Example 6.39 $L_1 = \{P, Q, R\}, L_2 = \{P, Q, S\}$. Let

$$\varphi = \forall x(Px \to Rx) \wedge \forall x(Rx \to Qx)$$

and

$$\psi = \forall x(Sx \to Px) \to \forall x(Sx \to Qx).$$

Now

$$\models \varphi \to \psi,$$

and indeed, if

$$\theta = \forall x(Px \to Qx),$$

then θ is an $L_1 \cap L_2$-sentence such that

$$\models \varphi \to \theta \text{ and } \models \theta \to \psi.$$

The Craig Interpolation Theorem is a consequence of the following remarkable *subformula property* of the Model Existence Game $\mathrm{MEG}(T, L)$: Player **II** never has to play anything but subformulas of sentences of T up to a substitution of terms for free variables.

Theorem 6.40 (Craig Interpolation Theorem)   *Suppose $\models \varphi \to \psi$, where φ is an L_1-sentence and ψ is an L_2-sentence. Then there is an $L_1 \cap L_2$-sentence θ such that $\models \varphi \to \theta$ and $\models \theta \to \psi$.*

Proof   We assume, for simplicity, that L_1 and L_2 are relational. This restriction can be avoided (see Exercise 6.97). Let us assume that the claim of the theorem is false and derive a contradiction. Since $\models \varphi \to \psi$, player **I** has a winning strategy in $\mathrm{MEG}(\{\varphi, \neg\psi\}, L_1 \cup L_2)$. Therefore to reach a contradiction it suffices to construct a winning strategy for player **II** in $\text{MEG}(\{\varphi, \neg\psi\}, L_1 \cup L_2)$. If φ alone is inconsistent, we can take any inconsistent L-sentence as θ. Likewise if $\neg\psi$ alone is inconsistent, we can take any valid L-sentence as θ. Let $L = L_1 \cap L_2$. Let us consider the following strategy of player **II**. Suppose $C = \{c_n : n \in \mathbb{N}\}$ is a set of new constant symbols. We denote $L \cup C$-sentences by $\theta(c_0, \ldots, c_{m-1})$ where $\theta(z_0, \ldots, z_{m-1})$ is assumed to be an L-formula. Suppose player **II** has played $Y = \{y_0, \ldots, y_{n-1}\}$ so far. While she plays, she maintains two subsets S_1^n and S_2^n of Y such that $S_1^n \cup S_2^n = Y$. The set S_1^n consists of all $L_1 \cup C$-sentences in Y, and S_2^n consists of all $L_2 \cup C$-sentences in Y. Let us say that an $L \cup C$-sentence θ *separates* S_1^n and S_2^n if $S_1^n \models \theta$ and $S_2^n \models \neg\theta$. Player **II** plays so that the following condition holds at all times:

($\star$)   There is no $L \cup C$-sentence θ that separates S_1^n and S_2^n.

Let us check that she can maintain this strategy: (There is no harm in assuming that player **I** plays φ and $\neg\psi$ first.)

Case 1. Player **I** plays φ. We let $S_1^0 = \{\varphi\}$ and $S_2^0 = \emptyset$. Condition ($\star$) holds, as S_1^n is consistent.

Case 2. Player **I** plays $\neg\psi$ having already played φ. We let $S_1^1 = \{\varphi\}$ and $S_2^1 = \{\neg\psi\}$. Suppose $\theta(c_0, \ldots, c_{m-1})$ separates S_1^1 and S_2^1. Then $\models \varphi \rightarrow \forall z_0 \ldots \forall z_{m-1} \theta(z_0, \ldots, z_{m-1})$ and $\models \forall z_0 \ldots \forall z_{m-1} \theta(z_0, \ldots, z_{m-1}) \rightarrow \psi$ contrary to assumption.

Case 3. Player **I** plays $\approx cc$, where, for example, $c \in L_1 \cup C$. We let $S_1^{n+1} = S_0^n \cup \{\approx cc\}$ and $S_2^{n+1} = S_1^n \cup \{\approx cc\}$. Suppose $\theta(c_0, \ldots, c_{m-1})$ separates S_1^{n+1} and S_2^{n+1}. Then clearly also $\theta(c_0, \ldots, c_{m-1})$ separates S_1^n and S_2^n, a contradiction.

Case 4. Player **I** plays $\varphi_0(c_1)$, where, for example, $\varphi_0(c_0), \approx c_0 c_1 \in S_1^n$. We let $S_1^{n+1} = S_1^n \cup \{\varphi_0(c_1)\}$ and $S_2^{n+1} = S_2^n$. Suppose $\theta(c_0, \ldots, c_m)$ separates S_1^{n+1} and S_2^{n+1}. Then as $S_1^n \models \varphi_0(c_1)$ clearly $\theta(c_0, \ldots, c_{m-1})$ separates S_1^n and S_2^n, a contradiction.

Case 5. Player **I** plays φ_i, where, for example, $\varphi_1 \wedge \varphi_1 \in S_1^n$. We let $S_1^{n+1} = S_1^n \cup \{\varphi_i\}$ and $S_2^{n+1} = S_2^n$. Suppose $\theta(c_0, \ldots, c_{m-1})$ separates S_1^{n+1} and S_2^{n+1}. Then, as $S_1^n \models \varphi_i$, clearly $\theta(c_0, \ldots, c_{m-1})$ separates S_1^n and S_2^n, a contradiction.

Case 6. Player **I** plays $\varphi_0 \vee \varphi_1$, where, for example, $\varphi_0 \vee \varphi_1 \in S_1^n$. We claim that for one of $i \in \{0, 1\}$ the sets $S_1^n \cup \{\varphi_i\}$ and S_2^n satisfy ($\star$). Otherwise there is for both $i \in \{0, 1\}$ some $\theta_i(c_0, \ldots, c_{m-1})$ that separates $S_1^n \cup \{\varphi_i\}$

and S_2^n. Let

$$\theta(c_0, \ldots, c_{m-1}) = \theta_0(c_0, \ldots, c_{m-1}) \vee \theta_1(c_0, \ldots, c_{m-1}).$$

Then, as $S_1^n \models \varphi_0 \vee \varphi_1$, clearly $\theta(c_0, \ldots, c_{m-1})$ separates S_1^n and S_2^n, a contradiction.

Case 7. Player **I** plays $\varphi(c_0)$, where, for example, $\forall x \varphi(x) \in S_1^n$. We claim that the sets $S_1^n \cup \{\varphi(c_0)\}$ and S_2^n satisfy $(\star)$. Otherwise there is $\theta(c_0, \ldots, c_{m-1})$ that separates $S_1^n \cup \{\varphi(c_0)\}$ and S_2^n. Let

$$\theta'(c_1, \ldots, c_{m-1}) = \forall x \theta(x, c_1, \ldots, c_{m-1}).$$

Then, as $S_1^n \models \forall x \varphi(x)$, we have $S_1^n \models \varphi(c_0)$, and hence $\theta'(c_0, c_1, \ldots, c_{m-1})$ separates S_1^n and S_2^n, a contradiction.

Case 8. Player **I** plays $\exists x \varphi(x)$, where, for example, $\exists x \varphi(x) \in S_1^n$. Let $c \in C$ be such that c does not occur in Y yet. We claim that the sets $S_1^n \cup \{\varphi(c)\}$ and S_2^n satisfy $(\star)$. Otherwise there is some $\theta(c, c_0, \ldots, c_{m-1})$ that separates $S_1^n \cup \{\varphi(c)\}$ and S_2^n. Let

$$\theta'(c_1, \ldots, c_{m-1}) = \exists x \theta(x, c_0, \ldots, c_{m-1}).$$

Then, as $S_1^n \models \exists x \varphi(x)$ and $S_1^n \models \varphi(c) \to \theta(c, c_0, \ldots, c_{m-1})$ we clearly have that $\theta'(c_1, \ldots, c_{m-1})$ separates S_1^n and S_2^n, a contradiction. $\square$

Example 6.41 The Craig Interpolation Theorem is false in finite models. To see this, let $L_1 = \{R\}$ and $L_2 = \{P\}$ where R and P are distinct binary predicates. Let φ say that R is an equivalence relation with all classes of size 2 and let ψ say P is not an equivalence relation with all classes of size 2 except one of size 1. Then $\mathcal{M} \models \varphi \to \psi$ holds for finite $\mathcal{M}$. If there were a sentence θ of the empty vocabulary such that $\mathcal{M} \models \varphi \to \theta$ and $\mathcal{M} \models \theta \to \psi$ for all finite $\mathcal{M}$, then θ would characterize even cardinality in finite models. It is easy to see with Ehrenfeucht–Fraïssé Games that this is impossible.

Theorem 6.42 (Beth Definability Theorem) *Suppose L is a vocabulary and P is a predicate symbol not in L. Let φ be an $L \cup \{P\}$-sentence. Then the following are equivalent:*

1. If $(\mathcal{M}, A) \models \varphi$ and $(\mathcal{M}, B) \models \varphi$, where $\mathcal{M}$ is an L-structure, then $A = B$.
2. There is an L-formula θ such that

$$\varphi \models \forall x_0 \ldots x_{n-1}(\theta(x_0, \ldots, x_{n-1}) \leftrightarrow P(x_0, \ldots, x_{n-1})).$$

If condition 1 holds we say that φ defines P *implicitly*. If condition 2 holds, we say that θ defines P *explicitly* relative to φ.

Proof   Let φ' be obtained from φ by replacing everywhere P by P' (another new predicate symbol). Then condition 1 implies

$$\models (\varphi \wedge Pc_0 \ldots c_{n-1}) \rightarrow (\varphi' \rightarrow P'c_0 \ldots c_{n-1}).$$

By the Craig Interpolation Theorem there is an L-formula $\theta(x_0, \ldots, x_{n-1})$ such that

$$\models (\varphi \wedge Pc_0 \ldots c_{n-1}) \rightarrow \theta(c_0, \ldots, c_{n-1})$$

and

$$\models \theta(c_0, \ldots, c_{n-1}) \rightarrow (\varphi' \rightarrow P'c_0 \ldots c_{n-1}).$$

It follows easily that θ is the formula we are looking for. $\square$

Example 6.43   The Beth Definability Theorem is false in finite models. Let φ be the conjunction of

1. "$<$ is a linear order".
2. $\exists x(Px \wedge \forall y(\approx xy \vee x < y))$.
3. $\forall x \forall y("y$ immediate successor of $x" \rightarrow (Px \leftrightarrow \neg Py))$.

Every finite linear order has a unique P with φ, but there is no $\{<\}$-formula $\theta(x)$ which defines P in models of φ. For then the sentence

$$\exists x(\theta(x) \wedge \forall y(\approx xy \vee y < x))$$

would characterize ordered sets of odd length among finite ordered sets, and it is easy to see with Ehrenfeucht–Fraïssé Games that no such sentence can exist. There are infinite linear orders (e.g. $(\mathbb{N} + \mathbb{Z}, <)$) where several different P satisfy φ.

Recall that the *reduct* of an L-structure $\mathcal{M}$ to a smaller vocabulary K is the structure $\mathcal{N} = \mathcal{M} \restriction K$ which has M as its universe and the same interpretations of all symbols of K as $\mathcal{M}$. In such a case we call $\mathcal{M}$ an *expansion* of $\mathcal{N}$ from vocabulary K to vocabulary L. Another useful operation on structures is the following. The *relativization* of an L-structure $\mathcal{M}$ to a set N is the structure $\mathcal{N} = \mathcal{M}^{(N)}$ which has N as its universe, $R^{\mathcal{M}} \cap N^{\#(R)}$ as the interpretation of any predicate symbol $R \in L$, $f^{\mathcal{M}} \restriction N^{\#(f)}$ as the interpretation of any function symbol $f \in L$, and $c^{\mathcal{M}}$ as the interpretation of any constant symbol $c \in L$. Relativization is only possible when the result actually *is* an L-structure. There is a corresponding operation on formulas: The *relativization* of an L-formula φ to a predicate $P \in L$ is defined by replacing every quantifier $\forall y \ldots$ in φ by $\forall y(Py \rightarrow \ldots)$ and every quantifier $\exists y \ldots$ in φ by $\exists y(Py \wedge \ldots)$. We denote the relativization by $\psi^{(P)}$.

Lemma 6.44 *Suppose L is a relational vocabulary and $P \in L$ is a unary predicate symbol. The following are equivalent for all L-formulas φ and all L-structures $\mathcal{M}$ such that $P^{\mathcal{M}} \neq \emptyset$:*

1. $\mathcal{M} \models \varphi^{(P)}$.
2. $\mathcal{M}^{(P^{\mathcal{M}})} \models \varphi$.

Proof   Exercise 6.101. □

Definition 6.45   Suppose L is a vocabulary. A class K of L-structures is an *EC-class* if there is an L-sentence φ such that

$$K = \{\mathcal{M} \in \operatorname{Str}(L) : \mathcal{M} \models \varphi\}$$

and a *PC-class* if there is an L'-sentence φ for some $L' \supseteq L$ such that

$$K = \{\mathcal{M} \upharpoonright L : \mathcal{M} \in \operatorname{Str}(L') \text{ and } \mathcal{M} \models \varphi\}.$$

Example 6.46   Let $L = \emptyset$. The class of infinite L-structures is a *PC*-class which is not an EC class. (Exercise 6.102.)

Example 6.47   Let $L = \emptyset$. The class of finite L-structures is not a *PC*-class. (Exercise 6.103.)

Example 6.48   Let $L = \{<\}$. The class of non-well-ordered L-structures is a *PC*-class which is not an *EC*-class. (Exercise 6.104.)

Suppose $\models \varphi \rightarrow \psi$, where φ is an L_1-sentence and ψ is an L_2-sentence. Let

$$K_1 = \{\mathcal{M} \upharpoonright (L_1 \cap L_2) : \mathcal{M} \models \varphi\}$$

and

$$K_2 = \{\mathcal{M} \upharpoonright (L_1 \cap L_2) : \mathcal{M} \models \neg\psi\}.$$

Now K_1 and K_2 are disjoint *PC*-classes. If there is an $L_1 \cap L_2$-sentence θ such that $\models \varphi \rightarrow \theta$ and $\models \theta \rightarrow \psi$, then the *EC*-class

$$K = \{\mathcal{M} : \mathcal{M} \models \theta\}$$

separates K_1 and K_2 in the sense that $K_1 \subseteq K$ and $K_2 \cap K = \emptyset$. On the other hand, if an *EC*-class K separates in this sense K_1 and K_2, then there is an $L_1 \cap L_2$-sentence θ such that $\models \varphi \rightarrow \theta$ and $\models \theta \rightarrow \psi$. Thus the Craig Interpolation Theorem can be stated as: disjoint *PC*-classes can always be separated by an *EC*-class.

Theorem 6.49 (Separation Theorem)   *Suppose K_1 and K_2 are disjoint PC-classes of models. Then there is an EC-class K that separates K_1 and K_2, i.e. $K_1 \subseteq K$ and $K_2 \cap K = \emptyset$.*

Proof   The claim has already been proved in Theorem 6.40, but we give here a different – model-theoretic – proof. This proof is of independent interest, being as it is, in effect, the proof of the so-called *Lindström's Theorem* (Lindström (1973)), which gives a model theoretic characterization of first order logic.

Case 1: There is an $n \in \mathbb{N}$ such that some union K of $\simeq_p^n$-equivalence classes of models separates K_1 and K_2. By Theorem 6.5 the model class K is an EC-class, so the claim is proved.

Case 2: There are, for any $n \in \mathbb{N}$, $L_1 \cap L_2$-models $\mathcal{M}_n$ and $\mathcal{N}_n$ such that $\mathcal{M}_n \in K_1$, $\mathcal{N}_n \in K_2$, and there is a back-and-forth sequence $(I_i : i \le n)$ for $\mathcal{M}_n$ and $\mathcal{N}_n$. Suppose K_1 is the class of reducts of models of φ, and K_2 respectively the class of reducts of models of ψ. Let T be the following set of sentences:

1. $\varphi^{(P_1)}$.
2. $\psi^{(P_2)}$.
3. $(R, <)$ is a non-empty linear order in which every element with a predecessor has an immediate predecessor.
4. $\forall z (Rz \to Q_0 z)$.
5. $\forall z \forall u_1 \ldots \forall u_m \forall v_1 \ldots \forall v_m ((Rz \wedge Q_n z u_1 \ldots u_m v_1 \ldots v_m) \to (\theta(u_1, \ldots, u_m) \leftrightarrow \theta(v_1, \ldots, v_m)))$ for all atomic $L_1 \cap L_2$-formulas θ.
6. $\forall z \forall u_1 \ldots \forall u_n \forall v_1 \ldots \forall v_m ((Rz \wedge Q_n z u_1 \ldots u_m v_1 \ldots v_m) \to \forall z' \forall x ((Rz' \wedge z' < z \wedge \forall w (w < z \to (w < z' \vee w = z')) \wedge P_1 x) \to \exists y (P_2 y \wedge Q_{n+1} z' u_1 \ldots u_m x v_1 \ldots v_m y)))$.
7. $\forall z \forall u_1 \ldots \forall u_m \forall v_1 \ldots \forall v_m ((Rz \wedge Q_n z u_1 \ldots u_m v_1 \ldots v_m) \to \forall z' \forall y ((Rz' \wedge z' < z \wedge \forall w (w < z \to (w < z' \vee w = z')) \wedge P_2 y) \to \exists x (P_1 x \wedge Q_{n+1} z' u_1 \ldots u_m x v_1 \ldots v_m y)))$.

For all $n \in \mathbb{N}$ there is a model $\mathcal{A}_n$ of T with $(R, <)$ of length n. The model $\mathcal{A}_n$ is obtained as follows. The universe A_n is the (disjoint) union of M_n, N_n, and $\{1, \ldots, n\}$. The L_1-structure $(\mathcal{A}_n \upharpoonright L_1)^{P_1^{\mathcal{A}_n}}$ is chosen to be a copy of the model $\mathcal{M}_n$ of φ. The L_2-structure $(\mathcal{A}_n \upharpoonright L_2)^{P_2^{\mathcal{A}_n}}$ is chosen to be a copy of the model $\mathcal{N}_n$ of ψ. The $2i + 1$-ary predicate Q_i is interpreted in $\mathcal{A}_n$ as the set

$$\{(n - i, u_1, \ldots, u_i, v_1, \ldots, v_i) : \{(u_1, v_1), \ldots, (u_i, v_i)\} \in I_{n-i}\}.$$

By the Compactness Theorem, there is a countable model $\mathcal{M}$ of T with $(R, <)$ non-well-founded (see Exercise 6.107). That is, there are a_n, $n \in \mathbb{N}$, in M such that a_{n+1} is an immediate predecessor of a_n in $\mathcal{M}$ for all $n \in \mathbb{N}$. Let $\mathcal{M}_1$ be the $L_1 \cap L_2$-structure $(\mathcal{M} \upharpoonright (L_1 \cap L_2))^{(P_1^{\mathcal{M}})}$. Let $\mathcal{M}_2$ be the $L_1 \cap L_2$-structure $(\mathcal{M} \upharpoonright (L_1 \cap L_2))^{(P_2^{\mathcal{M}})}$. Now $\mathcal{M}_1 \simeq_p \mathcal{M}_2$, for we have the back-and-forth set:

$$P = \{\{(u_1, v_1), \ldots, (u_n, v_n)\} : \mathcal{M} \models Q_n a_n u_1 \ldots u_n v_1 \ldots v_n, n \in \mathbb{N}\}.$$

Since $\mathcal{M}_1$ and $\mathcal{M}_2$ are countable, they are isomorphic. But $\mathcal{M}_1 \in K_1$ and $\mathcal{M}_2 \in K_2$, a contradiction. $\qquad\square$

6.9 Uncountable Vocabularies

So far we have concentrated on methods based on the assumption that vocabularies are countable. Several key methods work also for uncountable vocabularies. A typical application of uncountable vocabularies is the task of finding an elementary extension of an uncountable structure. In this case a new constant symbol is added to the vocabulary for each element of the model, and the vocabulary may become uncountable.

Strictly speaking, handling uncountable vocabularies does not require dealing with ordinals, but since we use the Axiom of Choice anyway, it is more natural to assume our vocabularies are well-ordered as in

$$L = \{R_\alpha : \alpha < \beta\} \cup \{f_\alpha : \alpha < \gamma\} \cup \{c_\alpha : \alpha < \delta\}.$$

We then allow also variable symbols $x_\alpha, \alpha < \epsilon$.

An important method throughout logic is the method of Skolem functions.

Definition 6.50 Suppose L is a vocabulary, $\mathcal{M}$ is an L-structure, and we have an L-formula $\varphi(x_0, \ldots, x_n)$ of first-order logic. A *Skolem function* for $\varphi(x_0, \ldots, x_n)$ in $\mathcal{M}$ is any function $f_\varphi : M^n \to M$ such that for all elements $a_0, \ldots, a_{n-1}$ of M:

$$M \models \exists x_n \varphi(a_0, \ldots, a_{n-1}, x_n) \to \varphi(a_0, \ldots, a_{n-1}, f_\varphi(a_0, \ldots, a_{n-1})).$$

The following simple but fundamental fact is very helpful in the applications of Skolem functions:

Proposition 6.51 (Tarski–Vaught criterion) *Suppose L is a vocabulary, $\mathcal{M}$ an L-structure, and $\mathcal{N} \subseteq \mathcal{M}$ such that for all L-formulas $\varphi(x_0, \ldots, x_n)$ the following holds:*

$$\text{If } a_0, \ldots, a_{n-1} \in N \text{ and } \mathcal{M} \models \varphi(a_0, \ldots, a_{n-1}, a_n) \text{ for some} \qquad (6.7)$$
$$a_n \in M, \text{ then } \mathcal{M} \models \varphi(a_0, \ldots, a_{n-1}, a'_n) \text{ for some } a'_n \in N.$$

Then $\mathcal{N} \prec \mathcal{M}$.

Proof Exercise 6.110. $\qquad\square$

Proposition 6.52 *Suppose L is a vocabulary, $\mathcal{M}$ an L-structure, and $\mathcal{F}$ a family of functions such that every L-formula has a Skolem function $\in \mathcal{F}$ in*

$\mathcal{M}$. *Suppose $M_0 \subseteq M$ is closed under all functions in $\mathcal{F}$. Then $\mathcal{M}_0$ is the universe of an elementary submodel $\mathcal{M}_0$ of $\mathcal{M}$.*

Proof The claim follows immediately from the Tarski–Vaught criterion. □

Theorem 6.53 (Downward Löwenheim–Skolem Theorem) *Suppose L is a vocabulary of cardinality $\leq \kappa$, $\mathcal{M}$ an L-structure, and $X \subseteq M$ such that $|X| \leq \kappa$. Then there is $\mathcal{M}' \prec \mathcal{M}$ such that $X \subseteq M'$ and $|M'| \leq \kappa$.*

Proof Apply Proposition 6.52 to a family $\mathcal{F}$ which has a Skolem function in $\mathcal{M}$ for each L-formula of first-order logic. We may assume $|\mathcal{F}| \leq \kappa$. Let M_0 be the smallest set containing X which is closed under all the functions in $\mathcal{F}$. Clearly, $|M_0| \leq \kappa$. By the above Proposition, M_0 is the universe of a submodel $\mathcal{M}_0$ of $\mathcal{M}$ such that $\mathcal{M}_0 \prec \mathcal{M}$. □

Definition 6.54 Suppose L is a vocabulary. The *Skolem expansion* of L is the extension L^* of L which has a function symbol

$$f_{\exists x_0 \varphi(x_0, \dots, x_n)}$$

for each formula $\psi = \exists x_0 \varphi(x_0, \dots, x_n)$ of the vocabulary L. The *Skolem expansion* T^* of an L-theory T is the extension of T by the axioms

$$(S) \qquad \forall x_1 \dots \forall x_n (\exists x_0 \varphi(x_0, \dots, x_n) \to \varphi(f_\psi(x_1, \dots, x_n), x_1, \dots, x_n))$$

for each L-formula $\psi = \exists x_0 \varphi(x_0, \dots, x_n)$. A Skolem expansion of an L-model $\mathcal{M}$ is any expansion of $\mathcal{M}$ to an L^*-model $\mathcal{M}^*$ such that $\mathcal{M}^* \models (S)$.

Lemma 6.55 *1. Every model has a Skolem expansion.*
2. If $\mathcal{M} \models T$, then $\mathcal{M}^* \models T^*$.
3. If $\mathcal{N}_1 \models T^*, \mathcal{N}_2 \models T^*$ and $\mathcal{N}_1 \subseteq \mathcal{N}_2$, then $\mathcal{N}_1 \restriction L \prec \mathcal{N}_2 \restriction L$.

Proof Exercise 6.111. □

The Skolem Hull $SH(X)$ of $X \subseteq N$ in $\mathcal{N} \models T^*$ is the smallest submodel of $\mathcal{N}$ which satisfies (S). Thus $SH(X) \models T$.

Lemma 6.56 $|SH(X)| \leq |X| + |L| + \aleph_0$.

Proof Exercise 6.112. □

Note that every element a of $SH(X)$ is of the form $\tau^{\mathcal{N}}(a_1, \dots, a_n)$ for some term $\tau(x_1, \dots, x_n)$ and some $a_1, \dots, a_n \in X$.

Definition 6.57 Suppose L is a vocabulary, $|L| \leq \kappa$, and C is a set of new constant symbols such that $|C| = \kappa$. The Model Existence Game $\mathrm{MEG}_\kappa(T, L)$ of a set T of first-order L-sentences in NNF is defined as $\mathrm{MEG}(T, L)$ (in

Definition 6.29) except that there are κ rounds in the game. The rounds are as in $\mathrm{MEG}(T, L)$.

Example 6.58 Let $L = \{P_\alpha : \alpha < \kappa\}$, $\#(P_\alpha) = 1$, and $C = \{c_\alpha : \alpha < \kappa\}$. Let T be the set of L-sentences

$$\exists x(P_\alpha x \wedge \neg P_\beta x)$$
$$\forall x(\neg P_\beta x \vee P_\alpha x)$$

where $\alpha < \beta < \kappa$. The game could proceed as follows:

I	**II**
$\exists x(P_0 x \wedge \neg P_1 x)$	
	$P_0 c_0 \wedge \neg P_1 c_0$
$\exists x(P_1 x \wedge \neg P_2 x)$	
$\vdots$	$P_1 c_1 \wedge \neg P_2 c_1$
$\exists x(P_n x \wedge \neg P_{n+1} x)$	
$\vdots$	$P_n c_n \wedge \neg P_{n+1} c_n$
$\neg P_\omega c_1 \vee P_0 c_1$	$\vdots$
	$\neg P_\omega c_1$
$\neg P_\omega c_2 \vee P_1 c_2$	
$\vdots$	$\neg P_\omega c_2$
$\exists x(P_0 x \wedge \neg P_\omega x)$	
	$P_0 c_\omega \wedge \neg P_\omega c_\omega$
$\vdots$	$\vdots$

It is clear that **II** has a winning strategy.

Theorem 6.59 (Model Existence Theorem) *Suppose L is a vocabulary of cardinality $\leq \kappa$ and T is a set of L-sentences of first-order logic. The following conditions are equivalent:*

1. *There is an L-structure $\mathcal{M}$ such that $\mathcal{M} \models T$.*
2. *Player **II** has a winning strategy in $\mathrm{MEG}_\kappa(T, L)$.*

Proof If $\mathcal{M} \models T$, then **II** has a winning strategy by just maintaining the condition that every played sentence is true in $\mathcal{M}$. This requires her to keep interpreting the constants $c \in C$ in $\mathcal{M}$ as they appear in the game. For the other direction, assume **II** has a winning strategy τ in $\mathrm{MEG}_\kappa(T, L)$. We now define an "enumeration strategy" for **I**. Let Trm be the set of all constant terms of $L \cup C$. Let

$$T = \{\varphi_\alpha : \alpha < \kappa\}$$
$$C = \{c_\alpha : \alpha < \kappa\}$$
$$\mathrm{Trm} = \{t_\alpha : \alpha < \kappa\}.$$

The strategy of **I** is the following: Suppose x_β and y_β have been played on round $\beta < \alpha$. We describe x_α. Let $\pi : \omega \times \kappa \times \kappa \times \kappa \times \kappa \to \kappa$ be a bijection.

1. If $\alpha = \pi(0, \beta, 0, 0, 0)$ then $x_\alpha = \varphi_\beta$.
2. If $\alpha = \pi(1, \beta, 0, 0, 0)$ then $x_\alpha = \approx c_\beta c_\beta$.
3. If $\alpha = \pi(2, \beta, \gamma, \delta, \epsilon), y_\beta = \approx c_\gamma c_\delta$ and $y_\epsilon = \varphi(c_\gamma)$, then $x_\alpha = \varphi(c_\delta)$.
4. If $\alpha = \pi(3, \beta, \gamma, 0, 0), y_\beta = \theta_0 \wedge \theta_1$ and $\gamma < 2$, then $x_\alpha = \theta_\gamma$.
5. If $\alpha = \pi(4, \beta, 0, 0, 0)$ and $y_\beta = \theta_0 \vee \theta_1$, then $x_\alpha = \theta_0 \vee \theta_1$.
6. If $\alpha = \pi(5, \beta, \gamma, 0, 0)$ and $y_\beta = \forall x \varphi(x)$, then $x_\alpha = \varphi(c_\gamma)$.
7. If $\alpha = \pi(6, \beta, 0, 0, 0)$ and $y_\beta = \exists x \varphi(x)$, then $x_\alpha = \exists x \varphi(x)$.
8. If $\alpha = \pi(7, \beta, 0, 0, 0)$ then $x_\alpha = t_\beta$.

As before, the idea of the enumeration strategy is that **I** tries all possible moves in a systematic way. Thus, if **II** plays her winning strategy against this enumeration strategy of **I** , and H is the set of all responses of **II**, then H is a Hintikka set for the vocabulary $L \cup C$ in the sense that

1. $\approx tt \in H$ for every constant $L \cup C$-term t.
2. If $\varphi(t) \in H$ and $\approx ct \in H$ then $\varphi(c) \in H$.
3. If $\varphi \wedge \psi \in H$, then $\varphi \in H$ and $\psi \in H$.
4. If $\varphi \vee \psi \in H$, then $\varphi \in H$ or $\psi \in H$.
5. If $\forall x \varphi \in H$, then $\varphi(c) \in H$ for all $c \in C$.
6. If $\exists x \varphi \in H$, then $\varphi(c) \in H$ for some $c \in C$.
7. For every constant $L \cup C$-term t there is $c \in C$ such that $\approx ct \in H$.
8. For no φ both $\varphi \in H$ and $\neg\varphi \in H$.

Now H gives rise, as before, to a model $\mathcal{M}$ such that the universe of M consists of equivalence classes $[c]$ under the equivalence relation

$$c \sim d \iff \approx cd \in H$$

and for all $L \cup C$-sentences φ

$$\varphi \in H \implies \mathcal{M} \models \varphi.$$

In particular, $\mathcal{M} \restriction L \models T$. $\square$

Theorem 6.60 (Compactness Theorem)   *Suppose L is a vocabulary and T is a set of first-order L-sentences. If every finite subset of T has a model, then T itself has a model.*

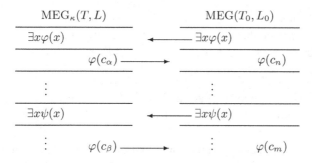

Figure 6.19 Transfer of plays.

Proof Let $\kappa = |L| + \aleph_0$. We show that Player **II** has a winning strategy in $\mathrm{MEG}_\kappa(T, L)$. The strategy is as follows: Suppose x_β and y_β were played during round $\beta < \alpha$ and then **I** plays x_α. Player **II** has made sure that $T \cup \{y_\beta : \beta < \alpha\}$ is finitely consistent. Now she plays y_α so that $T \cup \{y_\beta : \beta \le \alpha\}$ remains finitely consistent. Checking that this is possible for Player **II** involves no new tricks over and above the case that L is countable (Theorem 6.36). $\square$

We can now note that the game $\mathrm{MEG}_\kappa(T, L)$ is always determined: If T has a model, Player **II** has a winning strategy. If T has no models, there is a finite subset $T_0 \subseteq T$ which has no models. Let L_0 be the (finite) vocabulary of T_0. Now **I** has a winning strategy in $\mathrm{MEG}(T_0, L_0)$. If **I** uses this strategy in $\mathrm{MEG}_\kappa(T, L)$, he wins after a finite number of moves. To transfer moves of $\mathrm{MEG}_\kappa(T, L)$ to moves of $\mathrm{MEG}(T_0, L_0)$, Player **I** has to rewrite constants c_α played by **II** as constants c_n with $n < \omega$, as in Figure 6.19. As he is going to win after a finite number of moves, there is no difficulty in finding unused c_n.

Corollary (Upward Löwenheim–Skolem Theorem) *Suppose L is a vocabulary, $\mathcal{M}$ an infinite L-structure, and $\kappa \ge |L| + |M|$. There is an L-structure $\mathcal{M}'$ such that $\mathcal{M} \prec \mathcal{M}'$ and $|M'| = \kappa$.*

Proof Let $M = \{a_\alpha : \alpha < \lambda\}$, where $\lambda \le \kappa$. Let $L' = L \cup \{c_\alpha : \alpha < \kappa\}$. Let T be the theory

$\neg \approx c_\alpha c_\beta$, where $\alpha < \beta < \kappa$,

$\varphi(c_{\alpha_1}, \ldots, c_{\alpha_n})$, where $\alpha_1 < \cdots < \alpha_n < \lambda$ and $\mathcal{M} \models \varphi(a_{\alpha_1}, \ldots, a_{\alpha_n})$.

Clearly, $|T| \le \kappa$, and T is finitely consistent. By the Compactness Theorem T has a model $\mathcal{M}_0$. Since $\mathcal{M}_0 \models \neg \approx c_\alpha c_\beta$ for $\alpha < \beta < \kappa, |M_0| \ge \kappa$. The mapping

$$f(a_\alpha) = c_\alpha^{\mathcal{M}_0}$$

is an isomorphism from $\mathcal{M}$ onto an elementary substructure of $\mathcal{M}_0$. Thus there is $\mathcal{M}_1 \cong \mathcal{M}_0$ such that $\mathcal{M} \prec \mathcal{M}_1$. By the Downward Löwenheim–Skolem Theorem there is $\mathcal{M}' \prec \mathcal{M}_1$ such that $M \subseteq \mathcal{M}'$ and $|\mathcal{M}'| = \kappa$. Thus $\mathcal{M} \prec \mathcal{M}'$. □

Thus the standard model of arithmetic $(\mathbb{N}, +, \cdot, 0, 1)$ has elementary extensions of all infinite cardinalities, and more generally for any model $\mathcal{M}$ of countable vocabulary models $\mathcal{M}' \equiv \mathcal{M}$ of all infinite cardinalities exist. This is a strong absoluteness phenomenon in first-order logic. The so-called Lindström's Theorem (see Barwise and Feferman (1985)) asserts that no proper extension of first-order logic enjoys this property.

Theorem 6.61 (Omitting Types Theorem) *Suppose κ is an infinite cardinal, L is a vocabulary of cardinality $\leq \kappa$, T is a consistent first-order L-theory, and for each $\xi < \kappa$, Γ_ξ is a set of formulas $\varphi(x)$ of vocabulary L such that if $\Sigma(x)$ is any set of formulas $\psi(x)$ such that $|\Sigma(x)| < \kappa$ and $T \cup \Sigma(x)$ is consistent, then $T \cup \Sigma(x) \cup \{\neg\varphi(x)\}$ is consistent for some $\varphi(x) \in \Gamma_\xi$. Then there is a model $\mathcal{M}$ of T of cardinality $\leq \kappa$ which omits each Γ_ξ, $\xi < \kappa$, i.e. no element $a \in M$ satisfies in $\mathcal{M}$ all formulas of Γ_ξ.*

Proof Let $\Gamma_\xi = \{\varphi_\alpha^\xi(x) : \alpha < \kappa\}$. We know that Player **II** can win the game $\mathrm{MEG}_\kappa(T, L)$ with the strategy of keeping the set of played sentences finitely consistent with T. This is simply because T itself is consistent. But now we show that **II** can win also if player **I** is allowed to make the following additional move: if the round is $\alpha = \pi(8 + \xi, \beta, 0, 0, 0)$, then **I** can ask **II** to produce an ordinal $f^\xi(\beta) < \kappa$ and play

$$\neg\varphi_{f^\xi(\beta)}^\xi(c_\beta).$$

In this way eventually functions $f^\xi : \kappa \to \kappa$ are constructed. The strategy of **II** is again to make sure that after round α, the set of played sentences is consistent. Let us see how this is possible. Suppose $\alpha = \pi(8 + \xi, \beta, 0, 0, 0)$ and **II** tries to define $f^\xi(\alpha)$. Suppose the played sentences are $\psi_\beta, \beta < \alpha$. Thus we know at this stage that

$$T' = T \cup \{\psi_\beta : \beta < \alpha\}$$

is finitely consistent. Let us write ψ_β as

$$\psi_\beta = \psi_\beta(c_\alpha, c_\beta, c_i)_{i \in I_\beta}$$

where $I_\beta \subseteq \kappa$ is finite. We want to find $\gamma < \kappa$ such that $T' \cup \{\neg\varphi_\gamma^\xi(c_\alpha)\}$ is finitely consistent. If such a γ can be found, we let $f^\xi(\alpha) = \gamma$. Otherwise we can find for each $\gamma < \kappa$ a finite $J_\gamma \subseteq \alpha$ such that

$$T \cup \{\psi_\beta : \beta \in J_\gamma\} \cup \{\neg\varphi_\gamma^\xi(c_\alpha)\} \qquad (6.8)$$

is inconsistent. Let

$$\Sigma(x) = \left\{ \exists x_\beta \exists (x_i)_{i \in I_\beta, \beta \in J_\gamma} \bigwedge_{\beta \in J_\gamma} \psi_\beta(x, x_\beta, x_i)_{i \in I_\beta} : \gamma < \kappa \right\}.$$

Clearly, $|\Sigma(x)| < \kappa$, as each J_γ is a finite subset of α, and $T \cup \Sigma(x)$ is consistent. By assumption, $T \cup \Sigma(x) \cup \{\neg\varphi_\gamma^\xi(x)\}$ is consistent for some $\gamma < \kappa$. But this contradicts the fact that (6.8) is inconsistent. $\qquad\square$

The problem with this Omitting Types Theorem is that it is rather difficult to satisfy the assumption on Γ involving as it does infinite sets of formulas $\Sigma(x)$, seriously limiting the applicability of the result. In particular, Γ cannot be countable unless $\kappa = \aleph_0$.

Theorem 6.62 *Assume κ is an infinite cardinal. Let L be a vocabulary of cardinality $\leq \kappa$, T an L-theory, and for each $\xi < \kappa$, Γ_ξ is a set $\{\varphi_\alpha^\xi(x) : \alpha < \kappa\}$ of formulas in the vocabulary L. Assume that*

1. *If $\alpha \leq \beta < \kappa$, then $T \vdash \varphi_\beta^\xi(x) \to \varphi_\alpha^\xi(x)$.*
2. *For every L-formula $\psi(x)$, for which $T \cup \{\psi(x)\}$ is consistent, and for every $\xi < \kappa$, there is an $\alpha < \kappa$ such that $T \cup \{\psi(x)\} \cup \{\neg\varphi_\alpha^\xi(x)\}$ is consistent.*

Then T has a model which omits Γ.

Proof We show that the assumption of Theorem 6.61 is valid. Suppose therefore that $\Sigma(x)$ is a set of formulas such that $|\Sigma(x)| < \kappa$ and $T \cup \Sigma(x)$ is consistent. We claim that $T \cup \Sigma(x) \cup \{\neg\varphi_\alpha^\xi(s)\}$ is consistent for some $\alpha < \kappa$. Otherwise there is for every $\alpha < \kappa$ some finite $\Sigma_\alpha(x) \subseteq \Sigma(x)$ such that $T \cup \Sigma_\alpha(x) \cup \{\neg\varphi_\alpha^\xi(s)\}$ has no models. Since $|\Sigma(x)| < \kappa$, there is a fixed finite $\Sigma^*(x) \subseteq \Sigma(x)$ such that $T \cup \Sigma^*(x) \cup \{\neg\varphi_\alpha^\xi(s)\}$ has no models for κ different α. Because of our assumption 1 above, $T \cup \Sigma^*(x) \cup \{\neg\varphi_\alpha^\xi(s)\}$ has no models for any $\alpha < \kappa$ whatsoever. Let $\psi(x) = \bigwedge \Sigma^*(x)$. Now ψ contradicts assumption 2 above. $\qquad\square$

6.10 Ultraproducts

In this section we introduce a new operation on models: the ultraproduct. An ultraproduct is a way to combine a set A of models into one single model in such a way that the new model has all the first-order properties that "most" of the models in A have. A variety of results in model theory can be obtained by means of ultraproducts. We give a few examples below. Ultraproducts can be used to get new examples of partially isomorphic structures, as they are particularly suitable for the method of Ehrenfeucht–Fraïssé Games.

The basic idea of ultraproducts is the following. Suppose we consider infinite sequences

$$f \in \prod_n M_n$$

where $M_0, M_1, \ldots$ are some fixed sets. Suppose we wish to study the "limit behavior" of such sequences and therefore identify two sequences f and g if they agree from some m onwards:

$$f \sim g \iff \exists m \forall n \geq m(f(n) = g(n)). \tag{6.9}$$

This makes a lot of sense and leads us to the study of so-called *reduced products*. However, suppose we want to make the identification (6.9) in such a way that if two sequences f and g are not identified, then they differ from some m onwards. But this is impossible! For example, the sequences

$$f(n) = \begin{cases} a, & n \text{ even} \\ b, & n \text{ odd} \end{cases}$$

$$g(n) = a \text{ for all } n$$

where $a \neq b$ neither eventually agree nor eventually disagree. We have to change our identification (6.9). Let us fix $F \subseteq \mathcal{P}(\mathbb{N})$ and define

$$f \sim g \iff \{n \in \mathbb{N} : f(n) = g(n)\} \in F. \tag{6.10}$$

Suppose $f \not\sim g$. Then $\{n \in \mathbb{N} : f(n) = g(n)\} \notin F$. We want $\{n \in \mathbb{N} : f(n) \neq g(n)\} \in F$, so we assume that F has the property

$$\forall X \subseteq \mathbb{N}(X \notin F \iff \mathbb{N} \setminus X \in F). \tag{6.11}$$

Naturally we want that if $f \sim g$ and $g \sim h$, then $f \sim h$. This follows if

$$\forall X \in F \ \forall Y \in F(X \cap Y \in F). \tag{6.12}$$

Finally, if the intuition behind $f \sim g$ is that $f(n) = g(n)$ for a large number of $n \in \mathbb{N}$, then it makes sense to assume

$$\forall X \in F \ \forall Y \subseteq \mathbb{N}(X \subseteq Y \implies Y \in F), \tag{6.13}$$

i.e. that F is a filter. Now the question arises, whether there are any such wonderful sets F.

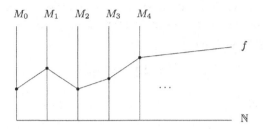

M_0 M_1 M_2 M_3 M_4

f

$\cdots$

$\mathbb{N}$

Figure 6.20 Cartesian product.

Lemma 6.63 *For any infinite set I there is $F \subseteq \mathcal{P}(I)$ satisfying (6.11), (6.12), and (6.13), and not containing any finite sets.*

Proof Let D be the set of all cofinite subsets of $\mathcal{I}$. Clearly, D satisfies (6.12) and (6.13), i.e. D is a filter. By Zorn's Lemma there is a maximal filter F containing D but not containing $\emptyset$ as an element. We show that F satisfies (6.11). Suppose $X^* \subseteq I$ is not in F. Let F' be the set of $X \subseteq I$ containing some $Y \cap X^*$ where $Y \in F$. Now F' satisfies (6.12) and (6.13). If $\emptyset \notin F'$, then $F' \subseteq F$ and $X^* \in F$, contrary to assumption. Thus $\emptyset \in F'$ whence $Y \subseteq I \setminus X^*$ for some $Y \in F$. Thus $I \setminus X^* \in F$ as desired. $\square$

Note that we did not construct a set $F \subseteq \mathcal{P}(I)$ satisfying (6.11), (6.12), and (6.13), but merely proved the existence of such. This is an essential feature of such F.

A set $F \subseteq \mathcal{P}(I)$ satisfying (6.12) and (6.13) is called a *filter on I*. A filter $F \subseteq \mathcal{P}(I)$ satisfying (6.11) is called an *ultrafilter on I*. An ultrafilter $F \subseteq \mathcal{P}(I)$ which is not of the form $\{X \subseteq I : i \in X\}$ for any $i \in I$ is called a *nonprincipal ultrafilter on I*. These concepts were introduced in Example 5.7.

Definition 6.64 Suppose M_i, $i \in I$, are sets and F is a filter on I. Let

$$f \sim g \iff \{i \in I : f(i) = g(i)\} \in F \tag{6.14}$$

for $f, g \in \prod_i M_i$. The set

$$\prod_i M_i / F = \{[f] : f \in \prod_i M_i\}$$

is called a *reduced product* of the sets M_i, $i \in I$. If F is an ultrafilter, it is called an *ultraproduct* of the sets M_i, $i \in I$.

We can picture the Cartesian product $\prod_i M_i$ as in Figure 6.20. It is more difficult to picture the ultraproduct, but Figure 6.20 is useful for it too.

Let L be a vocabulary, F a filter on I, and $\mathcal{M}_i$, $i \in I$ a collection of L-structures. We can define a new L-structure $\mathcal{M}$ with

$$M = \prod_i M_i/F \tag{6.15}$$

as universe as follows. For $R \in L, \#(R) = m$, we define

$$([f_1], \ldots, [f_m]) \in R^{\mathcal{M}} \iff \{i \in I : (f_1(i), \ldots, f_m(i)) \in R^{\mathcal{M}_i}\} \in F. \tag{6.16}$$

It should be checked that (6.16) is independent of the choice of $f_1, \ldots, f_m$ (see Exercise 6.115). For $h \in L, \#(h) = m$, we define

$$h^{\mathcal{M}}([f_1], \ldots, [f_m]) = [f], \text{ where } f(i) = h^{\mathcal{M}_i}(f_1(i), \ldots, f_m(i)). \tag{6.17}$$

Again, (6.17) is independent of the choice of $f_1, \ldots, f_m$, but this is easy (see Exercise 6.116). Finally, for $c \in L$ we define

$$c^{\mathcal{M}} = [f], \text{ where } f(i) = c^{\mathcal{M}_i}. \tag{6.18}$$

Definition 6.65 The *reduced product* $\prod_i \mathcal{M}_i/F$ of the L-structures $\mathcal{M}_i (i \in I)$ with respect to the filter F is the L-structure $\mathcal{M}$ defined by (6.15), (6.16), (6.17), and (6.18) above. If the filter is an ultrafilter, it is called the *ultraproduct* of the L-structures $\mathcal{M}_i (i \in I)$ with respect to the filter F.

Ultraproducts have many interesting properties and they are invariably based on the following lemma. First some notation: If s is an assignment into the set $\prod_i M_i/F$ then $s(x_i) = [f_i]$ for some function f_i. We then denote by s_n the induced assignment $s_n(x_i) = f_i(n)$ into M_n (see Figure 6.21).

Lemma 6.66 (Łoś Lemma) *If F is an ultrafilter and φ is a first-order formula, then*

$$\prod_i \mathcal{M}_i/F \models_s \varphi \iff \{i \in I : \mathcal{M}_i \models_{s_i} \varphi\} \in F. \tag{6.19}$$

Proof We use induction on φ. For atomic φ (6.19) is true by definition. For negation (6.19) follows from (6.11), for conjunction from (6.12). Let us then consider $\varphi = \exists x_m \psi$. If $\prod_i \mathcal{M}_n/F \models_s \varphi$, then there is f such that $\prod_i \mathcal{M}_i/F \models_{s[[f]/x_m]} \psi$. By the induction hypothesis $\{i \in I : \mathcal{M}_i \models_{s_i[f(i)/x_m]} \psi\} \in F$. Thus $\{i \in I : \mathcal{M}_i \models_{s_i} \varphi\} \in F$ by (6.13). Conversely, suppose $X = \{i \in I : \mathcal{M}_i \models_{s_i} \varphi\} \in F$. Let for $i \in X$ a value $f(i)$ be chosen such that $\mathcal{M}_i \models_{s_i[f(i)/x_m]} \psi$. For $i \notin X$ we can let $f(i)$ be anything and still the induction hypothesis gives $\prod_i \mathcal{M}_i/F \models_{s[[f]/x_m]} \psi$. Now $\prod_i \mathcal{M}_i/F \models_s \varphi$ follows. $\square$

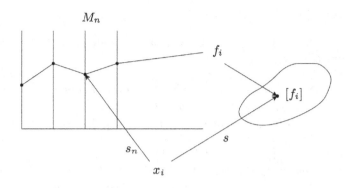

Figure 6.21

Among the many corollaries of the Łoś Lemma the simplest is: For all $\mathcal{M}$ and all ultrafilters F:

$$\prod_i \mathcal{M}/F \equiv \mathcal{M}.$$

This is remarkable as $\prod_i \mathcal{M}/F$ may be quite unlike $\mathcal{M}$. For example,

$$\prod_n (\mathbb{N}, +, \cdot, 0, 1)/F \not\cong (\mathbb{N}, +, \cdot, 0, 1)$$

(see Exercise 6.117). Note that also $\prod_n (\mathbb{R}, <, +, \cdot, 0, 1)/F \equiv (\mathbb{R}, <, +, \cdot, 0, 1)$, when F is an ultrafilter, whence $\prod_n (\mathbb{R}, <, +, \cdot, 0, 1)/F$ is an ordered field. This field is the basis of *non-standard analysis*.

We can also use the Łoś Lemma to give a new proof of the Compactness Theorem in any vocabulary, countable or uncountable:

Theorem 6.67 (Compactness Theorem) *Suppose L is a vocabulary (of any cardinality) and T is a set of first-order L-sentences. If every finite subset of T has a model, then T itself has a model.*

Proof Suppose T is a set of first-order sentences. Let $\mathcal{A}$ be the set of finite subsets of T. The assumption is that each $S \in \mathcal{A}$ has a model $\mathcal{M}_S$. Let $\widehat{\varphi} = \{S \in \mathcal{A} : \varphi \in S\}$. Let F be a non-principal ultrafilter on $\mathcal{A}$ extending the set

$$\{\widehat{\varphi} : \varphi \in T\}.$$

Let $\mathcal{M} = \prod_{S \in \mathcal{A}} \mathcal{M}_S/F$. For $\varphi \in T$:

$$\{S \in \mathcal{A} : \mathcal{M}_S \models \varphi\} \supseteq \widehat{\varphi} \in F.$$

Hence $\mathcal{M} \models \varphi$ for all $\varphi \in T$. $\qquad\qquad\qquad\qquad\qquad\qquad\qquad$ □

A filter is *countably incomplete* if it contains an infinite descending sequence with an empty intersection.

Theorem 6.68 *Suppose D is a countably incomplete filter on I. Suppose $\mathcal{M}_i$ and $\mathcal{N}_i$, $i \in I$, are structures of a countable vocabulary L. If $\mathcal{M}_i \equiv \mathcal{N}_i$ for each $i \in I$, then* **II** *has a winning strategy in $EF_\omega(\prod_i \mathcal{M}_i/D, \prod_i \mathcal{N}_i/D)$.*

Proof Let $\mathcal{M} = \prod_i \mathcal{M}_i/D$ and $\mathcal{N} = \prod_i \mathcal{N}_i/D$. Let $E_0 \supset E_1 \supset \cdots$ be a sequence of elements of D such that $\bigcap_n E_n = \emptyset$. Let $L = \bigcup_n L_n$ where $L_0 \subseteq L_1 \subseteq \cdots$ are all finite. If $i \in I$, there is a largest n such that $i \in E_n$; let us denote it by n_i. Let τ_i be a winning strategy of **II** in $EF_{n_i}(\mathcal{M}_i \restriction L_{n_i}, \mathcal{N}_i \restriction L_{n_i})$. We describe the winning strategy of **II** in $EF_\omega(\mathcal{M}, \mathcal{N})$. Player **II** makes sure that the following condition is maintained: Suppose the position in the game $EF_\omega(\mathcal{M}, \mathcal{N})$ is

$$
\begin{array}{c|c}
\mathcal{M} & \mathcal{N} \\
\hline
[a_0] & [b_0] \\
\cdots & \cdots \\
[a_{n-1}] & [b_{n-1}]
\end{array}
\qquad (6.20)
$$

and $i \in E_{n_i}$. Then

$$
\begin{array}{c|c}
\mathcal{M}_i & \mathcal{N}_i \\
\hline
a_0(i) & b_0(i) \\
\cdots & \cdots \\
a_{n-1}(i) & b_{n-1}(i)
\end{array}
\qquad (6.21)
$$

is a position in the game $EF_{n_i}(\mathcal{M}_i \restriction L_{n_i}, \mathcal{N}_i \restriction L_{n_i})$ while **II** uses τ_i. Suppose in such a position **I** makes a move $[a_n]$ and waits for **II**'s move. To specify **II**'s move $[b_n]$ we need only define $b_n(i)$ for i in a set which is in the filter. So consider $i \in E_{n+1}$. Then necessarily $n_i \geq n + 1$. So the game $EF_{n_i}(\mathcal{M}_i \restriction L_{n_i}, \mathcal{N}_i \restriction L_{n_i})$ has at least $n + 1$ moves and (6.21) is an initial segment of one play where **II** uses τ_i. We now submit $a_n(i)$ as the next move of **I** in this game. The strategy τ_i gives a move $b_n(i)$ for **II**. This is how $b_n(i)$ is defined for $i \in E_{n+1}$. This concludes the description of the strategy of **II**. To see that this is a winning strategy, suppose (6.20) is the position while **II** plays the above strategy. Let $\varphi(x_0, \ldots, x_{n-1})$ be an atomic formula such that $\mathcal{M} \models \varphi(a_0, \ldots, a_{n-1})$. Let us choose $m \geq n$ so that $\varphi(x_0, \ldots, x_{n-1})$ is a formula of the vocabulary L_m. By definition,

$$J = \{i \in I \mid \mathcal{M}_i \models \varphi(a_0(i), \ldots, a_{n-1}(i))\} \in D.$$

Suppose $i \in J \cap E_m$. Again, $n_i \geq m$. Now

$$\mathcal{M}_i \models \varphi(a_0(i), \ldots, a_{n-1}(i))$$

and (6.21) is a position in the game $EF_{n_i}(\mathcal{M}_i{\restriction}L_{n_i}, \mathcal{N}_i{\restriction}L_{n_i})$ while **II** uses τ_i. Since τ_i is a winning strategy of **II** and $\varphi(x_0, \ldots, x_{n-1})$ is a formula of the vocabulary L_{n_i},

$$\mathcal{N}_i \models \varphi(b_0(i), \ldots, b_{n-1}(i)).$$

Thus

$$\{i \in I \mid \mathcal{N}_i \models \varphi(b_0(i), \ldots, b_{n-1}(i))\} \supseteq J \cap E_m \in D,$$

whence $\mathcal{N} \models \varphi(b_0, \ldots, b_{n-1})$. $\square$

Theorem 6.68 is by no means the best in this direction (see Benda (1969)). A particularly beautiful stronger result is the following result of Shelah (1971): $\mathcal{M} \equiv \mathcal{N}$ if and only if there are I and an ultrafilter D on I such that $\prod_i \mathcal{M}/D \cong \prod_i \mathcal{N}/D$.

6.11 Historical Remarks and References

Basic texts in model theory are Chang and Keisler (1990) and Hodges (1993). For the history of model theory, see Vaught (1974). Characterization of elementary equivalence in terms of back-and-forth sequences (Theorem 6.5 and its Corollary) is due to Fraïssé (1955).

The concepts and results of Section 6.4 are due to Kueker (1972, 1977). Theorem 6.23 goes back to Löwenheim (1915) and Skolem (1923, 1970).

The idea of interpreting the quantifiers in terms of moves in a game, as in the Semantic Game, is due to Henkin (1961). Hintikka (1968) extended this from quantifiers to propositional connectives and emphasized its role in semantics in general. The roots of interpreting logic as a game go back, arguably, to Wittgenstein's language games. Lorenzen (1961) used a similar game in proof theory. For the close general connection between inductive definitions and games see Aczel (1977).

Our Model Existence Game is a game-theoretic rendering of the method of semantic tableaux of Beth (1955a,b), model sets of Hintikka (1955), and consistency properties of Smullyan (1963, 1968). Its roots are in the proof-theoretic method of natural deduction of Gentzen (1934, 1969). A good source for more advanced applications of the Model Existence Game is Hodges (1985). Theorem 6.42 is due to Beth (1953) and the stronger but related Theorem 6.40 to Craig (1957a). For the background of Theorem 6.40 see Craig (2008), and for its early applications Craig (1957b). The failure of Graig Interpolation in finite models was observed in Hájek (1976), see also Gurevich (1984), which has this and Example 6.43. The proof of Theorem 6.49 is modeled according to the proof in Barwise and Feferman (1985) of the main result of Lindström (1973), the so-called Lindström's Theorem, which characterizes first-order logic as a maximal logic which satisfies the Compactness Theorem and the Löwenheim–Skolem Theorem in the form: every sentence of the logic which has an infinite model has a countable model. The connection to Theorem 6.49 is the following: Suppose L^* were such a logic and $\varphi \in L^*$. We could treat the class of models of φ and the class of models of $\neg\varphi$ as we treat the disjoint PC-classes K_1 and K_2 in Theorem 6.49. The proof then shows that a first-order sentence θ can separate the class of models of φ and the class of models of $\neg\varphi$. This would clearly mean that φ would be logically equivalent to θ, hence first-order definable. Theorem 6.62 is from Keisler and Morley (1968).

Ultraproducts were introduced by Łoś (1955). For a survey of the use of them in model theory see Bell and Slomson (1969). Theorem 6.68 is from Benda (1969).

Exercise 6.9 is from Brown and Hoshino (2007), where more information about Ehrenfeucht-Fraïssé games for paths and cycles can be found. See also Bissell-Siders (2007). Exercise 6.114 is from Morley (1968).

Exercises

6.1   A finite connected graph is a *cycle* if every vertex has degree 2. Write a sentence of quantifier rank 2 which holds in a cycle if and only if the cycle has length 3. Show that no such sentence of quantifier rank 1 exists.

6.2   Write a sentence of quantifier rank 3 which holds in a cycle if and only if the cycle has length 4. Show that no such sentence of quantifier rank 2 exists. Do the same for the cycle of length 5.

6.3   Do the previous Exercise for the cycle of length 6.

6.4   Write a sentence of quantifier rank 4 which holds in a graph if and only if the cycle has length 7. Show that no such sentence of quantifier rank 3 exists. Do the same for the cycle of length 8.

6.5   Write a sentence of quantifier rank 4 which holds in a graph if and only if the cycle has length 9. Show that no such sentence of quantifier rank 3 exists.

6.6 Construct a sentence of quantifier rank 2 which is true in an ordered set $\mathcal{M}$ if and only if $\mathcal{M}$ has length 2.

6.7 Construct a sentence φ_n of quantifier rank 3 which is true in an ordered set $\mathcal{M}$ if and only if $\mathcal{M}$ has length n, where n is $3, 4, 5$, or 6.

6.8 Show that there is a sentence of quantifier rank 4 which is true in a graph if and only if the graph is a cycle, but no such sentence of quantifier rank 3 exists.

6.9 Show that if $n \geq 3$ and $\mathcal{M}$ and $\mathcal{N}$ are cycles of length $\geq 2^{n-1} + 3$, then $\mathcal{M} \simeq_p^n \mathcal{N}$.

6.10 Suppose $L = \emptyset$ and $n \in \mathbb{N}$. Into how many classes does $\equiv_n$ divide $\mathrm{Str}(L)$?

6.11 Suppose $L = \{c\}$ and $n \in \mathbb{N}$. Into how many classes does $\equiv_n$ divide $\mathrm{Str}(L)$?

6.12 Suppose $L = \{P\}, \#(P) = 1$, and $n \in \mathbb{N}$. Into how many classes does $\equiv_n$ divide $\mathrm{Str}(L)$?

6.13 Suppose $L = \{R\}, \#(R) = 2$. Into how many classes does $\equiv_1$ divide $\mathrm{Str}(L)$?

6.14 Suppose $L = \{R\}, \#(R) = 2$. Show that $\equiv_2$ divides $\mathrm{Str}(L)$ into at least 11 classes.

6.15 Construct for each $n > 0$ trees $\mathcal{T}$ and $\mathcal{T}'$ of height 2 such that $\mathcal{T} \simeq_p^n \mathcal{T}'$ but $\mathcal{T} \not\simeq_p^{n+1} \mathcal{T}'$.

6.16 Consider $\mathrm{EF}_3(\mathcal{T}, \mathcal{T}')$ where $\mathcal{T}$ and T' are the trees below. Show that

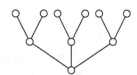 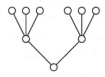

player **I** has a winning strategy. Then write a sentence of quantifier rank 3 which is true in $\mathcal{T}$ but false in $\mathcal{T}'$.

6.17 Suppose $\mathcal{M}$ is an equivalence relation with n classes of size 1 and $n + 1$ classes of size 2. Suppose, on the other hand, that $\mathcal{N}$ is an equivalence relation with $n + 1$ classes of size 1 and n classes of size 2. Show that **II** has a winning strategy in $\mathrm{EF}_{n+1}(\mathcal{M}, \mathcal{M}')$ but **I** has a winning strategy in $\mathrm{EF}_{n+2}(\mathcal{M}, \mathcal{M}')$.

6.18 Suppose $\mathcal{M}$ is an equivalence relation with n classes of size k and $n + 1$ classes of size $k+1$. Suppose, on the other hand, that $\mathcal{N}$ is an equivalence relation with $n + 1$ classes of size k and n classes of size $k + 1$. Show that **II** has a winning strategy in $\mathrm{EF}_{n+k}(\mathcal{M}, \mathcal{M}')$ but **I** has a winning strategy in $\mathrm{EF}_{n+k+1}(\mathcal{M}, \mathcal{M}')$.

6.19 Suppose $L = \{c, d\}$. Which of the following properties of L-structures $\mathcal{M}$ can be expressed in FO with a sentence of quantifier rank ≤ 1:

    (a) $M \neq \{c^{\mathcal{M}}, d^{\mathcal{M}}\}$.

    (b) $|M| \geq 2$.

    (c) $|M \setminus \{c^{\mathcal{M}}\}| \geq 1$.

    (d) $|M| = 2$.

6.20 Like Exercise 6.19 but $L = \{R\}$, $\#(R) = 2$, and the cases are:

    (a) There is $a \in M$ such that $(b, a) \in R^{\mathcal{M}}$ for all $b \in M \setminus \{a\}$.

    (b) $R^{\mathcal{M}}$ is symmetric.

    (c) $R^{\mathcal{M}}$ is reflexive.

6.21 Suppose $L = \{R\}$, $\#(R) = 2$. Which of the following properties of L-structures $\mathcal{M}$ can be expressed in FO with a sentence of quantifier rank ≤ 2.

    (a) $\mathcal{M}$ is an ordered set.

    (b) $\mathcal{M}$ is a partially ordered set.

    (c) $\mathcal{M}$ is an equivalence relation.

    (d) $\mathcal{M}$ is a graph.

6.22 Suppose $L = \{<\}$, $\#(<) = 2$. Which of the following properties of L-structures $\mathcal{M}$ can be expressed in FO with a sentence of quantifier rank ≤ 3:

    (a) $\mathcal{M}$ is a dense linear order.

    (b) $\mathcal{M}$ is an ordered set with at least eight elements.

    (c) $\mathcal{M}$ is a linear order with at least two limit points. (a is a limit point if a has predecessors but no immediate predecessor.)

6.23 Which of the following sentences are logically equivalent to a sentence of quantifier rank ≤ 1:

    (a) $\forall x_0 \exists x_1 (\neg R x_0 \vee P x_1)$.

    (b) $\exists x_0 \exists x_1 (R x_1 \wedge R x_0)$.

    (c) $\exists x_0 \exists x_1 (\neg \approx x_0 x_1 \wedge P x_0)$.

    (d) $\forall x_0 \exists x_1 \neg \approx x_0 x_1$.

6.24 Which of the following sentences are logically equivalent to a sentence of quantifier rank ≤ 2:

    (a) $\forall x_0 \exists x_1 \forall x_2 (R x_0 x_2 \vee S x_0 x_1)$.

    (b) $\exists x_0 \exists x_1 \forall x_2 (R x_0 x_2 \vee S x_1 x_2)$.

6.25 Suppose we are told of an ordered set $\mathcal{M}$ that $\mathcal{M} \equiv_2 (\mathbb{N}, <)$. Can we conclude that

(a) M is infinite?

(b) $\mathcal{M}$ has a smallest element?

(c) Every element has finitely many predecessors?

6.26 Show that if $\mathcal{M} \equiv_3 (\mathbb{Q}, <)$, then $\mathcal{M} \equiv (\mathbb{Q}, <)$.

6.27 Show that if $\mathcal{M} \equiv_3 (\mathbb{Z}, <)$, then $\mathcal{M} \equiv (\mathbb{Z}, <)$.

6.28 Show that for all n there are $\mathcal{M}$ and $\mathcal{N}$ such that $\mathcal{M} \equiv_n \mathcal{N}$ but $\mathcal{M} \not\equiv_{n+1} \mathcal{N}$.

6.29 Suppose L is a vocabulary and $\mathcal{A}$ an L-structure. $\mathcal{A}$ is ω-*saturated* if every type of the expanded structure $(\mathcal{A}, a_0, \ldots, a_{n-1})$, where the elements $a_0, \ldots, a_{n-1}$ are from A, is realized in $(\mathcal{A}, a_0, \ldots, a_{n-1})$. Suppose $\mathcal{A}$ and $\mathcal{B}$ are ω-saturated structures such that $\mathcal{A} \equiv \mathcal{B}$. Show that $\mathcal{A} \simeq_p \mathcal{B}$.

6.30 Let $\mathcal{C} = \{X \in \mathcal{P}_\omega(\mathbb{R}) : \sup(X) = 1000\}$. What is a good starting move for player **I** in $G_{cub}(\mathcal{C})$? Let $\mathcal{C}' = \{X \in \mathcal{P}_\omega(\mathbb{R}) : \inf(X) \leq 1000\}$. What is a good starting move for player **II** in $G_{cub}(\mathcal{C}')$?

6.31 Let $\mathcal{C} = \{X \in \mathcal{P}_\omega(\mathbb{R}) : X$ is dense (meets every non-empty open set) in $\mathbb{R}\}$. What is a good strategy for player **II** in $G_{cub}(\mathcal{C})$?

6.32 Let $\mathcal{C} = \{X \in \mathcal{P}_\omega(\mathbb{R}) :$ every point in X is a limit point of $X\}$. What is a good strategy for player **II** in $G_{cub}(\mathcal{C})$?

6.33 Let $\mathcal{C} = \{X \subseteq \mathbb{N} : \forall m \in \mathbb{N} \exists n \in \mathbb{N}(X \cap [n, n+m] = \emptyset)\}$. What is a good strategy for player **I** in $G_{cub}(\mathcal{C})$?

6.34 Decide which player has a winning strategy in the Cub Game of the following sets:

1. $\{X \in \mathcal{P}_\omega(A) : a \in X\}$, where $a \in A$.
2. $\{X \in \mathcal{P}_\omega(A) : B \cap X$ is finite$\}$, where $B \in \mathcal{P}(A)$.
3. $\{X \in \mathcal{P}_\omega(A) : B \cap X$ is countable$\}$, where $B \in \mathcal{P}(A)$.
4. $\{X \in \mathcal{P}_\omega(\mathbb{R}) : X$ is bounded$\}$.
5. $\{X \in \mathcal{P}_\omega(\mathbb{R}) : X$ is closed$\}$.

6.35 Compute $\triangle_{a \in A} \mathcal{C}_a$ if $\mathcal{C}_a = \{X \in \mathcal{P}_\omega(A) : a \in X\}$.

6.36 Let $f : A \to A$. Let $\mathcal{C}_a = \{X \in \mathcal{P}_\omega(A) : f(a) \in X\}$ and $\mathcal{C}'_a = \{X \in \mathcal{P}_\omega(A) : f(a) \notin X\}$. Compute $\triangle_{a \in A} \mathcal{C}_a$ and $\triangledown_{a \in A} \mathcal{C}'_a$.

6.37 Suppose $\mathcal{M}$ is an ordered set. For $a \in M$ let $\mathcal{C}_a$ be the set of $X \subseteq M$ which have an element above a in $\mathcal{M}$ and let $\mathcal{C}'_a$ be the set of $X \subseteq M$ which are bounded by a in $\mathcal{M}$. Describe the sets $\triangle_{a \in M} \mathcal{C}_a$ and $\triangledown_{a \in M} \mathcal{C}_a$.

6.38 Let L be a relational vocabulary. Suppose $f : \mathcal{M} \cong \mathcal{N}$, where $\mathcal{M}$ and $\mathcal{N}$ and L-structures such that $M = N$. If $A \subseteq M$, there is a unique submodel $\mathcal{M} \upharpoonright A$ of $\mathcal{M}$ with domain A. Show that player **II** has a winning strategy in $G_{cub}(\{X \in \mathcal{P}_\omega(M) : \mathcal{M} \upharpoonright A \cong \mathcal{N} \upharpoonright A\})$.

6.39 Suppose $\mathcal{G}$ is a connected graph. Describe a winning strategy for player
II in $G_{\text{cub}}(\mathcal{C})$, where $\mathcal{C} = \{X \in \mathcal{P}_\omega(G) : [X]_\mathcal{G}$ is connected$\}$.

6.40 Suppose $(A, <)$ is an ordered set and X has a last element for a stationary set of countable $X \subseteq A$. Show that $(A, <)$ itself has a last element.

6.41 Show that the set $\mathcal{CUB}_A$ of sets $\mathcal{C} \subseteq \mathcal{P}_\omega(A)$ which contain a cub is a
countably closed filter (i.e. (1) If $\mathcal{C} \in \mathcal{CUB}_A$ and $\mathcal{C} \subseteq \mathcal{D} \subseteq \mathcal{P}_\omega(A)$, then
$\mathcal{D} \in \mathcal{CUB}_A$. (2) If $\mathcal{C}_n \in \mathcal{CUB}_A$ for all $n \in \mathbb{N}$, then $\bigcap_{n\in\mathbb{N}} \mathcal{C}_n \in \mathcal{CUB}_A$).
In fact, $\mathcal{CUB}_A$ is a normal filter (i.e. if $\mathcal{C}_a \in \mathcal{CUB}_A$ for all $a \in A$, then
$\triangle_{a\in A}\mathcal{C}_a \in \mathcal{CUB}_A$).

6.42 Show that if $\mathcal{D}$ is stationary and $\mathcal{C}$ cub, then $\mathcal{D} \cap \mathcal{C}$ is stationary.

6.43 Show that if $\mathcal{D} = \bigcup_{n\in\mathbb{N}} \mathcal{D}_n$ is stationary, then there is $n \in \mathbb{N}$ such that
$\mathcal{D}_n$ is stationary.

6.44 Show that if $\mathcal{D} = \triangledown_{a\in A}\mathcal{D}_a$ is stationary, then there is $a \in A$ such that
$\mathcal{D}_a$ is stationary.

6.45 (Fodor's Lemma) Suppose $\mathcal{D}$ is stationary and $f(X) \in X$ for every
$X \in \mathcal{C}$. Show that there is a stationary $\mathcal{D} \subseteq \mathcal{C}$ such that f is constant on
$\mathcal{D}$. (Hint: Let $\mathcal{C}_a = \{X : f(X) = a\}$. Assume no $\mathcal{C}_a$ is stationary and
use Lemma 6.15 to derive a contradiction.)

6.46 Show that if A is an uncountable set, then there is a stationary set $\mathcal{C} \subseteq$
$\mathcal{P}_\omega(A)$ such that also $\mathcal{P}_\omega(A) \setminus \mathcal{C}$ is stationary. Such sets are called *bistationary*. Note that then $\mathcal{C} \notin \mathcal{CUB}_A$. (Hint: Write $X = \{a_n^X : n \in \mathbb{N}\}$
whenever $X \in \mathcal{P}_\omega(A)$. Apply the above Fodor's Lemma to the func-
tions $f_n(X) = a_n^X$ to find for each n a stationary $\mathcal{D}_n$ on which f_n is
constant. If each $\mathcal{P}_\omega(A) \setminus \mathcal{D}_n$ is non-stationary, there is for each n a cub
set $\mathcal{C}_n \subseteq \mathcal{D}_n$. Let $\mathcal{C} = \bigcap \mathcal{C}_n$ and show that $\mathcal{C}$ can have only one element,
which contradicts the fact that $\mathcal{C}$ is cub.)

6.47 Use the previous exercise to conclude that $\mathcal{CUB}_A$ is not an ultrafilter (i.e.
a maximal filter) if A is infinite.

6.48 Show that the set NS^A of sets $\mathcal{C} \subseteq \mathcal{P}_\omega(A)$ which are non-stationary is a
σ-ideal (i.e. (1) If $\mathcal{D} \in \text{NS}^A$ and $\mathcal{C} \subseteq \mathcal{D} \subseteq \mathcal{P}_\omega(A)$, then $\mathcal{C} \in \text{NS}^A$. (2)
If $\mathcal{D}_n \in \text{NS}^A$ for all $n \in \mathbb{N}$, then $\bigcup_{n\in\mathbb{N}} \mathcal{D}_n \in \text{NS}^A$). In fact, NS^A is a
normal ideal (i.e. if $\mathcal{D}_a \in \text{NS}^A$ for all $a \in A$, then $\triangledown_{a\in A}\mathcal{D}_a \in \text{NS}^A$).

6.49 Show that if a sentence is true in a stationary set of countable submodels
of a model then it is true in the model itself. More exactly: Let L be a
countable vocabulary, $\mathcal{M}$ an L-model, and φ an L-sentence. Suppose
$\{X \in \mathcal{P}_\omega(M) : [X]_\mathcal{M} \models \varphi\}$ is stationary. Show that $\mathcal{M} \models \varphi$.

6.50 In this and the following exercises we develop the theory of cub and
stationary subsets of a regular cardinal $\kappa > \omega$. A set $C \subseteq \kappa$ is *closed* if it
contains every non-zero limit ordinal $\delta < \kappa$ such that $C \cap \delta$ is unbounded
in δ, and *unbounded* if it is unbounded as a subset of κ. We call $C \subseteq \kappa$

a *closed unbounded* (*cub*) set if C is both closed and unbounded. Show that the following sets are cub

(i) κ.
(ii) $\{\alpha < \kappa : \alpha$ is a limit ordinal$\}$.
(iii) $\{\alpha < \kappa : \alpha = \omega^\beta$ for some $\beta\}$.
(iv) $\{\alpha < \kappa :$ if $\beta < \alpha$ and $\gamma < \alpha$, then $\beta + \gamma < \alpha\}$.
(v) $\{\alpha < \kappa :$ if $\alpha = \beta \cdot \gamma$, then $\alpha = \beta$ or $\alpha = \gamma\}$.

6.51 Show that the following sets are not cub:

(i) $\emptyset$.
(ii) $\{\alpha < \omega_1 : \alpha = \beta + 1$ for some $\beta\}$.
(iii) $\{\alpha < \omega_1 : \alpha = \omega^\beta + \omega$ for some $\beta\}$.
(iv) $\{\alpha < \omega_2 : \mathrm{cf}(\alpha) = \omega\}$.

6.52 Show that a set C contains a cub subset of ω_1 if and only if player **II** wins the game $G_\omega(W_C)$, where

$$W_C = \{(x_0, x_1, x_2, \ldots) : \sup_n x_n \in C\}.$$

6.53 A filter $\mathcal{F}$ on M is *λ-closed* if $A_\alpha \in \mathcal{F}$ for $\alpha < \beta$, where $\beta < \lambda$, implies $\bigcap_\alpha A_\alpha \in \mathcal{F}$. A filter $\mathcal{F}$ on κ is *normal* if $A_\alpha \in \mathcal{F}$ for $\alpha < \kappa$ implies $\triangle_\alpha A_\alpha \in \mathcal{F}$, where

$$\triangle_\alpha A_\alpha = \{\alpha < \kappa : \alpha \in A_\beta \text{ for all } \beta < \alpha\}.$$

Note that normality implies κ-closure. Show that if $\kappa > \omega$ is regular, then the set $\mathcal{F}$ of subsets of κ that contain a cub set is a proper normal filter on κ. The filter $\mathcal{F}$ is called the *cub-filter* on κ.

6.54 A subset of κ which meets every cub set is called *stationary*. Equivalently, a subset S of κ is stationary if its complement is not in the cubfilter. A set which is not stationary, is *non-stationary*. Show that all sets in the cub-filter are stationary. Show that

$$\{\alpha < \omega_2 : \mathrm{cof}(\alpha) = \omega\}$$

is a stationary set which is not in the cub-filter on ω_2.

6.55 (Fodor's Lemma, second formulation) Suppose $\kappa > \omega$ is a regular cardinal. If $S \subseteq \kappa$ is stationary and $f : S \to \kappa$ satisfies $f(\alpha) < \alpha$ for all $\alpha \in S$, then there is a stationary $S' \subseteq S$ such that f is constant on S'. (Hint: For each $\alpha < \kappa$ let $S_\alpha = \{\beta < \kappa : f(\beta) = \alpha\}$. Show that one of the sets S_α has to be stationary.)

6.56 Suppose κ is a regular cardinal $> \omega$. Show that there is a bistationary set $S \subseteq \kappa$ (i.e. both S and $\kappa \setminus S$ are stationary). (Hint: Note that $S = \{\alpha < \kappa : \mathrm{cf}(\alpha) = \omega\}$ is always stationary. For $\alpha \in S$ let $\delta_\alpha : \omega \to \alpha$ be strictly increasing with $\sup_n \delta_\alpha(n) = \alpha$. By the previous exercise there is for each $n < \omega$ a stationary $A_n \subseteq S$ such that the regressive function $f_n(\alpha) = \delta_\alpha(n)$ is constant δ_n on A_n. Argue that some $\kappa \setminus A_n$ must be stationary.)

6.57 Suppose κ is a regular cardinal $> \omega$. Show that $\kappa = \bigcup_{\alpha < \kappa} S_\alpha$ where the sets S_α are disjoint stationary sets. (Hint: Proceed as in Exercise 6.56. Find $n < \omega$ such that for all $\beta < \kappa$ the set $S_\beta = \{\alpha < \kappa : \delta_\alpha(n) \geq \beta\}$ is stationary. Find stationary $S'_\beta \subseteq S_\beta$ such that $\delta_\alpha(n)$ is constant for $\alpha \in S'_\beta$. Argue that there are κ different sets S'_β.)

6.58 Show that $S \subseteq \omega_1$ is bistationary if and only if the game $G_\omega(W_S)$ is non-determined.

6.59 Suppose κ is regular $> \omega$. Show that $S \subseteq \kappa$ is stationary if and only if every regressive $f : S \to \kappa$ is constant on an unbounded set.

6.60 Prove that $C \subseteq \omega_1$ is in the cub filter if and only if almost all countable subsets of ω_1 have their sup in C.

6.61 Suppose $S \subseteq \omega_1$ is stationary. Show that for all $\alpha < \omega_1$ there is a closed subset of S of order-type $\geq \alpha$. (Hint: Prove a stronger claim by induction on α.)

6.62 Decide first which of the following are true and then show how the winner should play the game $\mathrm{SG}(\mathcal{M}, T)$:

   1. $(\mathbb{R}, <, 0) \models \exists x \forall y (y < x \vee 0 < y)$.
   2. $(\mathbb{N}, <) \models \forall x \forall y (\neg y < x \vee \forall z (z < y \vee \neg z < x))$.

6.63 Prove directly that if **II** has a winning strategy in $\mathrm{SG}(\mathcal{M}, T)$ and $\mathcal{M} \simeq_p \mathcal{N}$, then **II** has a winning strategy in $\mathrm{SG}(\mathcal{N}, T)$.

6.64 The *Existential Semantic Game* $\mathrm{SG}_\exists(\mathcal{M}, T)$ differs from $\mathrm{SG}(\mathcal{M}, T)$ only in that the $\forall$-rule is omitted. Show that if **II** has a winning strategy in $\mathrm{SG}_\exists(\mathcal{M}, T)$ and $\mathcal{M} \subseteq \mathcal{N}$, then **II** has a winning strategy in $\mathrm{SG}_\exists(\mathcal{N}, T)$.

6.65 A formula in NNF is *existential* if it contains no universal quantifiers. (Then it is logically equivalent to one of the form $\exists x_1 \ldots \exists x_n \varphi$, where φ is quantifier free.) Show that if L is countable and T is a set of existential L-sentences, then $\mathcal{M} \models T$ if and only if player **II** has a winning strategy in the game $\mathrm{SG}_\exists(\mathcal{M}, T)$.

6.66 The *Universal-Existential Semantic Game* $\mathrm{SG}_{\forall\exists}(\mathcal{M}, T)$ differs from the game $\mathrm{SG}(\mathcal{M}, T)$ only in that player **I** has to make all applications of the $\forall$-rule before all applications of the $\exists$-rule. Show that if $\mathcal{M}_0 \subseteq \mathcal{M}_1 \subseteq$

I	II
$\neg Pc \lor Pfc$	
	Pfc

Figure 6.22

... and **II** has a winning strategy in each $SG_{\forall\exists}(\mathcal{M}_n, T)$, then **II** has a winning strategy in $SG_{\forall\exists}(\cup_{n=0}^{\infty}\mathcal{M}_n, T)$.

6.67 A formula in NNF is *universal-existential* if it is of the form

$$\forall y_1 \ldots \forall y_n \exists x_1 \ldots \exists x_m \varphi,$$

where φ is quantifier free. Show that if L is countable and T is a set of universal-existential L-sentences, then $\mathcal{M} \models T$ if and only if player **II** has a winning strategy in the game $SG_{\forall\exists}(\mathcal{M}, T)$.

6.68 The *Positive Semantic Game* $SG_{pos}(\mathcal{M}, T)$ differs from $SG(\mathcal{M}, T)$ only in that the winning condition "If player **II** plays the pair (φ, s), where φ is basic, then $\mathcal{M} \models_s \varphi$" is weakened to "If player **II** plays the pair (φ, s), where φ is atomic, then $\mathcal{M} \models_s \varphi$". Suppose $\mathcal{M}$ and $\mathcal{N}$ are L-structures. A surjection $h : M \to N$ is a *homomorphism* $\mathcal{M} \to \mathcal{N}$ if

$$\mathcal{M} \models \varphi(a_1, \ldots, a_n) \Rightarrow \mathcal{N} \models \varphi(f(a_1), \ldots, f(a_n))$$

for all atomic L-formulas φ and all $a_1, \ldots, a_n \in M$. Show that if **II** has a winning strategy in $SG_{pos}(\mathcal{M}, T)$ and $h : \mathcal{M} \to \mathcal{N}$ is a surjective homomorphism, then **II** has a winning strategy in $SG_{pos}(\mathcal{N}, T)$.

6.69 A formula in NNF is *positive* if it contains no negations. Show that if L is countable and T is a set of positive L-sentences, then $\mathcal{M} \models T$ if and only if player **II** has a winning strategy in the game $SG_{pos}(\mathcal{M}, T)$.

6.70 The game $MEG(T, L)$ is played with

$$T = \{Pc, \neg Qfc, \forall x_0(\neg Px_0 \lor Qx_0), \forall x_0(\neg Px_0 \lor Pfx_0)\}.$$

The game starts as in Figure 6.22. How does **I** play now and win?

6.71 Consider $T = \{\exists x_0 \forall x_1 Rx_0 x_1, \exists x_1 \forall x_0 \neg Rx_0 x_1\}$. Now we start the game $MEG(T, L)$ as in Figure 6.23. How does **I** play now and win?

6.72 Consider $T = \{\forall x_0(\neg Px_0 \lor Qx_0), \exists x_0(Qx_0 \land \neg Px_0)\}$. The game $MEG(T, L)$ is played. Player **I** immediately resigns. Why?

6.73 The game $MEG(T, L)$ is played with

$$T = \{\forall x_0 \neg x_0 Ex_0, \forall x_0 \forall x_1(\neg x_0 Ex_1 \lor x_1 Ex_0),$$
$$\forall x_0 \exists x_1 x_0 Ex_1, \forall x_0 \exists x_1 \neg x_0 Ex_1\}.$$

I	II
$\exists x_0 \forall x_1 R x_0 x_1$	
	$\forall x_1 R c_0 x_1$
$\exists x_1 \forall x_0 \neg R x_0 x_1$	
	$\forall x_0 \neg R x_0 c_1$

Figure 6.23

Player **I** immediately resigns. Why?

6.74 Use the game $\mathrm{MEG}(T, L)$ to decide whether the following sets T have a model:

1. $\{\exists x P x, \forall y(\neg P y \lor R y)\}$.
2. $\{\forall x P x x, \exists y \forall x \neg P x y\}$.

6.75 Prove the following by giving a winning strategy of player **I** in the appropriate game $\mathrm{MEG}(T \cup \{\neg\varphi\}, L)$:

1. $\{\forall x(P x \to Q x), \exists x P x\} \models \exists x Q x$.
2. $\{\forall x R x f x\} \models \forall x \exists y R x y$.

6.76 Suppose T is the following theory

$$\forall x_0 \neg x_0 < x_0$$
$$\forall x_0 \forall x_1 \forall x_2 (\neg(x_0 < x_1 \land x_1 < x_2) \lor x_0 < x_2)$$
$$\forall x_0 \forall x_1 (x_0 < x_1 \lor x_1 < x_0 \lor x_0 \approx x_1)$$
$$\exists x_0 (P x_0 \land \forall x_1 (\neg P x_1 \lor x_0 \approx x_1 \lor x_1 < x_0))$$
$$\exists x_0 (\neg P x_0 \land \forall x_1 (P x_1 \lor x_0 \approx x_1 \lor x_1 < x_0)).$$

Give a winning strategy for player **I** in $\mathrm{MEG}(T, L)$.

6.77 Prove that the relation $\sim$ is an equivalence relation on C in the proof of Lemma 6.33.

6.78 Prove that the relation $\sim$ in the proof of Lemma 6.33 has the properties:
(1) If $c_i \sim c_i'$ for $1 \le i \le n$ and $f \in L$, then $f c_1 \ldots c_n \sim f c_1' \ldots c_n'$.
(2) If $c_i \sim c_i'$ for $1 \le i \le n$ and $R \in L$ such that $R c_1 \ldots c_n \in H$, then $R c_1' \ldots c_n' \in H$.

6.79 Show in the proof of Lemma 6.33, that if $\varphi(x_1, \ldots, x_n)$ is an L-formula and $\varphi(d_1, \ldots, d_n) \in H$ for some $d_1 \ldots, d_n$, then $\mathcal{M} \models \varphi(d_1, \ldots, d_n)$.

6.80 Suppose L is a vocabulary and $\mathcal{M}$ an L-structure. Let $C = \{c_a : a \in M\}$ be a new set of constants, one for each element of M. There is a canonical expansion $\mathcal{M}^*$ of $\mathcal{M}$ to an $L \cup C$-structure where each constant c_a is interpreted as a. The *diagram* of $\mathcal{M}$ is the set $D(\mathcal{M})$ of basic $L \cup C$-sentences φ such that $\mathcal{M} \models \varphi$. Show that an L-structure $\mathcal{N}$ has a substructure isomorphic to $\mathcal{M}$ if and only if $\mathcal{N}$ can be expanded (by adding interpretations to the new constants) to a model of $D(\mathcal{M})$. You may assume M is countable although the claim is true for all M.

6.81 Show that a sentence φ is logically equivalent to an existential sentence if and only if for all $\mathcal{M} \subseteq \mathcal{N}$: If $\mathcal{M} \models \varphi$, then $\mathcal{N} \models \varphi$. (Hint: Let T be the set of existential sentences that logically imply φ. Show that a finite disjunction of sentences in T is logically equivalent to φ. Use the Compactness Theorem and the previous exercise.)

6.82 Show that a sentence φ is logically equivalent to a positive sentence if and only if for all $\mathcal{M}$ and $\mathcal{N}$: If $\mathcal{M} \models \varphi$ and $\mathcal{N}$ is a homomorphic image of $\mathcal{M}$, then $\mathcal{N} \models \varphi$.

6.83 Show that if $\mathcal{M} \equiv \mathcal{N}$, then there are $\mathcal{M}^*$ and $\mathcal{N}^*$ such that $\mathcal{M} \preceq \mathcal{M}^*$, $\mathcal{N} \preceq \mathcal{N}^*$ and $\mathcal{N}^* \cong \mathcal{M}^*$. (You may assume N and M are countable although the claim is true without this assumption.)

6.84 Suppose $\mathcal{M}$ is a structure in which $<^{\mathcal{M}}$ is a linear order of M without a last element. Show that there is $\mathcal{N}$ such that $\mathcal{M} \preceq \mathcal{N}$ and some element a of N satisfies $b <^{\mathcal{N}} a$ for all $b \in M$. (You may assume M is countable although the claim is true for all M.)

6.85 Suppose (M, R) is a partially ordered set. Prove that there is an ordered set (M, R') such that $R \subseteq R'$. (You may assume M is countable although the claim is true for all M.)

6.86 Prove using the Compactness Theorem that for every set M there is a relation $< \subseteq M \times M$ such that $(M, <)$ is an ordered set. Hint: Consider a vocabulary which has a constant symbol for each element of M. (You may assume M is countable although the claim is true for all M.)

6.87 Suppose T is a theory with an infinite model $\mathcal{M}$ in which $<^{\mathcal{M}}$ is a linear order. Show that T has a model $\mathcal{N}$ in which $<^{\mathcal{N}}$ is not well-ordered.

6.88 Suppose T is a theory which has for each $n > 0$ a model $\mathcal{M}_n$ such that $(M_n, E^{\mathcal{M}_n})$ is a graph in which there are two elements which are not connected by a path of length $\leq n$. Show that T has a model $\mathcal{N}$ in which $(N, E^{\mathcal{N}})$ is a disconnected graph.

6.89 Show that the function used in the proof of Theorem 6.38 really exists.

6.90 Suppose p is a type of T. Show that p is included in the type of some element of some model of T.

6.91 Let T be the theory of dense linear order without endpoints plus the axioms $c_n < c_m$ for natural numbers $n < m$. Show that the type $p = \{c_0 < x, c_1 < x, c_2 < x, \ldots\}$ of T is non-principal.

6.92 Suppose p and p' are types of the theory T. Does there have to be a model of T in which p is included in the type of some element and also p' is included in the type of some element? Does it make a difference if T is complete?

6.93 Suppose p and p' are types of the theory T. Under which conditions does T have a model which realizes p but omits p'? (Hint: Consider the condition: For every $\varphi(x, y)$ there is $\psi(x) \in p'$ such that for no $\delta_1(y), \ldots, \delta_n(y) \in p$ do we have both

$$T \vdash (\varphi(x, y) \wedge \delta_1(y) \wedge \ldots \wedge \delta_n(y)) \to \psi(x)$$

and

$$T \cup \{\exists x \exists y (\varphi(x, y) \wedge \delta_1(y) \wedge \ldots \wedge \delta_n(y))\} \text{ is consistent.)}$$

6.94 Let T be the theory of dense linear order without endpoints plus the axioms $c_i < c_j$ for positive and negative integers $i < j$. Let $p = \{c_0 < x, c_1 < x, c_2 < x, \ldots\}$ and $p' = \{y < c_0, y < c_{-1}, y < c_{-2}, \ldots\}$. Show that T has a model which realizes p but omits p'.

6.95 Show that if $p_0, p_1, \ldots$ are non-principal types of a countable theory T, then there is a model of T which omits each p_n.

6.96 Show that the predicate P is not explicitly definable relative to

1. $\neg \forall x P x \wedge \neg \forall x \neg P x$.
2. $\exists x \forall y ((P y \wedge Q y) \to \approx x y)$.

6.97 Deduce the Craig Interpolation Theorem for arbitrary vocabularies from the assumption that it holds for relational vocabularies.

6.98 A predicate symbol occurs *positively* in a formula if the formula is in NNF and there is a non-negated occurrence of the predicate symbol in the formula. A predicate symbol occurs *negatively* in a formula if the formula is in NNF and there is a negated occurrence of the predicate symbol in the formula. Show that the Craig Interpolation Theorem holds in the following form, known as the *Lyndon Interpolation Theorem*: If L is a relational vocabulary, φ and ψ are L-sentences and $\models \varphi \to \psi$, then there is θ such that $\models \varphi \to \theta$, $\models \theta \to \psi$, every predicate symbol occurring positively in θ occurs positively in φ and ψ, and every predicate symbol symbol occurring negatively in θ occurs negatively in φ and ψ.

6.99 Assume in the previous Exercise that the sentences φ and ψ have no occurrences of the identity symbol. Assume also $\not\models \neg\varphi$ and $\not\models \psi$. Show that θ can be chosen such that it does not contain identity.

6.100 Suppose L_1 and L_2 are vocabularies which contain no function symbols. Let φ be an L_1-sentence and ψ an L_2-sentence such that ψ is universal and $\models \varphi \to \psi$. Show that there is a universal $L_1 \cap L_2$-sentence θ such that $\models \varphi \to \theta$ and $\models \theta \to \psi$.

6.101 Prove Lemma 6.44.

6.102 Prove Example 6.46.

6.103 Prove Example 6.47.

6.104 Prove Example 6.48.

6.105 Suppose L is a finite vocabulary, P_1 and P_2 are unary predicate symbols in L. Show that the class of L-structures $\mathcal{M}$ such that

$$\mathcal{M}^{(P_1^{\mathcal{M}})} \text{ and } \mathcal{M}^{(P_2^{\mathcal{M}})} \text{ are well-defined and } \mathcal{M}^{(P_1^{\mathcal{M}})} \cong \mathcal{M}^{(P_2^{\mathcal{M}})}$$

is a PC-class.

6.106 Suppose L is a finite vocabulary, P_1 and P_2 are unary predicate symbols in L. Show that for all $n \in \mathbb{N}$ the class of L-structures $\mathcal{M}$ such that

$$\mathcal{M}^{(P_1^{\mathcal{M}})} \text{ and } \mathcal{M}^{(P_2^{\mathcal{M}})} \text{ are well-defined and } \mathcal{M}^{(P_1^{\mathcal{M}})} \simeq_p^n \mathcal{M}^{(P_2^{\mathcal{M}})}$$

is a PC-class.

6.107 Suppose L is a countable vocabulary containing the binary predicate symbol $<$. Suppose T is a set of L-sentences. Prove that if T has for each $n \in \mathbb{N}$ a model $\mathcal{M}$ in which $<^{\mathcal{M}}$ is infinite or finite of length at least n, then T has a model $\mathcal{M}$ in which $<^{\mathcal{M}}$ in non-well-ordered.

6.108 Show that every PC-class is closed under isomorphisms.

6.109 Show that the intersection and union of any two PC-classes is again a PC-class.

6.110 Prove Proposition 6.51, the Tarski–Vaught Criterion.

6.111 Prove Lemma 6.55.

6.112 Prove Lemma 6.56.

6.113 Show that the Omitting Types Theorem of first-order logic fails (in its original form) for uncountable vocabularies. (Hint: Let L be a vocabulary consisting of uncountably many constants c_α and countably many constants d_n. Let T say all the constants c_α denote different elements, and all the constants d_n likewise denote different elements. Let p be the type of an element different from each d_n. Then p is non-principal in the original sense of Theorem 6.38.)

6.114 Suppose L is a vocabulary (not necessarily countable). Show that if T has a countable model, then T has a model of cardinality $2^{\aleph_0}$.

6.115 Prove that equivalence (6.16) is independent of the choice of $f_1, \dots, f_m$.

6.116 Prove that Equation (6.17) is independent of the choice of $f_1, \dots, f_m$.

6.117 Prove $\prod_n (\mathbb{N}, +, \cdot, 0, 1)/F \not\cong (\mathbb{N}, +, \cdot, 0, 1)$, where F is a non-principal ultrafilter on $\mathbb{N}$.

6.118 Show that the ordered field $\prod_n (\mathbb{R}, <, +, \cdot, 0, 1)/F$, where F is a non-principal ultrafilter on $\mathbb{N}$, has "infinitely small" elements, i.e. elements that are greater than zero but smaller than $1/n$ for all $n \in \mathbb{N}$.

6.119 Let G_n be the graph consisting of a cycle of $n + 3$ elements, and $G = \prod_n G_n/F$. Show that G is disconnected.

6.120 Suppose L is a vocabulary and $\varphi(x)$ and $\psi(x)$ are first-order formulas. Suppose that for each n we have an L-model $\mathcal{M}_n$ such that

$$\{a \in M_n : \mathcal{M}_n \models \varphi(a)\} \sim \{a \in M_n : \mathcal{M}_n \models \psi(a)\}.$$

Let $\mathcal{M} = \prod_n \mathcal{M}_n / F$. Show that

$$\{a \in M : \mathcal{M} \models \varphi(a)\} \sim \{a \in M : \mathcal{M} \models \psi(a)\}.$$

(Recall: $A \sim B$ means that the sets A and B have the same cardinality.)

6.121 Use ultraproducts to show that every infinite structure has a proper elementary extension.

7

Infinitary Logic

7.1 Introduction

As the name indicates, infinitary logic has infinite formulas. The oldest use of infinitary formulas is the elimination of quantifiers in number theory:

$$\exists x \varphi(x) \leftrightarrow \bigvee_{n \in \mathbb{N}} \varphi(n)$$

$$\forall x \varphi(x) \leftrightarrow \bigwedge_{n \in \mathbb{N}} \varphi(n).$$

Here we leave behind logic as a study of sentences humans can write down on paper. Infinitary formulas are merely mathematical objects used to study properties of structures and proofs. It turns out that games are particularly suitable for the study of infinitary logic. In a sense games replace the use of the Compactness Theorem which fails badly in infinitary logic.

7.2 Preliminary Examples

The games we have encountered so far have had a fixed length, which has been either a natural number or ω (an infinite game). Now we introduce a game which is "dynamic" in the sense that it is possible for player **I** to change the length of the game during the game. He may first claim he can win in five moves, but seeing what the first move of **II** is, he may decide he needs ten moves. In these games player **I** is not allowed to declare he will need infinitely many moves, although we shall study such games, too, later.

Before giving a rigorous definition of the Dynamic Ehrenfeucht–Fraïssé Game we discuss some simple versions of it.

Definition 7.1 (Preliminary) Suppose $\mathcal{M}, \mathcal{M}'$ are L-structures such that L is a relational vocabulary and $M \cap M' = \emptyset$. The *Dynamic Ehrenfeucht–Fraïssé Game*, denoted $\mathrm{EFD}_\omega(\mathcal{M}, \mathcal{M}')$ is defined as follows: First player **I** chooses a natural number n and then the game $\mathrm{EF}_n(\mathcal{M}, \mathcal{M}')$ is played.

Note that $\mathrm{EFD}_\omega(\mathcal{M}, \mathcal{M}')$ is *not* a game of length ω. Player **II** has a winning strategy in $\mathrm{EFD}_\omega(\mathcal{M}, \mathcal{M}')$ if she has one in each $\mathrm{EF}_n(\mathcal{M}, \mathcal{M}')$. On the other hand, player **I** has a winning strategy in $\mathrm{EFD}_\omega(\mathcal{M}, \mathcal{M}')$ if he can envisage a number n so that he has a winning strategy in $\mathrm{EF}_n(\mathcal{M}, \mathcal{M}')$.

Example 7.2 If $\mathcal{M}$ and $\mathcal{M}'$ are L-structures such that M is finite and M' is infinite, then player **I** has a winning strategy in $\mathrm{EFD}_\omega(\mathcal{M}, \mathcal{M}')$. Suppose $|M| = n$. Player **I** has a winning strategy in $\mathrm{EF}_{n+1}(\mathcal{M}, \mathcal{M}')$. He first plays all n elements of M and then any unplayed element of M'. Player **II** is out of good moves, and loses the game.

Example 7.3 If $\mathcal{M}$ and $\mathcal{M}'$ are equivalence relations such that $\mathcal{M}$ has finitely many equivalence classes and $\mathcal{M}'$ infinitely many, then player **I** has a winning strategy in $\mathrm{EFD}_\omega(\mathcal{M}, \mathcal{M}')$. Suppose the equivalence classes of $\mathcal{M}$ are $[a_1], \ldots, [a_n]$. The strategy of **I** is to play first the elements $a_1, \ldots, a_n$. Then he plays an element from M' which is not equivalent to any element played so far. Player **II** is at a loss. She has to play an element of M equivalent to one of $a_1, \ldots, a_n$. She loses.

Definition 7.4 (Preliminary) Suppose $n \in \mathbb{N}$. The game $\mathrm{EFD}_{\omega+n}(\mathcal{M}, \mathcal{M}')$ is played as follows. First the game $\mathrm{EF}_\omega(\mathcal{M}, \mathcal{M}')$ is played for n moves. Then player **I** declares a natural number m and the game $\mathrm{EF}_\omega(\mathcal{M}, \mathcal{M}')$ is continued for m more moves. If **II** has not lost yet, she has won $\mathrm{EFD}_{\omega+n}(\mathcal{M}, \mathcal{M}')$. Otherwise player **I** has won.

Example 7.5 Suppose $\mathcal{G}$ and $\mathcal{G}'$ are graphs so that in $\mathcal{G}$ every vertex has a finite degree while in $\mathcal{G}'$ some vertex has infinite degree. Then player **I** has a winning strategy in $\mathrm{EFD}_{\omega+1}(\mathcal{G}, \mathcal{G}')$. Suppose $a \in G'$ has infinite degree. Player **I** plays first the element a. Let $b \in G$ be the response of player **II**. We know that every element of $\mathcal{G}$ has finite degree. Let the degree of b be n. Player **I** declares that we play $n+1$ more moves. Accordingly, he plays $n+1$ different neighbors of a. Player **II** cannot play $n+1$ different neighbors of b since b has degree n. She loses.

Example 7.6 Suppose $\mathcal{G}$ is a connected graph and $\mathcal{G}'$ a disconnected graph. Then player **I** has a winning strategy in $\mathrm{EFD}_{\omega+2}(\mathcal{G}, \mathcal{G}')$. Suppose a and b are elements of G' that are not connected by a path. Player **I** plays first elements a and b. Suppose the responses of player **II** are c and d. Since $\mathcal{G}$ is connected, there is a connected path $c = c_0, c_1, \ldots, c_n, c_{n+1} = d$ connecting c and d in $\mathcal{G}$.

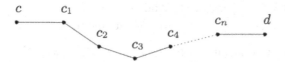

Now player **I** declares that he needs n more moves. He plays the elements $c_1, \ldots, c_n$ one by one. Player **II** has to play a connected path $a_1, \ldots, a_n$ in $\mathcal{G}'$. Now d is a neighbor of c_n in $\mathcal{G}$ but b is not a neighbor of a_n in $\mathcal{G}'$ (see Figure 7.1).

Example 7.7 An *abelian group* is a structure $\mathcal{G} = (G, +)$ with $+_\mathcal{G} : G \times G \to G$ satisfying the conditions

(1) $x +_\mathcal{G} (y +_\mathcal{G} z) = (x +_\mathcal{G} y) +_\mathcal{G} z$ for x, y, z.
(2) there is an element $0_\mathcal{G}$ such that $x +_\mathcal{G} 0_\mathcal{G} = 0_\mathcal{G} +_\mathcal{G} x = x$ for all x.
(3) for all x there is $-x$ such that $x +_\mathcal{G} (-x) = 0_\mathcal{G}$.
(4) for all x and $y : x +_\mathcal{G} y = y +_\mathcal{G} x$.

Examples of abelian groups are

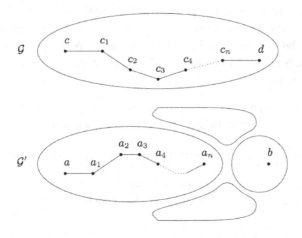

Figure 7.1

$(\mathbb{Z}, +)$ integers with addition.

$(\mathbb{Z}(n), +)$ integers modulo n with modular adddition:

$$x +_{\mathbb{Z}(n)} y = (x +_{\mathbb{Z}} y) \bmod n.$$

$(\mathbb{Q}, +)$ rationals with addition.

$(\mathbb{R}, +)$ reals with addition.

$(\mathbb{R}^+, \cdot)$ positive reals with multiplication.

Example 7.8 Consider the abelian groups $\mathbb{Z} = (\mathbb{Z}, +)$ and $\mathbb{Z}^2 = (\mathbb{Z} \times \mathbb{Z}, +)$ with

$$(m, n) + (p, q) = (m + p, n + q).$$

It is trivial that **II** has a winning strategy in $\mathrm{EFD}_1(\mathbb{Z}, \mathbb{Z}^2)$. But **I** has a winning strategy already in $\mathrm{EFD}_2(\mathbb{Z}, \mathbb{Z}^2)$: First he plays $x_0 = (1, 0)$ and $\alpha_0 = 1$. Suppose **II** responds with $y_0 \in \mathbb{Z}$. Then **I** plays $x_1 = (0, 1)$ and $\alpha_1 = 0$. Player **II** responds with $y_1 \in \mathbb{Z}$. Now

$$\sum_{i=1}^{y_1} y_0 = \sum_{i=1}^{y_0} y_1$$

but

$$\sum_{i=1}^{y_1} x_0 = (y_1, 0) \neq (0, y_0) = \sum_{i=1}^{y_0} x_1$$

unless $y_1 = y_0 = 0$, in which case **II** has lost anyway.

Example 7.9 Consider the structures $(\mathbb{Z}, +, 0)$ and $(\mathbb{Z}, +, 1)$. Player **II** cannot guarantee victory even in a zero-move game, as $0 + 0 = 0$, but $1 + 1 \neq 1$. If instead we have the structures $(\mathbb{Z}, +, 0)$ and $(\mathbb{Z}, \cdot, 1)$, then **II** wins the zero-move game, but if **I** has even just one move, he can play $x_0 = 0$ in $(\mathbb{Z}, \cdot, 1)$ and he wins. Namely, if **II** plays $y_0 \in \mathbb{Z}$ with $y_0 \neq 0$, we have $x_0 \cdot x_0 = x_0$ but $y_0 + y_0 \neq y_0$.

An element a of an abelian group $\mathcal{G}$ is a *torsion element* if there is an $n \in \mathbb{N}$ such that $\underbrace{a + \ldots + a}_{n} = 0$. In $\mathbb{Z}(n)$ every element is a torsion element because if $a < n$, then $\underbrace{a + \ldots + a}_{n} = na = 0 \bmod n$. A group in which every element is a torsion element is a *torsion group*. If no element is a torsion element, the group is *torsion-free*. Torsion-freeness can be axomatized with

$$\forall x (\underbrace{x + \ldots + x}_{n} = 0 \to x = 0), n = 1, 2, \ldots$$

Torsion groups cannot be axiomatized:

Proposition 7.10 *If T is a first-order theory in the vocabulary $\{+\}$ and $\mathbb{Z}(n) \models T$ for arbitrarily large $n \in \mathbb{N}$, then T has a model which is not a torsion group.*

Proof Let T' consist of axioms of abelian groups, T and the axioms

$$\underbrace{c + \ldots + c}_{n} \neq 0$$

for all $n \in \mathbb{N}, n > 0$. Any finite subtheory of T' is satisfied by $\mathbb{Z}(n)$ for large enough n, if we interpret c as 1. By the Compactness Theorem T' has a model $\mathcal{G}$. Let $c^{\mathcal{G}} = a$. Now in $\mathcal{G}$ we have $\underbrace{a + \ldots + a}_{n} \neq 0$ for all $n \in \mathbb{N}$. Thus a is not a torsion element of $\mathcal{G}$. $\square$

Lemma 7.11 *If $\mathcal{G}$ is an abelian torsion group and $\mathcal{G}'$ is a non-torsion abelian group, then* **I** *has a winning strategy in* $\mathrm{EFD}_1(\mathcal{G}, \mathcal{G}')$.

Proof We let **I** play x_0 as a non-torsion element of $\mathcal{G}'$. Suppose **II** plays $y_0 \in \mathcal{G}$. Now there is $n \in \mathbb{N}$ such that

$$\underbrace{y_0 + \ldots + y_0}_{n} = 0$$

but

$$\underbrace{x_0 + \ldots + x_0}_{n} \neq 0$$

so **I** wins. $\square$

We can construe abelian groups also as relational structures. Thus instead of a binary function $+ : G \times G \to G$ we have a ternary relation $R_+ \subseteq G \times G \times G$. Then the axioms of abelian groups are

(1) $\forall x \forall y \exists z R_+ xyz$.
(2) $\forall x \forall y \forall z \forall u ((R_+ xyz \wedge R_+ xyu) \to z = u)$.
(3) $\forall x \forall y \forall z \forall u \forall v \forall w ((R_+ xyu \wedge R_+ uzv \wedge R_+ yzw) \to R_+ xwv)$.
(4) $\exists x \forall y (R_+ xyy \wedge R_+ yxy \wedge \forall z \exists u (R_+ zux \wedge R_+ uzx))$.

In Ehrenfeucht–Fraïssé Games abelian groups behave quite differently depending on whether they are construed as relational structures or as algebraic structures.

Lemma 7.12 *If $\mathcal{G} = (G, R_+)$ is an abelian torsion group and $\mathcal{G}' = (G', R_+)$ is a non-torsion abelian group, then* **I** *has a winning strategy in the game* $\mathrm{EFD}_{\omega+1}(\mathcal{G}, \mathcal{G}')$.

Figure 7.2

Proof Let **I** play first $x_0 \in \mathcal{G}'$ which is not a torsion element. The response $y_o \in \mathcal{G}$ of **II** is a torsion element, so if we use algebraic notation, we have $z_1, \ldots, z_n$ such that

$$
\begin{aligned}
y_0 + y_0 &= z_1. \\
z_1 + y_0 &= z_2. \\
&\vdots \\
z_n + y_0 &= 0.
\end{aligned}
$$

Now **I** declares there are $n+2$ moves left, and plays $x_i = z_i$ for $i = 1, \ldots, n$. Let the responses of **II** be $y_1, \ldots, y_n$. Next **I** plays $x_{n+1} = 0_\mathcal{G}$, and **II** plays $y_{n+1} \in \mathcal{G}'$. Since $x_0 \in \mathcal{G}'$ is not a torsion element, **II** cannot have played $y_{n+1} = 0_{\mathcal{G}'}$ or else she loses. So there is x_{n+2} in $\mathcal{G}'$ with $x_{n+2} + y_{n+1} \neq x_{n+2}$. Now finally **I** plays this x_{n+2}, and **II** plays y_{n+2}. As $y_{n+2} + x_{n+1} = y_{n+2}$, **II** has now lost.

□

7.3 The Dynamic Ehrenfeucht–Fraïssé Game

From $\mathrm{EFD}_{\omega+n}(\mathcal{M}, \mathcal{M}')$ we could go on to define a game $\mathrm{EFD}_{\omega+\omega}(\mathcal{M}, \mathcal{M}')$ in which player **I** starts by choosing a natural number n and declaring that we are going to play the game $\mathrm{EFD}_{\omega+n}(\mathcal{M}, \mathcal{M}')$. But what is the general form of such games? We can have a situation where player **I** wants to decide that after n_0 moves he decides how many moves are left. At that point he decides that after n_1 moves he will decide how many moves now are left. At that point he decides that after n_2 moves he ... until finally he decides that the game lasts n_k more moves. A natural way of making this decision process of player **I** exact is to say that player **I** moves down an ordinal. For example, if he moves down the ordinal $\omega + \omega + 1$, he can move as in Figure 7.2.

So first he wants n_0 moves and after they have been played he decides on n_1. If he moves down on the ordinal $\omega \cdot \omega + 1$, he first chooses k and wants

Figure 7.3

n_0 moves and after they have been played he can still make k changes of mind about the length of the rest of the game (see Figure 7.3).

Definition 7.13 Let L be a relational vocabulary and $\mathcal{M}, \mathcal{M}'$ L-structures such that $M \cap M' = \emptyset$. Let α be an ordinal. The Dynamic Ehrenfeucht–Fraïssé Game $\text{EFD}_\alpha(\mathcal{M}, \mathcal{M}')$ is the game $G_\omega(M \cup M' \cup \alpha, W_{\omega,\alpha}(\mathcal{M}, \mathcal{M}'))$, where $W_{\omega,\alpha}(\mathcal{M}, \mathcal{M}')$ is the set of

$$p = (x_0, \alpha_0, y_0, \ldots, x_{n-1}, \alpha_{n-1}, y_{n-1})$$

such that

(D1) For all $i < n : x_i \in M \leftrightarrow y_i \in M'$.

(D2) $\alpha > \alpha_0 > \ldots > \alpha_{n-1} = 0$.

(D3) If we denote

$$v_i = \begin{cases} x_i \text{ if } x_i \in M \\ y_i \text{ if } y_i \in M \end{cases} \text{ and } v_i' = \begin{cases} x_i \text{ if } x_i \in M' \\ y_i \text{ if } y_i \in M' \end{cases}$$

then

$$f_p = \{(v_0, v_0'), \cdots, (v_{n-1}, v_{n-1}')\}$$

is a partial isomorphism $\mathcal{M} \to \mathcal{M}'$.

Note that $\text{EFD}_\alpha(\mathcal{M}, \mathcal{M}')$ is *not* a game of length α. Every play in the game $\text{EFD}_\alpha(\mathcal{M}, \mathcal{M}')$ is finite, it is just how the length of the game is determined during the game where the ordinal α is used. Compared to $\text{EF}_\omega(\mathcal{M}, \mathcal{M}')$, the only new feature in $\text{EFD}_\alpha(\mathcal{M}, \mathcal{M}')$ is condition (D2). Thus $\text{EFD}_\alpha(\mathcal{M}, \mathcal{M}')$ is more difficult for **I** to play than $\text{EF}_\omega(\mathcal{M}, \mathcal{M}')$, but – if $\alpha \geq \omega$ – easier than any $\text{EF}_n(\mathcal{M}, \mathcal{M}')$.

Lemma 7.14 *(1) If* **II** *has a winning strategy in* $\text{EFD}_\alpha(\mathcal{M}, \mathcal{M}')$ *and* $\beta \leq \alpha$, *then* **II** *has a winning strategy in* $\text{EFD}_\beta(\mathcal{M}, \mathcal{M}')$.

(2) If **I** *has a winning strategy in* $\text{EFD}_\alpha(\mathcal{M}, \mathcal{M}')$ *and* $\alpha \leq \beta$, *then* **I** *has a winning strategy in* $\text{EFD}_\beta(\mathcal{M}, \mathcal{M}')$.

Proof (1) Any move of **I** in $\mathrm{EFD}_\beta(\mathcal{M}, \mathcal{M}')$ is as it is a legal move of **I** in $\mathrm{EFD}_\alpha(\mathcal{M}, \mathcal{M}')$. Thus if **II** can beat **I** in EFD_α she can beat him in EFD_β.

(2) If **I** knows how to beat **II** in EFD_α, he can use the very same moves to beat **II** in EFD_β.   □

Lemma 7.15 *If α is a limit ordinal $\neq 0$ and **II** has a winning strategy in the game $\mathrm{EFD}_\beta(\mathcal{M}, \mathcal{M}')$ for each $\beta < \alpha$, then **II** has a winning strategy in the game $\mathrm{EFD}_\alpha(\mathcal{M}, \mathcal{M}')$.*

Proof In his opening move **I** plays $\alpha_0 < \alpha$. Now **II** can pretend we are actually playing the game $\mathrm{EFD}_{\alpha_0+1}(\mathcal{M}, \mathcal{M}')$. And she has a winning strategy for that game!   □

Back-and-forth sequences are a way of representing a winning strategy of player **II** in the game EFD_α.

Definition 7.16 A *back-and-forth sequence* $(P_\beta : \beta \leq \alpha)$ is defined by the conditions

$$\emptyset \neq P_\alpha \subseteq \ldots \subseteq P_0 \subseteq \mathrm{Part}(\mathcal{A}, \mathcal{B}) \tag{7.1}$$

$$\forall f \in P_{\beta+1} \forall a \in A \exists b \in B \exists g \in P_\beta (f \cup \{(a,b)\} \subseteq g) \text{ for } \beta < \alpha \tag{7.2}$$

$$\forall f \in P_{\beta+1} \forall b \in B \exists a \in A \exists g \in P_\beta (f \cup \{(a,b)\} \subseteq g) \text{ for } \beta < \alpha. \tag{7.3}$$

We write

$$\mathcal{A} \simeq_p^\alpha \mathcal{B}$$

if there is a back-and-forth sequence of length α for $\mathcal{A}$ and $\mathcal{B}$.

The following proposition shows that back-and-forth sequences indeed capture the winning strategies of player **II** in $\mathrm{EFD}_\alpha(\mathcal{A}, \mathcal{B})$:

Proposition 7.17 *Suppose L is a vocabulary and $\mathcal{A}$ and $\mathcal{B}$ are two L-structures. The following are equivalent:*

1. $\mathcal{A} \cong_p^\alpha \mathcal{B}$.
*2. **II** has a winning strategy in $\mathrm{EFD}_\alpha(\mathcal{A}, \mathcal{B})$.*

Proof Let us assume $A \cap B = \emptyset$. Let $(P_i : i \leq \alpha)$ be a back-and-forth sequence for $\mathcal{A}$ and $\mathcal{B}$. We define a winning strategy $\tau = (\tau_i : i \in \mathbb{N})$ for **II**. Suppose we have defined τ_i for $i < j$ and we want to define τ_j. Suppose player **I** has played $x_0, \alpha_0, \ldots, x_{j-1}, \alpha_{j-1}$ and player **II** has followed τ_i during round $i < j$. During the inductive construction of τ_i we took care to define also a partial isomorphism $f_i \in P_{\alpha_i}$ such that $\{v_0, \ldots, v_{i-1}\} \subseteq \mathrm{dom}(f_i)$. Now player **I** plays x_j and $\alpha_j < \alpha_{j-1}$. Note that $f_{j-1} \in P_{\alpha_j+1}$. By assumption there is $f_j \in P_{\alpha_j}$ extending f_{j-1} such that if $x_j \in A$, then $x_j \in \mathrm{dom}(f_j)$

and if $x_j \in B$, then $x_j \in \text{rng}(f_j)$. We let $\tau_j(x_0, \ldots, x_j) = f_j(x_j)$ if $x_j \in A$, and $\tau_j(x_0, \ldots, x_j) = f_j^{-1}(x_j)$ otherwise. This ends the construction of τ_j. This is a winning strategy because every f_p extends to a partial isomorphism $\mathcal{M} \to \mathcal{N}$.

For the converse, suppose $\tau = (\tau_n : n \in \mathbb{N})$ is a winning strategy of **II**. Let Q consist of all plays of $\text{EFD}_\alpha(\mathcal{A}, \mathcal{B})$ in which player **II** has used τ. Let P_β consist of all possible f_p where $p = (x_0, \alpha_0, y_0, \ldots, x_{i-1}, \alpha_{i-1}, y_{i-1})$ is a position in the game $\text{EFD}_\alpha(\mathcal{A}, \mathcal{B})$ with an extension in Q and $\alpha_{i-1} \geq \beta$. It is clear that $(P_\beta : \beta \leq \alpha)$ has the properties (7.1) and (7.2). $\qquad \square$

We have already learnt in Lemma 7.14 that the bigger the ordinal α in $\text{EFD}_\alpha(\mathcal{M}, \mathcal{M}')$ is, the harder it is for player **II** to win and eventually, in a typical case, her luck turns and player **I** starts to win. From that point on it is easier for **I** to win the bigger α is. Lemma 7.15, combined with the fact that the game is determined, tells us that there is a first ordinal where player **I** starts to win. So all the excitement concentrates around just one ordinal up to which player **II** has a winning strategy and starting from which player **I** has a winning strategy. It is clear that this ordinal tells us something important about the two models. This motivates the following:

Definition 7.18 An ordinal α such that player **II** has a winning strategy in $\text{EFD}_\alpha(\mathcal{M}, \mathcal{M}')$ and player **I** has a winning strategy in $\text{EFD}_{\alpha+1}(\mathcal{M}, \mathcal{M}')$ is called the *Scott watershed of $\mathcal{M}$ and $\mathcal{M}'$*.

By Lemma 7.14 the Scott watershed is uniquely determined, if it exists. In two extreme cases the Scott watershed does not exist. First, maybe **I** has a winning strategy even in $\text{EF}_0(\mathcal{M}, \mathcal{M}')$. Here $\text{Part}(\mathcal{M}, \mathcal{M}') = \emptyset$. Secondly, player **II** may have a winning strategy even in $\text{EF}_\omega(\mathcal{M}, \mathcal{M}')$, so **I** has no chance in any $\text{EFD}_\alpha(\mathcal{M}, \mathcal{M}')$, and there is no Scott watershed. In any other case the Scott watershed exists. The bigger it is, the closer $\mathcal{M}$ and $\mathcal{M}'$ are to being isomorphic. Respectively, the smaller it is, the farther $\mathcal{M}$ and $\mathcal{M}'$ are from being isomorphic. If the watershed is so small that it is finite, the structures $\mathcal{M}$ and $\mathcal{M}'$ are not even elementary equivalent.

General problem: Given $\mathcal{M}$ and $\mathcal{M}'$, find the Scott watershed!

How far afield do we have to go to find the Scott watershed? It is very natural to try first some small ordinals. But if we try big ordinals, it would be nice to know how high we have to go. There is a simple answer given by the next proposition: If the models have infinite cardinality κ, and the Scott watershed exists, then it is $< \kappa^+$. Thus for countable models we only need to check

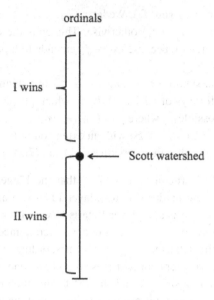

Figure 7.4

countable ordinals. For finite models this is not very interesting: if the models
have at most n elements, and there is a watershed, then it is at most n.

Proposition 7.19 *If* **II** *has a winning strategy in* $\mathrm{EFD}_\alpha(\mathcal{M}, \mathcal{M}')$ *for all* $\alpha <$
$(|M| + |M'|)^+$ *then* **II** *has a winning strategy in* $\mathrm{EF}_\omega(\mathcal{M}, \mathcal{M}')$.

Proof Let $\kappa = |M| + |M'|$. The idea of **II** is to make sure that

($\star$) If the game $\mathrm{EF}_\omega(\mathcal{M}, \mathcal{M}')$ has reached a position

$$p = (x_0, y_0, \ldots, x_{n-1}, y_{n-1}) \text{ with } f_p = \{(v_0, v_0'), \ldots, (v_{n-1}, v_{n-1}')\}$$

then **II** has a winning strategy in

$$\mathrm{EFD}_{\alpha+1}((\mathcal{M}, v_0, \ldots, v_{n-1}), (\mathcal{M}', v_0', \ldots, v_{n-1}')) \qquad (7.4)$$

for all $\alpha < \kappa^+$.

In the beginning $n = 0$ and condition ($\star$) holds. Let us suppose **II** has been
able to maintain ($\star$) and then **I** plays x_n in $\mathrm{EF}_\omega(\mathcal{M}, \mathcal{M}')$. Let us look at the
possibilities of **II**: She has to play some y_n and there are $\leq \kappa$ possibilities. Let
Ψ be the set of them. Assume none of them works. Then for each legal move
y_n there is $\alpha_{y_n} < \kappa^+$ such that

(1) **II** does not have a winning strategy in

$$\text{EFD}_{\alpha_{y_n}}((\mathcal{M}, v_0, \ldots, v_n), (\mathcal{M}', v_0', \ldots, v_n'))$$

where

$$v_n = \left\{ \begin{array}{l} x_n \text{ if } x_n \in M \\ y_n \text{ if } x_n \in M' \end{array} \right. \text{ and } v_n' = \left\{ \begin{array}{l} y_n \text{ if } y_n \in M' \\ x_n \text{ if } y_n \in M. \end{array} \right.$$

Let $\alpha = \sup_{y_n \in \Psi} \alpha_{y_n}$. As $|\Psi| \leq \kappa$, we have $\alpha < \kappa^+$. By the induction hypothesis, **II** has a winning strategy in the game (7.4). So, let us play this game. We let **I** play x_n and α. The winning strategy of **II** gives $y_n \in \Psi$. Let v_n and v_n' be determined as above. Now

(2) **II** has a winning strategy in $\text{EFD}_\alpha((\mathcal{M}, v_0, \ldots, v_n), (\mathcal{M}', v_0', \ldots, v_n'))$.

We have a contradiction between (1), (2), $\alpha_{y_n} < \alpha$ and Lemma 7.14. $\qquad\square$

The above theorem is particularly important for countable models since countable partially isomorphic structures are isomorphic. Thus the countable ordinals provide a complete hierarchy of thresholds all the way from not being even elementary equivalent to being actually isomorphic. For uncountable models the hierarchy of thresholds reaches only to partial isomorphism which may be far from actual isomorphism.

We list here two structural properties of $\simeq_p^\alpha$, which are very easy to prove. There are many others and we will meet them later.

Lemma 7.20 *(Transitivity) If* $\mathcal{M} \simeq_p^\alpha \mathcal{M}'$ *and* $\mathcal{M}' \simeq_p^\alpha \mathcal{M}''$, *then* $\mathcal{M} \simeq_p^\alpha \mathcal{M}''$.

Proof Exercise 7.14. $\qquad\square$

Lemma 7.21 *(Projection) If* $\mathcal{M} \simeq_p^\alpha \mathcal{M}'$, *then* $\mathcal{M} \restriction L \simeq_p^\alpha \mathcal{M}' \restriction L$.

Proof Exercise 7.15. $\qquad\square$

We shall now introduce one of the most important concepts in infinitary logic, namely that of a Scott height of a structure. It is an invariant which sheds light on numerous aspects of the model.

Definition 7.22 The *Scott height* $\text{SH}(\mathcal{M})$ of a model $\mathcal{M}$ is the supremum of all ordinals $\alpha + 1$, where α is the Scott watershed of a pair

$$(\mathcal{M}, a_1, \ldots, a_n) \not\simeq_p (\mathcal{M}, b_1, \ldots, b_n)$$

and $a_1, \ldots, a_n, b_1, \ldots, b_n \in M$.

Lemma 7.23 $\mathrm{SH}(\mathcal{M})$ *is the least* α *such that if* $a_1, \ldots, a_n, b_1, \ldots, b_n \in M$
and

$$(\mathcal{M}, a_1, \ldots, a_n) \simeq_p^\alpha (\mathcal{M}, b_1, \ldots, b_n)$$

then

$$(\mathcal{M}, a_1, \ldots, a_n) \simeq_p^{\alpha+1} (\mathcal{M}, b_1, \ldots, b_n).$$

Proof Exercise 7.16. $\square$

Theorem 7.24 *If* $\mathcal{M} \simeq_p^{\mathrm{SH}(\mathcal{M})+\omega} \mathcal{M}'$, *then* $\mathcal{M} \simeq_p \mathcal{M}'$.

Proof Let $\mathrm{SH}(\mathcal{M}) = \alpha$. The strategy of **II** in $\mathrm{EF}_\omega(\mathcal{M}, \mathcal{M}')$ is to make sure
that if the position is

$$(1) \qquad p = (x_0, y_0, \ldots, x_{n-1}, y_{n-1})$$

then

$$(2) \qquad (\mathcal{M}, v_0, \ldots, v_{n-1}) \simeq_p^\alpha (\mathcal{M}', v_0', \ldots, v_{n-1}').$$

In the beginning of the game (2) holds by assumption. Let us then assume
we are in the middle of the game $\mathrm{EF}_\omega(\mathcal{M}, \mathcal{M}')$, say in position p, and (2)
holds. Now player **I** moves x_n, say $x_n = v_n \in M$. We want to find a move
$y_n = v_n' \in M'$ of **II** which would yield

$$(3) \qquad (\mathcal{M}, v_0, \ldots, v_n) \simeq_p^\alpha (\mathcal{M}', v_0', \ldots, v_n').$$

Now we use the assumption $\mathcal{M} \simeq_p^{\alpha+\omega} \mathcal{M}'$. We play a sequence of rounds of
an auxiliary game $G = \mathrm{EFD}_{\alpha+n+1}(\mathcal{M}, \mathcal{M}')$ in which player **II** has a winning
strategy τ. First player **I** moves the elements $v_0', \ldots, v_{n-1}'$. Let the responses
of player **II** according to τ be $u_0, \ldots, u_{n-1}$. We get

$$(4) \qquad (\mathcal{M}', v_0', \ldots, v_{n-1}') \simeq_p^{\alpha+1} (\mathcal{M}, u_0, \ldots, u_{n-1}).$$

By transitivity,

$$(\mathcal{M}, v_0, \ldots, v_{n-1}) \simeq_p^\alpha (\mathcal{M}, u_0, \ldots, u_{n-1}).$$

See Figure 7.5.
  By Lemma 7.23,

$$(\mathcal{M}, v_0, \ldots, v_{n-1}) \simeq_p^{\alpha+1} (\mathcal{M}, u_0, \ldots, u_{n-1}).$$

Now we apply the definition of $\simeq_p^{\alpha+1}$ and find $a \in M$ such that

$$(5) \qquad (\mathcal{M}, v_0, \ldots, v_{n-1}, v_n) \simeq_p^\alpha (\mathcal{M}, u_0, \ldots, u_{n-1}, a).$$

Finally we play one more round of the auxiliary game G using (4) so that

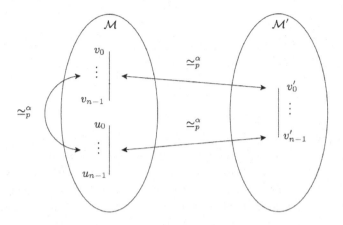

Figure 7.5

player **I** moves $a \in M$ and **II** moves according to τ an element $y_n = v'_n \in M'$. Again

$$(\mathcal{M}', v'_0, \ldots, v'_{n-1}, v'_n) \simeq^\alpha_p (\mathcal{M}, u_0, \ldots, u_{n-1}, a),$$

which together with (5) gives (3).

$\square$

Note, that for countable models we obtain the interesting corollary:

Corollary *If $\mathcal{M}$ is countable, then for any other countable $\mathcal{M}'$ we have*

$$\mathcal{M} \simeq^{\mathrm{SH}(\mathcal{M})+\omega}_p \mathcal{M}' \iff \mathcal{M} \cong \mathcal{M}'.$$

The *Scott spectrum* $\mathrm{ss}(T)$ of a first-order theory is the class of Scott heights of its models:

$$\mathrm{ss}(T) = \{\mathrm{SH}(\mathcal{M}) : \mathcal{M} \models T\}.$$

It is in general quite difficult to determine what the Scott spectrum of a given theory is. For some theories the Scott spectrum is bounded from above. An extreme case is the case of the empty vocabulary, where the Scott height of any model is zero. It follows from Example 7.29 below that the Scott spectrum of the theory of linear order is unbounded in the class of all ordinals. A *gap* in a Scott spectrum $\mathrm{ss}(T)$ is an ordinal which is missing from $\mathrm{ss}(T)$.

Vaught's Conjecture: If T is a countable first-order theory, then T has, up to isomorphism, either $\leq \aleph_0$ or exactly $2^{\aleph_0}$ countable models.

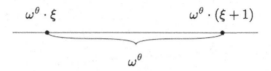

It can be proved that any first-order theory can have only $\leq \aleph_0$ or exactly $2^{\aleph_0}$ countable models of a fixed Scott height Morley (1970). Thus, since there are $\aleph_1$ Scott heights of countable models, any first-order theory can have $\leq \aleph_1$ or exactly $2^{\aleph_0}$ countable models, up to isomorphism, all in all. To prove Vaught's Conjecture it would suffice to prove that for every first-order theory T there is an upper bound $\alpha < \omega_1$ for the Scott heights of its countable models or else there are $2^{\aleph_0}$ countable models of some fixed Scott height. This leads to the following concept: a first-order theory is *scattered* if it has at most $\aleph_0$ countable models of any fixed Scott height. Vaught's Conjecture now has the following equivalent form: *If T is scattered, then the Scott heights of its countable models have a countable upper bound.*

We now prove that there are for arbitrarily large α models with Scott height α. First we prove that for any α there are non-isomorphic models $\mathcal{M}$ and $\mathcal{M}'$ such that $\mathcal{M} \simeq_p^\alpha \mathcal{M}'$. For this we need the following useful concept:

Definition 7.25 If $\mathcal{M} = (M, <)$ and $\mathcal{M}' = (M', <')$ are ordered sets, their *product* $M \times M'$ is the ordered set $(M \times M', <^*)$ where

$$(x, x') <^* (y, y') \iff x' <' y' \text{ or } (x' = y' \text{ and } x < y).$$

Every ordinal α determines canonically a well-ordered set $(\alpha, <)$ which we denote also by α.

Theorem 7.26 *Suppose δ satisfies the condition*

$$\alpha < \delta \implies \omega^\alpha < \delta$$

and $\mathcal{M}$ is any linear order with a first element. Then $\delta \simeq_p^\delta \delta \times \mathcal{M}$.

Proof An ω^θ-interval of δ is any set of the form

$$I_\xi^\theta = \{\alpha : \omega^\theta \cdot \xi \leq \alpha < \omega^\theta \cdot (\xi + 1)\}.$$

An ω^θ-interval of $\delta \times \mathcal{M}$ is any set of the form $I_\xi^\theta \times \{a\}$, where $a \in M$.

We shall define a back-and-forth sequence $P_\delta \subseteq \ldots \subseteq P_0$ as follows: A partial isomorphism f is put into P_θ if f is a finite subfunction of a partial isomorphism g from δ to $\delta \times \mathcal{M}$ such that

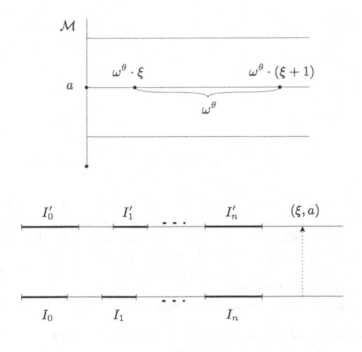

Figure 7.6

(1) $\mathrm{dom}(g)$ is a union of finitely many ω^θ-intervals $I_0, \dots, I_n$ of δ.
(2) $\mathrm{rng}(g)$ is a union of finitely many ω^θ-intervals $I'_0, \dots, I'_n$ of $\delta \times \mathcal{M}$.
(3) $g(0) = (0, \min(\mathcal{M}))$.
(4) $g \restriction I_j : I_j \cong I'_j$.

The empty function is in P_δ. If $\eta < \theta$, then $P_\theta \subseteq P_\eta$ for every ω^θ-interval is a union of ω^η-intervals. To prove the back-and-forth property, suppose $f \in P_{\beta+1}$, where $\beta < \delta$ and $(\xi, a) \in \delta \times \mathcal{M}$. Suppose f is a finite subfunction of g satisfying (1)–(4). If (ξ, a) happens to be in the range of g, it is clear how to proceed: we simply extend f inside g. Let us assume that (ξ, a) is not in the range of g. Let $I_0, \dots, I_n$ be the $\omega^{\beta+1}$-intervals in increasing order containing elements of the domain of f. Let the corresponding $\omega^{\beta+1}$-intervals in $\delta \times \mathcal{M}$ be $I'_0, \dots, I'_n$. Let m be the largest m such that (ξ, a) is above the interval I'_m. If $m = n$, we have Figure 7.6.

Let k be an isomorphism between an $\omega^{\beta+1}$-interval above I'_n and an $\omega^{\beta+1}$-interval of $\delta \times \mathcal{M}$ above I'_n. Then $g \cup k$ satisfies (1)–(4) and the restriction of $g \cup k$ to $\mathrm{dom}(f) \cup \{k^{-1}(\xi, a)\}$ is the extension of f in P_β we are looking

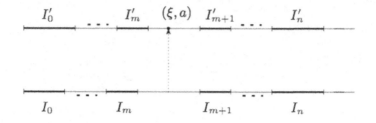

Figure 7.7

for. If on the other hand $m < n$ (Figure 7.7), we argue differently. We may not have a whole new $\omega^{\beta+1}$-interval, but we only need ω^{β}-intervals. So we break I_m into ω copies of ω^{β}-intervals $J_i (i \in \mathbb{N})$ and find a J_i which is above the finitely many elements of $\mathrm{dom}(f)$. Now we just have to choose an ω^{β}-interval J' containing (ξ, a) and choose an isomorphism $k : J_i \to J'$. Clearly, the restriction of $g \cup k$ to $\mathrm{dom}(f) \cup k^{-1}(\xi, a)$ is in P_β.

The other half of the back-and-forth condition is symmetric.   □

Before drawing conclusions from the above important theorem we need to introduce some operations on linear orders.

The *sum* $\mathcal{M} + \mathcal{M}'$ of two linear orders $\mathcal{M}$ and $\mathcal{M}'$ is defined as the linear order consisting of $\mathcal{M}$ and $\mathcal{M}'$ one after the other, $\mathcal{M}$ first then $\mathcal{M}'$. More technically:

Definition 7.27   Suppose $\mathcal{M} = (M, <)$ and $\mathcal{M}' = (M', <')$ are linear orders. Their *sum* $\mathcal{M} + \mathcal{M}'$ is the linear order $(M'', <'')$ where

(1) $M'' = M \times \{0\} \cup M' \times \{1\}$.
(2) $(x, i) <'' (y, j) \iff i < j$ or $(i = j = 0$ and $x < y)$ or $(i = j = 1$ and $x <' y)$.

The *inverse* of a linear order $\mathcal{M} = (M, <)$ is the linear order $\mathcal{M}^* = (M, >)$. Note that if $\mathcal{M}$ is an infinite well-order, then $\mathcal{M}^*$ is necessarily non-well-ordered.

Example 7.28

$$(\mathbb{Z}, <) \cong \omega^* + \omega \not\cong (\mathbb{Z}, <) + 1 + (\mathbb{Z}, <)$$
$$(\mathbb{Q}, <) \cong (\mathbb{Q}, <) + (\mathbb{Q}, <) \cong (\mathbb{Q}, <) + 1 + (\mathbb{Q}, <)$$
$$(\mathbb{R}, <) \cong (\mathbb{R}, <) + 1 + (\mathbb{R}, <) \not\cong (\mathbb{R}, <) + (\mathbb{R}, <).$$

Example 7.29 Let $\alpha_0 = \omega, \alpha_{n+1} = \omega^{\alpha_n}$, and $\epsilon_0 = \sup_{n<\omega} \alpha_n$. Then

$$\epsilon_0 \sim_p^{\epsilon_0} \epsilon_0 \times (1 + \omega^*).$$

More generally, if $\alpha = \sup_{\beta<\alpha} \omega^\beta$, then $\alpha \sim_p^\alpha \alpha \times (1 + \omega^*)$.

The above example shows that there is *no* ordinal α such that

$$\forall \mathcal{M}((\mathcal{M} \text{ well-order } \& \ \mathcal{M} \sim_p^\alpha \mathcal{M}') \to \mathcal{M}' \text{ well-order}).$$

This should be compared with the *fact*

$$\forall \mathcal{M}((\mathcal{M} \text{ well-order } \& \ \mathcal{M} \simeq_p \mathcal{M}') \to \mathcal{M}' \cong \mathcal{M}').$$

The above example also shows that Scott heights can be arbitrarily large and the Scott spectra of first-order theories can be unbounded in the class of all ordinals.

We now prove a result of D. Kueker about the number of automorphisms of countable models.

Lemma 7.30 *Suppose $\mathcal{M} \simeq_p \mathcal{M}'$ where $|M| < |M'|$. Then there are $a \neq a'$ in M and $b \in M'$ such that $(\mathcal{M}, a) \simeq_p (\mathcal{M}, a') \simeq_p (\mathcal{M}', b)$.*

Proof For any $b \in M'$ there is $a \in M$ such that $(\mathcal{M}, a) \simeq_p (\mathcal{M}', b)$. Since there are $|M'|$ many different b but only $|M|$ many different a, there has to be one $a_0 \in M$ such that $(\mathcal{M}, a_0) \simeq_p (\mathcal{M}', b_0)$ and $(\mathcal{M}, a_0) \simeq_p (\mathcal{M}', b_1)$ for some $b_0 \neq b_1$. Let $a_1 \in M$ such that

$$(\mathcal{M}, a_0, a_1) \simeq_p (\mathcal{M}', b_0, b_1).$$

Thus

$$(\mathcal{M}, a_0) \simeq_p (\mathcal{M}', b_1) \simeq_p (\mathcal{M}, a_1).$$

Clearly $a_0 \neq a_1$. $\square$

Theorem 7.31 *If $\mathcal{M} \simeq_p \mathcal{M}'$ where $\mathcal{M}$ is countable and $\mathcal{M}'$ is uncountable, then $\mathcal{M}$ has $2^{\aleph_0}$ automorphisms.*

Proof We construct an automorphism π_s of $\mathcal{M}$ for each $s : \mathbb{N} \to 2$ such that if $s \neq s'$, then $\pi_s \neq \pi_{s'}$. To this end let $M = \{b_n : n \in \mathbb{N}\}$. We define π_s as the union of finite partial mappings $\pi_{s\restriction n}, n \in \mathbb{N}$. Let $\pi_\emptyset = \emptyset$. Suppose $\pi_{s\restriction n}$ has been defined and we want to define $\pi_{s\restriction n+1}$. As an induction hypothesis we assume that if

$$\pi_{s\restriction n} = \{(x_i, y_i) : i < m\}$$

then
$$(\mathcal{M}, x_0, \ldots, x_{m-1}) \simeq_p (\mathcal{M}, y_0, \ldots, y_{m-1})$$
$$\simeq_p (\mathcal{M}', z_0, \ldots, z_{m-1}).$$

By Lemma 7.30 there are $a \neq a' \in M$ and $b \in M'$ such that
$$(\mathcal{M}, x_0, \ldots, x_{m-1}, a) \simeq_p (\mathcal{M}, x_0, \ldots, x_{m-1}, a')$$
$$\simeq_p (\mathcal{M}', z_0, \ldots, z_{m-1}, b).$$

Let $c \neq c' \in M$ such that
$$(\mathcal{M}, x_0, \ldots, x_{m-1}, a) \simeq_p (\mathcal{M}, y_0, \ldots, y_{m-1}, c)$$

and
$$(\mathcal{M}, x_0, \ldots, x_{m-1}, a, a') \simeq_p (\mathcal{M}, y_0, \ldots, y_{m-1}, c, c').$$

Then
$$(\mathcal{M}, x_0, \ldots, x_{m-1}, a) \simeq_p (\mathcal{M}, x_0, \ldots, x_{m-1}, a')$$
$$\simeq_p (\mathcal{M}, y_0, \ldots, y_{m-1}, c').$$

Let $x_m = a$ and
$$y_m = \begin{cases} c \text{ if } s(n) = 0 \\ c' \text{ if } s(n) = 1. \end{cases}$$

Let $c_n, d_n \in M$ such that
$$(\mathcal{M}, x_0, \ldots, x_m, b_n, c_n) \simeq_p (\mathcal{M}, y_0, \ldots, y_m, d_n, b_n)$$

and
$$\pi_{s\upharpoonright n+1} = \{(x_i, y_i) : i \leq m\} \cup \{(b_n, d_n), (c_n, b_n)\}.$$

Two more applications of the back-and-forth property of $\simeq_p$ guarantee that the induction condition remains valid. Let
$$\pi_s = \bigcup_{n=0}^{\infty} \pi_{s\upharpoonright n}.$$

If $s \neq s'$, say $s(n) \neq s'(n)$, then $\pi_{s\upharpoonright n+1} \neq \pi_{s'\upharpoonright n+1}$, so $\pi_s \neq \pi_{s'}$. Clearly each π_s is an automorphism of $\mathcal{M}$. $\qquad\square$

Corollary *If $\mathcal{M}$ is a countable model with only countably many automorphisms, then for all $\mathcal{M}'$*
$$\mathcal{M} \sim_p^{\text{SH}(\mathcal{M})+\omega} \mathcal{M}' \Longleftrightarrow \mathcal{M} \cong \mathcal{M}'.$$

Proof If $\mathcal{M} \simeq_p \mathcal{M}'$, then $\mathcal{M}'$ must be countable by the previous theorem. Then $\mathcal{M} \cong \mathcal{M}'$ by Proposition 5.16. $\qquad\square$

Example 7.32 The following structures have only countably many automorphisms:

$$(\mathbb{N}, <, \cdot, 0, 1), \ (\alpha, <), \ (\mathbb{Z}, <), \ (\mathbb{Z}, +), \ (\mathbb{Q}, +).$$

7.4 Syntax and Semantics of Infinitary Logic

The syntax and semantics of the infinitary logic $L_{\infty\omega}$ that we now introduce are very much like the syntax and semantics of first-order logic. The logical symbols are $\approx, \neg, \bigwedge, \bigvee, \forall, \exists, (,), x_0, x_1, \ldots$. Terms and atomic formulas are defined as usual. Formulas of $L_{\infty\omega}$ are of the form

$$\approx tt'$$

$$Rt_1 \ldots t_n$$

$$\neg \varphi$$

$$\bigwedge_{i \in I} \varphi_i, \bigvee_{i \in I} \varphi_i$$

$$\forall x_n \varphi, \exists x_n \varphi$$

where $t, t', t_1, \ldots, t_n$ are L-terms, $R \in L$ with $\#_l(R) = n$, and φ and all φ_i, $i \in I$, where I is an arbitrary set, are formulas of $L_{\infty\omega}$, and the formulas φ_i have altogether only finitely many free variables.[1] We regard $\varphi \wedge \psi$, $\varphi \vee \psi$, $(\varphi \to \psi)$ and $(\varphi \leftrightarrow \psi)$ as abbreviations.

In first-order logic we can think of formulas as finite strings of symbols. In infinitary logic it is customary to consider formulas as sets. Then we have the following more exact albeit more cumbersome definition:

Definition 7.33 Suppose L is a vocabulary. The class of L-formulas of $L_{\infty\omega}$ is defined as follows:

(1) If t and t' are L-terms, then $(0, t, t')$ is an L-formula denoted by $\approx tt'$.
(2) If $t_1, \ldots, t_n$ are L-terms, then $(1, R, t_1, \ldots, t_n)$ is an L-formula denoted by $Rt_1 \ldots t_n$.
(3) If φ is an L-formula, so is $(2, \varphi)$, and we denote it by $\neg \varphi$.
(4) If Φ is a set of L-formulas with a fixed finite set of free variables, then $(3, \Phi)$ is an L-formula and we denote it by $\bigwedge_{\varphi \in \Phi} \varphi$.
(5) If Φ is a set of L-formulas with a fixed finite set of free variables, then $(4, \Phi)$ is an L-formula and we denote it by $\bigvee_{\varphi \in \Phi} \varphi$.
(6) If φ is an L-formula and $n \in \mathbb{N}$, then $(5, \varphi, n)$ is an L-formula and we denote it by $\forall x_n \varphi$.

---

[1] This restriction makes it possible to quantify all free variables in a formula.

(7) If φ is an L-formula and $n \in \mathbb{N}$, then $(6, \varphi, n)$ is an L-formula and we denote it by $\exists x_n \varphi$.

Every formula of $L_{\infty\omega}$ is now a finite sequence of sets and the first element of the sequence is one of $\{0, 1, 2, 3, 4, 5, 6\}$. With this definition it is easy to write exact inductive definitions for various concepts related to infinitary logic.

A formula of $L_{\infty\omega}$ can be thought of as a tree, too. In this tree the formula itself is the root and the set $\mathrm{ISub}(\varphi)$ of immediate successors of a node φ of the tree are:

(1) $\mathrm{ISub}((0, t, t')) = \emptyset$.
(2) $\mathrm{ISub}((1, t_1, \ldots, t_n)) = \emptyset$.
(3) $\mathrm{ISub}((2, \varphi)) = \{\varphi\}$.
(4) $\mathrm{ISub}((3, \Phi)) = \Phi$.
(5) $\mathrm{ISub}((4, \Phi)) = \Phi$.
(6) $\mathrm{ISub}((5, \varphi, n)) = \{\varphi\}$.
(7) $\mathrm{ISub}((6, \varphi, n)) = \{\varphi\}$.

The tree thus consists of the elements of

$$\mathrm{Sub}(\varphi) = \bigcup_{n=0}^{\infty} \mathrm{Sub}_n(\varphi)$$

where

$$\mathrm{Sub}_0(\varphi) = \{\varphi\}$$
$$\mathrm{Sub}_{n+1}(\varphi) = \cup\{\mathrm{ISub}(\psi) : \psi \in \mathrm{Sub}_n(\varphi)\},$$

and the order is

$$\psi <_{Sub} \theta \iff \theta \in \mathrm{Sub}_n(\psi) \text{ for some } n > 0.$$

The tree $(\mathrm{Sub}(\varphi), <_{Sub})$ is a well-founded tree.

The quantifier rank of a formula of $L_{\infty\omega}$ is defined by induction as follows:

(1) $\mathrm{QR}(\approx t t') = 0$.
(2) $\mathrm{QR}(R t_1 \ldots t_n) = 0$.
(3) $\mathrm{QR}(\neg \varphi) = \mathrm{QR}(\varphi)$.
(4) $\mathrm{QR}(\bigwedge \Phi) = \sup\{\mathrm{QR}(\psi) : \psi \in \Phi\}$.
(5) $\mathrm{QR}(\bigvee \Phi) = \sup\{\mathrm{QR}(\psi) : \psi \in \Phi\}$.
(6) $\mathrm{QR}(\forall x_n \varphi) = \mathrm{QR}(\varphi) + 1$.
(7) $\mathrm{QR}(\exists x_n \varphi) = \mathrm{QR}(\varphi) + 1$.

Example 7.34

$$QR(\exists x_0 \ldots \exists x_n (\bigwedge_{0 \le i < j \le n} \neg \approx x_i x_j)) = QR(\bigwedge_{0 \le i < j \le n} \neg \approx x_i x_j) + n + 1 = n + 1.$$

Example 7.35   Let

$$\theta_0 = \neg \exists x_1 (x_1 < x_0)$$

$$\theta_\alpha = \forall x_1 \left(x_1 < x_0 \leftrightarrow \exists x_0 \left(\approx x_0 x_1 \wedge \left(\bigvee_{\beta < \alpha} \theta_\beta \right) \right) \right)$$

All formulas θ_α are built up from two variables x_0 and x_1, and have just x_0 free. With appropriate agreements about the exchange of bound variables in substitution, these formulas could be written more succinctly as

$$\theta_\alpha(x_0) = \forall x_1 \left(x_1 < x_0 \leftrightarrow \left(\bigvee_{\beta < \alpha} \theta_\beta(x_1) \right) \right).$$

Note that

$$QR \left(\forall x_1 \left(x_1 < x_0 \leftrightarrow \left(\bigvee_{\beta < \alpha} \theta_\beta(x_1) \right) \right) \right) = (\sup_{\beta < \alpha} QR(\theta_\beta(x_1))) + 1.$$

Thus $QR(\theta_\alpha) = \alpha + 1$.

The truth-definition of $L_{\infty\omega}$ is standard:

Definition 7.36   The concept of an assignment $s : \mathbb{N} \to M$ *satisfying* a formula φ in a model $\mathcal{M}$, $\mathcal{M} \models_s \varphi$ is defined as follows:

$\mathcal{M} \models_s \approx t_1 t_2$	iff	$t_1^{\mathcal{M}}(s) = t_2^{\mathcal{M}}(s)$
$\mathcal{M} \models_s Rt_1 \ldots t_n$	iff	$(t_1^{\mathcal{M}}(s), \ldots, t_n^{\mathcal{M}}(s)) \in \mathrm{Val}_{\mathcal{M}}(R)$
$\mathcal{M} \models_s \neg\varphi$	iff	$\mathcal{M} \not\models_s \varphi$
$\mathcal{M} \models_s \bigwedge_{i \in I} \varphi_i$	iff	$\mathcal{M} \models_s \varphi_i$ for all $i \in I$
$\mathcal{M} \models_s \bigvee_{i \in I} \varphi_i$	iff	$\mathcal{M} \models_s \varphi_i$ for some $i \in I$
$\mathcal{M} \models_s \forall x_n \varphi$	iff	$\mathcal{M} \models_{s[a/x_n]} \varphi$ for all $a \in M$
$\mathcal{M} \models_s \exists x_n \varphi$	iff	$\mathcal{M} \models_{s[a/x_n]} \varphi$ for some $a \in M$.

An alternative definition can be given in terms of games:

Definition 7.37   Suppose L is a vocabulary, $\mathcal{M}$ is an L-structure, φ^* is an L-formula, and s^* is an assignment for M. The game $\mathrm{SG}^{\mathrm{sym}}(\mathcal{M}, \varphi^*)$ is defined as follows. In the beginning player **II** holds (φ^*, s^*). The rules of the game are as follows:

I	**II**
x_0	
	y_0
x_1	
	y_1
$\vdots$	$\vdots$

Figure 7.8 The game $G_\omega(W)$.

1. If φ is atomic, and s satisfies it in $\mathcal{M}$, then the player who holds (φ, s) wins the game, otherwise the other player wins.
2. If $\varphi = \neg\psi$, then the player who holds (φ, s), gives (ψ, s) to the other player.
3. If $\varphi = \bigwedge_{i\in I}\varphi_i$, then the player who holds (φ, s) switches to hold some (φ_i, s) and the other player decides which.
4. If $\varphi = \bigvee_{i\in I}\varphi_i$, then the player who holds (φ, s) switches to hold some (φ_i, s) and can himself or herself decide which.
5. If $\varphi = \forall x_n\psi$, then the player who holds (φ, s) switches to hold some $(\psi, s[a/x_n])$ and the other player chooses $a \in M$.
6. If $\varphi = \exists x_n\psi$, then the player who holds (φ, s) switches to hold some $(\psi, s[a/x_n])$ and can himself or herself choose $a \in M$.

As was pointed out in Section 6.5, $\mathcal{M} \models_s \varphi$ if and only if player **II** has a winning strategy in the above game, starting with (φ, s). Why? If $\mathcal{M} \models_s \varphi$, then the winning strategy of player **II** is to play so that if she holds (φ', s'), then $\mathcal{M} \models_{s'} \varphi'$, and if player **I** holds (φ', s'), then $\mathcal{M} \not\models_{s'} \varphi'$.

The *negation normal form* NNF is defined for $L_{\infty\omega}$ exactly as for first-order logic by requiring that negations occur in front of atomic formulas only.

Definition 7.38   The *Semantic Game* $\text{SG}(\mathcal{M}, T, s)$ of the set T of L-sentences of $L_{\infty\omega}$ in NNF is the game $G_\omega(W)$ (see Figure 7.8), where W consists of sequences $(x_0, y_0, x_1, y_1, \ldots)$ such that player **II** has followed the rules of Figure 7.9, and moreover, if ψ_i is a basic formula and player **II** plays the pair (ψ_i, s) then $\mathcal{M} \models_s \psi_i$.

Proposition 7.39 $\mathcal{M} \models_s T$ *iff* **II** *has a winning strategy in* $\text{SG}(\mathcal{M}, T, s)$.

Proof   Exercise 7.37. □

Example 7.40   Let ψ_n be the sentence $\exists x_0 \ldots \exists x_n (\bigwedge_{0\leq i<j\leq n} \neg\approx x_i x_j)$. Then $\mathcal{M} \models (\bigvee_{n\in\mathbb{N}} \neg\psi_n)$ iff M is finite. Thus

$$\mathcal{M} \models (\bigwedge_{n\in\mathbb{N}} \psi_n) \quad \text{iff} \quad |M| \geq \aleph_0.$$

x_n	y_n	Explanation	Rule
$(\varphi, \emptyset)$		**I** enquires about $\varphi \in T$.	
	$(\varphi, \emptyset)$	**II** confirms.	Axiom rule
(φ_i, s)		**I** tests a played $(\bigwedge_{i \in I} \varphi_i, s)$ by choosing $i \in I$.	
	(φ_i, s)	**II** confirms.	$\wedge$-rule
$(\bigvee_{i \in I} \varphi_i, s)$		**I** enquires about a played disjunction.	
	(φ_i, s)	**II** makes a choice of $i \in I$.	$\vee$-rule
$(\varphi, s[a/x])$		**I** tests a played $(\forall x \varphi, s)$ by choosing $a \in M$.	
	$(\varphi, s[a/x])$	**II** confirms.	$\forall$-rule
$(\exists x \varphi, s)$		**I** enquires about a played existential statement.	
	$(\varphi, s[a/x])$	**II** makes a choice of $a \in M$.	$\exists$-rule

Figure 7.9 The game $\mathrm{SG}(\mathcal{M}, T, s)$.

Example 7.41   Let

$$\psi_0 = \approx x_0 x_1$$
$$\psi_{n+1} = \exists x_2 (x_0 E x_2 \wedge \exists x_0 (\approx x_0 x_2 \wedge \psi_n)).$$

Then for graphs $\mathcal{G}$ we have

$$\mathcal{G} \models \forall x_0 \forall x_1 \left(\bigvee_{n \in \mathbb{N}} \psi_n \right) \quad \text{iff} \quad \mathcal{G} \text{ is connected.}$$

Note that the sentence $\forall x_0 \forall x_1 (\bigvee_{n \in \mathbb{N}} \psi_n)$ uses just the variables x_0, x_1, and x_2.

Example 7.42   Consider the vocabulary $\{+, 0\}$ of abelian groups. Let us introduce the notation

$$x_i \cdot 0 = 0$$
$$x_i \cdot (n+1) = x_i \cdot n + x_i.$$

Thus

$$x_i \cdot n = \underbrace{x_i + \cdots + x_i}_{n}.$$

A group $\mathcal{G}$ is torsion-free iff

$$\mathcal{G} \models \forall x_0(\approx 0x_0 \vee \bigwedge_{n>0} \neg\approx 0x_0 \cdot n)$$

and $\mathcal{G}$ is torsion if

$$\mathcal{G} \models \forall x_0(\bigvee_{n\geq 0} \approx 0x_0 \cdot n).$$

Example 7.43 Consider the vocabulary $\{+, \cdot, 0, 1\}$ of arithmetic. Let $x_i \cdot n$ be defined as above. Then for models $\mathcal{M}$ of Peano's axioms we have

$$\mathcal{M} \cong (\mathbb{N}, +, \cdot, 0, 1) \quad \text{iff} \quad \mathcal{M} \models \forall x_0 \left(\bigvee_{n\geq 0} \approx x_0 1 \cdot n \right).$$

Example 7.44 Suppose (M, d) is a metric space. For each positive rational r let $D_r = \{(x, y) \in M \times M : d(x, y) < r\}$ and $\mathcal{M} = (M, (D_r)_{r>0})$. We can now actually define the original metric:

$$d(s(n), s(m)) = z \iff \mathcal{M} \models_s \bigwedge_{r>z>r'} (D_r x_n x_m \wedge \neg D_{r'} x_n x_m).$$

We can express the continuity of a function $f : M \to M$ with

$$(\mathcal{M}, f) \models_s \forall x_0 \left(\bigwedge_\epsilon \bigvee_\delta \forall x_1(D_\delta x_0 x_1 \to D_\epsilon f x_0 f x_1) \right).$$

Example 7.45 Consider the formulas θ_α of Example 7.35. Then

$$\mathcal{M} \models_s \theta_\alpha \quad \text{iff} \quad (\leftarrow, s(0))^{\mathcal{M}} \cong \alpha$$

where $(\leftarrow, x)^{\mathcal{M}} = (\{y \in M : y <^{\mathcal{M}} x\}, <^{\mathcal{M}})$. We prove this by induction on α. Suppose first $f : (\leftarrow, s(0))^{\mathcal{M}} \cong \alpha$. The winning strategy of **II** in $SG((\leftarrow, s(0))^{\mathcal{M}}, \theta_\alpha, s)$ is: if **I** chooses $a \in (\leftarrow, s(0))^{\mathcal{M}}$ and enquires about $\beta < \alpha$, **II** chooses $\beta = f(a)$ and plays $(\theta_\beta, s[0/a])$. By the induction hypothesis, as $(\leftarrow, a)^{\mathcal{M}} \cong \beta$, she has a winning strategy in the new position. Conversely, suppose $\mathcal{M} \models_s \theta_\alpha$. We show that $(\leftarrow, s(0))^{\mathcal{M}} \simeq_p \alpha$, from which $(\leftarrow, s(0))^{\mathcal{M}} \cong \alpha$ follows. The back-and-forth set for $(\leftarrow, s(0))^{\mathcal{M}}$ and α is the set P of finite partial isomorphisms

$$f = \{(x_0, \alpha_0), \ldots, (x_{n-1}, \alpha_{n-1})\}$$

such that for all $i < n : \mathcal{M} \models_{s[0/x_i]} \theta_{\alpha_i}$. By the induction hypothesis $(\leftarrow$

, $x_i)^\mathcal{M} \cong \alpha_i$. Note that isomorphisms between well-ordered sets are unique. To prove the back-and-forth property for P, suppose first $f \in P$ and $a \in (\leftarrow , s(0))^\mathcal{M}$. We play $SG((\leftarrow, s(0))^\mathcal{M}, \theta_\alpha, s)$ such that player **I** enquires about $(\bigvee_{\beta<\alpha} \theta_\beta, s[0/a])$. The winning strategy of **II** yields $\beta < \alpha$ such that she plays $(\theta_\beta, s[0/a])$. By the induction hypothesis $(\leftarrow, a)^\mathcal{M} \cong \beta$. So $f \cup \{(a, \beta)\} \in P$. The other half of back-and-forth is proved similarly.

Example 7.46 Let θ_α be as above. Then

$$\mathcal{M} \models \left(\forall x_0 \bigvee_{\beta<\alpha} \theta_\beta\right) \wedge \left(\bigwedge_{\beta<\alpha} \exists x_0 \theta_\beta\right) \quad \text{iff} \quad \mathcal{M} \cong \alpha.$$

The proof is just as above (see Exercise 7.52).

We write $\mathcal{M} \equiv_{\infty\omega} \mathcal{M}'$ if $\mathcal{M}$ and $\mathcal{M}'$ satisfy the same $L_{\infty\omega}$-sentences and $\mathcal{M} \equiv_\alpha \mathcal{M}'$ if they satisfy the same $L_{\infty\omega}$-sentences of quantifier rank $\leq \alpha$.

We now extend an important leg of the Strategic Balance of Logic, namely the equivalence of the Semantic Game and the Ehrenfeucht–Fraïssé Game, from first-order logic to infinitary logic:

Theorem 7.47 *The following are equivalent:*

(i) $\mathcal{A} \equiv_\alpha \mathcal{B}$.
(ii) $\mathcal{A} \sim_p^\alpha \mathcal{B}$.

Proof $(ii) \rightarrow (i)$ Suppose $(P_\beta : \beta \leq \alpha)$ is a back-and-forth sequence for $\mathcal{A}$ and $\mathcal{B}$. We use induction on $\beta \leq \alpha$ to prove:

Claim: If $f \in P_\beta$ and $a_1, \ldots, a_k \in \text{dom}(f)$, then

$$(\mathcal{A}, a_1, \ldots, a_k) \equiv_\beta (\mathcal{B}, fa_1, \ldots, fa_k).$$

We use induction on φ of quantifier rank $\leq \beta$ to prove the claim

$$(\mathcal{A}, a_1, \ldots, a_k) \models \varphi \Rightarrow (\mathcal{B}, fa_1, \ldots, fa_k) \models \varphi.$$

The only non-trivial case is that $\varphi = \exists x_n \psi(x_n)$ and $\gamma = \text{QR}(\psi) < \beta$. By assumption, $f \in P_{\gamma+1}$. Since $(\mathcal{A}, a_1, \ldots, a_k) \models \varphi$, there is $a \in A$ such that $(\mathcal{A}, a_1, \ldots, a_k, a) \models \psi(c)$, where c is a new constant symbol, a name for a. Since $f \in P_{\gamma+1}$, there is $b \in B$ such that $f \cup \{(a, b)\} \in P_\gamma$. By the induction hypothesis $(\mathcal{B}, fa_1, \ldots, fa_k, b) \models \psi(c)$. Thus $(\mathcal{B}, fa_1, \ldots, fa_k) \models \varphi$.

$(i) \rightarrow (ii)$ Let P_β consist of such finite $f \in \text{Part}(\mathcal{A}, \mathcal{B})$ that if $\text{dom}(f) = \{a_0, \ldots, a_{n-1}\}$, then

$$(\mathcal{A}, a_0, \ldots, a_{n-1}) \equiv_\beta (\mathcal{B}, fa_0, \ldots, fa_{n-1}).$$

By assumption (i), $\emptyset \in P_\alpha$, so $P_\alpha \neq \emptyset$. Certainly $\beta < \gamma$ implies $P_\gamma \subseteq P_\beta$. To prove the back-and-forth criterion, suppose $f \in P_{\beta+1}, a \in A$ and there is no $b \in B$ with

$$(\mathcal{A}, a_0, \ldots, a_{n-1}, a) \equiv_\beta (\mathcal{B}, fa_0, \ldots, fa_{n-1}, b). \tag{7.5}$$

Then for each $b \in B$ there is some φ_b of quantifier rank $\leq \beta$ such that

$$(\mathcal{A}, a_0, \ldots, a_{n-1}, a) \models \varphi_b(c)$$

and

$$(\mathcal{B}, fa_0, \ldots, fa_{n-1}, b) \models \neg\varphi_b(c)$$

where c is a name for a in $\mathcal{A}$ and b in $\mathcal{B}$. Thus

$$(\mathcal{A}, a_0, \ldots, a_{n-1}) \models \exists x_0 \bigwedge_{b \in B} \varphi_b(x_0).$$

Since $\mathrm{QR}(\exists x_0 \bigwedge_{b \in B} \varphi_b(x_0)) \leq \beta + 1$ and $f \in P_{\beta+1}$ we may conclude

$$(\mathcal{B}, fa_0, \ldots, fa_{n-1}) \models \exists x_0 \bigwedge_{b \in B} \varphi_b(x_0).$$

Let $z \in B$ with $(\mathcal{B}, fa_0, \ldots, fa_{n-1}, z) \models \bigwedge_{b \in B} \varphi_b(c)$. We get the contradiction

$$(\mathcal{B}, fa_0, \ldots, fa_{n-1}, z) \models \neg\varphi_z(c) \wedge \varphi_z(c).$$

Thus $a, b \in B$ with (7.5) must exist. The other half of the back-and-forth criterion is similar. □

By combining the above theorem with our previous results about the relation $\simeq_p^\alpha$, we obtain many interesting facts about $L_{\infty\omega}$:

Proposition 7.48 *The following are equivalent for all $\mathcal{A}$ and $\mathcal{B}$:*

1. $\mathcal{A} \equiv_{\infty\omega} \mathcal{B}$.
2. $\mathcal{A} \simeq_p \mathcal{B}$ i.e. there is a back-and-forth set for $\mathcal{A}$ and $\mathcal{B}$.
3. II has a winning strategy in $\mathrm{EF}_\omega(\mathcal{A}, \mathcal{B})$.

Example 7.49 (1) There is no $L_{\infty\omega}$-sentence ψ in the empty vocabulary such that for all $\mathcal{M}$:

$$\mathcal{M} \models \psi \quad \text{iff} \quad |\mathcal{M}| \leq \aleph_0,$$

because all infinite models in this vocabulary are partially isomorphic.

(2) There is no $L_{\infty\omega}$-sentence ψ in the vocabulary $\{\sim\}$ of equivalence relations such that any equivalence relation satisfies

$$\mathcal{M} \models \psi \quad \text{iff} \quad \mathcal{M} \text{ has only countably many equivalence classes.}$$

(3) There is no $L_{\infty\omega}$-sentence ψ of the vocabulary $\{E\}$ of graph theory such that for all graphs $\mathcal{G}$:

$$\mathcal{G} \models \psi \quad \text{iff} \quad \mathcal{G} \text{ has an uncountable clique.}$$

The following consequence of Karp's Theorem (Theorem 7.26) is of fundamental importance for understanding $L_{\infty\omega}$:

Corollary   *There is no $L_{\infty\omega}$-sentence of the vocabulary $\{<\}$ such that for all linear orders $\mathcal{M}$:*

$$\mathcal{M} \models \psi \quad \text{iff} \quad \mathcal{M} \text{ is a well-order.}$$

This should be contrasted with the fact that for all α and all $\mathcal{M}$

$$\mathcal{M} \models \theta_\alpha \quad \text{iff} \quad \mathcal{M} \text{ is a well-order of type } \alpha.$$

If we could take the disjunction of all θ_α, $\alpha \in On$, we could characterize well-order, but On is a proper class, so the disjunction cannot be formed in $L_{\infty\omega}$.

Definition 7.50   Let L be a vocabulary, $\mathcal{M}$ an L-structure, and $a_0, \ldots, a_{n-1} \in M$. Then we define

$$\sigma^0_{\mathcal{M},a_0,\ldots,a_{n-1}} = \bigwedge \{\varphi(x_0,\ldots,x_{n-1}) : \varphi(x_0,\ldots,x_{n-1})$$
$$\text{is a basic } L\text{-formula and } \mathcal{M} \models \varphi(a_0,\ldots,a_{n-1})\}$$

$$\sigma^{\alpha+1}_{\mathcal{M},a_0,\ldots,a_{n-1}} = \left(\forall x_n \bigvee_{a_n \in M} \sigma^\alpha_{\mathcal{M},a_0,\ldots,a_n}\right) \wedge \left(\bigwedge_{a_n \in M} \exists x_n \sigma^\alpha_{\mathcal{M},a_0,\ldots,a_n}\right)$$

$$\sigma^\nu_{\mathcal{M},a_0,\ldots,a_{n-1}} = \bigwedge_{\alpha<\nu} \sigma^\alpha_{\mathcal{M},a_0,\ldots,a_{n-1}}, \text{ for limit } \nu$$

$$\sigma^\alpha_{\mathcal{M}} = \sigma^\alpha_{\mathcal{M},\emptyset}.$$

Lemma 7.51   *1.* $\mathcal{M} \models \sigma^\alpha_{\mathcal{M},a_0,\ldots,a_{n-1}}(a_0,\ldots,a_{n-1})$.
2. $\sigma^\alpha_{\mathcal{M},a_0,\ldots,a_k}(x_0,\ldots,x_k) \models \sigma^\alpha_{\mathcal{M},a_0,\ldots,a_{n-1}}(x_0,\ldots,x_{n-1})$ *for* $n \leq k+1$.
3. If $\alpha < \beta$, *then*

$$\sigma^\beta_{\mathcal{M},a_0,\ldots,a_{n-1}}(x_0,\ldots,x_{n-1}) \models \sigma^\alpha_{\mathcal{M},a_0,\ldots,a_{n-1}}(x_0,\ldots,x_{n-1}).$$

Proof   Exercise 7.44.   □

Proposition 7.52   *The following are equivalent:*

(1) $\mathcal{M}' \models \sigma^\alpha_{\mathcal{M},a_0,\ldots,a_{n-1}}(b_0,\ldots,b_{n-1})$.
(2) $(\mathcal{M},a_0,\ldots,a_{n-1}) \simeq^\alpha_p (\mathcal{M}',b_0,\ldots,b_{n-1})$.

Proof Note that the quantifier rank of the formula $\sigma^\alpha_{\mathcal{M},a_0,\ldots,a_{n-1}}$ is α. Since $\mathcal{M} \models \sigma^\alpha_{\mathcal{M},a_0,\ldots,a_{n-1}}(a_0,\ldots,a_{n-1})$ by Lemma 7.51, the implication (2) $\to$ (1) follows from Proposition 7.47. Next we prove (1) $\to$ (2). Intuitively, the winning strategy of **II** in EFD_α on $(\mathcal{M}, a_0,\ldots,a_{n-1})$ and $(\mathcal{M}', b_0,\ldots,b_{n-1})$ is written into the structure of $\sigma^\alpha_{\mathcal{M},a_0,\ldots,a_{n-1}}$. More exactly, we can define a back-and-forth sequence $(P_\beta : \beta \leq \alpha)$ by letting P_β consist of finite mappings

$$f = \{(a_0,b_0),\ldots,(a_{n-1},b_{n-1}),\ldots,(a_m,b_m)\}$$

such that

$$\mathcal{M}' \models \sigma^\beta_{\mathcal{M},a_0,\ldots,a_{n-1},\ldots,a_m}(b_0,\ldots,b_{n-1},\ldots,b_m)\}.$$

By the definition of the formulas $\sigma^\beta_{\mathcal{M},a_0,\ldots,a_{n-1}}$, the sequence $(P_\beta : \beta \leq \alpha)$ is indeed a back-and-forth sequence. Note that (1) implies $P_\alpha \neq \emptyset$, as $\{(a_0,b_0),\ldots,(a_{n-1},b_{n-1})\} \in P_\alpha$. $\qquad\square$

Definition 7.53 The *Scott sentence* of a structure $\mathcal{M}$ is the $L_{\infty\omega}$-sentence

$$\sigma_{\mathcal{M}} = \sigma^{\text{SH}(\mathcal{M})}_{\mathcal{M},\emptyset} \wedge \bigwedge_{\substack{a_0,\ldots,a_{n-1}\in M \\ n \in \mathbb{N}}} \forall x_0 \ldots \forall x_{n-1}(\sigma^{\text{SH}(\mathcal{M})}_{\mathcal{M},a_0,\ldots,a_{n-1}} \to \sigma^{\text{SH}(\mathcal{M})+1}_{\mathcal{M},a_0,\ldots,a_{n-1}}).$$

Proposition 7.54 *The following are equivalent:*

(1) $\mathcal{M}' \models \sigma_{\mathcal{M}}$.
(2) $\mathcal{M}' \simeq_p \mathcal{M}$.

Proof (2) $\to$ (1): Lemma 7.51 gives $\mathcal{M} \models \sigma_{\mathcal{M}}$. The implication follows now from Proposition 7.48. (1) $\to$ (2): Suppose $\mathcal{M}' \models \sigma_{\mathcal{M}}$. We prove $\mathcal{M}' \simeq_p \mathcal{M}$ by giving a winning strategy for player **II** in the game $\text{EFD}_{\text{SH}(\mathcal{M})}(\mathcal{M},\mathcal{M}')$. The strategy of **II** is to make sure that if the position is

$$p = (x_0,y_0,\ldots,x_{n-1},y_{n-1})$$

then

$$(\star) \qquad \mathcal{M}' \models \sigma^{\text{SH}(\mathcal{M})}_{v_0,\ldots,v_{n-1}}(v'_0,\ldots,v'_{n-1}).$$

In the beginning of the game $(\star)$ holds by assumption. Let us then assume we are in the middle of the game $\text{EF}_\omega(\mathcal{M},\mathcal{M}')$, say in position p, and $(\star)$ holds. Now player **I** moves x_n, say $x_n = v_n \in M$. Now we use the assumption $\mathcal{M}' \models \sigma_{\mathcal{M}}$. It gives

$$\mathcal{M}' \models \sigma^{\text{SH}(\mathcal{M})+1}_{\mathcal{M},v_0,\ldots,v_{n-1}}(v'_0,\ldots,v'_{n-1})),$$

whence

$$\mathcal{M}' \models \bigwedge_{a \in M} \exists y \sigma^{\alpha}_{\mathcal{M}, v_0, \ldots, v_{n-1}, a}(v'_0, \ldots, v'_{n-1}, y).$$

By choosing $a = v_n$ we find a move $y_n = v'_n \in M'$ of **II** which yields

$$\mathcal{M}' \models \sigma^{\mathrm{SH}(\mathcal{M})}_{v_0, \ldots, v_n}(v'_0, \ldots, v'_n).$$

□

Note that if $\mathcal{M}$ is a well-ordered set then by the above result it is, up to isomorphism, the only model of $\sigma_{\mathcal{M}}$.

Corollary (Scott Isomorphism Theorem) *Suppose $\mathcal{M}$ is a countable model. Then for all countable $\mathcal{M}'$*

$$\mathcal{M}' \models \sigma_{\mathcal{M}} \quad \Longleftrightarrow \quad \mathcal{M}' \cong \mathcal{M}.$$

This is a remarkable result. It puts countable models on levels of a well-ordered hierarchy according to their Scott height. On each level there is an invariant, the Scott sentence of the model, that characterizes the model up to isomorphism. These invariants need not, of course, be simple in any way, but they have a uniform tree-structure, the differences occurring only at the leaves of the tree. The invariants provide a way to systematize and classify countable models according to the syntactic properties of the Scott sentence.

For the next result we have to compute an upper bound for the number of non-equivalent infinitary formulas of a given quantifier rank. As in the finite case (see Propositions 6.3 and 4.15), the upper bound is an exponential tower, only this time we deal with infinite cardinals rather than natural numbers. These cardinal numbers look very big, but at this point the only relevant thing is that they exist. We want to be sure that there is not a proper class of non-equivalent formulas of a fixed quantifier rank. Note that if we do not limit the quantifier rank, there is a proper class of non-equivalent formulas, namely the Scott sentences of different ordinals. Recall that

$$\begin{cases} \beth_0(\lambda) = \lambda \\ \beth_{\alpha+1}(\lambda) = 2^{\beth_\alpha(\lambda)} \\ \beth_\nu(\lambda) = \sup_{\alpha < \nu} \beth_\alpha(\lambda). \end{cases}$$

Lemma 7.55 *Suppose L is a vocabulary of size μ and $\alpha = \nu + n$ where ν is a limit ordinal. There are at most $\beth_{\nu+2n+2}(\mu + \aleph_0)$ non-equivalent formulas of $L_{\infty\omega}$ of quantifier rank $\leq \alpha$.*

Proof There are at most $\mu + \aleph_0$ atomic formulas and therefore at most $\beth_2(\mu + \aleph_0)$ non-equivalent formulas of quantifier rank 0. Suppose then $\alpha = \nu + n + 1$ where ν is a limit ordinal. Formulas of quantifier rank $\leq \alpha$ are of the form $\forall x_n \varphi$ or of the form $\exists x_n \varphi$, where $\mathrm{QR}(\varphi) < \alpha$, and what can be built from them by means of $\neg$, $\bigwedge_{i \in I}$ and $\bigvee_{i \in I}$. Thus their number (up to logical equivalence) is at most $\beth_2(\beth_{\nu+2n+2}(\mu + \aleph_0)) = \beth_{\nu+2(n+1)+2}(\mu + \aleph_0))$. If ν is a limit ordinal, the number of non-equivalent formulas of quantifier rank $< \nu$ is $\leq \sup_{\alpha < \nu} \beth_\alpha(\mu + \aleph_0) = \beth_\nu(\mu + \aleph_0)$. Therefore, the number of non-equivalent formulas of quantifier rank $\leq \nu$ is at most $\beth_2(\beth_\nu(\mu + \aleph_0)) = \beth_{\nu+2}(\mu + \aleph_0)$. $\square$

Thus, for example, for any α there is only a set of non-equivalent sentences $\sigma_{\mathcal{M}}^\alpha$, while there is a proper class of non-equivalent sentences $\sigma_{\mathcal{M}}$.

Corollary *Suppose L is a vocabulary. Then for all ordinals α the equivalence relation*

$$\mathcal{A} \equiv_\alpha \mathcal{B}$$

divides the class $\mathrm{Str}(L)$ of all L-structures into a set of equivalence classes $C_i^\alpha, i \in I$, such that if we choose any representatives $\mathcal{M}_i \in C_i^\alpha$, then:

1. *For all L-structures $\mathcal{M}$: $\mathcal{M} \in C_i^\alpha \iff \mathcal{M} \models \sigma_{\mathcal{M}_i}^\alpha$.*
2. *If φ is an L-sentence of $L_{\infty\omega}$ of quantifier rank $\leq \alpha$, then there is a set $I_0 \subseteq I$ such that $\models \varphi \leftrightarrow \bigvee_{i \in I_0} \sigma_{\mathcal{M}_i}^\alpha$.*

Proof For any L-structure $\mathcal{M}$ let $\mathrm{Th}_\alpha(\mathcal{M})$ be the set of $L_{\infty\omega}$-sentences of quantifier rank $\leq \alpha$ (up to logical equivalence) which are true in $\mathcal{M}$. Thus

$$\mathcal{M} \equiv_\alpha \mathcal{M}' \iff \mathrm{Th}_\alpha(\mathcal{M}) = \mathrm{Th}_\alpha(\mathcal{M}').$$

Let $\mathrm{Th}_\alpha(\mathcal{M}_i), i \in I$, be a complete list of all $\mathrm{Th}_\alpha(\mathcal{M})$. The claim follows. For the second claim let I_0 consist of such $i \in I$ that $\mathcal{M}_i \models \varphi$. If $\mathcal{M} \models \varphi$ and $\mathrm{Th}_\alpha(\mathcal{M}) = \mathrm{Th}_\alpha(\mathcal{M}_i)$, then $i \in I_0$ and $\mathcal{M} \models \sigma_{\mathcal{M}_i}^\alpha$. Conversely, if $\mathcal{M} \models \sigma_{\mathcal{M}+i}^\alpha, i \in I_0$, then $\mathcal{M} \equiv_\alpha \mathcal{M}_i$ and $\mathcal{M} \models \varphi$ follows. $\square$

Note again that if we tried to prove the above corollary for the finer relation $\simeq_p$, we would run into the difficulty that there is a proper class of equivalence classes.

Corollary *Suppose L is an arbitrary vocabulary and K is a class of L-structures. Then the following are equivalent:*

(i) K is definable in $L_{\infty\omega}$.
(ii) K is closed under $\simeq_p^\alpha$ for some α.

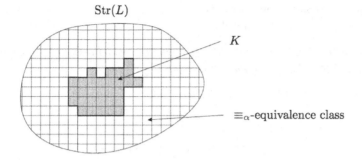

Figure 7.10 Model class K definable in $L_{\infty\omega}$.

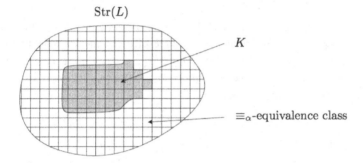

Figure 7.11 Model class K not definable in $L_{\infty\omega}$.

These equivalent conditions are strictly stronger than

(iii) *K is closed under $\simeq_p$.*

The above theorem gives a kind of normal form for sentences of $L_{\infty\omega}$: every sentence is a disjunction of sentences $\sigma_{\mathcal{M}}^{\alpha}$, which in turn have a very canonical form. For finite α and finite relational vocabulary the formulas $\sigma_{\mathcal{M}}^{\alpha}$ are first-order.

Definition 7.56 Suppose κ is a regular cardinal. $L_{\kappa\omega}$ is the fragment of $L_{\infty\omega}$ which obtains if in the definition of the syntax of $L_{\infty\omega}$ we modify condition (4) and (5) by requiring that $|I| < \kappa$.

First-order logic is in this notation $L_{\omega\omega}$. The most important non-first-order case is $L_{\omega_1\omega}$, the extension of first-order logic obtained by allowing countable

disjunctions and conjunctions. Note that in a countable vocabulary the Scott
sentence of a countable model is in $L_{\omega_1\omega}$.

Proposition 7.57 *Suppose $\mathcal{M}$ is a countable model in a countable vocabulary and $P \subseteq M^n$. Then the following are equivalent:*

(i) *P is closed under automorphisms of $\mathcal{M}$.*
(ii) *There is a formula $\varphi(x_0, \ldots, x_n)$ in $L_{\omega_1\omega}$ such that for all $a_0, \ldots, a_n \in$
       M*

$$(a_0, \ldots, a_n) \in P \iff \mathcal{M} \models \varphi(a_0, \ldots, a_n).$$

Proof $(ii) \rightarrow (i)$ is trivial because automorphisms preserve truth. To prove
$(i) \rightarrow (ii)$ consider

$$\varphi(x_0, \ldots, x_n) = \bigvee \{\sigma_{(\mathcal{M}, a_0, \ldots, a_n)}(x_0, \ldots, x_n) : (a_0, \ldots, a_n) \in P\},$$

where $\sigma_{(\mathcal{M}, a_0, \ldots, a_n)}(x_0, \ldots, x_n)$ denotes the formula obtained from the sentence $\sigma_{(\mathcal{M}, a_0, \ldots, a_n)}$ by replacing the name of a_i by the variable symbol x_i.
Since M (and hence P) is countable, $\varphi(x_0, \ldots, x_n) \in L_{\omega_1\omega}$. If we now
have $(a_0, \ldots, a_n) \in P$, then $\mathcal{M} \models \sigma_{(\mathcal{M}, a_0, \ldots, a_n)}(a_0, \ldots, a_n)$. Thus $\mathcal{M} \models$
$\varphi(a_0, \ldots, a_n)$. Conversely, suppose

$$\mathcal{M} \models \sigma_{(\mathcal{M}, a_0, \ldots, a_n)}(b_0, \ldots, b_n) \text{ and } (a_0, \ldots, a_n) \in P.$$

Then $(\mathcal{M}, b_0, \ldots, b_n) \simeq_p (\mathcal{M}, a_0, \ldots, a_n)$. Thus there is an automorphism π
of $\mathcal{M}$ such that $\pi(b_i) = a_i$ for $i \leq n$. Since P is closed under automorphisms,
$(b_0, \ldots, b_n) \in P$. $\square$

If we want to show that a relation on a countable structure is not definable in
$L_{\omega_1\omega}$, a natural approach is to show that the relation is not preserved by automorphisms of the structure. The above theorem demonstrates that this natural
approach is as good as any other.

Example 7.58 Let $\mathcal{M} = (\mathbb{Z}, <)$. The only subsets of $\mathbb{Z}$ that are closed under automorphisms of $\mathcal{M}$ are $\emptyset$ and $\mathbb{Z}$. Thus they are the only subsets of $\mathcal{M}$
definable in $L_{\omega_1\omega}$.

Corollary *If $\mathcal{M}$ is a rigid countable model in a countable vocabulary, then
every relation on M is $L_{\omega_1\omega}$-definable on $\mathcal{M}$.*

7.5 Historical Remarks and References

Infinitary languages were introduced in propositional calculus in Scott and
Tarski (1958) and in predicate logic in Tarski (1958). An early book on infinitary

languages is Karp (1964). More recent books are Keisler (1971), Dickmann (1975), and Barwise (1975). A good source is the survey article Makkai (1977).

The back-and-forth sets, and thereby in effect the Ehrenfeucht–Fraïssé Game was introduced to infinitary logic in Karp (1965), where Proposition 7.48 and Proposition 7.26 appear. A good survey article on back-and-forth sets is Kueker (1975). Propositions 7.54 and 7.57 and their corollaries are from Scott (1965). Definition 7.16 is from Karp (1965). Theorem 7.31 is from Kueker (1968).

Exercises

7.1 Show that if **II** has a winning strategy in $\mathrm{EFD}_\omega(\mathcal{M}, \mathcal{M}')$ and **M** is finite, then $\mathcal{M} \cong \mathcal{M}'$.

7.2 Let $\mathcal{M} = (\mathbb{Z}, <)$ and $\mathcal{M}' = (\mathbb{Z} + \mathbb{Z}, <)$ (i.e. two copies of $\mathcal{M}$ one after the other). For which n does **I** have a winning strategy in the game $\mathrm{EFD}_{\omega+n}(\mathcal{M}, \mathcal{M}')$, and for which does **II**?

7.3 Suppose $\mathcal{G}$ and $\mathcal{G}'$ are graphs such that player **II** has a winning strategy in $\mathrm{EFD}_\omega(\mathcal{G}, \mathcal{G}')$. Show that if $\mathcal{G}$ has a cycle path, then so does $\mathcal{G}'$.

7.4 Suppose $\mathcal{G}$ and $\mathcal{G}'$ are graphs and player **II** has a winning strategy in $\mathrm{EFD}_\omega(\mathcal{G}, \mathcal{G}')$. Show that if $\mathcal{G}$ has infinitely many edges, also $\mathcal{G}'$ has.

7.5 Suppose $\mathcal{M} \equiv \mathcal{N}$ where $\mathcal{N} = (\mathbb{N}, +, \cdot, 0, 1)$ but $\mathcal{M} \not\cong \mathcal{N}$. Show that **I** has a winning strategy in $\mathrm{EFD}_1(\mathcal{M}, \mathcal{N})$.

7.6 Player **I** wants to play $\mathrm{EFD}_\alpha(\mathcal{M}, \mathcal{M}')$ but cannot decide which α to choose. He wants to play as follows:

1. First **I** wants to play 10 moves.
2. Then, depending on how **II** has played, **I** wants to play 2^n moves for some n that he chooses.
3. Then **I** wants to play five additional moves.
4. Then, depending on how **II** has played, **I** wants to play $5n+1$ moves for some n that he chooses.
5. Finally **I** wants to play 15 additional moves, whereupon the game should end.

Can you help him choose α?

7.7 Show that player **I** (**II**) has a winning strategy in $\mathrm{EF}_n(\mathcal{M}, \mathcal{M}')$ iff he (she) has a winning strategy in $\mathrm{EFD}_n(\mathcal{M}, \mathcal{M}')$.

7.8 Let the game $\mathrm{EFD}_\alpha^*(\mathcal{A}, \mathcal{B})$ be like the game $\mathrm{EFD}_\alpha(\mathcal{A}, \mathcal{B})$ except that **I** has to play $x_{2n} \in A$ and $x_{2n+1} \in B$ for all $n \in \mathbb{N}$. Show that if ν is a limit ordinal, then player **II** has a winning strategy in $\mathrm{EFD}_{\nu+2n}^*(\mathcal{A}, \mathcal{B})$ if and only if she has a winning strategy in $\mathrm{EF}_{\nu+n}(\mathcal{A}, \mathcal{B})$.

7.9 Suppose $B = \{b_n : n \in \mathbb{N}\}$. Let the game $\text{EFD}_\alpha^{**}(\mathcal{A}, \mathcal{B})$ be like the game $\text{EFD}_\alpha(\mathcal{A}, \mathcal{B})$ except that **I** has to play $x_{2n} \in A$ and $x_{2n+1} = b_n$ for all $n \in \mathbb{N}$. Show that if ν is a limit ordinal, then player **II** has a winning strategy in $\text{EFD}_{\nu+2n}^{**}(\mathcal{A}, \mathcal{B})$ if and only if she has a winning strategy in $\text{EFD}_{\nu+n}(\mathcal{A}, \mathcal{B})$.

7.10 Find the Scott watershed for

   (a) $(\mathbb{N}, +, \cdot, 0, 1)$ and $(\mathbb{Q}, +, \cdot, 0, 1)$.
   (b) $(\mathbb{Z} + \mathbb{Z}, <)$ and $(\mathbb{Z} + \mathbb{Z} + \mathbb{Z}, <)$.

7.11 What is the Scott watershed of $(\mathbb{Z}^2, +)$ and $(\mathbb{Z}^3, +)$?

7.12 What is the Scott watershed of $(\mathbb{Q}, +)$ and $(\mathbb{R}, +)$?

7.13 Prove $\mathbb{Z}(15) \cong \mathbb{Z}(3) \times \mathbb{Z}(5)$.

7.14 Prove Lemma 7.20.

7.15 Prove Lemma 7.21.

7.16 Prove Lemma 7.23.

7.17 Show that if $(\mathcal{M}, v_0, \ldots, v_{n-1}) \simeq_p^{\text{SH}(\mathcal{M})} (\mathcal{M}, v_0', \ldots, v_{n-1}')$, then

$$(\mathcal{M}, v_0, \ldots, v_{n-1}) \simeq_p (\mathcal{M}, v_0', \ldots, v_{n-1}').$$

7.18 A model $\mathcal{M}$ is $\aleph_0$-*homogeneous* if the following holds for all $v_0, \ldots, v_n$ and $v_0', \ldots, v_{n-1}'$ in M: If $(\mathcal{M}, v_0, \ldots, v_{n-1}) \equiv (\mathcal{M}, v_0', \ldots, v_{n-1}')$ then there is v_n' in M such that $(\mathcal{M}, v_0, \ldots, v_n) \equiv (\mathcal{M}, v_0', \ldots, v_n')$. Show that the Scott height of an $\aleph_0$-homogeneous model is $\leq \omega$. Show that if $\mathcal{M}$ is a countable $\aleph_0$-homogeneous model and

$$(\mathcal{M}, v_0, \ldots, v_{n-1}) \equiv (\mathcal{M}, v_0', \ldots, v_{n-1}'),$$

then there is an automorphism of $\mathcal{M}$ which maps each v_i to v_i'.

7.19 Show that there are, up to isomorphism, exactly three countable $\aleph_0$-homogeneous models $\mathcal{M}$ such that $\mathcal{M} \simeq_p^\omega (\omega, <)$.

7.20 Show that if $\mathcal{M}$ and $\mathcal{M}'$ are well-orderings, then so is $\mathcal{M} \times \mathcal{M}'$.

7.21 Prove $\alpha \cdot \beta \cong \alpha \times \beta$ starting from the inductive definition of multiplication in Exercise 2.22.

7.22 Prove $\alpha < \kappa \implies \omega^\alpha < \kappa$ for uncountable cardinals κ. Recall the inductive definition of exponentiation in Exercise 2.26.

7.23 Prove $\mathcal{M} \times (\mathcal{M}' \times \mathcal{M}'') \cong (\mathcal{M} \times \mathcal{M}') \times \mathcal{M}''$ for linear orders $\mathcal{M}, \mathcal{M}'$ and $\mathcal{M}''$. Show also that it is possible that $\mathcal{M} \times \mathcal{M}' \not\cong \mathcal{M}' \times \mathcal{M}$.

7.24 Show that $\omega_1 \simeq_p^{\omega_1} \omega_1 \cdot 2$, and $(\mathbb{Q}, <) \times \omega_1 \simeq_p^{\omega_1} (\mathbb{R}, <) \times (\omega_1 \cdot 2)$.

7.25 Show that $\epsilon_0 \times (\mathbb{R}^{\geq 0}, <) \simeq_p^{\epsilon_0} \omega_1 \times (\mathbb{Q}^{\geq 0}, <)$.

7.26 Find for all α a scattered $\mathcal{M}$ and a non-scattered $\mathcal{M}'$ such that $\mathcal{M} \simeq_p^\alpha \mathcal{M}'$.

7.27 Prove the claims of Example 7.28.

7.28 Suppose $\mathcal{M} = (M, <)$ is a linear order and $\mathcal{M}'$ is the set of initial segments of $\mathcal{M}$ ordered by proper inclusion. Show that $\mathcal{M} \not\cong \mathcal{M}'$.

7.29 Show that there is a countable model $\mathcal{M}$ such that $\mathcal{M}$ has $2^{\aleph_0}$ automorphisms but there is no uncountable $\mathcal{M}'$ such that $\mathcal{M} \simeq_p \mathcal{M}'$. (Hint: Let $\mathcal{M} = (M, \omega, <, R)$ where $<$ is the usual ordering of ω, $M = \omega \cup \{(n, i) : n \in \mathbb{N}, i \in \{0, 1\}\}$ and $R = \{(n, (n, i)) : n \in \mathbb{N}, i \in \{0, 1\}\}$.)

7.30 Write a sentence of $L_{\infty\omega}$, as simple as possible, which holds in a finite graph iff

   (a) the number of vertices is even.

   (b) the number of edges is even.

   (c) the graph has a cycle path.

7.31 Write a sentence of $L_{\infty\omega}$, as simple as possible, which holds in a finite graph iff the graph is 3-colorable. Use this to prove that there is a sentence of $L_{\infty\omega}$ which holds in a graph iff the graph is 3-colorable. (Hint: use the Compactness Theorem of propositional logic to reduce the second part to the finite case.)

7.32 An ordered field $(K, +, \cdot, 0, 1, <)$ is *Archimedian* if for all $r_1 > 0$ and r_2 in K there is a natural number n so that $\underbrace{r_1 + \cdots + r_1}_{n} > r_2$. Show that the Archimedian property can be expressed in $L_{\infty\omega}$.

7.33 Let $L^n_{\infty\omega}$ denote the fragment of $L_{\infty\omega}$ consisting of formulas in which only variables $x_0, \ldots, x_{n-1}$ occur. Show that if $\mathcal{G}$ and $\mathcal{G}'$ are graphs so that player **II** has a winning strategy in the n-Pebble Game, then they satisfy the same sentences of $L^n_{\infty\omega}$. Hence, if $\mathcal{G}$ and $\mathcal{G}'$ satisfy the extension axiom E_n, then they satisfy the same sentences of $L^n_{\infty\omega}$. (See Exercise 4.17 for the definition of the n-Pebble Game.)

7.34 Use Exercise 7.33 to conclude that if two graphs satisfy E_n for all $n \in \mathbb{N}$, then the graphs are partially isomorphic. (See Exercise 4.16 for the definition of E_n.) Conclude also that if φ is a first-order sentence, then $E_n \models \varphi$ for some $n \in \mathbb{N}$, or else $E_n \models \neg\varphi$ for some $n \in \mathbb{N}$.

7.35 Show that there is no $L_{\infty\omega}$-sentence ψ of the vocabulary of linear order such that for all linear orders $\mathcal{M}$: $\mathcal{M} \models \psi$ iff $\mathcal{M}$ has cofinality ω (i.e. has a countable unbounded subset.)

7.36 Show that there is no $L_{\infty\omega}$-sentence ψ of the vocabulary $\{P, Q\}$ of two unary predicates such that

$$(M, P^{\mathcal{M}}, Q^{\mathcal{M}}) \models \psi \quad \text{iff} \quad |P^{\mathcal{M}}| = |Q^{\mathcal{M}}|.$$

7.37 Prove Proposition 7.39.

7.38 Show that there is no $L_{\infty\omega}$-sentence ψ of the vocabulary $\{<\}$ such that for all linear orders $\mathcal{M}$:

$$\mathcal{M} \models \psi \quad \text{iff} \quad |M| \text{ is countable.}$$

Show that such a ψ exists if "linear order" is replaced by "well-order".

7.39 Let (M, d) be a metric space and $\mathcal{M} = (M, (D_r)_{r>0})$ as in Example 7.44. Write a sentence φ of $L_{\infty\omega}$ such that

   (1) $(\mathcal{M}, p_0, p_1, \dots) \models \varphi$ iff the sequence $p_0, p_1, \dots$ converges in (M, d). (Expand the vocabulary to include names for the points p_n.)

   (2) $(\mathcal{M}, f_0, f_1, \dots, f) \models \varphi$ iff the sequence $f_0, f_1, \dots$ of functions $f : M \to M$ converges uniformly to f. (Expand the vocabulary to include names for the functions f_n and for the function f.)

   (3) $(\mathcal{M}, A) \models \varphi$ iff the set A is closed. (Expand the vocabulary to include a name for A.)

7.40 Let (M, d) and $\mathcal{M}$ be as above. Is there a sentence φ of $L_{\infty\omega}$ such that $\mathcal{M} \models \varphi$ iff (M, d) is compact?

7.41 Let V be a $\mathbb{Q}$-vector space. Let $\mathcal{M}_V = (V, +_V, 0_V, (f_r)_{r\geq 0})$, where for each non-negative rational r,

$$f_r(v) = r \cdot_V v.$$

Write a sentence φ of $L_{\infty\omega}$ such that

   (1) $\mathcal{M}_V \models \varphi$ iff $\dim(V) = n$.

   (2) $\mathcal{M}_V \models \varphi$ iff $\dim(V)$ is infinite.

   (3) $(\mathcal{M}_V, f) \models \varphi$ iff $f : V \times V \to V$ is a linear mapping. (Expand the vocabulary to include a name for f.)

7.42 Let V be an $\mathbb{R}$-vector space with a norm $\| \cdot \|_V : V \to \mathbb{R}$. Let $\mathcal{N}_V = (V, +_V, 0_V, D_V, (f_r)_{r\geq 0})$ where $D_V = \{v \in V : \|v\| < 1\}$ and for non-negative rational r, f_r is as above. Write a sentence φ of $L_{\infty\omega}$ such that

   (1) $(\mathcal{N}_V, f) \models \varphi$ iff $f : V \to V$ is continuous.

   (2) $(\mathcal{N}_V, f) \models \varphi$ iff $f : V \to V$ is differentiable.

   (In both (1) and (2), expand the vocabulary to include a name for f.)

7.43 Let $\mathcal{M} = (\mathbb{R}, +, \cdot, 0, 1, <)$ and L a vocabulary which extends the vocabulary of $\mathcal{M}$ by a name for a function $f : M \to M$. Write a sentence φ of $L_{\infty\omega}$ such that $(\mathcal{M}, f) \models \varphi$ iff

   (1) $f \restriction [0, 1]$ has bounded variation, i.e. there exists an M such that $\sum_{i=0}^{n} |f(x_{i+1}) - f(x_i)| \leq M$ for all $0 = x_0 < x_1 < \cdots < x_n = 1$.

(2) f is homogeneous, i.e. there is $n \in \mathbb{N}$ such that $f(ax) = a^n f(x)$ for all $a \in \mathbb{R}$.

(3) $f \restriction [0,1]$ is Riemann integrable, i.e. for each $\epsilon > 0$ there are $0 = x_0 < x_1 < \cdots < x_n = 1$ such that

$$\sum_{i=0}^{n} (x_{i+1} - x_i) \left(\sup_{x_i < x \leq x_{i+1}} f(x) - \inf_{x_i < x \leq x_{i+1}} f(x) \right) < \epsilon.$$

7.44 Prove Lemma 7.51.

7.45 Prove that if $\mathcal{M}$ is a well-ordered set, then $\mathcal{M}$ is, up to isomorphism, the only model of $\sigma_{\mathcal{M}}$.

7.46 Suppose $\mathcal{M}$ is a countable model in a countable vocabulary. Suppose $a \in M$ is fixed by all automorphisms of $\mathcal{M}$. Show that a is definable in $\mathcal{M}$ by a formula of $L_{\omega_1 \omega}$.

7.47 Suppose $\mathcal{M}$ is a countable model in a countable vocabulary. Show that $\mathcal{M}$ is rigid if and only if every element of $\mathcal{M}$ is definable in $\mathcal{M}$ by a formula of $L_{\omega_1 \omega}$.

7.48 Suppose $\mathcal{M}$ is a countable model in a countable vocabulary. Suppose there are $a_0, \ldots, a_{n-1} \in M$ such that $(\mathcal{M}, a_0, \ldots, a_{n-1})$ is rigid. Show that $\mathcal{M}$ can have at most countably many automorphisms.

7.49 Suppose $\mathcal{M}$ is a countable model in a countable vocabulary. Suppose $\mathcal{M}$ has $< 2^\omega$ many automorphisms. Show that there are $a_0, \ldots, a_{n-1} \in M$ such that $(\mathcal{M}, a_0, \ldots, a_{n-1})$ is rigid.

7.50 Let us write $\mathcal{M} <^\zeta_{\infty\omega} \mathcal{N}$ if $\mathcal{M} \subseteq \mathcal{N}$ and if $a_0, \ldots, a_{n-1} \in M$, then $(\mathcal{M}, a_0, \ldots, a_{n-1}) \cong^\zeta_p (\mathcal{N}, a_0, \ldots, a_{n-1})$. Suppose $\vec{\mathcal{M}} = (\mathcal{M}_\xi : \xi < \gamma)$ is a $<^\zeta_{\infty\omega}$-chain, i.e. $\mathcal{M}_\xi <^\zeta_{\infty\omega} \mathcal{M}_\eta$ for $\xi < \eta < \gamma$. Let $\mathcal{M}$ be the union of $\vec{\mathcal{M}}$, i.e.

$$M = \bigcup_{\xi < \gamma} M_\xi, \quad R^{\mathcal{M}} = \bigcup_{\xi < \gamma} R^{\mathcal{M}_\xi}, \quad f^{\mathcal{M}} = \bigcup_{\xi < \gamma} f^{\mathcal{M}_\xi}, c^{\mathcal{M}} = c^{\mathcal{M}_0}.$$

Show that $\mathcal{M}_\xi <^\zeta_{\infty\omega} \mathcal{M}$ for all $\xi < \gamma$.

7.51 Suppose $\mathcal{M}$ is a countable model for a countable vocabulary. Suppose there are formulas φ_n and ψ^n_m of $L_{\infty\omega}$ such that $\mathcal{M}$ satisfies the sentence:

$$\forall x_0 (\bigvee_{n<\omega} \varphi_n(x_0)) \wedge$$

$$\bigwedge_{n<\omega} \exists x_3 \ldots \exists x_{k_n} \forall x_1 (\varphi_n(x_1) \rightarrow$$
$$\bigvee_{m<\omega} \forall x_2 (\approx x_1 x_2 \leftrightarrow \psi^n_m(x_2, x_3, \ldots, x_{k_n}))).$$

Show that $\mathcal{M}$ is, up to isomorphism, the only model of $\sigma_{\mathcal{M}}$.

7.52 Prove the claim made in Example 7.46.

8

Model Theory of Infinitary Logic

8.1 Introduction

The model theory of $L_{\omega_1\omega}$ is dominated by the Model Existence Theorem. It more or less takes the role of the Compactness Theorem which can be rightfully called the cornerstone of model theory of first-order logic. The Model Existence Theorem is used to prove the Craig Interpolation Theorem and the important undefinability of the concept of well-order. When we move to the stronger logics $L_{\kappa^+\omega}$, $\kappa > \omega$, the Model Existence Theorem in general fails. However, we use a union of chains argument to prove the undefinability of well-order. In the final section we introduce game quantifiers. Here we cross the line to logics in which well-order is definable. Game quantifiers permit an approximation process which leads to the Covering Theorem, a kind of Interpolation Theorem.

8.2 Löwenheim–Skolem Theorem for $L_{\infty\omega}$

In Section 6.4 we saw that if a first-order sentence is true in a model it is true in "almost" every countable approximation of that model. We now extend this to $L_{\infty\omega}$ but of course with some modification because $L_{\infty\omega}$ has consistent sentences without any countable models. We show that if a sentence φ of $L_{\infty\omega}$ is true in a structure $\mathcal{M}$, a countable "approximation" of φ is true in a countable "approximation" of $\mathcal{M}$, and even more, there are this kind of approximations of φ and $\mathcal{M}$ in a sense "everywhere". To make this statement precise we employ the Cub Game introduced in Definition 6.10. We say

$$\ldots X \ldots \text{ for almost all } X \in \mathcal{P}_\omega(A)$$

if

$$\text{player II has a winning strategy in } G_{\text{cub}}(\mathcal{P}_\omega(A)).$$

Recall the following facts:

1. If $X_0 \in \mathcal{P}_\omega(A)$, then $X_0 \subseteq X$ for almost all $X \in \mathcal{P}_\omega(A)$.
2. If $X \in \mathcal{C}$ for almost all $X \in \mathcal{P}_\omega(A)$ and $\mathcal{C} \subseteq \mathcal{C}'$, then $X \in \mathcal{C}'$ for almost all $X \in \mathcal{P}_\omega(A)$.
3. If for all $n \in \mathbb{N}$ we have $X \in \mathcal{C}_n$ for almost all $X \in \mathcal{P}_\omega(A)$, then $X \in \bigcap_{n\in\mathbb{N}} \mathcal{C}_n$ for almost all $X \in \mathcal{P}_\omega(A)$.
4. If for all $a \in A$ we have $X \in \mathcal{C}_a$ for almost all $X \in \mathcal{P}_\omega(A)$, then $X \in \triangle_{a\in A}\mathcal{C}_a$ for almost all $X \in \mathcal{P}_\omega(A)$.

In other words, the set of subsets of $\mathcal{P}_\omega(A)$ which contain almost all $X \in \mathcal{P}_\omega(A)$ is a countably complete filter.

Now that approximations extend not only to models but also to formulas we assume that models and formulas have a common universe V, which is supposed to be a transitive[1] set. As the following lemma demonstrates, the exact choice of this set V is not relevant:

Lemma 8.1 *Suppose $\emptyset \neq A \subseteq V$ and $\mathcal{C} \subseteq \mathcal{P}_\omega(A)$. Then the following are equivalent:*

1. $X \in \mathcal{C}$ for almost all $X \in \mathcal{P}_\omega(A)$.
2. $X \cap A \in \mathcal{C}$ for almost all $X \in \mathcal{P}_\omega(V)$.

Proof (1) implies (2): Let $a \in A$. Player **II** applies her winning strategy in $G_{\text{cub}}(\mathcal{C})$ in the game $G_{\text{cub}}(\{X \in \mathcal{P}_\omega(V) : X \cap A \in \mathcal{C}\})$ as follows: If **I** plays his element in A, player **II** interprets it as a move in $G_{\text{cub}}(\mathcal{C})$, where she has a winning strategy. If **I** plays x_n outside A, player **II** plays $y_n = a$. (2) implies (1): player **II** interprets all moves of **I** in A as his moves in V and then uses her winning strategy in $G_{\text{cub}}(\{X \in \mathcal{P}_\omega(V) : X \cap A \in \mathcal{C}\})$. $\qquad\square$

Definition 8.2 Suppose $\varphi \in L_{\infty\omega}$ and X is a countable set. The approximation φ^X of φ is defined by induction as follows:

(1) $(\approx tt')^X = \approx tt'$.
(2) $(Rt_1 \ldots t_n)^X = Rt_1 \ldots t_n$.
(3) $(\neg\varphi)^X = \neg\varphi^X$.
(4) $(\bigwedge \Phi)^X = \bigwedge\{\varphi^X : \varphi \in \Phi \cap X\}$.
(5) $(\bigvee \Phi)^X = \bigvee\{\varphi^X : \varphi \in \Phi \cap X\}$.
(6) $(\forall x_n\varphi)^X = \forall x_n(\varphi^X)$.

---

[1] A set A is *transitive* if $y \in x \in A$ implies $y \in A$ for all x and y.

(7) $(\exists x_n \varphi)^X = \exists x_n(\varphi^X)$.

Note that φ^X is always in $L_{\omega_1\omega}$, whatever countable set X is.

Example 8.3   Suppose $X \cap \{\varphi_\alpha : \alpha < \omega_1\} = \{\varphi_{\alpha_0}, \varphi_{\alpha_1}, \ldots\}$. Then

$$\left(\forall x_0 \bigvee_{\alpha < \omega_1} \varphi_\alpha(x_0)\right)^X = \forall x_0 \bigvee_n \varphi_{\alpha_n}^X(x_0)$$

Example 8.4   Suppose $X, \mathcal{M}, \theta_\delta \in V$, V transitive, and δ is the order type of $X \cap On$. Then for all $\alpha \geq \delta$ we have $\mathcal{M} \models \forall x_0(\theta_\alpha^X \leftrightarrow \theta_\delta)$ (Exercise 8.4).

Lemma 8.5   *If $\varphi \in L_{\omega_1\omega}$, then player* **II** *has a winning strategy in the game* $G_{cub}(\{X \in \mathcal{P}_\omega(V) : \varphi^X = \varphi\})$. *That is, almost all approximations of $\varphi \in L_{\omega_1\omega}$ are equal to φ.*

Proof   We use induction on φ. If φ is atomic, the claim is trivial since $\varphi^X = \varphi$ holds for all X. Also negation and the cases of $\forall x_n \varphi$ and $\exists x_n \varphi$ are immediate. Let us then assume $\varphi = \bigwedge_{n \in \mathbb{N}} \varphi_n$ and the claim holds for each φ_n, that is, player **II** has a winning strategy in $G_{cub}(\{X \in \mathcal{P}_\omega(V) : \varphi_n^X = \varphi_n\})$ for each n. By Lemma 6.14 player **II** has a winning strategy in the Cub Game for the set

$$\bigcap_{n \in \mathbb{N}} \{X : \varphi_n^X = \varphi_n\} \cap \{X : \varphi_n \in X \text{ for all } n \in \mathbb{N}\}.$$

$\square$

Definition 8.6   Suppose L is a vocabulary and $\mathcal{M}$ an L-structure. Suppose φ is a first-order formula in NNF and s an assignment for the set M the domain of which includes the free variables of φ. We define the set $\mathcal{D}_{\varphi,s}$ of countable subsets of M as follows: If φ is basic, $\mathcal{D}_{\varphi,s}$ contains as an element any countable $X \subseteq V$ such that $X \cap M$ is the domain of a countable submodel $\mathcal{A}$ of $\mathcal{M}$ such that $\text{rng}(s) \subseteq A$ and:

- If φ is $\approx tt'$, then $t^\mathcal{A}(s) = t'^\mathcal{A}(t)$.
- If φ is $\neg\approx tt'$, then $t^\mathcal{A}(s) \neq t'^\mathcal{A}(t)$.
- If φ is $Rt_1 \ldots t_n$, then $(t_1^\mathcal{A}(s), \ldots, t_n{}^\mathcal{A}(t)) \in R^\mathcal{A}$.
- If φ is $\neg Rt_1 \ldots t_n$, then $(t_1^\mathcal{A}(s), \ldots, t_n{}^\mathcal{A}(t)) \notin R^\mathcal{A}$.

For non-basic φ we define

- $\mathcal{D}_{\bigwedge \Phi,s} = \triangle_{\varphi \in \Phi} \mathcal{D}_{\varphi,s}$.
- $\mathcal{D}_{\bigvee \Phi,s} = \triangledown_{\varphi \in \Phi} \mathcal{D}_{\varphi,s}$
- $\mathcal{D}_{\forall x \varphi,s} = \triangle_{a \in M} \mathcal{D}_{\varphi,s[a/x]}$.
- $\mathcal{D}_{\exists x \varphi,s} = \triangledown_{a \in M} \mathcal{D}_{\varphi,s(a/x)}$.

If φ is a sentence, we denote $\mathcal{D}_{\varphi,s}$ by $\mathcal{D}_\varphi$. If φ is not in NNF, we define $\mathcal{D}_{\varphi,s}$ and $\mathcal{D}_\varphi$ by first translating φ into a logically equivalent NNF formula.

Intuitively, $\mathcal{D}_\varphi$ is the collection of countable sets X, which *simultaneously* give an $L_{\omega_1\omega}$-approximation φ^X of φ and a countable approximation $\mathcal{M}^X$ of $\mathcal{M}$ such that $\mathcal{M}^X \models \varphi^X$.

Proposition 8.7 *Suppose $\mathcal{A}$ is an L-structure and $X \in \mathcal{D}_{\varphi,s}$. Then $[X \cap A]_\mathcal{A} \models_t \varphi^X$.*

Proof This is trivial for basic φ. For the induction step for $\bigwedge \Phi$ suppose $X \in \mathcal{D}_{\bigwedge \Phi,s}$. Suppose $\varphi \in X \cap \Phi$. Then $X \in \mathcal{D}_{\varphi,s}$. By the induction hypothesis $[X \cap A]_\mathcal{A} \models_t \varphi^X$. Thus $[X]_\mathcal{A} \models_t (\bigwedge \Phi)^X$. The other cases are as in the proof of Proposition 6.21. $\square$

Proposition 8.8 *Suppose L is a countable vocabulary and $\mathcal{M}$ an L-structure such that $\mathcal{M} \models \varphi$. Then player **II** has a winning strategy in $G_{cub}(\mathcal{D}_\varphi)$.*

Proof We use induction on φ to prove that if $\mathcal{M} \models_s \varphi$, then **II** has a winning strategy in $G_{cub}(\mathcal{D}_{\varphi,s})$. Most steps are as in the proof of Proposition 6.22. Let us look at the induction step for $\bigwedge \Phi$. We assume $\mathcal{M} \models_s \bigwedge \varphi$. It suffices to prove that **II** has a winning strategy in $G_{cub}(\mathcal{D}_{\varphi,s})$ for each $\varphi \in \Phi$. But this follows from the induction hypothesis. $\square$

Theorem 8.9 (Löwenheim–Skolem Theorem) *Suppose L is a countable vocabulary, $\mathcal{M}$ an arbitrary L-structure, and φ an $L_{\infty\omega}$-sentence of vocabulary L, and V a transitive set containing $\mathcal{M}$ and φ such that $M \cap TC(\varphi) = \emptyset$. Suppose $\mathcal{M} \models \varphi$. Let*

$$\mathcal{C} = \{X \in \mathcal{P}_\omega(V) : [X \cap M]_\mathcal{M} \models \varphi^X\}.$$

*Then player **II** has a winning strategy in the game $G_{cub}(\mathcal{C})$.*

Proof The claim follows from Propositions 8.7 and 8.8. $\square$

Theorem 8.10 *1. $\mathcal{M} \equiv_{\infty\omega} \mathcal{N}$ if and only if $\mathcal{M}^X \cong \mathcal{N}^X$ for almost all X.*
2. $\mathcal{M} \not\equiv_{\infty\omega} \mathcal{N}$ if and only if $\mathcal{M}^X \not\cong \mathcal{N}^X$ for almost all X.

8.3 Model Theory of $L_{\omega_1\omega}$

The Model Existence Game $\mathrm{MEG}(T, L)$ of first-order logic (Definition 6.35) can be easily modified to $L_{\omega_1\omega}$.

x_n	y_n	Explanation
φ		**I** enquires about φ.
	φ	**II** confirms.
$\approx tt$		**I** enquires about an equation.
	$\approx tt$	**II** confirms.
$\varphi(t)$		**I** chooses played $\varphi(c)$ and $\approx ct$ with φ basic and enquires about substituting t for c in φ.
	$\varphi(t)$	**II** confirms.
φ_i		**I** tests a played $\bigwedge_{i \in I} \varphi_i$ by choosing $i \in I$.
	φ_i	**II** confirms.
$\bigvee_{i \in I} \varphi_i$		**I** enquires about a played disjunction.
	φ_i	**II** makes a choice of $i \in I$.
$\varphi(c)$		**I** tests a played $\forall x \varphi(x)$ by choosing $c \in C$.
	$\varphi(c)$	**II** confirms.
$\exists x \varphi(x)$		**I** enquires about a played existential statement.
	$\varphi(c)$	**II** makes a choice of $c \in C$.
t		**I** enquires about a constant $L \cup C$-term t.
	$\approx ct$	**II** makes a choice of $c \in C$.

Figure 8.1 The game $\mathrm{MEG}(T, L)$.

Definition 8.11 The Model Existence Game $\mathrm{MEG}(\varphi, L)$ for a countable vocabulary L and a sentence φ of $L_{\omega_1 \omega}$ is the game $G_\omega(W)$ where W consists of sequences $(x_0, y_0, x_1, y_1, \ldots)$ where player **II** has followed the rules of Figure 8.1 and for no atomic $L \cup C$-sentence ψ both ψ and $\neg\psi$ are in $\{y_0, y_1, \ldots\}$.

We now extend the first leg of the Strategic Balance of Logic, the equivalence between the Semantic Game and the Model Existence Game, from first-order logic to infinitary logic:

Theorem 8.12 (Model Existence Theorem for $L_{\omega_1\omega}$) *Suppose L is a countable vocabulary and φ is an L-sentence of $L_{\omega_1\omega}$. The following are equivalent:*

(1) There is an L-structure $\mathcal{M}$ such that $\mathcal{M} \models \varphi$.
*(2) Player **II** has a winning strategy in $\mathrm{MEG}(\varphi, L)$.*

Proof The implication $(1) \rightarrow (2)$ is clear as **II** can keep playing sentences that are true in $\mathcal{M}$. For the other implication we proceed as in the proof of Theorem 6.35. Let $C = \{c_n : n \in \mathbb{N}\}$ and $Trm = \{t_n : n \in \mathbb{N}\}$. Let $(x_0, y_0, x_1, y_1, \dots)$ be a play in which player **II** has used her winning strategy and player **I** has maintained the following conditions:

1. If $n = 0$, then $x_n = \varphi$.
2. If $n = 2 \cdot 3^i$, then x_n is $\approx c_i c_i$.
3. If $n = 4 \cdot 3^i \cdot 5^j \cdot 7^k \cdot 11^l$, y_i is $\approx c_j t_k$, and y_l is $\varphi(c_j)$, then x_n is $\varphi(c_i)$.
4. If $n = 8 \cdot 3^i \cdot 5^j$ and y_i is $\bigwedge_{m \in \mathbb{N}} \varphi_m$, then x_n is φ_j.
5. If $n = 16 \cdot 3^i$ and y_i is $\bigvee_{m \in \mathbb{N}} \varphi_m$, then x_n is $\bigvee_{m \in \mathbb{N}} \varphi_m$.
6. If $n = 32 \cdot 3^i \cdot 5^j$, y_i is $\forall x \varphi(x)$, then x_n is $\varphi(c_j)$.
7. etc.

The rest of the proof is exactly as in the proof of Theorem 6.35. □

Our success in the above proof is based on the fact that even if we deal with infinitary formulas we can still manage to let player **I** list all possible formulas that are relevant for the consistency of the starting formula. If even one uncountable conjunction popped up, we would be in trouble.

It suffices to consider in $\mathrm{MEG}(\varphi, L)$ such constant terms t that are either constants or contain no other constants than those of C. Moreover, we may assume that if player **I** enquires about $\approx tt$, then $t = c_n$ for some $n \in \mathbb{N}$.

Corollary *Let L be a countable vocabulary. Suppose φ and ψ are sentences of $L_{\omega_1\omega}$. The following are equivalent:*

(1) $\varphi \models \psi$.
*(2) Player **I** has a winning strategy in $\mathrm{MEG}(\varphi \wedge \neg\psi, L)$.*

The proof of the Compactness Theorem does not go through, and should not, because there are obvious counter-examples to compactness in $L_{\omega_1\omega}$. In many proofs where one would like to use the Compactness Theorem one can instead use the Model Existence Theorem. The non-definability of well-order in $L_{\infty\omega}$ was proved already in Theorem 7.26 but we will now prove a stronger version for $L_{\omega_1\omega}$:

Theorem 8.13 (Undefinability of Well-Order) *Suppose L is a countable vocabulary containing a unary predicate symbol U and a binary predicate symbol $<$, and $\varphi \in L_{\omega_1 \omega}$. Suppose that for all $\alpha < \omega_1$ there is a model $\mathcal{M}$ of φ such that $(\alpha, <) \subseteq (U^{\mathcal{M}}, <^{\mathcal{M}})$. Then φ has a model $\mathcal{N}$ such that $(\mathbb{Q}, <) \subseteq (U^{\mathcal{N}}, <^{\mathcal{N}})$.*

Proof Let $D = \{d_r : r \in \mathbb{Q}\}$ be a set of new constant symbols. Let us call them d-constants. Let $\theta = \bigwedge_{r < s} (d_r < d_s)$. We show that player **II** has a winning strategy in

$$\text{MEG}(\varphi \wedge \theta, L \cup D).$$

This clearly suffices. The strategy of **II** is the following: Suppose she has played $\{y_0, \ldots, y_{n-1}\}$ so far and $y_i = \theta$ or

$$y_i = \varphi_i(c_0, \ldots, c_m, d_{r_1}, \ldots, d_{r_l}),$$

where $d_{r_1}, \ldots, d_{r_l}$ are the d-constants appearing in $\{y_0, \ldots, y_{n-1}\}$ except in θ. She maintains the following condition:

($\star$) For all $\alpha < \omega_1$ there is a model $\mathcal{M}$ of φ and $b_1, \ldots, b_l \in U^{\mathcal{M}} \subseteq \omega_1$ such that

$$\mathcal{M} \models \exists x_0 \ldots \exists x_m \bigwedge_{i < n} \varphi_i(x_0, \ldots, x_m, b_1, \ldots, b_l)$$

and

$$\alpha \leq b_1, b_1 + \alpha \leq b_2, \ldots, b_{l-1} + \alpha \leq b_l.$$

We show that player **II** can indeed maintain this condition.

For most moves of player **I** the move of **II** is predetermined and we just have to check that ($\star$) remains valid. For a start, if **I** plays φ, condition ($\star$) holds by assumption. If **I** enquires about substitution or plays a conjunct of a played conjunction, no new constants are introduced, so ($\star$) remains true. Also, if **I** tests a played $\forall x \varphi(x)$ or enquires about a played $\exists x \varphi(x)$, no new constants of D are introduced, so ($\star$) remains true. We may assume that **I** enquires about $\approx tt$ only if $t = c_n$ and so ($\star$) holds by the induction hypothesis. Let us then

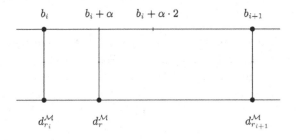

Figure 8.2

assume $(\star)$ holds and **I** enquires about a played disjunction $\bigvee_{i \in I} \psi_i$. For each $\alpha < \omega_1$ we have a model $\mathcal{M}_\alpha$ as in $(\star)$ and some $i_\alpha \in I$ such that $\mathcal{M}_\alpha \models \psi_{i_\alpha}$. Since **I** is countable, there is a fixed $i \in I$ such that for uncountably many $\alpha < \omega_1$: $\mathcal{M}_\alpha \models \psi_i$. If **II** plays this ψ_i, condition $(\star)$ is still true.

The remaining case is that **I** enquires about a constant term t. We may assume $t = d_r$ as otherwise there is nothing to prove. The constants of D occurring so far in the game are $d_{r_1}, \ldots, d_{r_l}$. Let us assume $r_i < r < r_{i+1}$. To prove $(\star)$, assume $\alpha < \omega_1$ and let $\beta = \alpha \cdot 2$. By the induction hypothesis there is $\mathcal{M}$ as in $(\star)$ such that $b_i + \beta \le b_{i+1}$. Let d_r be interpreted in $\mathcal{M}$ as $b_i + \alpha$. Now $\mathcal{M}$ satisfies the condition $(\star)$ (see Figure 8.3). $\qquad\square$

The following corollary is due to Lopez-Escobar (1966b).

Corollary *If φ is a sentence of $L_{\omega_1\omega}$ in a vocabulary which contains the unary predicate U and the binary predicate $<$, and $(U^\mathcal{M}, <^\mathcal{M})$ is well-ordered in every model of φ, then there is $\alpha < \omega_1$ such that the order type of the structure $(U^\mathcal{M}, <^\mathcal{M})$ is $< \alpha$ for every model $\mathcal{M}$ of φ.*

Corollary *The class of well-orderings is not a PC-class of $L_{\omega_1\omega}$.*

The undefinability of well-ordering as a PC-class of $L_{\infty\omega}$ will be established later. We now prove the Craig Interpolation Theorem for $L_{\omega_1\omega}$. There are several different proofs of this theorem, some of which employ the above corollary directly. Our proof is like the original proof by Lopez-Escobar, except that we operate with Model Existence Games instead of Gentzen systems.

Theorem 8.14 (Separation Theorem) *Suppose L_1 and L_2 are vocabularies. Suppose φ is an L_1-sentence of $L_{\omega_1\omega}$ and ψ is an L_2-sentence of $L_{\omega_1\omega}$ such* *that* $\models \varphi \to \psi$. *Then there is an* $L_1 \cap L_2$-*sentence* θ *of* $L_{\omega_1\omega}$ *such that* $\models \varphi \to$ θ *and* $\models \theta \to \psi$.

Proof This is similar to the proof of Theorem 6.40. We assume, w.l.o.g., that L_1 and L_2 are relational. Let $L = L_1 \cap L_2$. We describe, assuming that no such θ exists, a winning strategy of **II** in MEG($\varphi \wedge \neg\psi, L_1 \cup L_2$). We now follow closely the proof of Theorem 6.40, where the strategy of **II** was to divide the set Ψ of her moves into two parts S_1^n and S_2^n such that S_1^n consists of the $L_1 \cup C$-sentences of Ψ and S_2^n consists of the $L_2 \cup C$-sentences of Ψ. In addition it is assumed that

(*) There is no $L \cup C$-sentence θ that separates S_1^n and S_2^n.

There are two new cases over and above those of Theorem 6.40:

Case 5′. Player **I** plays φ_i where for example $\bigwedge_{i \in I} \varphi_i \in S_1^n$. Let $S_1^{n+1} = S_1^n \cup \{\varphi_i\}$ and $S_2^{n+1} = S_2^n$. If θ separates S_1^{n+1} and S_2^{n+1}, then clearly θ also separates S_1^n and S_2^n.

Case 6′. Player **I** plays $\bigvee_{i \in I} \varphi_i$, where for example $\bigvee_{i \in I} \varphi_i \in S_1^n$. We claim that for some $i \in I$ the sets $S_1^n \cup \{\varphi_i\}$ and S_2^n satisfy (*). Otherwise there is for each $i \in I$ some θ_i that separates $S_1^n \cup \{\varphi_i\}$ and S_2^n. Let $\theta = \bigvee_{i \in I} \theta_i$. Then θ separates S_1^n and S_2^n contrary to assumption.

□

8.4 Large Models

Suppose Γ is an L-fragment of $L_{\kappa^+\omega}$ of size κ and $\mathcal{M}$ is an L-structure of size $> \kappa$. If we define on M

$$a \sim b \quad \Longleftrightarrow \quad \text{for every } \varphi(x) \in \Gamma : \mathcal{M} \models \varphi(a) \Longleftrightarrow \mathcal{M} \models \varphi(b)$$

then by the Pigeonhole Principle there is a subset I of M of size $> \kappa$ such that for $a, b \in I$ and $\varphi(x) \in \Gamma$:

$$\mathcal{M} \models \varphi(a) \quad \Longleftrightarrow \quad \mathcal{M} \models \varphi(b).$$

We say that the set I is *indiscernible* in $\mathcal{M}$ with respect to Γ. If we want indiscernibility relative to formulas with more than one free variable, we have to use *Ramsey theory*.

Definition 8.15 Suppose L is a vocabulary, $\mathcal{M}$ is an L-structure, and Γ is an L-fragment. A linear order $(I, <)$, where $I \subseteq M$, is Γ-*indiscernible* in $\mathcal{M}$ if for all $a_1 < \ldots < a_n, b_1 < \ldots < b_n$ in I and any $\varphi(x_1, \ldots, x_n)$ in Γ :

$$\mathcal{M} \models \varphi(a_1, \ldots, a_n) \quad \Longleftrightarrow \quad \mathcal{M} \models \varphi(b_1, \ldots, b_n).$$

Example 8.16 If $L = \emptyset$, any linear order is Γ-indiscernible in any L-structure. If $L = \{P\}$, P a unary predicate, and $\mathcal{M}$ is an L-structure, then any linear order $(I, <)$, where $I \subseteq P^{\mathcal{M}}$ or $I \subseteq M \setminus P^{\mathcal{M}}$ is Γ-indiscernible in $\mathcal{M}$ for any Γ. If $\mathcal{M}$ is a dense linear order without endpoints, then any suborder of $\mathcal{M}$ is Γ-indiscernible for any Γ.

Example 8.17 Suppose $\mathcal{M} = (M, \cdot, +, 0, \mathbb{R}, \cdot_{\mathbb{R}}, +_{\mathbb{R}}, 0_{\mathbb{R}}, 1_{\mathbb{R}})$ is a vector space over $\mathbb{R}$, where $(\mathbb{R}, \cdot_{\mathbb{R}}, +_{\mathbb{R}}, 0_{\mathbb{R}}, 1_{\mathbb{R}})$ is the field of reals, $+$ is the vector sum in $M \setminus \mathbb{R}$, and $\cdot$ is the scalar product $\mathbb{R} \times (M \setminus \mathbb{R}) \to (M \setminus \mathbb{R})$. If X is a basis of $\mathcal{M}$, then X is Γ-indiscernible in $\mathcal{M}$ for any Γ.

Proposition 8.18 *Suppose L is a vocabulary, Γ an L-fragment, and $\mathcal{M}$ is an L-structure such that for each $\varphi \in \Gamma$ there is $f \in L$ such that $f^{\mathcal{M}}$ is a Skolem function for φ in $\mathcal{M}$. If $X \subseteq M$, then*

$$[X]_{\mathcal{M}} \prec_{\Gamma} \mathcal{M}.$$

Proof We use induction on $\varphi \in \Gamma$ to prove for all $a_1, \ldots, a_n \in [X]_{\mathcal{M}}$

$$[X]_{\mathcal{M}} \models \varphi(a_1, \ldots, a_n) \quad \Longleftrightarrow \quad \mathcal{M} \models \varphi(a_1, \ldots, a_n).$$

The only interesting case is the following: Suppose

$$\mathcal{M} \models \exists x_{n+1} \varphi(a_1, \ldots, a_n, x_{n+1}).$$

Let $f \in L$ such that $f^{\mathcal{M}}$ is a Skolem function for φ. Thus if

$$b = f^{\mathcal{M}}(a_1, \ldots, a_n)$$

then

$$\mathcal{M} \models \varphi(a_1, \ldots, a_n, b).$$

Since $[X]_{\mathcal{M}}$ is closed under $f^{\mathcal{M}}$, $b \in [X]_{\mathcal{M}}$ and we can use the induction hypothesis to conclude

$$[X]_{\mathcal{M}} \models \varphi(a_1, \ldots, a_n, b)$$

from which

$$[X]_{\mathcal{M}} \models \exists x_{n+1} \varphi(a_1, \ldots, a_n, x_{n+1})$$

follows. □

Suppose $(I, <)$ is Γ-indiscernible in $\mathcal{M}$. Thus for any $\varphi(x_1, \ldots, x_n)$ in Γ the truth-value of $\varphi(a_1, \ldots, a_n)$ in $\mathcal{M}$ is independent of $a_1, \ldots, a_n \in I$ as long as $a_1 < \ldots < a_n$. Owing to this situation that $\mathcal{M}$ does not recognize any difference between the elements of I, we can "smuggle" more elements into $\mathcal{M}$ without changing properties of $\mathcal{M}$ expressible in Γ.

Theorem 8.19 *Suppose L is a vocabulary, Γ an L-fragment, and $\mathcal{M}$ is an L-structure such that for each $\varphi \in \Gamma$ there is $f \in L$ for which $f^{\mathcal{M}}$ is a Skolem function for φ in $\mathcal{M}$. Assume furthermore that $(I, <_I)$ is an infinite linear order which is Γ-indiscernible in $\mathcal{M}$. Then for any linear order $(J, <_J)$ there is an L-structure $\mathcal{N}$ such that for any $\varphi(x_1, \ldots, x_n) \in \Gamma, a_1 <_I \ldots <_I a_n$ in I and $b_1 <_J \ldots <_J b_n$ in J*

$$\mathcal{M} \models \varphi(a_1, \ldots, a_n) \quad \Longleftrightarrow \quad \mathcal{N} \models \varphi(b_1, \ldots, b_n).$$

In particular, $\mathcal{M} \equiv_\Gamma \mathcal{N}$.

Proof Let N' be the set of tuples

$$(t, b_0, \ldots, b_{n-1})$$

where t has at most $x_0, \ldots, x_{n-1}$ as its variables and $b_0 <_J \ldots <_J b_{n-1}$. If we have two elements

$$(t, b_0, \ldots, b_{n-1}), (t', b'_0, \ldots, b'_{n'-1})$$

of N', we make the following construction: Let $c_0 <_J \ldots <_J c_{m-1}$ be any sequence (since $(J, <)$ is infinite, such exist) containing the sequences $b_0, \ldots, b_{n-1}, b'_0, \ldots, b'_{n'-1}$, i.e.

$$b_i = c_{r_i}, b'_j = c_{s_j}$$

for $i < n$ and $j < n'$. Let t_0 be obtained from t by replacing x_i everywhere by x_{r_i}, and t'_0 from t' by replacing everywhere x_j by x_{s_j}. Now we define

$$(t, b_0, \ldots, b_{n-1}) \sim (t', b'_0, \ldots, b'_{n'-1})$$

if there are $a_0 <_I \ldots <_I a_{m-1}$ in I such that

$$t_0(a_0, \ldots, a_{m-1})^{\mathcal{M}} = t'_0(a_0, \ldots, a_{m-1})^{\mathcal{M}}.$$

Note that this definition of $\sim$ is independent of the sequences $c_0 <_J \ldots <_J c_{m-1}$ and $a_0 <_I \ldots <_I a_{m-1}$.

Claim $\sim$ is an equivalence relation on N'.

Clearly, $\sim$ is reflexive and symmetric. To prove transitivity, assume

(*) $(t, b_0, \ldots, b_{n-1}) \sim (t', b'_0, \ldots, b'_{n'-1}) \sim (t'', b''_0, \ldots, b''_{n''-1}).$

Since $(J, <)$ is infinite, there is $c_0 <_J \ldots <_J c_{m-1}$ in J such that

$$b_i = c_{r_i}, b'_j = c_{s_j}, b''_k = c_{u_k}$$

for $i < n, j < n'$ and $k < n''$. Let $a_0 <_I \ldots <_I a_{m-1}$ in I. Let t_0 be gotten from t by replacing x_i by x_{r_i}, t_0' from t' by replacing x_j by x_{s_j}, and t_0'' from t'' by replacing x_k by x_{u_k}. Then (*) implies

$$t_0(a_0, \ldots, a_{m-1})^{\mathcal{M}} = t_0'(a_0, \ldots, a_{m-1})^{\mathcal{M}} = t_0''(a_0, \ldots, a_{m-1})^{\mathcal{M}}.$$

For $z \in N'$ let $[z]$ be the equivalence class of z under $\sim$. Let N be the set of all $[z]$, $z \in N'$. We make N into an L-structure by defining for $b_0 <_J \ldots <_J b_{m-1}$ in J:

$$([[(t_1, b_0, \ldots, b_{m-1})]], \ldots [(t_n, b_0, \ldots, b_{m-1})]]) \in R^{\mathcal{N}}$$

iff for some (all) $a_0 <_I \ldots <_I a_{m-1}$ in I :

$$(t_1(b_0, \ldots, b_{m-1})^{\mathcal{M}}, \ldots, t_n(b_0, \ldots, b_{m-1})^{\mathcal{M}}) \in R^{\mathcal{M}}.$$

For $f \in L$ we define

$$f^{\mathcal{N}}([[(t_1, b_0, \ldots, b_{m-1})]], \ldots [(t_n, b_0, \ldots, b_{m-1})]]) =$$

$$[(ft_1 \ldots t_n, b_0, \ldots, b_{m-1})].$$

Finally, for $c \in L$, we let

$$c^{\mathcal{N}} = [(c, \emptyset)].$$

We can identify the element b of J with $[(v_0, b)]$.

Claim The following are equivalent for $\varphi \in \Gamma$:

(1) $\mathcal{M} \models \varphi(a_0, \ldots, a_{n-1})$ for all $a_0 <_I \ldots <_I a_{n-1}$ in I.
(2) $\mathcal{N} \models \varphi(b_0, \ldots, b_{n-1})$ for all $b_0 <_J \ldots <_J b_{n-1}$ in J.
(3) $\mathcal{N} \models \varphi(b_0, \ldots, b_{n-1})$ for some $b_0 <_J \ldots <_J b_{n-1}$ in J.

To prove this claim we use induction on φ. For atomic and negated atomic φ this follows from the definition of $\mathcal{N}$. Let us prove the case of the existential quantifier. Suppose $\mathcal{M} \models \exists x_n \varphi(a_0, \ldots, a_{n-1}, x_n)$ for some (equivalently, for all) $a_0 <_I \ldots <_I a_{n-1}$ in I. There is $f \in L$ such that $f^{\mathcal{M}}$ is a Skolem function for φ in $\mathcal{M}$. Thus

$$\mathcal{M} \models \varphi(a_0, \ldots, a_{n-1}, f^{\mathcal{M}}(a_0, \ldots, a_{n-1})).$$

Let φ' be obtained from φ by replacing everywhere x_n by $fx_0 \ldots x_{n-1}$. Then

$$\mathcal{M} \models \varphi'(a_0, \ldots, a_{n-1}).$$

Let $b_0 <_J \ldots <_J b_{n-1}$ in Ψ. By the induction hypothesis,

$$\mathcal{N} \models \varphi'(b_0, \ldots, b_{n-1})$$

whence

$$(**) \quad \mathcal{N} \models \exists x_n \varphi(b_0, \ldots, b_{n-1}, x_n).$$

Conversely, suppose (**) holds for some (equivalently, for all) $b_0 <_J \ldots <_J b_{n-1}$. Let $z \in N$ so that $\mathcal{N} \models \varphi(b_0, \ldots, b_{n-1}, z)$. Let $z = [(t, b'_0, \ldots, b'_{n'-1})]$. Let $c_0 <_J \ldots <_J c_{m-1}$ be such that

$$b_i = c_{r_i}, \; b'_j = c_{s_j}$$

for $i < n$ and $j < n'$. Let t_0 be obtained from t by replacing everywhere x_j by x_{s_j}. Let φ_0 be obtained from φ by replacing everywhere x_i by x_{r_i} and x_n by $t_0(x_0, \ldots, x_{m-1})$. Then

$$\mathcal{N} \models \varphi_0(c_0, \ldots, c_{m-1}).$$

By the induction hypothesis

$$\mathcal{M} \models \varphi_0(a_0, \ldots, a_{m-1})$$

for all $a_0 <_I \ldots <_I a_{m-1}$. Fix some $a_0 <_I \ldots <_I a_{n-1}$. Then there is $a_n \in M$ such that

$$\mathcal{M} \models \varphi(a_0, \ldots, a_{n-1}, a_n).$$

Thus

$$\mathcal{M} \models \exists x_n \varphi(a_0, \ldots, a_{n-1}).$$

$\square$

Corollary *Suppose L is a vocabulary, Γ an L-fragment, and $\mathcal{M}$ is an L-structure such that for each $\varphi \in \Gamma$ there is $f \in L$ for which $f^{\mathcal{M}}$ is a Skolem function for φ in $\mathcal{M}$, and $(I, <_I)$ is an infinite linear order which is Γ-indiscernible in $\mathcal{M}$. Then φ has arbitrarily large models.*

Proof Let $(J, <_J)$ be an arbitrary linear order. By the previous theorem there is a model $\mathcal{N}$ containing J such that $\mathcal{M} \equiv_\Gamma \mathcal{N}$. Then $|N| \geq |J|$. $\square$

Definition 8.20 Suppose κ, λ, and μ are cardinals and $n \in \mathbb{N}$. We write

$$\kappa \to (\lambda)^n_\mu$$

if for every function ("coloring")

$$f : [\kappa]^n \to \mu$$

there is $I \subseteq \kappa$, $|I| = \lambda$, such that for all $a_1 < \ldots < a_n, b_1 < \ldots < b_n$ in I

$$f(\{a_1, \ldots, a_n\}) = f(\{b_1, \ldots, b_n\}).$$

Example 8.21 Ramsey's Theorem says $\omega \to (\omega)^n_m$, The Erdös–Rado Theorem says $(\beth_n(\kappa))^+ \to (\kappa^+)^{n+1}_\kappa$. We will not prove these results of set theory here, but refer to such set theory texts as (Jech, 1997, Theorem 9.6).

For finite Γ we could use the Ramsey Theorem $\omega \to (\omega)^n_m$ to get models with infinite sets of indiscernibles. In infinitary logic we cannot expect Γ to be finite, and we have to use the Erdös–Rado Theorem instead.

Theorem 8.22 *Suppose $\varphi \in \Gamma \subseteq L_{\omega_1\omega}$, $|\Gamma| \le \aleph_0$, and φ has for each $\alpha < \omega_1$ a model of size $\beth_\alpha$. Then φ has a model with an infinite Γ-indiscernible linear order, and hence arbitrary large models.*

Proof Let L be the countable vocabulary of φ. Let $D = \{d_n : n < \omega\}$ be a new set of constant symbols. It suffices to prove that **II** has a winning strategy in

$$\text{MEG}(\varphi \wedge \theta, L \cup D),$$

where θ is the conjunction of

$$\varphi(d_{i_1}, \ldots, d_{i_n}) \leftrightarrow \varphi(d_{j_1}, \ldots, d_{j_n})$$
$$d_i < d_j$$

for $i_1 < \ldots < i_n < \omega, j_1 < \ldots < j_n < \omega, i < j < \omega$ and $\varphi(x_1, \ldots, x_n) \in \Gamma$. The winning strategy of **II** is the following: Suppose $\{y_0, \ldots, y_{n-1}\}$ is her play so far and $y_i = \theta$ or

$$y_i = \varphi_i(c_0, \ldots, c_m, d_{r_1}, \ldots, d_{r_l}).$$

She maintains the following condition: For all $\alpha < \omega_1$ there is a model $\mathcal{M}$ of φ and a linear order $(I, <_I)$ such that $I \subset M, |I| = \beth_\alpha$ and for $a_0 <_I \ldots <_I a_l$ in I :

$$\mathcal{M} \models \exists x_0 \ldots \exists x_n \bigwedge_{i<n} \varphi_i(x_0, \ldots, x_m, a_1, \ldots, a_l).$$

Note that if $n = 0$, i.e. the game has not started yet, the condition holds as for all $\alpha < \omega_1$ there is a model $\mathcal{M}$ of φ with $|M| = \beth_\alpha$.

Let us assume **I** enquires about a played disjunction $\bigvee_{n<\omega} \psi_n$. For each $\alpha < \omega_1$ we have $\mathcal{M}_\alpha \models \varphi$ and a linear order $(I_\alpha, <_\alpha)$ such that $|I_\alpha| = \beth_\alpha$ and for $a_1 <_\alpha \ldots <_\alpha a_l$ in I_α

$$\mathcal{M}_\alpha \models \exists x_0 \ldots \exists x_m [\bigwedge_{i<n} \varphi_i(x_0, \ldots, x_m, a_1, \ldots, a_l) \wedge$$
$$\bigvee_{n<\omega} \psi_n(x_0, \ldots, x_m, a_1, \ldots, a_l)].$$

Fix $\alpha < \omega_1$. If $a_1 <_{\alpha+l} \ldots <_{\alpha+l} a_l \in I_{\alpha+l}$, let $\chi(a_1, \ldots, a_l) = k < \omega$ be such that

$$\mathcal{M}_{\alpha+l} \models \exists x_0 \ldots \exists x_m [\ \bigwedge_{i<n} \varphi_i(x_0, \ldots, x_m, a_1, \ldots, a_l) \wedge$$
$$\psi_k(x_0, \ldots, x_m, a_1, \ldots, a_l)].$$

By the Erdös–Rado Theorem there is a set $J_\alpha \subseteq I_{\alpha+l}$ such that $\chi(a_1, \ldots, a_l)$ is a fixed n_α for all $a_1 <_{\alpha+l} \ldots <_{\alpha+l} a_l$ in J_α and $|I| = \beth_\alpha$. For arbitrarily large $\alpha < \omega_1$ we have n_α fixed. Let that fixed value be n^*. Then for all $\alpha < \omega_1$ there is a model $\mathcal{N}_\alpha (= \mathcal{M}_{\alpha+l})$ and a linear order $(J_\alpha, <_\alpha)$ such that $|J_\alpha| = \beth_\alpha$ and for all $a_1 <_\alpha \ldots <_\alpha a_l$ in J_α

$$\mathcal{N}_\alpha \models \exists x_0 \ldots \exists x_m [\ \bigwedge_{i<n} \varphi_i(x_0, \ldots, x_m, a_1, \ldots, a_l) \wedge$$
$$\psi_{n^*}(x_0, \ldots, x_m, a_1, \ldots, a_l)].$$

Now **II** plays the sentence

$$\psi_{n^*}(c_0, \ldots, c_m, d_{r_1}, \ldots, d_{r_l}).$$

To prove that φ has arbitrarily large models we first expand the vocabulary L to L^* and the countable L-fragment Γ to a countable L^*-fragment Γ^* so that for every L^*-formula $\psi = \psi(x_1, \ldots, x_{n_\psi})$ in Γ^* there is a function symbol f_ψ in $L^* \setminus L$. We replace φ by

$$\varphi^* = \varphi \wedge \bigwedge_{f_\psi \in L^* \setminus L} \forall x_{1_\psi} \ldots \quad \forall x_{n_\psi - 1}(\exists x_n \psi(x_1, \ldots, x_{n_\psi}) \rightarrow$$
$$\psi(x_1, \ldots, x_{n_\psi - 1}, f_\psi(x_1, \ldots, x_{n_\psi}))).$$

Note that φ^* has still for each $\alpha < \omega_1$ a model of size $\beth_\alpha$. Then the model $\mathcal{N}$ constructed above has a Skolem function $f_\psi^{\mathcal{N}}$ for each $\psi \in \Gamma$. Thus we can use Theorem 8.22 to construct arbitrarily large models for φ^* and hence for φ. □

Proposition 8.23 *Suppose L is a vocabulary, Γ an L-fragment, $\mathcal{M}$ an L-structure which has Skolem functions for all $\varphi \in \Gamma$, and $(I, <)$ a linear order which is Γ-indiscernible in $\mathcal{M}$. Then every automorphism of $(I, <)$ can be extended to an automorphism of $[I]_{\mathcal{M}}$.*

Proof Suppose π is an automorphism of $(I, <)$. If $a \in [I]_{\mathcal{M}}$, then $a = t^{\mathcal{M}}(a_1, \ldots, a_n)$ for some L-term t and some $a_1 < \cdots < a_n$ in I.

Claim If $t^{\mathcal{M}}(a_1, \ldots, a_n) = u^{\mathcal{M}}(b_1, \ldots, b_m)$, where $a_1 < \cdots < a_n$ in I and $b_1 < \cdots < b_m$ in I, then $t^{\mathcal{M}}(\pi a_1, \ldots, \pi a_n) = u^{\mathcal{M}}(\pi b_1, \ldots, \pi b_m)$.

To prove the claim let $c_1 < \cdots < c_k$ in I be such that

$$a_i = c_{r_i}, b_j = c_{s_j}$$

for $1 \leq i \leq n, 1 \leq j \leq m$. Let t_0 be obtained from t by replacing x_i everywhere by x_{r_i}. Let u_0 be obtained from u by replacing x_j everywhere by x_{s_j}.

Thus

$$\mathcal{M} \models \approx t_0 u_0 (c_1, \ldots, c_k).$$

By indiscernibility,

$$\mathcal{M} \models \approx t_0 u_0 (\pi c_1, \ldots, \pi c_k).$$

Thus $t^{\mathcal{M}}(\pi a_1, \ldots, \pi a_n) = u^{\mathcal{M}}(\pi b_1, \ldots, \pi b_m)$. The claim is proved. Now we can define

$$\pi(t^{\mathcal{M}}(a_1, \ldots, a_n)) = t^{\mathcal{M}}(\pi a_1, \ldots, \pi a_n)$$

and thereby extend π to an automorphism of $[I]_{\mathcal{M}}$. $\square$

Corollary *Suppose $\varphi \in L_{\omega_1\omega}$ has for each $\alpha < \omega_1$ a model of size $\beth_\alpha$. Then φ has for each infinite cardinal κ a model of cardinality κ with 2^κ automorphisms.*

Proof Let $(I, <)$ be a linear order of size κ with 2^κ automorphisms (see Exercise 8.36). Let Γ be a countable fragment of $L_{\omega_1\omega}$ such that $\varphi \in \Gamma$. Let $\mathcal{M}$ be a model of φ with $(I, <)$ Γ-indiscernible in $\mathcal{M}$. Without loss of generality, $\mathcal{M}$ has Skolem functions for each $\psi \in \Gamma$. Now $[I]_{\mathcal{M}}$ is a model of φ of size κ with 2^κ automorphisms. $\square$

8.5 Model Theory of $L_{\kappa^+\omega}$

New phenomena arise when we pass from $L_{\omega_1\omega}$ to $L_{\kappa^+\omega}$ with $\kappa > \omega$. For an example consider the sentence ψ:

$$\bigwedge_{A \subseteq \mathbb{N}} \exists x_0 \bigwedge_{n \in A} c_n E x_0 \wedge \bigwedge_{n \notin A} \neg c_n E x_0.$$

$\mathcal{M}$ satisfies ψ iff every subset X of $\{c_n^{\mathcal{M}} : n \in \mathbb{N}\}$ is coded by some element $a \in M$:

$$X = \{c_n^{\mathcal{M}} : \mathcal{M} \models c_n E a\}.$$

Thus we can talk about all subsets of $\{c_n^{\mathcal{M}} : n \in \mathbb{N}\}$. This is new since a priori we cannot talk about subsets of the model in $L_{\infty\omega}$, only about elements. However, this trick of coding subsets by elements works only as far as we can explicitly refer to each element of a part of the model, e.g. by having enough constant symbols, or by being able to define a long well-ordering. One of the main results of this section shows that even in $L_{\kappa^+\omega}$ there is a bound to how long well-orderings we can define.

Definition 8.24   Let L be a vocabulary. An *L-fragment* of $L_{\infty\omega}$ is any set $\mathcal{T}$ of formulas of $L_{\infty\omega}$ in the vocabulary L such that

(1) $\mathcal{T}$ contains the atomic L-formulas.
(2) $\varphi \in \mathcal{T}$ if and only if $\neg\varphi \in \mathcal{T}$.
(3) $\wedge\Phi \in \mathcal{T}$ implies $\Phi \subseteq \mathcal{T}$.
(4) $\Phi \subseteq \mathcal{T}$ implies $\wedge\Phi \in \mathcal{T}$ for finite Φ.
(5) $\vee\Phi\mathcal{T}$ implies $\Phi \subseteq \mathcal{T}$.
(6) $\Phi \subseteq \mathcal{T}$ implies $\vee\Phi \in \mathcal{T}$ for finite Φ.
(7) $\forall x_n\varphi \in \mathcal{T}$ if and only if $\varphi \in \mathcal{T}$.
(8) $\exists x_n\varphi \in \mathcal{T}$ if and only if $\varphi \in \mathcal{T}$.

Note that a fragment is necessarily closed under subformulas. The main use of fragments is the following fact:

Lemma 8.25   *Suppose $\varphi \in L_{\kappa^+\omega}$ with a vocabulary $L, |L| \leq \kappa$. There is a fragment $\mathcal{T}$ such that $\varphi \in \mathcal{T}$ and $|\mathcal{T}| = \kappa$.*

Proof   Let $\mathcal{T}_0$ consist of atomic L-formulas and the formula φ. Let

$$
\begin{aligned}
\mathcal{T}_{n+1} = \mathcal{T}_n &\cup \{\psi : \neg\psi \in \mathcal{T}_n\} \\
&\cup \{\neg\psi : \psi \in \mathcal{T}_n\} \\
&\cup \{\psi : \psi \in \Phi, \wedge\Phi \in \mathcal{T}_n \text{ or } \vee\Phi \in \mathcal{T}_n\} \\
&\cup \{\psi : \forall x_n\psi \in \mathcal{T}_n \text{ or } \exists x_n\psi \in \mathcal{T}_n\} \\
&\cup \{\forall x_n\psi : \psi \in \mathcal{T}_n\} \\
&\cup \{\exists x_n\psi : \psi \in \mathcal{T}_n\} \\
&\cup \{\wedge\Phi : \Phi \subseteq \mathcal{T}_n \text{ finite}\} \\
&\cup \{\vee\Phi : \Phi \subseteq \mathcal{T}_n \text{ finite}\}.
\end{aligned}
$$

Finally, let $\mathcal{T} = \bigcup_n \mathcal{T}_n$. $\square$

We quite often use induction on formulas. In such situations it often suffices to use induction on formulas in a fragment. Since the fragment is smaller than the set of all formulas, preparations for a successful induction are easier to make. The following Löwenheim-Skolem Theorem is an example of this. First we review the nice method of Skolem functions:

Definition 8.26   Suppose L is a vocabulary, $\mathcal{M}$ is an L-structure, and $\mathcal{T}$ is an L-fragment of $L_{\infty\omega}$. A *Skolem function* for a formula $\varphi(x_0, \ldots, x_n)$ is any function $f_\varphi : M^n \to M$ such that for all $a_0, \ldots, a_{n-1} \in M$:

$$
\mathcal{M} \models \exists x_n\varphi(a_0, \ldots, a_{n-1}, x_n) \to \varphi(a_0, \ldots, a_{n-1}, f_\varphi(a_0, \ldots, a_{n-1})).
$$

By the Axiom of Choice it is clear that Skolem functions exist in any model for any formula.

Proposition 8.27 *Suppose T is an L-fragment, $\mathcal{M}$ an L-structure, and $\mathcal{F}$ a family of functions such that every formula in T has a Skolem function in $\mathcal{F}$. Suppose $M_0 \subseteq M$ is closed under all functions in $\mathcal{F}$. Then M_0 is the universe of a substructure $\mathcal{M}_0$ of $\mathcal{M}$ such that for all $\varphi \in T$ and s:*

$$(*) \qquad \mathcal{M}_0 \models_s \varphi \iff \mathcal{M} \models_s \varphi.$$

Proof Let us first observe that M_0 is closed under the functions of $\mathcal{M}$ and contains $c^{\mathcal{M}}$ for each $c \in M$. For an n-ary function symbol g we use the Skolem function of the formula

$$\exists x_n \approx x_n g x_0, \ldots, x_{n-1}$$

and for any constant symbol c in L we use the Skolem function of the formula $\exists x_n \approx x_n c$. Now the induction. For atomic φ claim (*) follows from the fact that $\mathcal{M}_0 \subseteq \mathcal{M}$. Claim (*) is trivially closed under negation, $\wedge$ and $\vee$, bearing in mind that if $\wedge \Phi$ or $\vee \Phi$ is in T, then $\Phi \subseteq T$. Finally, assume $\varphi(x_0, \ldots, x_{n-1})$ is of the form $\exists x_n \psi(x_0, \ldots, x_{n-1})$. If $\mathcal{M}_0 \models_s \varphi$, then by the induction hypothesis $\mathcal{M} \models_s \varphi$. On the other hand, assume $\mathcal{M} \models_s \varphi$. By assumption there is a Skolem function f_φ for φ in $\mathcal{F}$. Thus

$$\mathcal{M} \models_{s[a/n]} \psi, a = f_\varphi(s(0), \ldots, s(n-1)).$$

By assumption $a \in M_0$. Thus $\mathcal{M}_0 \models_{s[a/n]} \psi$ whence $\mathcal{M}_0 \models_s \varphi$. $\qquad\square$

Theorem 8.28 (Löwenheim–Skolem Theorem) *Suppose L is a vocabulary of cardinality $\leq \kappa$, φ is a sentence of $L_{\kappa^+\omega}$ in the vocabulary L, and $\mathcal{M}$ is an L-structure such that $\mathcal{M} \models \varphi$. Then there is $\mathcal{M}_0 \subseteq \mathcal{M}$ such that $\mathcal{M}_0 \models \varphi$ and $|M_0| \leq \kappa$.*

Proof Let T be an L-fragment of cardinality κ such that $\varphi \in T$. Let $\mathcal{F}$ be a set of cardinality κ of finitary functions on M containing a Skolem function for each $\psi \in T$. Let M_0 be a subset of M of cardinality κ closed under the functions of $\mathcal{F}$. By Proposition 8.27 M_0 is the universe of a substructure $\mathcal{M}_0$ of $\mathcal{M}$ such that $\mathcal{M}_0 \models \varphi$. $\qquad\square$

Proposition 8.29 (Union Property I) *Suppose $\mathcal{M}_n \prec^\alpha_{\infty\omega} \mathcal{M}_{n+1}$ for all $n \in \mathbb{N}$, i.e. $\mathcal{M}_n \subseteq \mathcal{M}_{n+1}$ and*

$$(\mathcal{M}_n, a_0, \ldots, a_{m-1}) \simeq^\alpha_p (\mathcal{M}_{n+1}, a_0, \ldots, a_{m-1}) \qquad (8.1)$$

for all $n, m \in \mathbb{N}$ and $a_0, \ldots, a_{m-1} \in M_n$. Let $\mathcal{M} = \bigcup_n \mathcal{M}_n$. Then

$$\mathcal{M}_n \prec^\alpha_{\infty\omega} \mathcal{M}$$

for all $n \in \mathbb{N}$.

Proof In this proof we violate the principle that in an Ehrenfeucht–Fraïssé Game the domains of the two models should be disjoint. This is not a problem, we just have to be more careful. Let $a_1, \ldots, a_m \in M_n$. We describe the winning strategy of **II** in

$$\mathrm{DEF}_\alpha((\mathcal{M}_n, a_1, \ldots, a_m), (\mathcal{M}, a_1, \ldots, a_m)).$$

For brevity, we denote

$$\mathcal{M}'_n = (\mathcal{M}_n, a_1, \ldots, a_m)$$
$$\mathcal{M}' = (\mathcal{M}, a_1, \ldots, a_m).$$

The strategy of **II** is the following: Suppose after round s the players have played $\alpha_n < \alpha$, $v_n, \ldots, v_{s-1}$ in $\mathcal{M}'_n$ and $v'_n, \ldots, v'_{s-1}$ in $\mathcal{M}'$. During the game player **II** has formed a sequence

$$n = k_0 \leq k_1 \leq \cdots \leq k_s < \omega$$

such that the elements v'_i played in $\mathcal{M}$ are all contained in M_{k_s}. The idea of player **II** is to maintain the condition

$$(\mathcal{M}'_n, v_0, \ldots, v_{s-1}) \simeq_p^{\alpha_n} (\mathcal{M}'_{k_s}, v'_0, \ldots, v'_{s-1}) \tag{8.2}$$

(Figure 8.3).

    Round 0: Suppose player **I** plays $\alpha_0 < \alpha$ and either $v_0 \in M_n$ or $v'_0 \in M$. Player **II** chooses k_1 sufficiently large in order that $v'_0 \in M_{k_1}$, and then her move such that after round 0 we have

$$(\mathcal{M}'_n, v_0) \simeq_p^{\alpha_0} (\mathcal{M}'_{k_1}, v'_0). \tag{8.3}$$

    Round 1: Suppose player **I** plays $\alpha_1 < \alpha_0$ and either $v_1 \in M_n$ or $v'_1 \in M$. If **I** plays his v_1 in $M_n \cup M_{k_1}$, then **II** lets $k_2 = k_1$ and uses fact (8.3). Otherwise, player **II** chooses k_2 sufficiently large in order that $v'_1 \in M_{k_2}$, and she also uses the assumption

$$(\mathcal{M}'_{k_1}, v'_0) \simeq_p^{\alpha_0} (\mathcal{M}'_{k_2}, v'_0) \tag{8.4}$$

to find u such that

$$(\mathcal{M}'_{k_1}, v'_0, u) \simeq_p^{\alpha_1} (\mathcal{M}'_{k_2}, v'_0, v'_1) \tag{8.5}$$

and then (8.3) to find v_1 such that

$$(\mathcal{M}'_n, v_0, v_1) \simeq_p^{\alpha_1} (\mathcal{M}'_{k_1}, v'_0, u). \tag{8.6}$$

Summa summarum, (8.5) and (8.6) give

$$(\mathcal{M}'_n, v_0, v_1) \simeq_p^{\alpha_1} (\mathcal{M}'_{k_2}, v'_0, v'_1). \tag{8.7}$$

Round 2: Suppose player **I** plays $\alpha_2 < \alpha_1$ and either $v_2 \in M_n$ or $v'_2 \in M$. If **I** plays his v_2 in $M_n \cup M_{k_2}$, then **II** lets $k_3 = k_2$ and uses fact (8.4). Otherwise, player **II** chooses k_3 sufficiently large in order that $v'_2 \in M_{k_3}$, and she also uses the assumption

$$(\mathcal{M}'_{k_2}, v'_0, v'_1) \simeq_p^{\alpha_1} (\mathcal{M}'_{k_3}, v'_0, v'_1) \tag{8.8}$$

to find u such that

$$(\mathcal{M}'_{k_2}, v'_0, v'_1, u) \simeq_p^{\alpha_2} (\mathcal{M}'_{k_3}, v'_0, v'_1, v'_2) \tag{8.9}$$

and then (8.7) to find v_2 such that

$$(\mathcal{M}'_n, v_0, v_1, v_2) \simeq_p^{\alpha_2} (\mathcal{M}'_{k_2}, v'_0, v'_1, u). \tag{8.10}$$

Summa summarum, (8.9) and (8.10) give

$$(\mathcal{M}'_n, v_0, v_1, v_2) \simeq_p^{\alpha_2} (\mathcal{M}'_{k_3}, v'_0, v'_1, v'_2). \tag{8.11}$$

Round s: We assume (8.2). Suppose player **I** plays $\alpha_s < \alpha_{s-1}$ ($\alpha_{s-1} = \alpha$, if $s = 0$) and either $v_s \in M_n$ or $v'_s \in M$. If **I** plays his v_s in $M_n \cup M_{k_s}$, then **II** simply lets $k_{s+1} = k_s$ and uses fact (8.2). Otherwise, player **II** chooses k_{s+1} sufficiently large in order that $v'_s \in M_{k_{s+1}}$, and she then uses the assumption (8.1) to find u such that

$$(\mathcal{M}'_{k_s}, v'_0, \ldots, v'_{s-1}, u) \simeq_p^{\alpha_2} (\mathcal{M}'_{k_{s+1}}, v'_0, \ldots, v'_s) \tag{8.12}$$

and then (8.2) to find v_s such that

$$(\mathcal{M}'_n, v_0, \ldots, v_{s-1}, v_s) \simeq_p^{\alpha_s} (\mathcal{M}'_{k_{s+1}}, v'_0, \ldots, v'_{s-1}, u). \tag{8.13}$$

Putting (8.12) and (8.13) together gives

$$(\mathcal{M}'_n, v_0, \ldots, v_{s-1}, v_s) \simeq_p^{\alpha_2} (\mathcal{M}'_{k_{s+1}}, v'_0, \ldots, v'_{s-1}, v'_s). \tag{8.14}$$

If **II** plays following this strategy she wins as the mapping $v_i \mapsto v'_i$ is clearly a partial isomorphism. $\qquad\square$

Proposition 8.30 (Union Property II) *Suppose Γ is a set of formulas closed under subformulas. Suppose $\mathcal{M}_n \prec_\Gamma \mathcal{M}_{n+1}$ for all $n \in \mathbb{N}$, i.e. $\mathcal{M}_n \subseteq \mathcal{M}_{n+1}$ and*

$$\mathcal{M}_n \models \varphi(a_0, \ldots, a_{m-1}) \iff \mathcal{M}_{n+1} \models \varphi(a_0, \ldots, a_{m-1}) \tag{8.15}$$

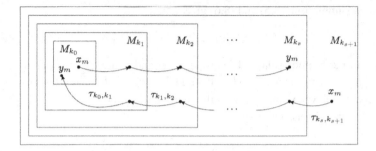

Figure 8.3

for all $\varphi \in \Gamma$ and $a_0, \ldots, a_{m-1} \in M_n$. Let $\mathcal{M} = \bigcup_n \mathcal{M}_n$. Then $\mathcal{M}_n \prec_\Gamma \mathcal{M}$ for all $n \in \mathbb{N}$.

Proof   We use induction on $\varphi \in \Gamma$ to prove for all $n \in \mathbb{N}$ and all $\ldots, a_{m-1} \in M_n$

$$\mathcal{M}_n \models \varphi(a_0, \ldots, a_{m-1}) \iff \mathcal{M} \models \varphi(a_0, \ldots, a_{m-1}) \qquad (8.16)$$

for $a_0, \ldots, a_{m-1} \in M_n$. The only non-trivial case is the case of the quantifiers. Suppose therefore $\varphi(x_0, \ldots, x_{m-1})$ is the formula $\exists x_m \psi(x_0, \ldots, x_{m-1}, x_m)$, where $a_1, \ldots, a_{m-1} \in M_n$, and

$$\mathcal{M}_n \models \varphi(a_0, \ldots, a_{m-1}).$$

Then

$$\mathcal{M}_n \models \psi(a_0, \ldots, a_{m-1}, a)$$

for some $a \in M_n$. By the induction hypothesis

$$\mathcal{M} \models \psi(a_0, \ldots, a_{m-1}, a).$$

Thus

$$\mathcal{M} \models \varphi(a_0, \ldots, a_{m-1}). \qquad (8.17)$$

Conversely, suppose (8.17). Let $a \in M$ be such that

$$\mathcal{M} \models \psi(a_0, \ldots, a_{m-1}, a).$$

There is $m \geq n$ such that $a \in M_m$. By the induction hypothesis

$$\mathcal{M}_m \models \psi(a_0, \ldots, a_{m-1}, a).$$

Thus

$$\mathcal{M}_m \models \varphi(a_0, \ldots, a_{m-1}).$$

By assumption (8.16), $\mathcal{M}_n \models \varphi(a_0, \ldots, a_{m-1})$. $\square$

Union Property I actually follows from Union Property II by letting Γ be the set of all formulas of quantifier rank $\leq \alpha$.

Theorem 8.31 *Suppose L is a vocabulary of cardinality κ and $\varphi \in L_{\kappa^+\omega}$ has vocabulary L. Suppose φ has a model $\mathcal{M}$ with $((2^\kappa)^+, <) \subseteq (U^{\mathcal{M}}, <^{\mathcal{M}})$. Then φ has a model $\mathcal{M}^*$ in which $(U^{\mathcal{M}^*}, <^{\mathcal{M}^*})$ is non-well-ordered.*

Proof Let $\mu = (2^\kappa)^+$. Let Γ be an L-fragment containing φ. Let

$$L_n = L \cup \{c_0, \ldots, c_{n-1}\}$$
$$L^* = \bigcup_n L_n.$$

We construct for each n an L_n-structure $\mathcal{M}_n$ such that

(i) $\mathcal{M}_n \models \bigwedge_{i<n} Uc_i \wedge \bigwedge_{i<j<n} c_j < c_i$.
(ii) $\mathcal{M}_n \prec_\Gamma \mathcal{M}_{n+1} \restriction L_n$.
(iii) $\mathcal{M}_n \models \varphi$.

(see Figure 8.4). There is clearly a unique L^*-structure $\mathcal{M}^*$ such that

$$M^* = \bigcup_n M_n$$
$$\mathcal{M}_n \subseteq \mathcal{M}^* \restriction L_n.$$

Then $\mathcal{M}^* \restriction L = \bigcup_n (\mathcal{M}_n \restriction L)$, so Proposition 8.29 gives $\mathcal{M}^* \models \varphi$. Also

$$\mathcal{M}^* \models \bigwedge_{i \in \mathbb{N}} Uc_i \wedge \bigwedge_{i<j<\omega} c_j < c_i.$$

So $\mathcal{M}^*$ is the non-well-ordered model of φ we seek.

In order to obtain the models $\mathcal{M}_n$ we define an auxiliary double sequence

$$\mathcal{M}_\alpha^n, n < \omega, \alpha < \mu$$

of models meeting the following conditions:

(1) $\mathcal{M}_\alpha^n$ is an L_n-structure, $\mathcal{M}_\alpha^n \restriction L \subseteq \mathcal{M}, |M_\alpha^n| = \kappa$.
(2) $\mathcal{M}_\alpha^n \models \bigwedge_{i<n} Uc_i \wedge \bigwedge_{i<j<n} c_j < c_i$.
(3) $\mathcal{M}_\alpha^n \cong \mathcal{M}_\beta^n$ for all n, α and β.
(4) Every $\mathcal{M}_\alpha^{n+1}$ is a supermodel of some $\mathcal{M}_\beta^n$.
(5) $\alpha \leq c_{n-1}^{\mathcal{M}_\alpha^n}$ for all α and n.

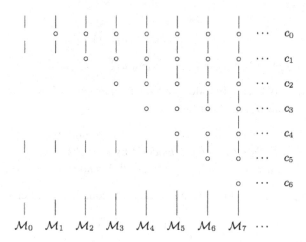

Figure 8.4

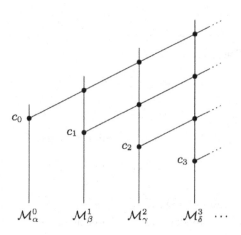

Figure 8.5

For a start, let $\mathcal{M}_0^0 \subseteq \mathcal{M}$ such that $|M_0^0| = \kappa$ and $M_0^0 \models \varphi$. Let then $\mathcal{M}_\alpha^0 = \mathcal{M}_0^0$ for $0 < \alpha < \mu$. For the inductive step, suppose $\mathcal{M}_\alpha^n$ has been constructed for each $\alpha < \mu$. For $\alpha < \mu$ let $\mathcal{M}_{\alpha+1}^n \prec_\Gamma \mathcal{N}_\alpha^{n+1} \subseteq (\mathcal{M}, c_0^{\mathcal{M}_{\alpha+1}^n}, \ldots, c_{n-1}^{\mathcal{M}_{\alpha+1}^n})$

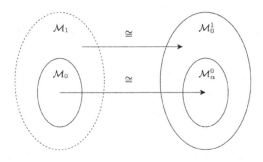

Figure 8.6

such that $|N_\alpha^{n+1}| = \kappa$, and $\alpha \in N_\alpha^{n+1}$. Since

$$\alpha + 1 \leq c_{n-1}^{\mathcal{M}_{\alpha+1}^n} = c_{n-1}^{\mathcal{N}_\alpha^{n+1}}$$

we can make $\mathcal{N}_\alpha^{n+1}$ an L_{n+1}-structure satisfying $c_n < c_{n+1}$ by defining $c_n^{\mathcal{N}_\alpha^{n+1}} = \alpha$. There are, up to isomorphism, only 2^κ L_{n+1}-structures of size κ. Thus there is $I \subseteq \mu$ of size μ such that $\mathcal{N}_\alpha^{n+1} \cong \mathcal{N}_\beta^{n+1}$ for $\alpha, \beta \in I$. Let I be $\{\gamma_\alpha : \alpha < \mu\}$ in increasing order, and $\mathcal{M}_\alpha^{n+1} = \mathcal{N}_{\gamma_\alpha}^{n+1}$. Note that $\gamma_\alpha \geq \alpha$, so $c_n^{\mathcal{M}_\alpha^{n+1}} = c_n^{\mathcal{M}_{\gamma_\alpha}^{n+1}} = \gamma_\alpha \geq \alpha$. The models $\mathcal{M}_\alpha^n$ satisfy conditions (1)–(5) (see Figure 8.5).

Now we return to the construction of the models $\mathcal{M}_n, n < \omega$. Let $\mathcal{M}_0 = \mathcal{M}_0^0$. There is $\alpha < \mu$ such that $\mathcal{M}_\alpha^0 \subseteq \mathcal{M}_0^1 \upharpoonright L_0$. As $\mathcal{M}_0^0 \cong \mathcal{M}_\alpha^0$ (in fact $\mathcal{M}_0^0 = \mathcal{M}_\alpha^0$), there is $\mathcal{M}_1$ such that $\mathcal{M}_0 \subseteq \mathcal{M}_1 \upharpoonright L_0$ and $\mathcal{M}_1 \cong \mathcal{M}_0^1$ (in fact $\mathcal{M}_1 = \mathcal{M}_0^1$) (Figure 8.6).

Next we find $\beta < \mu$ such that $\mathcal{M}_\beta^1 \subseteq \mathcal{M}_0^2 \upharpoonright L_1$. As $\mathcal{M}_1 \cong \mathcal{M}_0^1 \cong \mathcal{M}_\beta^1$, there is $\mathcal{M}_2$ such that $\mathcal{M}_1 \subseteq \mathcal{M}_2 \upharpoonright L_1$ and $\mathcal{M}_2 \cong \mathcal{M}_0^2$.

Then we find $\gamma < \mu$ such that $\mathcal{M}_\gamma^2 \subseteq \mathcal{M}_0^3 \upharpoonright L_2$. As $\mathcal{M}_2 \cong \mathcal{M}_0^2 \cong \mathcal{M}_\gamma^2$, there is $\mathcal{M}_3$ such that $\mathcal{M}_2 \subseteq \mathcal{M}_3 \upharpoonright L_2$ and $\mathcal{M}_3 \cong \mathcal{M}_0^3$.

Continuing in this way we get all $\mathcal{M}_n, n < \omega$, satisfying conditions (i)–(iii). $\qquad\square$

In Definition 6.45 we defined the concept of a PC-class. A more general concept is the following: A model class K with vocabulary L is an RPC-class ("relativized pseudo-elementary class") in a logic if there is a sentence φ in the logic, with a unary predicate U, such that K consists of the relativizations to U of reducts to L of models of φ.

Proposition 8.32 *The class of well-ordered models is not RPC-definable in*

$L_{\infty\omega}$, that is, if $\varphi \in L_{\infty\omega}$ and $(U^{\mathcal{M}}, <^{\mathcal{M}})$ is well-ordered in every model of φ, then there is an ordinal α such that $(U^{\mathcal{M}}, <^{\mathcal{M}})$ has order type $< \alpha$ for every model $\mathcal{M}$ of φ.

Proposition 8.32 can be used to prove the non-definability of various concepts in $L_{\infty\omega}$. Here are some examples:

Example 8.33 The class K of well-founded trees is not definable in $L_{\infty\omega}$. To see why, suppose $\varphi \in L_{\infty\omega}$ defines K. Let K' be the class of structures $(M, U, <, T, <', f)$ where $(T, <')$ is a tree in K, $(U, <)$ is a linear order, and $f : U \to T$ such that $x < y \to f(y) <' f(x)$ for all $x, y \in U$. Clearly, K' is definable in $L_{\infty\omega}$. Now $(U^{\mathcal{M}}, <^{\mathcal{M}})$ is a well-order in every model $\mathcal{M}$ of φ, but on the other hand, there is no upper bound for the order types of $(U^{\mathcal{M}}, <^{\mathcal{M}})$ in models $\mathcal{M}$ of φ. This contradicts Proposition 8.32.

Example 8.34 Let K be the class of structures $(M, <, A)$, where $(M, <)$ is a linear order and if $x, y \in A$ with $x < y$, then $|\{a \in M : a < x\}| < |\{a \in M : a < y\}|$. The class K is not definable in $L_{\infty\omega}$. Why? Suppose it is definable by $\varphi \in L_{\infty\omega}$. Let K' be the class of structures $(M, <, A, U, <', f)$ where $(M, <, A) \models \varphi$, $(U, <')$ is a linear order, and $f : U \to A$ is such that $x <' y \to f(y) < f(x)$ for all $x, y \in U$. Clearly, K' is definable in $L_{\infty\omega}$. Now $(U^{\mathcal{M}}, <'^{\mathcal{M}})$ is a well-order in every model $\mathcal{M}$ in K', but on the other hand, there is no upper bound for the order types of $(U^{\mathcal{M}}, <^{\mathcal{M}})$ in models $\mathcal{M}$ in K'. This contradicts Proposition 8.32.

Example 8.35 Let ZFC_n be a finite part of the Zermelo–Fraenkel axioms for set theory. Let K be the class of models (M, ε) of ZFC_n which are well-founded, i.e. there is no infinite sequence $s = \{a_0, a_1, \ldots\}$ of elements of M such that $a_{n+1}\varepsilon a_n$ for all $n \in \mathbb{N}$ (note that s need not be in M). The class K is not definable in $L_{\infty\omega}$. Why? Suppose it is definable by $\varphi \in L_{\infty\omega}$. Let K' be the class of structures $(M, \varepsilon, U, <, f)$ where $(M, \varepsilon) \models \varphi$, $(U, <)$ is a linear order, and $f : U \to M$ is such that $x < y \to f(x)\varepsilon f(y)$ for all $x, y \in U$. Clearly, K' is definable in $L_{\infty\omega}$. Now $(U^{\mathcal{M}}, <^{\mathcal{M}})$ is a well-order in every model $\mathcal{M}$ in K', but there is no upper bound for the order types of $(U^{\mathcal{M}}, <^{\mathcal{M}})$ in models $\mathcal{M}$ in K' since there are transitive models of ZFC_n of arbitrary large height (even of the form $(V_\kappa, \in)$ by the Levy Reflection Principle). This contradicts Proposition 8.32.

I	II	Winning condition
$a_0 \in A$		
	$b_0 \in A$	$\varphi_0(a_0, b_0)$
$a_1 \in A$		
	$b_1 \in A$	$\varphi_1(a_0, b_0, a_1, b_1)$
$\vdots$	$\vdots$	

Figure 8.7 The game quantifier.

8.6 Game Logic

In this section we sketch the basic properties and applications of the so-called closed game quantifier of length ω

$$\forall x_0 \exists y_0 \forall x_1 \exists y_1 \ldots \bigwedge_{n < \omega} \varphi_n(x_0, y_0 \ldots, x_n, y_n) \tag{8.18}$$

and its generalization, the so-called closed Vaught-formula of length ω

$$\forall x_0 \bigvee_{i_0 \in I_0} \bigwedge_{j_0 \in J_0} \exists y_0 \forall x_1 \bigvee_{i_1 \in I_1} \bigwedge_{j_1 \in J_1} \exists y_1 \ldots \bigwedge_{n < \omega} \varphi^{i_0 j_0 \ldots i_n j_n}(x_0, y_0 \ldots, x_n, y_n)$$

$$\tag{8.19}$$

as well as their open counterparts. We use the general term *game quantification* to cover expressions of the above type. The semantics of these expressions is defined below by reference to a proper version of a *Semantic Game*.

The first application of game quantifiers in model theory was Svenonius's Theorem (1965) to the effect that every PC-definable class of models is recursively axiomatizable in countable models. Moschovakis (1972) introduced the game quantifier to descriptive set theory showing that inductive relations on countable acceptable structures are definable by the game quantifier. Vaught (1973) applied game expressions to develop a general definability theory for $L_{\omega_1 \omega}$ including a Covering Theorem. Subsequently game quantifiers have become a standard tool in model theory.

Closed game formulas

We shall first discuss the simpler case (8.18) and show how it can be used as technical tool for an analysis of PC-definability in first-order logic.

Definition 8.36 (Game Quantifier)   The truth of a game expression (8.18) in a model $\mathcal{A}$ means the existence of a winning strategy of player **II** in the game of length ω of Figure 8.7. Player **II** wins this game if

$$\mathcal{A} \models \varphi_n(a_0, b_0, \ldots, a_n, b_n)$$

for all $n \in \mathbb{N}$. A winning strategy of **II** is a sequence $\{\tau_n : n < \omega\}$ of functions on A such that

$$\mathcal{A} \models \varphi_n(a_0, \tau_0(a_0), a_1, \tau_1(a_0, a_1), \ldots, a_n, \tau_n(a_0, \ldots, a_n))$$

for all $a_0, a_1, \ldots$ in A.

Example 8.37 Examples of formulas of the form (8.18) are

1. $\exists x_0 \exists x_1 \exists x_2 \ldots \bigwedge_n x_{n+1} < x_n$, which in a linearly ordered model $(A, <)$ says that the linear order is not a well-order.
2. $\exists y \exists z \forall x_0 \forall x_1 \forall x_2 \ldots \bigwedge_n ((yEx_0 \wedge x_0Ex_1 \wedge \ldots x_{n-1}Ex_n) \rightarrow \neg \approx x_n z)$, which in a graph says that the graph is not connected.
3. $\forall x_0 \forall x_1 \forall x_2 \ldots \bigwedge_{n>2}((x_0Ex_1 \wedge \ldots \wedge x_{n-1}Ex_n \wedge \bigwedge_{0 \le i < j < n} \neg \approx x_i x_j) \rightarrow \neg \approx x_0 x_n)$, which in a graph says that the graph is cycle-free.

As the above examples show, game expressions are more powerful than $L_{\infty \omega}$. In fact, we shall see below that they can express even things that go beyond $L_{\infty \infty}$. Therefore we cannot expect the model theory of game expressions to be as nice as that of $L_{\omega_1 \omega}$. However, the game expressions permit one very useful technique. This is the method of approximations, originally due in model theory to Keisler and then extensively used by Makkai, Vaught, and others.

We use $\bar{x}_i, \bar{y}_i$ or just $\bar{x}, \bar{y}$, when the length of the sequences is clear from the context, to denote $x_0, y_0, \ldots, x_{i-1}, y_{i-1}$.

Definition 8.38 Suppose Φ is the closed game formula (8.18). We shall associate Φ with a sequence $\Phi_\gamma^n, \gamma \in On$, of $L_{\infty \omega}$-formulas, called *approximations*, as follows:

$$
\begin{aligned}
\Phi_0^n(\bar{x}_n, \bar{y}_n) &= \bigwedge_{j<n} \varphi_{j-1}(\bar{x}_j, \bar{y}_j) \\
\Phi_{\gamma+1}^n(\bar{x}_n, \bar{y}_n) &= \forall x_n \exists y_n \Phi_\gamma^{n+1}(\bar{x}_{n+1}, \bar{y}_{n+1}) \\
\Phi_\nu^n(\bar{x}, \bar{y}) &= \bigwedge_{\gamma < \nu} \Phi_\gamma^n(\bar{x}, \bar{y}) \text{ for limit } \nu.
\end{aligned}
$$

The trivial properties of these approximations are proved easily by transfinite induction:

- $\Phi_\gamma^n(\bar{x}, \bar{y}) \in L_{\kappa \omega}$ for $\gamma < \kappa$.
- $QR(\Phi_\nu^n(\bar{x}, \bar{y})) = \nu$ for limit $\nu > 0$.
- $\models \Phi \rightarrow \Phi_\gamma^0$ for all γ.
- $\models \Phi_\gamma^n(\bar{x}, \bar{y}) \rightarrow \Phi_\beta^{\bar{n}}(\bar{x}, \bar{y})$ for $\beta \le \gamma$.

Less trivial is the following important and characteristic property of the approximations:

Proposition 8.39 *If $|A| = \kappa$ and $\mathcal{A} \models \Phi_\alpha^0$ for all $\alpha < \kappa^+$, then $\mathcal{A} \models \Phi$.*

Proof We define a winning strategy $\{\tau_n : n \in \mathbb{N}\}$ of **II** in the game (8.7) as follows: Suppose $a_0, b_0, \ldots, a_{n-1}, b_{n-1}$ have been played. The strategy of **II** is to maintain the property

$$\text{For all } \alpha < \kappa^+ : \mathcal{A} \models \Phi_\alpha^n(a_0, b_0, \ldots, a_{n-1}, b_{n-1}). \tag{8.20}$$

In the beginning this condition holds by assumption. Suppose the condition holds after $a_0, b_0, \ldots, a_{n-1}, b_{n-1}$ have been played. Now **I** plays a_n. If there is no b_n such that

$$\text{for all } \alpha < \kappa^+ : \mathcal{A} \models \Phi_\alpha^{n+1}(a_0, b_0, \ldots, a_n, b_n), \tag{8.21}$$

then for every $b_n \in A$ there is $\alpha(b_n) < \kappa^+$ such that

$$\mathcal{A} \not\models \Phi_{\alpha(b_n)}^{n+1}(a_0, b_0, \ldots, a_n, b_n).$$

Let $\delta = \sup_{b_n \in A} \alpha(b_n)$. Note that $\delta < \kappa^+$. Hence by assumption $\mathcal{A} \models \Phi_{\delta+1}^n(a_0, b_0, \ldots, a_{n-1}, b_{n-1})$. We obtain immediately a contradiction. Thus there must be a b_n such that (8.21). $\qquad\square$

Corollary *In countable models the game formula Φ and the $L_{\omega_2\omega}$-sentence $\bigwedge_{\alpha<\omega_1} \Phi_\alpha^0$ are logically equivalent.*

Thus as far as countable models are concerned, the only thing that the closed game formulas (8.18) add to $L_{\omega_1\omega}$ is an uncountable conjunction. When we move to bigger models, longer and longer conjunctions are needed, but that is all.

Definition 8.40 A structure in a countable recursive vocabulary is *recursively saturated* if it satisfies

$$\forall x_1 \ldots x_n \left(\left(\bigwedge_{n<\omega} \exists y \bigwedge_{m<n} \varphi_m(x_1, \ldots, x_n, y) \right) \to \exists y \bigwedge_{n<\omega} \varphi_n(x_1, \ldots, x_n, y) \right)$$

for all recursive sequences $\{\varphi_m(x_1, \ldots, x_n, y) : m < \omega\}$ of first-order formulas.

Examples of recursively saturated structures are the linear order $(\mathbb{Q}, <)$, and the field $(\mathbb{C}, +, \cdot)$ of complex numbers. Every infinite model of a recursive vocabulary has a countable recursively saturated elementary extension (see (Chang and Keisler, 1990, Section 2.4)).

Proposition 8.41 *Suppose $\mathcal{A}$ is recursively saturated. Then $\mathcal{A} \models \Phi \leftrightarrow \bigwedge_{n<\omega} \Phi_n^0$.*

Proof We proceed as in the proof of Proposition 8.39. Suppose $\mathcal{A}$ satisfies $\bigwedge_{n<\omega} \Phi_n^0$. We define a winning strategy $\{\tau_n : n \in \mathbb{N}\}$ of **II** in the game (8.7) as follows. Suppose $a_0, b_0, \ldots, a_{n-1}, b_{n-1}$ have been played. The strategy of **II** is to maintain the property

$$\text{For all } m < \omega \colon \mathcal{A} \models \Phi_m^n(a_0, b_0, \ldots, a_{n-1}, b_{n-1}). \tag{8.22}$$

In the beginning this condition holds by assumption. Suppose the condition holds after $a_0, b_0, \ldots, a_{n-1}, b_{n-1}$ have been played. Now **I** plays a_n. We look for b_n such that

$$\mathcal{A} \models \bigwedge_{m<\omega} \Phi_m^{n+1}(a_0, b_0, \ldots, a_n, b_n), \tag{8.23}$$

i.e. we want to show

$$\mathcal{A} \models \exists y_n \bigwedge_{m<\omega} \Phi_m^{n+1}(a_0, b_0, \ldots, a_n, y_n). \tag{8.24}$$

Since $\mathcal{A}$ is ω-saturated, it suffices to prove

$$\mathcal{A} \models \bigwedge_{m<\omega} \exists y_n \bigwedge_{k<m} \Phi_k^{n+1}(a_0, b_0, \ldots, a_n, y_n). \tag{8.25}$$

Suppose $m < \omega$ is given. By (8.22) we have

$$\mathcal{A} \models \forall x_n \exists y_n \Phi_k^{n+1}(a_0, b_0, \ldots, x_n, y_n).$$

By choosing the value of x_n to be a_n we get b such that

$$\mathcal{A} \models \Phi_k^{n+1}(a_0, b_0, \ldots, a_n, b).$$

We have proved (8.25). $\qquad\square$

What about structures that are not recursively saturated? To conclude $\mathcal{A} \models \Phi$ we have to assume $\mathcal{A} \models \Phi_\alpha^0$ for some infinite ordinals α, too. We refer to Barwise (1975) for details.

Barwise (1976) observed that game formulas can be used to "straighten" partially ordered quantifiers. Consider the so-called *Henkin quantifier*

$$\begin{pmatrix} \forall x & \exists y \\ \forall u & \exists v \end{pmatrix} \varphi(x, y, u, v, \bar{z}) \tag{8.26}$$

the meaning of which is

There are f and g such that for all a and b $\varphi(a, f(a), b, g(b), \bar{z})$. (8.27)

We call formulas of the form (8.26), with $\varphi(x, y, u, v)$ first-order, *Henkin formulas*. Indeed, they were introduced in Henkin (1961). An alternative notation for Henkin formulas is offered by *dependence logic* Väänänen (2007):

$$\forall x \exists y \forall u \exists v (=(u, \bar{z}, v) \wedge \varphi(x, y, u, v, \bar{z})),$$

where the intuitive interpretation of $=(u, \bar{z}, v)$ is "v depends only on u and $\bar{z}$". Let us compare (8.26) with the game formula

$$\forall x_0 \exists y_0 \forall u_0 \exists v_0 \forall x_1 \exists y_1 \forall u_1 \exists v_1 \ldots$$
$$\bigwedge_{i,j,k,l}((\approx x_i x_j \wedge \approx u_k u_l) \rightarrow (\approx y_i y_j \wedge \approx v_k v_l \wedge \varphi(x_i, y_i, u_i, v_i, \bar{z}))).$$
$$(8.28)$$

Clearly, (8.26), or rather (8.27), implies (8.28) as **II** can let $y_n = f(x_n)$ and $v_n = g(u_n)$. In a countable model the converse is true: Suppose $\mathcal{A}$ is a countable model and s is an assignment. Let (a_n, b_n), $n \in \mathbb{N}$, list all pairs of elements of A. Let us play the game associated with the formula (8.28) in $\mathcal{A}$ so that **I** plays $x_n = a_n$ and $u_n = b_n$. Let the responses of **II** be $y_n = a_n^*$ and $v_n = b_n^*$. Let $f(a_n) = a_n^*$ and $g(b_n) = b_n^*$. It is easy to see that $\mathcal{A} \models_s \varphi(a_n, f(a_n), b_m, g(b_m), \bar{z})$ for all n and m. Thus (8.26) holds in $\mathcal{A}$ under the assignment s. We have proved:

Proposition 8.42 *The formulas (8.26) and (8.28) are equivalent in all countable models.*

In consequence, in a countable recursively saturated model (8.26) is, by Proposition 8.41, equivalent to $\bigwedge_{n<\omega} \Phi_n^0$.

Let us say that $\{\varphi_n : n < \omega\}$ is a *first-order axiomatization* of a class K of models, if the following are equivalent for all first-order ψ:

1. ψ is true in all models in K.
2. $\{\varphi_n : n < \omega\} \models \psi$.

Proposition 8.43 $\{\Phi_n^0 : n < \omega\}$ *is a first-order axiomatization of the class of models of (8.26).*

Proof Suppose a first-order sentence ψ follows from (8.26). We show that it follows from $\{\Phi_n^0 : n < \omega\}$. If not, then there is a model $\mathcal{A}$ of $\{\Phi_n^0 : n < \omega\}$ which satisfies $\neg\psi$. Take a countable recursively saturated model of the first-order theory $\{\Phi_n^0 : n < \omega\} \cup \{\neg\psi\}$. We get a contradiction. $\square$

Example 8.44 (Models with an involution) Suppose L is the vocabulary $\{R\}$, where R is (for simplicity) binary. The class of L-models with an involution (non-trivial automorphism of order two) can be axiomatized by the Henkin sentence

$$\Phi = \exists z \begin{pmatrix} \forall x & \exists y \\ \forall u & \exists v \end{pmatrix} \varphi(x, y, u, v, z),$$

where $\varphi(x, y, u, v, z)$ is the conjunction of $(\approx xu \to \approx yv)$, $(\approx xv \to \approx yu)$, $(Rxu \leftrightarrow Ryv)$, and $\approx xz \to \neg \approx xy$. In countable models this Henkin sentence is equivalent to

$$\exists z \forall x_0 \exists y_0 \forall u_0 \exists v_0 \forall x_1 \exists y_1 \forall u_1 \exists v_1 \ldots$$
$$\bigwedge_{i,j,k,l}((\approx x_i x_j \wedge \approx u_k u_l) \to (\approx y_i y_j \wedge \approx v_l v_k \wedge \varphi(x_i, y_i, u_i, v_i, z))).$$

By inspecting the approximations Φ_n^0 of Φ we see that a first-order sentence has a model with an involution if and only if it is consistent with the set of the first-order sentences

$$\exists z \forall x_0 \exists y_0 \forall x_1 \exists y_1 \ldots \forall x_m \exists y_m$$
$$\bigwedge_{i,j,k,l \leq m}((\approx x_i x_j \to \approx y_i y_j) \wedge (\approx x_i y_j \to \approx x_j y_i) \wedge$$
$$(\approx x_i z \to \neg \approx x_i y_i) \wedge (Rx_i x_j \leftrightarrow Ry_i y_j)),$$

where $m \in \mathbb{N}$.

The above results about Henkin formulas are not limited to the particular form of (8.26). The meaning of the more general formula

$$\begin{pmatrix} \forall x_{i_1^1} \ldots \forall x_{i_{m_1}^1} \; \exists y_1 \\ \vdots \quad\quad \vdots \quad\quad \vdots \\ \forall x_{i_1^n} \ldots \forall x_{i_{m_n}^n} \; \exists y_n \end{pmatrix} \varphi(x_{i_1^1}, \ldots, x_{i_{m_1}^1}, y_1, \ldots, x_{i_1^n}, \ldots, x_{i_{m_n}^n}, y_n, \bar{z}) \quad (8.29)$$

is simply: There are $f_1, \ldots, f_n$ such that for all $a_{i_1^1}, \ldots, a_{i_{m_1}^1}$ $(i = 1, \ldots, n)$,

$$\varphi(a_{i_1^1}, \ldots, a_{i_{m_1}^1}, b_1, \ldots, a_{i_1^n}, \ldots, a_{i_{m_n}^n}, b_n, \bar{z}),$$

where

$$b_j = f_j(a_{i_1^j}, \ldots, a_{i_{m_j}^j}), \text{ for } j = 1, \ldots, n.$$

Note that (8.29) makes perfect sense even if the rows of the quantifier prefix are of different lengths, as in

$$\begin{pmatrix} \forall x_1 \forall x_2 & \exists y \\ \forall u & \exists v \end{pmatrix} \varphi(x_1, x_2, y, u, v, \bar{z}). \quad (8.30)$$

We call all formulas of the form (8.29) *Henkin formulas*. Let $\bar{\Phi}$ be obtained from (8.29) as (8.28) was obtained from (8.26). The following proposition is proved mutatis mutandis as in Proposition 8.42:

Proposition 8.45 *The formulas (8.29) and $\bar{\Phi}$ are equivalent in all countable models.*

In consequence, in a countable recursively saturated model (8.29) is, by Proposition 8.41, equivalent to $\bigwedge_{n<\omega} \bar{\Phi}_n^0$.

Enderton (1970) and Walkoe (1970) observed that any PC-class can be defined by a Henkin-formula:

Theorem 8.46 *For every PC-class K there is a Henkin sentence $\bar{\bar{\Phi}}$ such that for all $\mathcal{M}$:*

$$\mathcal{M} \in K \iff \mathcal{M} \models \bar{\bar{\Phi}}.$$

Proof Suppose K is the class of reducts of a first-order sentence φ. We may assume that φ is of the form

$$\forall x_1 \ldots \forall x_m \psi, \tag{8.31}$$

where ψ is quantifier-free but contains new function symbols $f_1, \ldots, f_n$. (This is the so called *Skolem Normal Form* of φ.) We will perform some reductions on (8.31) in order to make it more suitable for the construction of φ.

Step 1: If ψ contains nesting of the function symbols $f_1, \ldots, f_n$ or of the function symbols of the vocabulary, we can remove them one by one by using the equivalence of

$$\models \theta(f_i(t_1, \ldots, t_m))$$

and

$$\forall x_1 \ldots \forall x_m((t_1 = x_1 \wedge \ldots \wedge t_m = x_m) \rightarrow \theta(f_i(x_1, \ldots, x_m)))$$

for any first-order θ. Thus we may assume that all terms occurring in ψ are of the form x_i or $f_i(x_{i_1}, \ldots, x_{i_k})$.

Step 2: If ψ contains an occurrence of a function symbol $f_i(x_{i_1}, \ldots, x_{i_k})$ with the same variable occurring twice, e.g. $i_s = i_r$, $1 < r < k$, we can remove such by means of a new variable x_l and the equivalence

$$\models \forall x_1 \ldots \forall x_m \theta(f_i(x_{i_1}, \ldots, x_{i_k})) \leftrightarrow$$
$$\forall x_1 \ldots \forall x_m \forall x_l(x_l = x_r \rightarrow \theta(f_i(x_{i_1}, \ldots, x_{i_{r-1}}, x_l, x_{i_{r+1}}, \ldots, x_{i_k})))$$

for any first-order θ. Thus we may assume that if a term such as $f_i(x_{i_1}, \ldots, x_{i_k})$ occurs in ψ, its variables are all distinct.

Step 3: If ψ contains two occurrences of the same function symbol but with different variables or with the same variables in different order, we can remove such by using appropriate equivalences. If $\{i_1, \ldots, i_k\} \cap \{j_1, \ldots, j_k\} = \emptyset$, we have the equivalence

$$\models \forall x_1 \ldots \forall x_m \theta(f_i(x_{i_1}, \ldots, x_{i_k}), f_i(x_{j_1}, \ldots, x_{j_k})) \leftrightarrow$$

$$\exists f_i' \forall x_1 \ldots \forall x_m (\theta(f_i(x_{i_1}, \ldots, x_{i_k}), f_i'(x_{j_1}, \ldots, x_{j_k})) \wedge$$
$$((x_{i_1} = x_{j_1} \wedge \ldots \wedge x_{i_k} = x_{j_k}) \to$$
$$f_i(x_{i_1}, \ldots, x_{i_k}) = f_i'(x_{j_1}, \ldots, x_{j_k})))$$

for any first-order θ. We can reduce the more general case, where $\{i_1, \ldots, i_k\} \cap \{j_1, \ldots, j_k\} \neq \emptyset$, to this case by introducing new variables, as in Step 2. Thus we may assume that for each function symbol f_i occurring in ψ there are $j_1^i, \ldots, j_{n_i}^i$ such that *all* occurrences of f_i are of the form $f_i(x_{j_1^i}, \ldots, x_{j_{m_i}^i})$ and $j_1^i, \ldots, j_{m_i}^i$ are all different from each other.

In sum we may assume the function terms that occur in ψ are of the form $f_i(x_{j_1^i}, \ldots, x_{j_{m_i}^i})$ and for each i the variables $x_{j_1^i}, \ldots, x_{j_{m_i}^i}$ and their order is the same. Let N be greater than all the $x_{j_k^i}$. Let $\bar{\Phi}$ be the Henkin-sentence

$$\begin{pmatrix} \forall x_{j_1^1} & \cdots & \forall x_{j_{m_1}^1} & \exists x_{N+1} \\ \vdots & & \vdots & \vdots \\ \forall x_{j_1^n} & \cdots & \forall x_{j_{m_n}^n} & \exists x_{N+n} \end{pmatrix} \psi'$$

where ψ' is obtained from ψ by replacing $f_i(x_{j_1^i}, \ldots, x_{j_{m_i}^i})$ everywhere by x_{N+i}. This is clearly the desired Henkin-sentence. In the notation of dependence logic (Väänänen (2007)) this would look like:

$$\forall x_1 \ldots \forall x_m \exists x_{N+1} \ldots \exists x_{N+n} \ (=(x_{j_1^1}, \ldots, x_{j_{m_1}^1}, x_{N+1}) \wedge$$
$$\cdots$$
$$=(x_{j_1^n}, \ldots, x_{j_{m_n}^n}, x_{N+n}) \wedge \psi').$$

$\square$

By combining the above observations we get the following result of Svenonius (1965):

Theorem 8.47 *For every PC-class K there is a closed game sentence Φ and a sequence φ_n of first-order sentences such that for all structures $\mathcal{M}$:*

1. *If $\mathcal{M} \in K$, then $\mathcal{M} \models \Phi$ and $\mathcal{M} \models \varphi_n$ for all $n \in \mathbb{N}$.*
2. *If $\mathcal{M} \models \Phi$ and M is countable, then $\mathcal{M} \in K$.*
3. *If $\mathcal{M} \models \bigwedge_n \varphi_n$ and $\mathcal{M}$ is countable recursively saturated, then $\mathcal{M} \in K$.*
4. *If ψ is any first-order sentence, then ψ has a model in K if and only if ψ is consistent with $\{\varphi_n : n < \omega\}$.*

Moreover, the sequence $\{\varphi_n : n \in \mathbb{N}\}$ (or rather the set of Gödel numbers of the φ_n) is recursive.

I	II
$a_0 \in A$	
	$i_0 \in I_0$
$j_0 \in J_0$	
	$b_0 \in A$
$a_1 \in A$	
	$i_1 \in I_1$
$j_1 \in J_1$	
	$b_1 \in A$
$\vdots$	$\vdots$

Figure 8.8 The game quantifier.

Example 8.44 is but one of the many applications of the above theorem. One application is an immediate proof of the Craig Interpolation Theorem (see Exercise 8.45).

Closed Vaught formulas

Closed Vaught formulas (8.19) can be used to analyze $PC(L_{\omega_1\omega})$-classes of models much like closed game formulas were used above to analyze PC-classes. The details are a little different for reasons having to do with delicate discrepancies between properties of first-order logic and $L_{\omega_1\omega}$. However, the results for $L_{\omega_1\omega}$ are in a sense more canonical and interestingly analogous to *descriptive set theory*. In fact, Vaught calls this topic "descriptive set theory in $L_{\omega_1\omega}$" in his seminal work Vaught (1973).

Let us now define the semantics of the formulas (8.19):

Definition 8.48 The truth of (8.19) in a model $\mathcal{A}$ means the existence of a winning strategy of player **II** in the game of length ω of Figure 8.8. Player **II** wins this game if

$$\mathcal{A} \models \varphi^{i_0 j_0 \ldots i_n j_n}(a_0, b_0, \ldots, a_n, b_n)$$

for all $n \in \mathbb{N}$. Equivalently, there are functions τ_1 and τ_2 such that for all $a_0, \ldots, a_i \in A$ and all $m_0, \ldots, m_i \in \omega$ we have

$$\mathcal{A} \models \varphi^{n_0 m_0 \ldots n_i m_i}(a_0, b_0, \ldots, a_i, b_i), \tag{8.32}$$

where $n_i = \tau_1(x_0, m_0, \ldots, x_i)$ and $b_i = \tau_2(x_0, m_0, \ldots, x_i, m_i)$.

Technically the game of Figure 8.8 is exactly the same game as what we denoted in Chapter 3 by $G(C, \omega, W_0, W_1)$, where $C = A \cup \omega$, $W_1 = C^{\mathbb{N}} \setminus W_0$, and W_0 is the set of sequences

$$(a_0, n_0, m_0, b_0, \ldots, a_i, n_i, b_i, m_i, \ldots) \in \mathbb{N}$$

such that if $a_i \in A$ and $m_i \in \omega$ for $i < \omega$, then $n_i \in \omega$ and $b_i \in A$ for $i < \omega$, and (8.32) holds for all $i < \omega$. This game is determined by Theorem 3.12 since it is a closed game.

Example 8.49 Examples of formulas of the form (8.19) are

1. The closed game formula (8.18) is of course a special case of (8.19).
2. Souslin-formulas

$$\bigvee_{m_0 < \omega} \bigvee_{m_1 < \omega} \cdots \bigwedge_{n < \omega} \psi^{m_0 \cdots m_n},$$

where $\psi^{m_0 \cdots m_n}$ are first-order (or in $L_{\omega_1 \omega}$). The logic of such formulas was introduced inEllentuck (1975). These formulas are not expressible in $L_{\omega_1 \omega}$ alone (Exercise 8.46). However, they can be can be expressed in $L_{\infty \omega}$ as follows:

$$\bigvee_{f : \omega \to \omega} \bigwedge_n \varphi^{f(0) \ldots f(n)}.$$

3. A generalization of Souslin-formulas is the following propositional game formula:

$$\bigvee_{m_0 < \omega} \bigwedge_{k_0 < \omega} \bigvee_{m_1 < \omega} \bigwedge_{k_1 < \omega} \cdots \bigwedge_{n < \omega} \psi^{m_0 k_0 \cdots m_n k_n},$$

where $\psi^{m_0 k_0 \cdots m_n k_n}$ are first-order (or in $L_{\omega_1 \omega}$). The logic of such formulas was introduced in Green (1979). These formulas are not expressible in $L_{\omega_1 \omega}$ alone (see above). However, they can be can be expressed in $L_{\infty \omega}$ (Exercise 8.47).

4. A slight simplification of (8.19) is the following:

$$\exists x_0 \forall y_0 \bigvee_{n_0} \exists x_1 \forall y_1 \bigvee_{n_1} \cdots \bigwedge_i \varphi^{n_0 \cdots n_i}(x_0, y_0, \ldots, x_i, y_i), \qquad (8.33)$$

where $\varphi^{n_0 \cdots n_i}(x_0, y_0, \ldots, x_i, y_i)$ are first-order formulas. The truth of (8.33) in a model $\mathcal{A}$ means the existence of functions τ_n and τ'_n such that for any sequence $c_0, \ldots, c_i$ from A we have: $\mathcal{A} \models \varphi^{n_0 \cdots n_i}(a_0, c_0, \ldots, a_i, c_i)$, where $a_i = \tau(c_0, \ldots, c_{i-1})$ and $n_i = \tau'_n(c_0, \ldots, c_i)$.

We shall now consider at some length an important example of the use of the special case (8.33). Let L be a finite vocabulary, $\mathcal{B}$ a countable L-model, and $\{b_n : n < \omega\}$ an enumeration of the domain B of $\mathcal{B}$. Let

$$\Phi_{\mathcal{B}}$$

be the sentence (8.33), where $\varphi^{n_0 \cdots n_i}(x_0, y_0, \ldots, x_i, y_i)$ is the conjunction of all atomic and negated atomic L-formulas $\theta(x_0, y_0, \ldots, x_i, y_i)$ such that $\mathcal{B} \models \theta(b_0, b_{n_0}, \ldots, b_i, b_{n_i})$.

Lemma 8.50 *Suppose $\mathcal{A}$ is a countable L-model. Then the following are equivalent:*

(1) $\mathcal{A} \simeq_p \mathcal{B}$.
(2) $\mathcal{A} \models \Phi_{\mathcal{B}}$.

Proof (1) $\rightarrow$ (2). Suppose $\{\tau_n : n \in \mathbb{N}\}$ is a winning strategy of **II** in $\mathrm{EF}_\omega(\mathcal{A}, \mathcal{B})$. We define a winning strategy $\{(\bar{\tau}_n, \bar{\tau}'_n) : n \in \mathbb{N}\}$ of **II** in the game (2):

$$\bar{\tau}_0() = \tau_0(b_0)$$
$$\bar{\tau}'_0(c_0) = n_0, b_{n_0} = \tau_1(b_0, c_0)$$
$$\bar{\tau}_{i+1}(c_0, \ldots, c_i) = \tau_{2i+1}(b_0, c_0, \ldots, b_{i+1})$$
$$\bar{\tau}'_{i+1}(c_0, \ldots, c_{i+1}) = n_{i+1}$$
$$b_{n_{i+1}} = \tau_{2i+2}(b_0, c_0, \ldots, b_{i+1}, c_{i+1}).$$

(2) $\rightarrow$ (1). Suppose $\{(\bar{\tau}_n, \bar{\tau}'_n) : n \in \mathbb{N}\}$ is a winning strategy of **II** in the game (2). By Exercise 5.69 it suffices to give a winning strategy $\{\tau_n : n \in \mathbb{N}\}$ of **II** in $\mathrm{EF}^{**}_\omega(\mathcal{A}, \mathcal{B})$. We define the strategy as follows:

$$\tau_0(a_0) = \bar{\tau}_0(b_0)$$
$$\tau'_0(c_0) = n_0, b_{n_0} = \tau_1(b_0, c_0)$$
$$\tau_{i+1}(c_0, \ldots, c_i) = \tau_{2i+1}(b_0, c_0, \ldots, b_{i+1})$$
$$\tau'_{i+1}(c_0, \ldots, c_{i+1}) = n_{i+1}$$
$$b_{n_{i+1}} = \tau_{2i+2}(b_0, c_0, \ldots, b_{i+1}, c_{i+1}).$$

$\square$

The point of expressing condition (1) of Lemma 8.50 in the form of the game formula $\Phi_{\mathcal{B}}$ is that there is a procedure of *approximating* closed Vaught formulas, just as there was one for closed game formulas. By analyzing the approximations we will get useful information about the model $\mathcal{B}$.

Definition 8.51 We shall associate the Vaught formula (8.33) Φ with a sequence $\Phi_\gamma, \gamma \in On$, of $L_{\omega_1\omega}$-formulas as follows:

$$\Phi_0^{\bar{n}_i}(\bar{x}_i, \bar{y}_i) = \bigwedge_{j<i} \varphi^{\bar{n}_j}(\bar{x}_j, \bar{y}_j)$$
$$\Phi_{\gamma+1}^{\bar{n}_{i-1}}(\bar{x}_{i-1}, \bar{y}_{i-1}) = \exists x_i \forall y_i \bigvee_{n_i} \Phi_\gamma^{\bar{n}_i}(\bar{x}_i, \bar{y}_i)$$
$$\Phi_\nu^{\bar{n}_i}(\bar{x}_i, \bar{y}_i) = \bigwedge_{\gamma<\nu} \Phi_\gamma^{\bar{n}_i}(\bar{x}_i, \bar{y}_i) \text{ for limit } \nu.$$

The following properties of the approximations $\Phi_\gamma^{\bar{n}}(\bar{x}, \bar{y})$ are proved easily by transfinite induction:

- $\Phi_\gamma^{\bar{n}}(\bar{x}, \bar{y}) \in L_{\kappa\omega}$ for $\gamma < \kappa$.
- $qr(\Phi_\nu^{\bar{n}}(\bar{x}, \bar{y})) = \nu$ for limit ν.
- $\models \Phi \to \Phi_\gamma^{\emptyset}$ for all γ.
- $\models \Phi_\gamma^{\bar{n}}(\bar{x}, \bar{y}) \to \Phi_\beta^{\bar{n}}(\bar{x}, \bar{y})$ for $\beta \leq \gamma$.

The formula $(\Phi_{\mathcal{M}})_\alpha^{\bar{n}}(\bar{x}, \bar{y})$ is mutatis mutandis exactly what we denoted by $\sigma_{\mathcal{M}, a_0, \ldots, a_{n-1}}^\alpha(a_0, \ldots, a_{n-1})$ in Definition 7.50. Recall the definition of the game $\text{EFD}_\gamma^{**}(\mathcal{A}, \mathcal{B})$ from Exercise 7.9. Now:

Lemma 8.52 *The following are equivalent:*

(1) Player **II** *has a winning strategy in* $\text{EFD}_{\nu+2n}^{**}(\mathcal{A}, \mathcal{B})$
(2) $\mathcal{A} \models (\Phi_{\mathcal{B}})_{\nu+n}^{\emptyset}$.

Proof One can easily prove by induction the more general equivalence of the following two statements:
(1)′ Player **II** has a winning strategy in the game $\text{EFD}_{\nu+2n}(\mathcal{A}, \mathcal{B})$ in the position $\{(a_0, b_{k_0}), \ldots, (a_{n-1}, b_{k_{n-1}})\}$.
(2)′ $\mathcal{A} \models (\Phi_{\mathcal{B}})_{\nu+n}^{k_0 \ldots k_{n-1}}(a_0, \ldots, a_{n-1})$.   □

Thus we can think of all the formulas $\sigma_{\mathcal{M}, a_0, \ldots, a_{n-1}}^\alpha(a_0, \ldots, a_{n-1})$, $\alpha < \omega_1$, as arising from one single game formula $\Phi_{\mathcal{M}}$ by the process of approximation.

After the above examples of using simple forms of Vaught-formulas to study the isomorphism type of a countable model, we now turn our attention to the general concept.

Recall the definition of a *PC*-class from Definition 6.45. The concept of a $PC(L_{\omega_1\omega})$-class is defined similarly: A class K of L-structures is a $PC(L_{\omega_1\omega})$-*class* if there is an L'-sentence $\varphi \in L_{\omega_1\omega}$ for some $L' \supseteq L$ such that $K = \{\mathcal{M} \restriction L : \mathcal{M} \models \varphi\}$. We now prove that all properties expressed by closed Vaught formulas are in fact $PC(L_{\omega_1\omega})$ properties.

Lemma 8.53 *Suppose Ψ is a closed Vaught sentence of $L_{\omega_1\omega}$ in the vocabulary L. There is an $L_{\omega_1\omega}$-sentence φ of vocabulary $L \cup \{R_i : i < \omega\}$ so that the following are equivalent:*

(1) $\mathcal{A} \models \Psi$.
(2) $\mathcal{A} \models \exists R_0 \exists R_1 \ldots \varphi$.

Proof Let us take new predicate symbols $R^{\bar{n}, \bar{m}}(\bar{x}, \bar{y})$ for all $\bar{n}, \bar{m} \in \omega$. We let φ be the conjunction of the following sentences:

1. $R^{\emptyset, \emptyset}(\ )$.

2. $\forall \bar{x}_i, \bar{y}_i \bigwedge_{\bar{n}_i, \bar{m}_i} (R^{\bar{n}_i, \bar{m}_i}(\bar{x}_i, \bar{y}_i)$
$$\rightarrow \bigwedge_{j \leq i} \varphi^{n_0, m_0, \dots, n_j, m_j}(x_0, y_0, \dots, x_j, y_j)).$$

3. $\forall \bar{x}_{i-1}, \bar{y}_{i-1} \bigwedge_{\bar{n}_{i-1}, \bar{m}_{i-1}} (R^{\bar{n}_{i-1}, \bar{m}_{i-1}}(\bar{x}_{i-1}, \bar{y}_{i-1})$
$$\rightarrow \forall x_i \bigvee_{n_i} \bigwedge_{m_i} \exists y_i R^{\bar{n}_i, \bar{m}_i}(\bar{x}_i, \bar{y}_i)).$$

It is easy to see that this φ works. □

Proposition 8.54 *Suppose $\varphi \in L_{\omega_1 \omega}$. There is a closed Vaught sentence Ψ such that the following are equivalent for all countable L-models $\mathcal{A}$:*

(1) $\mathcal{A} \models \exists R_0 \exists R_1 \dots \varphi$
(2) $\mathcal{A} \models \Psi$.

Proof We shall first express the condition $\mathcal{A} \models \exists R_0 \exists R_1 \dots \varphi$ in a form which is easier to transform into a game form. Let $L' = L \cup \{R_0, \dots, R_n\}$ and $L'' = L' \cup \{d_n : n < \omega\}$, where the constants d_n are all new. Let S be the smallest set of $L_{\omega_1 \omega}$-formulas of the vocabulary L'' which contains all atomic formulas and all subformulas of φ and is closed under operations of first-order logic including substituting constants into free variables in a formula. We assume for simplicity that formulas in S have been built from atomic and negated atomic formulas using $\bigwedge, \bigvee, \exists$, and $\forall$.

As an intermediate step, we shall prove that the following are equivalent:

(1) $\mathcal{A} \models \exists R_0 \exists R_1 \dots \varphi$.
(2) There is an enumeration $\{a_n : n < \omega\}$ of the domain of $\mathcal{A}$ and a winning strategy of **II** in the Model Existence Game of $L_{\omega_1 \omega}$ in the extended vocabulary $L \cup \{R_0, R_1, \dots\}$ with new constants $\{c_0, c_1, \dots\}$ such that the following extra condition is satisfied: Player **II** plays an atomic formula $\theta(c_0, \dots, c_{n-1})$ only if $\mathcal{A} \models \theta(a_0, \dots, a_{n-1})$.

Suppose (1) holds. Let $P_0, \dots, P_n$ be predicates on the domain A of $\mathcal{A}$ so that $(\mathcal{A}, P_0, \dots, P_n) \models \varphi$. Let $\{a_n : n < \omega\}$ be an enumeration of A. Now **II** can simply always play formulas $\psi(c_0, \dots, c_{m-1})$ for which $(\mathcal{A}, P_0, \dots, P_n) \models \psi(a_0, \dots, a_{m-1})$. The extra condition is automatically satisfied. Conversely, if (2) holds then the proof of the Model Existence Theorem of $L_{\omega_1 \omega}$ (Theorem 8.12) gives a model $(\mathcal{A}', P_0, \dots, P_m)$ such that $\mathcal{A} \cong \mathcal{A}'$. Hence (1) holds. □

To end the proof of Proposition 8.54 one has to write condition (2) in the form of a closed Vaught formula Ψ. This involves judicious coding of all possible moves of **I**. We omit the details and refer to Vaught (1973) and Makkai (1977) for details.

Approximations of an arbitrary closed Vaught formula (8.19) are defined as above in the simpler case (8.33):

$$\Phi_0^{\bar{n}_i,\bar{m}_i}(\bar{x}_i,\bar{y}_i) = \bigwedge_{j<i} \varphi^{\bar{n}_j,\bar{m}_j}(\bar{x}_j,\bar{y}_j)$$
$$\Phi_{\gamma+1}^{\bar{n}_{i-1},\bar{m}_{i-1}}(\bar{x}_{i-1},\bar{y}_{i-1}) = \forall x_i \bigvee_{n_i} \bigwedge_{m_i} \exists y_i \Phi_\gamma^{\bar{n}_i,\bar{m}_i}(\bar{x}_i,\bar{y}_i)$$
$$\Phi_\nu^{\bar{n}_i,\bar{m}_i}(\bar{x}_i,\bar{y}_i) = \bigwedge_{\gamma<\nu} \Phi_\gamma^{\bar{n}_i,\bar{m}_i}(\bar{x}_i,\bar{y}_i).$$

The following properties of the approximations $\Phi_\gamma^{\bar{n},\bar{m}}(\bar{x},\bar{y})$ are proved easily by transfinite induction. The last claim is proved just as Proposition 7.19.

- $\Phi_\gamma^{\bar{n},\bar{m}}(\bar{x},\bar{y}) \in L_{\kappa\omega}$ for $\gamma < \kappa$.
- $qr(\Phi_\gamma^{\bar{n},\bar{m}}(\bar{x},\bar{y})) = \nu$ for limit ν.
- $\models \Phi \to \Phi_\gamma^\emptyset$ for all γ.
- $\models \Phi_\gamma^{\bar{n},\bar{m}}(\bar{x},\bar{y}) \to \Phi_\beta^{\bar{n},\bar{m}}(\bar{x},\bar{y})$ for $\beta < \gamma$.
- If $\mathcal{A}$ is countable and $\mathcal{A} \models \Phi_\gamma^\emptyset$ for all $\gamma < \omega_1$, then $\mathcal{A} \models \Phi$.

The approximations are most useful in an analysis of PC-definability, thanks to Proposition 8.54. Note that if

$$\Psi = \exists y_0 \cdots \exists y_i \cdots \bigwedge_n y_{i+1} < y_i$$

is the closed Vaught sentence (in fact a closed game sentence) defining non-well-founded relations, then mutatis mutandis

$$\Psi_{\alpha+1}^n(y_0,\ldots,y_{n-1}) = \exists y_n \Psi_\alpha^{n+1}(y_0,\ldots,y_n)$$

and

$$\Psi_0^n(y_0,\ldots,y_{n-1}) = y_{n-1} < y_{n-2} \wedge \cdots \wedge y_1 < y_0,$$

so up to logical equivalence

$$\Psi_\alpha^0 = \bigwedge_{\alpha_0<\alpha} \exists y_0 \left(\top \wedge \bigwedge_{\alpha_1<\alpha_0} \exists y_1 \left(y_1 < y_0 \wedge \bigwedge_{\alpha_2<\alpha_1} \exists y_2(y_2 < y_1 \wedge \cdots) \right) \right),$$

where $\top$ is a logical truth (e.g. $\forall x_0(\approx x_0 x_0)$). From this is is easy to see that $\Psi_\gamma^\emptyset$ is true in a linear order if and only if its order-type is $\geq \gamma$. As a more interesting application we shall now use the approximations to prove the Interpolation Theorem for $L_{\omega_1\omega}$. Note that we have previously used the Model Existence Theorem to prove Theorem 8.14, which is another formulation of the Craig Interpolation Theorem.

Theorem 8.55 (*Craig Interpolation Theorem for $L_{\omega_1\omega}$*) *Suppose $\mathcal{K}_1$ and $\mathcal{K}_2$ are disjoint $PC(L_{\omega_1\omega})$-classes. Then there is an $L_{\omega_1\omega}$-definable class $\mathcal{K}$ so that $\mathcal{K}_1 \subseteq \mathcal{K}$ and $\mathcal{K} \cap \mathcal{K}_2 = \emptyset$.*

Proof By Proposition 8.54 there is a closed Vaught formula Φ that defines K_1. Let $\Phi_\gamma^\emptyset$ be the sequence of approximations of Φ.

Case 1. For some $\gamma < \omega_1$ we have $Mod(\Phi_\gamma^\emptyset) \cap K_2 = \emptyset$. In this case we simply let $K = Mod(\Phi_\gamma^\emptyset)$ and we are done.

Case 2. For all $\gamma < \omega_1$ we have $Mod(\Phi_\gamma^\emptyset) \cap K_2 \neq \emptyset$. We shall see that this case contradicts the fact (Theorem 8.13) that the class of well-orders is not a $PC(L_{\omega_1\omega})$-class. Let $<$ and $F^{\bar{n},\bar{m}}(v,\bar{x},\bar{y})$ be new predicate symbols for $\bar{n}, \bar{m} \in \omega$. Let $\Phi_<$ be the conjunction of the universal closures of the formulas

1. "$(U,<)$ is a linear order".
2. $U(v) \to F^\emptyset(v)$.
3. $F^{\bar{n},\bar{m}}(v,\bar{x},\bar{y}) \to \bigwedge_{j \leq i} \Phi^{n_0,m_0,\ldots,n_j m_j}(x_0, y_0, \ldots, x_j, y_j)$.
4. $(u < v \wedge F^{\bar{n},\bar{m}}(v,\bar{x},\bar{y})) \to \forall x_i \bigvee_{n_i} \bigwedge_{m_i} \exists y_i F^{\bar{n},\bar{m},n_i,m_i}(u,\bar{x},\bar{y},x_i,y_i)$.
5. $F^{\bar{n},\bar{m}}(v,\bar{x},\bar{y}) \to \bigwedge_{\gamma<\nu} F_\gamma^{\bar{n},\bar{m}}(\bar{x},\bar{y})$ for limit ν.

Claim 1. If $\mathcal{A}$ is countable, $(\mathcal{A}, U, <) \models \Phi_<$ and $(U,<)$ is non-well-founded, then $\mathcal{A} \in K_1$.

To prove this, assume $e_0 > e_1 > \ldots$ is a descending chain in $\mathcal{A}$. Player **II** wins the game that defines the truth of $\mathcal{A} \models \Phi$ with the following strategy: Suppose **I** has played the elements $a_0, m_0, \ldots, a_{i-1}, m_{i-1}$ and **II** has played the elements $b_0, n_0, \ldots, b_{i-1}, n_{i-1}$. We assume as an induction hypothesis that

$$\mathcal{A} \models \forall x_i \bigvee_{n_i} \bigwedge_{m_i} \exists y_i F^{\bar{n},\bar{m},n_i,m_i}(u,\bar{a},\bar{b},x_i,y_i).$$

Now **I** plays a_i and then **II** chooses m_i such that

$$\mathcal{A} \models \bigwedge_{m_i} \exists y_i F^{\bar{n},\bar{m},n_i,m_i}(u,\bar{a},\bar{b},a_i,y_i).$$

Now **I** chooses m_i and then **II** chooses b_i so that

$$\mathcal{A} \models F^{n_0,m_0,\ldots,n_i,m_i}(e_{i+1},a_0,b_0,\ldots,a_i,b_i).$$

Since we have

$$\mathcal{A} \models \Phi^{n_0,m_0,\ldots,n_i,m_i}(e_{i+1},a_0,b_0,\ldots,a_i,b_i),$$

this is a winning strategy of **II**. Claim 1 is proved.

Claim 2. Suppose $\mathcal{A}$ is an infinite model and γ is a countable ordinal so that $\mathcal{A} \models \Phi_\gamma^\emptyset$. If $<$ is a well-ordering of a subset U of A so that the order-type of $(U,<)$ is at most γ, then $(\mathcal{A}, U, <) \models \Phi_<$.

To prove this, assume $\mathcal{A} \models \Phi_\gamma^\emptyset$. Since A is infinite and $\gamma < \omega_1$, we may assume $\gamma \subseteq A$. Let $\mathcal{U} = \gamma$ and define $F^{\bar{n},\bar{m}}(\beta,\bar{a},\bar{b})$ to hold in $\mathcal{A}$ if and only if $\mathcal{A} \models \Phi_\beta^{\bar{n},\bar{m}}(\bar{a},\bar{b})$. It is easy to see that this choice satisfies $\Phi_<$. Claim 2 is proved.

To end the proof we observe that we have proved the equivalence of the following two conditions:

(W1) $(U, <)$ is a countable well-ordering.
(W2) There is a countably infinite structure $\mathcal{A} \in \mathcal{K}_2$ so that $U \subseteq A$ and $(\mathcal{A}, U, <) \models \Phi_<$.

This contradicts Theorem 8.13.   □

Theorem 8.56 *(Covering Theorem) Suppose Φ is a closed Vaught sentence and φ a sentence of $L_{\omega_1\omega}$. If $\models \Phi \to \varphi$, then $\models \Phi_\gamma^\emptyset \to \varphi$ for some $\gamma < \omega_1$.*

Proof This proof is a replica of the proof of the Craig Interpolation Theorem. Therefore we omit some details.
Case 1. For some $\gamma < \omega_1$ we have $\models \Phi_\gamma^\emptyset \to \varphi$. In this case we are done.
Case 2. For all $\gamma < \omega_1$ we have a model for $\Phi_\gamma^\emptyset \wedge \neg\varphi$. Let $\Phi_<$ be as in the proof of the Craig Interpolation Theorem. We have again the equivalence of the following two conditions:

(1) $(U, <)$ is a countable well-ordering.
(2) There is a countably infinite structure $\mathcal{A}$ so that $U \subseteq A$ and $(\mathcal{A}, U, <) \models \Phi_< \wedge \neg\varphi$.

This contradicts Theorem 8.13.   □

The dual of a closed Vaught formula is an open Vaught formula:

Definition 8.57 Formulas Φ of the form

$$\exists x_0 \bigwedge_{n_0} \bigvee_{m_0} \forall y_0 \exists x_1 \bigwedge_{n_1} \bigvee_{m_1} \forall y_1 \dots \bigvee_i \varphi^{n_0 m_0 \dots n_i m_i}(x_0, y_0, \dots, x_i, y_i),$$

where the formulas $\varphi^{n_0 m_0 \dots n_i m_i}$ are from $L_{\omega_1\omega}$, are called *open Vaught formulas* of $L_{\omega_1\omega}$. If $\mathcal{A}$ is a model of the vocabulary of Φ, then we say that Φ is *true* in $\mathcal{A}$, $\mathcal{A} \models \Phi$, provided that there are functions τ_1 and τ_2 such that for all $b_0, b_1, \dots \in A$ and all $n_0, n_1, \dots \in \omega$ there is an $i < \omega$ such that

$$\mathcal{A} \models \varphi^{n_0 m_0 \dots n_i m_i}(a_0, b_0, \dots, a_i, b_i), \tag{8.34}$$

where $m_i = \tau_1(n_0, b_0, \dots, n_i)$ and $a_i = \tau_2(b_0, n_0, \dots, b_{i-1}, n_{i-1})$.

The following lemma is an immediate consequence of Lemma 8.53. Recall that the Semantic Games associated with Vaught formulas are determined. Thus if **II** does not have a winning strategy in the closed game arising from a closed Vaught formula, player **II** has a winning strategy in the open game arising from the open Vaught formula obtained by switching existential and universal quantifiers, disjunctions and conjunctions, and atomic formulas and their negations. Hence the negation of a closed Vaught formula is equivalent to an open Vaught formula, and vice versa.

Lemma 8.58 *Suppose Ψ is an open Vaught sentence of $L_{\omega_1\omega}$ of vocabulary L. There is an $L_{\omega_1\omega}$-sentence φ of vocabulary $L \cup \{R_i : i < \omega\}$ so that the following are equivalent:*

(1) $\mathcal{A} \models \Psi$.
(2) $\mathcal{A} \models \forall R_0 \forall R_1 \ldots \varphi$.

Proposition 8.59 *The class of well-ordered models $(A, <)$ is definable by an open game quantifier in $L_{\omega_1\omega}$ but not by a closed one. In fact, it is not definable by any expression of the form $\exists R_0 \exists R_1 \ldots \Phi$, where Φ is a closed Vaught formula.*

Proof To define the class of well-ordered models one adds the axiom

$$\forall x_0 \forall x_1 \ldots \bigvee_{i<\omega} \neg(x_{i+1} < x_i)$$

to the axioms of linear order. If this class were definable by an expression of the form $\exists R_0 \exists R_1 \ldots \Phi$, where Φ is a closed Vaught formula, it would by Lemma 8.53 be $PC(L_{\omega_1\omega})$-definable, contrary to Proposition 8.13. □

We now build a logic in which we combine closed and open Vaught formulas with other formulas and with each other.

Definition 8.60 The class of formulas of the logic $L_{\infty V}$ is the smallest set containing atomic formulas and closed under the following rules:

(1) If t and t' are L-terms, then $(0, t, t')$ is an L-formula denoted by $\approx tt'$.
(2) If $t_1, \ldots, t_n$ are L-terms, then $(1, R, t_1, \ldots, t_n)$ is an L-formula denoted by $Rt_1 \ldots t_n$.
(3) If φ is an L-formula, so is $(2, \varphi)$, and we denote it by $\neg\varphi$.
(4) If Φ is a set of L-formulas with a fixed finite set of free variables, then $(3, \Phi)$ is an L-formula and we denote it by $\bigwedge_{\varphi\in\Phi} \varphi$.
(5) If Φ is a set of L-formulas with a fixed finite set of free variables, then $(4, \Phi)$ is an L-formula and we denote it by $\bigvee_{\varphi\in\Phi} \varphi$.
(6) If φ is an L-formula and $n \in \mathbb{N}$, then $(5, \varphi, n)$ is an L-formula and we denote it by $\forall x_n \varphi$.
(7) If φ is an L-formula and $n \in \mathbb{N}$, then $(6, \varphi, n)$ is an L-formula and we denote it by $\exists x_n \varphi$.

(8) If $\varphi^{\bar{i},\bar{j}}(\bar{x},\bar{y})$ are L-formulas for $i_0 \in I_0,\ldots,i_{n-1} \in I_{n-1}$ and $j_0 \in J_0,\ldots,j_{n-1} \in J_{n-1}$, then so is $(7,f)$, where

$$f(\bar{i},\bar{j}) = \varphi^{\bar{i},\bar{j}}(\bar{x},\bar{y},\bar{z})$$

for all $\bar{i},\bar{j}$. We denote $(7,f)$ by

$$\forall x_0 \bigvee_{i_0} \bigwedge_{j_0} \exists y_0 \forall x_1 \bigvee_{i_1} \bigwedge_{j_1} \exists y_1 \ldots \bigwedge_k \varphi^{\bar{i},\bar{j}}(\bar{x}_i,\bar{y}_i,\bar{z}). \qquad (8.35)$$

(9) If $\varphi^{\bar{i},\bar{j}}(\bar{x},\bar{y})$ are L-formulas for $i_0 \in I_0,\ldots,i_{n-1} \in I_{n-1}, j_0 \in J_0,\ldots,$ $j_{n-1} \in J_{n-1}$, then so is $(8,f)$, where

$$f(\bar{i},\bar{j}) = \varphi^{\bar{i},\bar{j}}(\bar{x},\bar{y},\bar{z})$$

for all $\bar{i},\bar{j}$. We denote $(8,f)$ by

$$\exists x_0 \bigwedge_{i_0} \bigvee_{j_0} \forall y_0 \exists x_1 \bigwedge_{i_1} \bigvee_{j_1} \forall y_1 \ldots \bigvee_k \varphi^{\bar{i},\bar{j}}(\bar{x}_i,\bar{y}_i,\bar{z}). \qquad (8.36)$$

The logic $L_{\omega_1 V}$ is defined to consist of formulas of $L_{\infty V}$ that have only countable conjunctions and disjunctions in formulas of the form $\bigwedge \Phi$, $\bigvee \Phi$, $(7,f)$ and $(8,f)$.

The truth definition of $L_{\infty V}$ is as in Definition 7.36 with the addition

$$\mathcal{M} \models_s \forall x_0 \bigvee_{i_0} \bigwedge_{j_0} \exists y_0 \forall x_1 \bigvee_{i_1} \bigwedge_{j_1} \exists y_1 \ldots \bigwedge_k \varphi^{\bar{i},\bar{j}}(\bar{x}_i,\bar{y}_i,\bar{z})$$

if and only if

There are functions $f_i(x_0,w_0,\ldots,x_i)$ and $g_i(x_0,w_0,\ldots,x_i,w_i)$, $i \in \mathbb{N}$, such that for all $\bar{a}$ and $\bar{j}$ we have $\mathcal{M} \models_s \varphi^{\bar{i},\bar{j}}(\bar{x}_i,\bar{y}_i,\bar{z})$, where
$i_0 = f_0(a_0), b_0 = g_0(a_0,j_0)$
$i_1 = f_1(a_0,j_0,a_1), b_1 = g_1(a_0,j_0,a_1,j_1)$
$\ldots$

There is a certain intrinsic finiteness condition involved in the definition of the formulas of $L_{\infty\omega}$. To see this as clearly as possible it is useful to think of the formula (8.35) in the logically equivalent form

$$\forall x_0 \bigvee_{i_0} \bigwedge_{j_0} \exists y_0 (\varphi^{i_0,j_0}(x_0,y_0,\bar{z}) \wedge$$
$$\forall x_1 \bigvee_{i_1} \bigwedge_{j_1} \exists y_1 (\varphi^{i_0 i_1, j_0 j_1}(x_0,x_1,y_0,y_1,\bar{z}) \wedge$$
$$\ldots \qquad (8.37)$$
$$\forall x_{n-1} \bigvee_{i_{n-1}} \bigwedge_{j_{n-1}} \exists y_{n-1} (\varphi^{\bar{i},\bar{j}}(\bar{x},\bar{y},\bar{z}) \wedge$$
$$\ldots$$

Alternatively, we may define the semantics of $L_{\infty V}$ by means of a Semantic Game: Suppose L is a vocabulary, $\mathcal{M}$ is an L-structure, φ^* is an L-formula, and s^* is an assignment for M. The game $\text{SG}^{\text{sym}}(\mathcal{M}, \varphi^*)$ is defined as follows. In the beginning player **II** holds (φ^*, s^*). The rules of the game are as follows:

1. If φ is atomic, $\varphi = \neg\psi$, $\varphi = \bigwedge_{i \in I} \varphi_i$, $\varphi = \bigvee_{i \in I} \varphi_i$, $\varphi = \forall x_n \psi$, $\varphi = \exists x_n \psi$, then the game proceeds as in Definition 7.37

2. If $\varphi = (7, f)$ is the formula (8.35), then the player who holds (φ, s) switches to hold $(\varphi', s[a/x_0, b/y_0])$, where φ' is the conjunction

   $$\varphi^{i_0,j_0}(x_0, y_0, \bar{z}) \wedge$$
   $$\forall x_1 \bigvee_{i_1} \bigwedge_{j_1} \exists y_1 \forall x_2 \bigvee_{i_2} \bigwedge_{j_2} \exists y_2 \cdots \bigwedge_k \varphi^{i_0,\bar{i},j_0,\bar{j}}(x_0, y_0, \bar{x}_i, \bar{y}_i, \bar{z}),$$

   and the elements a, b, i_0, and j_0 are chosen as follows: Suppose the player who holds the formula is α and the other player is β. Then β chooses a, α chooses i_0, β chooses j_0, and then finally α chooses b. Note that the latter conjunct is again of the form $(7, g)$.

3. If $\varphi = (8, f)$ is the formula (8.36) then the player who holds (φ, s) switches to hold $(\varphi', s[a/x_0, b/y_0])$, where φ' is the disjunction

   $$\varphi^{i_0,j_0}(x_0, y_0, \bar{z}) \vee$$
   $$\exists x_1 \bigwedge_{i_1} \bigvee_{j_1} \forall y_1 \exists x_2 \bigwedge_{i_2} \bigvee_{j_2} \forall y_2 \cdots \bigvee_k \varphi^{i_0,\bar{i},j_0,\bar{j}}(x_0, y_0, \bar{x}_i, \bar{y}_i, \bar{z}),$$

   where the elements a, b, i_0, and j_0 are chosen as follows: Suppose the player who holds the formula is α and the other player is β. Then α chooses a, β chooses i_0, α chooses j_0, and then finally β chooses b. Note that the latter disjunction is again of the form $(8, g)$.

The definition of the logic $L_{\omega_1 V}$ may look complicated but its basic idea is very simple: in addition to having the usual logical operations $\bigwedge, \bigvee, \exists, \forall$ we are also allowed to iterate them in a sense indefinitely. Not all iterations are allowed. For example, the rules of forming formulas of $L_{\omega_1 V}$ do not allow the following formula:

$$\forall x_1 \forall x_2 \cdots \forall x_n \cdots \exists y \bigwedge_{n < \omega} \neg \approx y x_n.$$

That this formula, which clearly says that the universe is uncountable, is not expressible in $L_{\omega_1 V}$ follows from the following Downward Löwenheim–Skolem Theorem. We formulate this theorem for $L_{\omega_1 V}$ only but there is an obvious generalization to $L_{\infty V}$ based on approximations of models and formulas, as in Theorem 8.9.

Theorem 8.61 *(Downward Löwenheim–Skolem Theorem) Suppose L is a countable vocabulary, $\mathcal{A}$ is an L-model, and φ is a sentence of $L_{\omega_1 V}$. If $\mathcal{A} \models$*

φ, then $\mathcal{B} \models \varphi$ for some countable substructure $\mathcal{B}$ of $\mathcal{A}$. In fact there is a set S of countable substructures of $\mathcal{A}$ so that every model in S satisfies φ and the set of domains of models in S is a closed unbounded set of countable subsets of A.

Proof This theorem could be proved with the method of Theorem 8.9, but we use the different albeit related method of Theorem 8.28. We associate every formula $\psi(x_0, \dots, x_n)$ of $L_{\omega_1 V}$ with a set of functions on A, denoted $\mathcal{F}_{\psi(x_0, \dots, x_n)}$, so that the following condition holds:

If $B \subseteq A$ contains all the constants of $\mathcal{A}$ and is closed under the functions in $\mathcal{F}_{\psi(x_0, \dots, x_n)}$ and the interpretations of the function symbols of $\mathcal{A}$, then

$$\mathcal{A} \cap B \models \psi(a_0, \dots, a_n) \iff \mathcal{A} \models \psi(a_0, \dots, a_n)$$

for all $a_0, \dots, a_n \in B$.

The definition of $\mathcal{F}_{\psi(x_0, \dots, x_n)}$ is done by induction on the structure of the formula $\varphi(x_0, \dots, x_n)$.

1. If $\psi(x_0, \dots, x_n)$ is an atomic formula, then $\mathcal{F}_{\psi(x_0, \dots, x_n)} = \emptyset$.
2. $\mathcal{F}_{\neg\psi(x_0, \dots, x_n)} = \mathcal{F}_{\psi(x_0, \dots, x_n)}$.
3. $\mathcal{F}_{\bigwedge_{n<\omega} \psi_n(x_0, \dots, x_n)} = \bigcup_{n<\omega} \mathcal{F}_{\psi_n(x_0, \dots, x_n)}$.
4. $\mathcal{F}_{\bigvee_{n<\omega} \psi_n(x_0, \dots, x_n)} = \bigcup_{n<\omega} \mathcal{F}_{\psi_n(x_0, \dots, x_n)}$.
5. To define $\mathcal{F}_{\exists x_{n+1} \psi(x_0, \dots, x_{n+1})}$ define a function $f(x_0, \dots, x_n)$ in A as follows: If $\mathcal{A} \models \psi(a_0, \dots, a_n, a_{n+1})$ for some a_{n+1}, let $f(a_0, \dots, a_n)$ be such an a_{n+1}. Otherwise $f(a_0, \dots, a_n)$ is given a fixed but arbitrary value. Now we let $\mathcal{F}_{\exists x_{n+1} \psi(x_0, \dots, x_{n+1})}$ be $\mathcal{F}_{\psi(x_0, \dots, x_n)} \cup \{f\}$.
6. $\mathcal{F}_{\forall x_{n+1} \psi(x_0, \dots, x_{n+1})} = \mathcal{F}_{\exists x_{n+1} \neg\psi(x_0, \dots, x_{n+1})}$.
7. Suppose $\Psi(\bar{z})$ is obtained from the formulas $\varphi^{i,j}(\bar{x}, \bar{y}, \bar{z})$ as in (8.35). For any $\bar{a}$, if $\mathcal{A} \models \Psi(\bar{a})$, let $f_i^{\bar{a}}(x_0, w_0, \dots, x_i)$, $g_i^{\bar{a}}(x_0, w_0, \dots, x_i, w_i)$, $i \in \mathbb{N}$, be the winning strategy of **II**. If $\mathcal{A} \not\models \Psi(\bar{a})$, we let $f_i^{\bar{a}}(x_0, w_0, \dots, x_i) = x_0$, $g_i^{\bar{a}}(x_0, w_0, \dots, x_i, w_i) = x_0$ for $i \in \mathbb{N}$. Let $f_i(x_0, w_0, \dots, x_i, \bar{a}) = f_i^{\bar{a}}(x_0, w_0, \dots, x_i)$, $g_i(x_0, w_0, \dots, x_i, w_i, \bar{a}) = g_i^{\bar{a}}(x_0, w_0, \dots, x_i, w_i)$, $i \in \mathbb{N}$, Finally, let $\mathcal{F}_{\Psi(\bar{z})}$ be the union of the sets $\mathcal{F}_{\varphi^{i,j}(\bar{x}, \bar{y}, \bar{z})}$ and the functions $f_i(\bar{x}, \bar{w}, \bar{z})$ and $g_i(\bar{x}, \bar{w}, \bar{z})$ for $i \in \mathbb{N}$.
8. The case of open game quantifier is entirely similar to the previous case.

It is now easy to see that the sets $\mathcal{F}_{\psi(x_0, \dots, x_n)}$ satisfy the required condition. Let $\mathcal{C}$ be the set of those countable subsets of A that are closed under the functions of $\mathcal{F}_{\varphi}$. Let S be the set of structures $\mathcal{A} \cap B$, $B \in \mathcal{C}$. Then S is the required set. $\qquad\square$

Another important model-theoretic property of $L_{\infty V}$ is that its elementary equivalence is not more expressive than that of $L_{\infty\omega}$. A clear indication that this may be the case is given by the existence of the approximations of closed Vaught formulas.[2]

Theorem 8.62 *Let L be a countable vocabulary and A and A' two L-models. Then the following are equivalent:*

(1) $A \cong_p A'$.
(2) $A \equiv_{L_{\infty V}} A'$.

Proof Assume first (2). Then $A \equiv_{\infty\omega} B$, so by Proposition 7.48 (1) follows. Suppose then (1). Let us assume that **II** has a winning strategy τ in the game $E = EF_\omega(A, A')$. We show by induction on the structure of $\varphi(z_0, \dots, z_n)$: If $c_0, c_0', \dots, c_n, c_n'$ is a position in the game $EF_\omega(A, A')$ while **II** plays τ, then

$$A \models \varphi(c_0, \dots, c_n) \iff A' \models \varphi(c_0', \dots, c_n').$$

We need only consider the case that $\varphi(z_0, \dots, z_n)$ is of the form (6) of the above Definition 8.60. Let us suppose $A \models \varphi(c_0, \dots, c_n)$ and prove $A' \models \varphi(c_0', \dots, c_n')$. To this end we assume **II** has a winning strategy σ in the game $G = SG^{sym}(A, \varphi(c_0, \dots, c_n))$, and we have to describe a winning strategy ρ of **II** in $H = SG^{sym}(A', \varphi(c_0', \dots, c_n'))$.

The strategy is basically very straightforward, although it takes a few lines to write it down. The strategy consists of repeating the below steps 1–7 for $i = 0, 1, \dots$. Figure 8.9 presents these steps in schematic form.

1. **I** decides to play $a_i' \in A'$ in H and we let **I** play this a_i' in E.
2. Strategy τ directs **II** to play a_i in E and we let **I** play this a_i in G.
3. Strategy σ directs **II** to play n_i in G and we define the next move of **II** in H according to ρ to be this n_i.
4. **I** decides to play m_i in H and we let **I** play this m_i in G.
5. Strategy σ directs **II** to play b_i in G and we let **I** play this b_i in E.
6. Strategy τ directs **II** to play b_i' in E and we define the next move of **II** in H according to ρ to be this b_i'.
7. Score: Since σ is a winning strategy of **II**, $A \models \varphi^{\bar{n},\bar{m}}(\bar{a}, \bar{b}, \mathbf{c})$. The induction hypothesis together with the fact that τ is a winning strategy of **II** in E imply $A' \models \varphi^{\bar{n},\bar{m}}(\bar{a}', \bar{b}', \mathbf{c}')$. Therefore **II** has not lost yet.

$\square$

---

[2] Approximations can be defined for the entire $L_{\infty V}$.

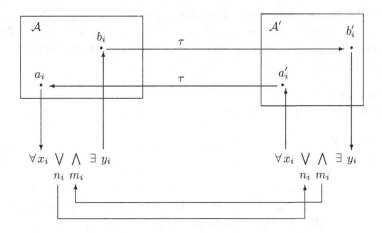

Figure 8.9 Strategy ρ.

Although elementary equivalence relative to $L_{\infty G}$ coincides with elementary equivalence relative to $L_{\infty \omega}$, there are properties expressible in $L_{\infty G}$ which are not expressible even in $L_{\infty \infty}$, as we shall prove in Proposition 9.38.

8.7 Historical Remarks and References

Good sources for the model theory of infinitary logic are Keisler (1971) and Makkai (1977). Theorem 8.10 is from Kueker (1972, 1977). Theorem 8.12 is from Keisler (1971), where the method of consistency properties is first presented in the context of infinitary logic. Subsequently it was extensively used in infinitary logic in Makkai (1969b,a), Harnik and Makkai (1976), and Green (1975). Theorem 8.13 and its two vorollaries are from Lopez-Escobar (1966b,a). The strong formulation of Theorem 8.13 is from Keisler (1971). Theorem 8.14 is from Lopez-Escobar (1965).

By means of the Model Existence Game (or consistency properties) many variations of the Craig Interpolation Theorem can be proved for $L_{\omega_1 \omega}$, as demonstrated convincingly in Keisler (1971). We have collected some of them in Exercises 8.10–8.17.

Theorem 8.22 is from Morley (1968). Theorem 8.31 goes back to Morley (1968), but the present proof is due to Shelah. Proposition 8.32 is from Lopez-Escobar (1966b,a). Proposition 8.41 is essentially due to Keisler (1965). A useful source for game quantifiers is Burgess (1977). Approximations of game formulas were first considered in set theory in Kuratowski (1966). For more on Souslin–formulas, see Burgess (1978) and Green (1978).

Exercises 8.10–8.12 and 8.16 are from Lopez-Escobar (1965). Exercises 8.13-8.14 are from Malitz (1969). Exercise 8.15 is from Barwise (1969). Exercise 8.17 is from Makkai (1969b), which is a good source for all the Exercises 8.10–8.17. Exercise 8.30 is from Sierpiński (1933). Exercise 8.46 is from Burgess (1978), where also related results are proved. Exercise 8.47 is from Green (1979).

Exercises

8.1 Find an uncountable model $\mathcal{M}$ such that there is no countable $\mathcal{N}$ with $\mathcal{N} \equiv_{\infty\omega} \mathcal{M}$.

8.2 Let L be the vocabulary $\{E\} \cup \{c_n : n \in \mathbb{N}\}$, where each c_n is a constant symbol and E is a binary relation symbol. Let for $A \subseteq \mathbb{N}$ the sentence φ_A be

$$\left(\bigwedge_{n<m<\omega} \neg \approx c_n c_m \right) \wedge \exists x_0 \forall x_1 \left(x_1 E x_0 \leftrightarrow \bigvee_{n \in A} \approx c_n x_1 \right).$$

Let $\Phi = \{\varphi_A : A \subseteq \mathbb{N}\}$. Show that the sentence $\bigwedge\{\varphi : \varphi \in \Phi \cap X\}$ has a model whatever X is, but it has a countable model if and only if X is countable.

8.3 Suppose $\mathcal{M} = (\alpha + \alpha, <)$. Show that $\mathcal{M}^X \cong (\beta_X + \beta_X, <)$ for some β_X for almost all X.

8.4 Prove the claim of Example 8.4.

8.5 A class K of models is *closed* if $\mathcal{M} \in K$ if and only if $\mathcal{M}^X \in K$ for almost all X. Suppose a closed class contains, up to isomorphism, only countably many countable models. Show that it is definable in $L_{\omega_1\omega}$.

8.6 Show that the class of models $(\mathcal{M}, P)$, where P is $L_{\infty\omega}$-definable on $\mathcal{M}$, is a closed class.

8.7 Show that the class of countable well-orders is the union of two disjoint closed classes.

8.8 Suppose $(A, <)$ is an uncountable linear order. Show that $(A, <)$ contains a copy of ω_1, a copy of ω_1^*, or a copy of the rationals. (Hint: Assume not. Prove first that there is $a \in A$ such that both $(\leftarrow, a]$ and $[a, \rightarrow)$ are uncountable in $(A, <)$.)

8.9 Show that if φ is a sentence of $L_{\omega_1\omega}$ in a vocabulary which contains the unary predicate U and the binary predicate $<$ and φ has a model $\mathcal{M}$ with

$(U^{\mathcal{M}}, <^{\mathcal{M}})$ an uncountable linear order, then φ has a model $\mathcal{N}$ such that $(U^{\mathcal{N}}, <^{\mathcal{N}})$ contains a copy of the rationals.

8.10 (Lyndon Interpolation Theorem) Suppose φ and ψ are sentences in $L_{\omega_1\omega}$ and $\models \varphi \to \psi$. Show that there is θ in $L_{\omega_1\omega}$ such that $\models \varphi \to \theta$, $\models \theta \to \psi$, every relation symbol occurring positively (negatively) in θ occurs positively (negatively) in φ and ψ.

8.11 Assume in Exercise 8.10 the sentences φ and ψ have no function or constant symbols, and no identity. Assume also $\not\models \neg\varphi$ and $\not\models \psi$. Show that θ can be chosen so that it does not contain identity.

8.12 Suppose φ and ψ are sentences of $L_{\omega_1\omega}$ such that if $\mathcal{M}$ and $\mathcal{N}$ are models of φ, $\mathcal{N}$ is a homomorphic image of $\mathcal{M}$, and $\mathcal{M} \models \psi$, then $\mathcal{N} \models \psi$. Show that there is a positive $L_{\omega_1\omega}$- sentence θ such that $\varphi \models \psi \leftrightarrow \theta$.

8.13 Suppose L_1 and L_2 are vocabularies which contain no function symbols. Let φ be an L_1-sentence and ψ an L_2-sentence of $L_{\omega_1\omega}$ such that ψ is universal and $\models \varphi \rightarrow \psi$. Show that there is a universal $L_1 \cap L_2$-sentence θ of $L_{\omega_1\omega}$ such that $\models \varphi \to \theta$ and $\models \theta \to \psi$.

8.14 Suppose φ and ψ are sentences of $L_{\omega_1\omega}$ such that if $\mathcal{M}$ and $\mathcal{N}$ are models of φ, $\mathcal{N}$ is a submodel of $\mathcal{M}$, and $\mathcal{M} \models \psi$, then $\mathcal{N} \models \psi$. Show that there is a universal sentence θ of $L_{\omega_1\omega}$ such that $\varphi \models \psi \leftrightarrow \theta$.

8.15 Suppose φ and ψ are sentences of $L_{\omega_1\omega}$. Show that every countable model of φ can be embedded in some countable model of ψ if and only if every universal logical consequence of ψ in $L_{\omega_1\omega}$ is a logical consequence of φ.

8.16 Suppose φ and ψ are sentences of $L_{\omega_1\omega}$. Show that every homomorphic image of a model of φ is a model of ψ if and only if there is a positive sentence θ in $L_{\omega_1\omega}$ such that $\models \varphi \to \theta$ and $\models \theta \to \psi$.

8.17 The class K of countable structures $\mathcal{B}$ such that $\mathcal{B}$ is isomorphic to a substructure of some model $\mathcal{A}$ of $\varphi \in L_{\omega_1\omega}$ is identical to the class of all countable models of the set of universal sentences $\theta \in L_{\omega_1\omega}$ such that $\varphi \models \psi$.

8.18 Consider the following game $G^\kappa_{\mathrm{cub}}(\mathcal{C})$ where $\mathcal{C}$ is a set of subsets of M of cardinality $\leq \kappa$. Players **I** and **II** play as in $G_{\mathrm{cub}}(\mathcal{C})$ but the game goes on for κ rounds producing a set $X = \{x_\alpha : \alpha < \kappa\} \cup \{y_\alpha : \alpha < \kappa\}$. Player **II** wins if $X \in \mathcal{C}$. Show that if **II** has a winning strategy in $G^\kappa_{\mathrm{cub}}(\mathcal{C})$ and $G^\kappa_{\mathrm{cub}}(\mathcal{D})$, then she has a winning strategy in $G^\kappa_{\mathrm{cub}}(\mathcal{C} \cap \mathcal{D})$.

8.19 Show that if $\mathcal{F}$ is a set of finitary functions on M, $|\mathcal{F}| = \kappa$, and $\mathcal{C}$ is the set of $X \subseteq M$ of cardinality κ closed under each function in $\mathcal{F}$, then **II** has a winning strategy in $G^\kappa_{\mathrm{cub}}(\mathcal{C})$. Use this to conclude that if $\mathcal{M}$ is an L-structure, $|L| \leq \kappa$, $\varphi \in L_{\kappa^+\omega}$, $\mathcal{M} \models \varphi$ and $\mathcal{C}$ is the set of domains of $\mathcal{M}_0 \subseteq \mathcal{M}$ with $\mathcal{M}_0 \models \varphi$, then **II** has a winning strategy in $G^\kappa_{\mathrm{cub}}(\mathcal{C})$.

8.20 Show that we can extend each vocabulary L to L^*, translate each $L_{\kappa^+\omega}$-sentence φ in the vocabulary L to an $L_{\kappa^+\omega}$-sentence φ^* in the vocabulary L^*, expand each L-structure $\mathcal{M}$ to an L^*-structure $\mathcal{M}^*$, and extend any L-fragment $\mathcal{T}$ to a set $\mathcal{T}^*$ of cardinality κ of $L_{\kappa^+\omega}$-sentences in the vocabulary L^* such that for all L-structures $\mathcal{M}$, L^*- structures $\mathcal{N}$, L-fragments $\mathcal{T}$ and $L_{\kappa^+\omega}$-sentences $\varphi \in \mathcal{T}$ in the vocabulary L:

  (1) $\mathcal{M} \models \varphi \implies \mathcal{M}^* \models \mathcal{T}^* \cup \{\varphi^*\}$.
  (2) $\mathcal{N} \models \mathcal{T}^* \cup \{\varphi^*\} \implies \mathcal{N} \restriction L \models \varphi$.
  (3) φ^* is quantifier-free.
  (4) $\mathcal{T}^*$ is a set of universal sentences.

8.21 Suppose L is a vocabulary of cardinality $\leq \kappa$, φ is a sentence of $L_{\kappa^+\omega}$ in the vocabulary L, $\mathcal{M}$ is an L-structure such that $\mathcal{M} \models \varphi$, and $\kappa \leq \mu \leq |M|$. Show that there is $\mathcal{M}_0 \subseteq \mathcal{M}$ such that $\mathcal{M}_0 \models \varphi$ and $|M_0| = \mu$.

8.22 Find a counter-example to the following claim: Suppose L is a vocabulary of cardinality $\leq \kappa$, $\alpha < \kappa^+$, and $\mathcal{M}$ is an L-structure. Then there is $\mathcal{M}_0 \subseteq \mathcal{M}$ such that $\mathcal{M}_0 \simeq^\alpha_p \mathcal{M}$ and $|M_0| \leq \kappa$.

8.23 Prove that the class of models (M, A, B), where $A, B \subseteq M$ and $|A| < |B|$ is not RPC in $L_{\infty\omega}$. In fact, prove the following statement: If $\varphi \in L_{\kappa^+\omega}$ has a model $\mathcal{M}$ such that $\kappa \leq |U^{\mathcal{M}}| < |V^{\mathcal{M}}|$, then φ has a model $\mathcal{N}$ such that $|U^{\mathcal{N}}| = |V^{\mathcal{N}}|$.

8.24 Show that the class of linear orders with uncountable cofinality is not PC-definable in $L_{\infty\omega}$. In fact, prove the following statement: If $\varphi \in L_{\kappa^+\omega}$ has a model $\mathcal{M}$ such that $(U^{\mathcal{M}}, <^{\mathcal{M}})$ has cofinality κ^+, then φ has a model $\mathcal{N}$ such that $(U^{\mathcal{N}}, <^{\mathcal{N}})$ has cofinality $\aleph_0$. (A linear order $(M, <)$ has *cofinality* κ (or is κ-cofinal) if κ is the smallest cardinal for which there is $A \subseteq M$ such that $|A| = \kappa$ and A has no upper bound in $(M, <)$.)

8.25 Prove that the class of graphs (G, E) which have no countable cover (i.e. there is no countable subset C such that for every $a \in G$ there is some $y \in C$ with xEy) is not RPC in $L_{\infty\omega}$. In fact, prove the following statement: If $\varphi \in L_{\kappa^+\omega}$ has, for every graph (G, E) of cardinality κ^+ without a countable cover, a model $\mathcal{M}$ such that $(U^{\mathcal{M}}, R^{\mathcal{M}}) \cong (G, E)$, then φ has a model $\mathcal{N}$ such that $(U^{\mathcal{M}}, R^{\mathcal{M}})$ has a countable cover.

8.26 Show that there is a sentence in $L_{\kappa^+\omega}$ which has a model of size κ^+ but none of size $> \kappa^+$.

8.27 Show that there is a sentence in $L_{\kappa^+\omega}$ which has a model of size κ^{++} but none of size $> \kappa^{++}$.

8.28 Show that there is a sentence in $L_{\kappa^+\omega}$ which has a model of size 2^κ but none of size $> 2^\kappa$.

8.29 Show that there is a sentence in $L_{\kappa^+\omega}$ which has a model of size 2^{2^κ} but none of size $> 2^{2^\kappa}$.

8.30 Show $2^\omega \not\to (\omega_1)^2_2$. (Hint: Take a well-ordering $\lhd$ of $\mathbb{R}$. Then define a coloring of pairs $\{x, y\}$ of reals by reference to $\lhd$ and the standard ordering of $\mathbb{R}$.)

8.31 Show that if $\kappa \to (\kappa)^2_2$, then κ is regular.

8.32 Prove directly $6 \to (3)^2_2$.

8.33 Show $15 \not\to (4)^2_2$ (it is probably just as easy to show $17 \not\to (4)^2_2$.)

8.34 Deduce the undefinability of well-order in $L_{\omega_1\omega}$ directly from Theorem 8.22. (Hint: Assume well-order could be defined in $L_{\omega_1\omega}$. Show that then there is a sentence in $L_{\omega_1\omega}$ with a model of size $\beth_{\omega_1}$ but none bigger.)

8.35 Show that for every vocabulary L, every L-fragment Γ of $L_{\infty\omega}$ and every L-structure $\mathcal{M}$ there is $L' \supseteq L$, an L'-fragment Γ' such that $\Gamma' \supseteq \Gamma$, and an expansion $\mathcal{M}'$ of $\mathcal{M}$ to an L'-structure such that $|L'| = |L| + \aleph_0$, $|\Gamma'| = |\Gamma| + \aleph_0$ and $\mathcal{M}'$ has Skolem functions for all $\varphi \in \Gamma'$.

8.36 Construct for each infinite κ a linear order with 2^κ automorphisms.

8.37 An element a of a Boolean algebra $\mathcal{M}$ is an *atom* if for all $b \in M : 0 \leq b \leq a$ implies $0 = b$ or $b = a$. Show that if $\mathcal{M}$ is a Boolean algebra and $(I, <)$ is a linear order such that I is a set of atoms of $\mathcal{M}$, then $(I, <)$ is Γ-indiscernible in $\mathcal{M}$ for any fragment Γ of $L_{\infty\omega}$.

8.38 Suppose $\mathcal{M}$ is an equivalence relation and $(I, <)$ is a linear order such that I is included in one of the equivalence classes of $\mathcal{M}$. Show that $(I, <)$ is Γ-indiscernible in $\mathcal{M}$ for all fragments Γ of $L_{\infty\omega}$. Is the same true if I is a set of non-$\mathcal{M}$-equivalent elements of M?

8.39 Suppose L is a vocabulary, $\mathcal{M}$ and $\mathcal{M}'$ L-structures, Γ an L-fragment of $L_{\infty\omega}$, $(I, <_I)$ Γ-indiscernible in $\mathcal{M}$, $(J, <_J)$ Γ-indiscernible in $\mathcal{N}$, and $\mathcal{M} \models \varphi(a_1, \ldots, a_n) \iff \mathcal{N} \models \varphi(b_1, \ldots, b_n)$ for all atomic $\varphi, a_1 <_I \cdots <_I a_n$ and $b_1 <_J \cdots <_J b_n$. Suppose $(I, <_I) \simeq_p (J, <_J)$. Show that $[I]_{\mathcal{M}} \simeq_p [J]_{\mathcal{N}}$.

8.40 Suppose $\varphi \in L_{\omega_1\omega}$ has for each $\alpha < \omega_1$ a model of size $\beth_\alpha$. Show that φ has arbitrarily large models all of which are partially isomorphic. (Hint: Use the previous exercise.)

8.41 Show that the Scott rank of $(\alpha, <)$ is always α.

8.42 Show that the Scott rank of $(\mathbb{Q}, <)$ is ω.

8.43 Show that the two truth definitions of game logic are equivalent.

8.44 Show that the conjunction and disjunction of two formulas of the form (8.18) is again of the form (8.18), up to logical equivalence.

8.45 Use Theorem 8.47 to give a quick proof of the Craig Interpolation Theorem.

8.46 Show that Souslin formulas (see Example 8.49) can express the property of well-foundedness of a binary relation on the structure $(\omega, <, R^{\mathcal{A}})$, where $<$ is the usual ordering of ω.

8.47 Show that propositional game formulas (see Example 8.49) can be expressed in $L_{\omega_1\omega}$.

9

Stronger Infinitary Logics

9.1 Introduction

The infinitary logics $L_{\kappa\omega}$, $L_{\infty\omega}$, and $L_{\infty G}$ of the previous chapter had one important feature in common with first-order logic: the truth predicate of these logics is absolute[1] in set theory. We now move on to logics which do not have this property. We lose something but we also gain something else. For example, we lose the last remnants of the Completeness Theorem of first-order logic. On the other hand, we can express deeper properties of models, such as uncountability, completeness of a separable order, and other properties, too. Perhaps surprisingly, some methods, such as the method of Ehrenfeucht–Fraïssé Games, still work perfectly even with these strong logics.

9.2 Infinite Quantifier Logic

First-order logic and the infinitary logic $L_{\infty\omega}$ are able to express

$$\exists V \varphi \text{ and } \forall V \varphi$$

when V is any finite set of variables. In the infinite quantifier logics of this section we can express this even when V is an infinite set of variables.

Before actually defining the infinite quantifier logics, we first define the appropriate version of the Ehrenfeucht–Fraïssé Game. In this game the players play sequences of a given length. Each round consists of a choice of a sequence by **I** followed by a choice of a sequence by **II**. The goal of **II** is to make sure the played sequences form, element by element, a partial isomorphism. Thus

---

[1] More exactly, if M is a transitive model of ZFC containing $\mathcal{A}$ and φ as elements, then $\mathcal{A}$ is a model of φ if and only if the set-theoretical statement "$\mathcal{A} \models \varphi$" holds in the model M.

if **I** plays a sequence

$$x_0 = (x_0(0), \dots, x_0(n), \dots)$$

which is a descending sequence relative to a linear order $<$ in one of the models, player **II** tries to play likewise a sequence

$$y_0 = (y_0(0), \dots, y_0(n), \dots)$$

which constitutes a descending sequence relative to $<$ in the other model. If that other model is well-ordered by $<$, she loses right away. Note that players have made so far just one move each, albeit a move with infinitely many components.

For another example, suppose one of the models is countable while the other is uncountable. If player **I** is allowed to play countable sequences he can immediately let x_0 enumerate the countable model. Whichever countable sequence **II** plays, **I** wins during the next round by playing an element from the uncountable model which is different from all the elements played by **II**.

To define the new game more exactly, we fix some notation. A function $s : \alpha \to M$ is called a *sequence* of *length* $\text{len}(s) = \alpha$. The set of all sequences of length α of elements of M is denoted by M^α. We define

$$M^{<\alpha} = \bigcup_{\beta < \alpha} M^\beta$$

and

$$\text{Part}_\kappa(\mathcal{A}, \mathcal{B}) = \{p \in \text{Part}(\mathcal{A}, \mathcal{B}) : |p| < \kappa\}.$$

Now we can define the new Ehrenfeucht–Fraïssé Game:

Definition 9.1 Suppose κ is a cardinal. The *Ehrenfeucht–Fraïssé Game with moves of size* $< \kappa$ on $\mathcal{M}$ and $\mathcal{M}'$, denoted $\text{EF}_\omega^\kappa(\mathcal{M}, \mathcal{M}')$, is the game in which player **I** plays

$$x_n \in M^{<\kappa} \cup (M')^{<\kappa}$$

and **II** responds with

$$y_n \in M^{<\kappa} \cup (M')^{<\kappa}$$

for all $n \in \mathbb{N}$. Player **II** wins if for all n,

(1) $\text{len}(x_n) = \text{len}(y_n)$;
(2) $x_n \in M^{<\kappa} \leftrightarrow y_n \in (M')^{<\kappa}$;

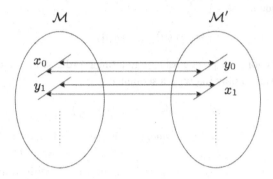

Figure 9.1 The Ehrenfeucht–Fraïssé Game with moves of size $< \kappa$.

(3) If we denote

$$v_i = \begin{cases} x_i & \text{if } x_i \in M^{<\kappa} \\ y_i & \text{if } y_i \in M^{<\kappa} \end{cases} \quad v_i' = \begin{cases} x_i & \text{if } x_i \in (M')^{<\kappa} \\ y_i & \text{if } y_i \in (M')^{<\kappa}, \end{cases}$$

then $\{(v_n(i), v_n'(i)) : i < \mathrm{len}(x_n), n \in \mathbb{N}\} \in \mathrm{Part}(\mathcal{M}, \mathcal{M}')$. (See Figure 9.1.)

We immediately introduce also the *dynamic version of* $\mathrm{EF}^{\kappa}_{\omega}$, denoted $\mathrm{EFD}^{\kappa}_{\alpha}$, which is easier for player **II** as there are only finitely many moves to play. It is defined by requiring that the moves of player **I** are pairs (x_n, α_n) where

$$x_n \in M^{<\kappa} \cup (M')^{<\kappa}, \quad \alpha_n < \alpha.$$

To win, player **I** has to play $\alpha > \alpha_0 > \alpha_1 > \ldots$ just as in EFD_{α}.

Example 9.2 Let $\mathcal{M} = (\mathbb{Q}, <)$ and $\mathcal{M}' = (\mathbb{R}, <)$. Player **I** has a winning strategy in $\mathrm{EFD}_3^{\omega_1}(\mathcal{M}, \mathcal{M}')$. His strategy is to play first in $\mathcal{M}$

$$x_0 = s_0$$

where

$$s_0 : \omega \to \mathbb{Q}, \quad \lim_n s(n) = \sqrt{2}, \quad s \text{ increasing.}$$

Suppose **II** responds with $y_0 : \omega \to \mathbb{R}$. We may assume y_0 is increasing, for otherwise **II** has lost already. If $\sup_n y_0(n) = \infty$, player **I** wins by playing $x_1 = (2)$ (i.e. a sequence of length 1). So let us assume $\sup_n y_0(n) = z \in \mathbb{R}$. Now **I** plays $x_1 = (z)$. Player **II** would like to play $(\sqrt{2})$ but $\sqrt{2}$ is not in

$\mathbb{Q}$ so she plays something bigger, say (a). Now **I** wins by playing $x_2 = (b)$ in $\mathcal{M}$, where $\sqrt{2} < b < a$. Actually, **I** has a winning strategy already in $\mathrm{EFD}_2^{\omega_1}(\mathcal{M}, \mathcal{M}')$ (Exercise 9.3). Note that $\mathcal{M} \simeq_p \mathcal{M}'$, so **II** has a winning strategy in $\mathrm{EF}_\omega(\mathcal{M}, \mathcal{M}')$.

Example 9.3 Suppose $\mathcal{M}$ is a countable L-structure and $\mathcal{M}'$ an uncountable L-structure. Now **I** has a winning strategy in $\mathrm{EFD}_2^{\omega_1}(\mathcal{M}, \mathcal{M}')$. He plays x_0 so that $x_0 : \mathbb{N} \to M$ enumerates M. After **II** has played $y_0 : \mathbb{N} \to M'$, **I** plays $x_1 = (a)$, where $a \notin \mathrm{rng}(y_0)$, and wins as **II** cannot play a new element in M. This should be compared with the fact that $L_{\infty\omega}$ is not able to express countability. In fact, the models $\mathcal{M}$ and $\mathcal{M}'$ could very well be partially isomorphic. This is the case in particular if $L = \emptyset$.

Example 9.4 Let $\mathcal{M}$ be the equivalence relation $(M, \sim)$, where $M = \omega_1 \times \omega_1$ and $(x, y) \sim (x', y')$ iff $y = y'$. Let $\mathcal{M}'$ be the equivalence relation $(M', \sim')$, where $M' = \omega_1 \times \omega$ and $(x, y) \sim (x', y')$ iff $y = y'$. Clearly $\mathcal{M} \simeq_p \mathcal{M}'$, but player **I** has a winning strategy in $\mathrm{EFD}_2^{\omega_1}(\mathcal{M}, \mathcal{M}')$. He plays in $\mathcal{M}'$

$$x_0 = s_0, \text{ where } s_0(n) = (0, n).$$

After **II** plays y_0 with $y_0(n) = (\gamma_n, \delta_n)$, **I** plays $x_1 = ((0, \delta))$, where $\delta \notin \{\delta_n : n \in \mathbb{N}\}$, and wins as **II** has no good moves remaining.

Example 9.5 A *vector space* (over $\mathbb{R}$) is a structure of the form

$$\mathcal{M} = (M, \cdot, +, 0, \mathbb{R}, \cdot_\mathbb{R}, +_\mathbb{R}, 0_\mathbb{R}, 1_\mathbb{R})$$

where $(\mathbb{R}, \cdot_\mathbb{R}, +_\mathbb{R}, 0_\mathbb{R}, 1_\mathbb{R})$ is the field of reals and $\cdot$ and $+$ are the scalar product

$$\cdot : \mathbb{R} \times M \to M$$

and the vector addition

$$+ : M \times M \to M$$

which satisfy the usual axioms of vector spaces (with 0 as the zero vector). There is just one such vector space (up to isomorphism) in any dimension. If $\mathcal{M}$ and $\mathcal{M}'$ have a different infinite dimension, say κ and $\kappa' > \kappa$, then player **I** has a winning strategy in $\mathrm{EFD}_3^{\kappa^+}(\mathcal{M}, \mathcal{M}')$. Note that $\mathcal{M} \simeq_p \mathcal{M}'$.

Definition 9.6 A linear order $\mathcal{M}$ is $\aleph_1$-*like* if every initial segment of $\mathcal{M}$ is countable but M itself is uncountable.

Lemma 9.7 *If $\mathcal{M}$ is an $\aleph_1$-like linear order, then $|M| = \aleph_1$.*

Proof By definition, $|M| \geq \aleph_1$. Let $\{a_\alpha : \alpha < \beta\}$ be a cofinal sequence in M defined by the transfinite recursion

$$a_\alpha \in M, a_\alpha > a_\gamma \text{ for all } \gamma < \alpha.$$

Since each $\{a_\alpha : \alpha < \gamma\}, \gamma < \beta$, is countable, being below a_γ in M, it must be the case that $\beta \leq \omega_1$. Thus M is the union of $\leq \aleph_1$ countable sets:

$$M = \bigcup_{\alpha < \beta} \{x \in M : x < a_\alpha\}.$$

Thus M has cardinality $\aleph_1$. $\square$

$\aleph_1$-like linear orders can be quite complicated but the *dense* ones among them are a bit easier to grasp:

Definition 9.8 Let η denote the order-type of the rationals. For $A \subseteq \omega_1$ let

$$r_\alpha^A = \begin{cases} 1 + \eta & \text{if} \quad \alpha \in A \setminus \{0\} \\ \eta & \text{if} \quad \alpha \notin A \setminus \{0\} \end{cases}$$

and

$$\Phi(A) = \sum_{\alpha < \omega_1} (r_\alpha^A \times \{\alpha\}).$$

$\Phi(A)$ is an $\aleph_1$-like dense linear order. (See Example 5.6 for the definition of Σ.)

Lemma 9.9 $\Phi(\emptyset) \not\cong \Phi(\omega_1)$.

Proof Suppose $f : \Phi(\emptyset) \cong \Phi(\omega_1)$. For each $a \in \Phi(A)$ let $\rho(a)$ be the unique α such that $a = (q, \alpha)$ for some q. Let us define a sequence $\{a_n : n \in \mathbb{N}\}$ as follows (Figure 9.2):

$$a_0 \in \Phi(\emptyset) \text{ arbitrary}$$
$$a_1 \in \Phi(\omega_1) \text{ such that } \rho(a_1) > \rho(f(a_0))$$
$$a_2 \in \Phi(\emptyset) \text{ such that } \rho(a_2) > \rho(f^{-1}(a_1))$$
$$a_3 \in \Phi(\omega_1) \text{ such that } \rho(a_3) > \rho(f(a_2))$$

etc.

Let $\delta = \sup_n \rho(a_n)$. Let a be the smallest element in $\Phi(\omega_1)$ with $\rho(a) = \delta$. Now a is the supremum of the set $\{a_{2n+1} : n \in \mathbb{N}\}$ in $\Phi(\omega_1)$, but $f^{-1}(a)$ cannot be the supremum of $\{a_{2n} : n \in \mathbb{N}\}$ in $\Phi(\emptyset)$ as $r_\delta^\emptyset$ has no first element. This shows that f cannot be an isomorphism. $\square$

Lemma 9.10 *Player* II *has a winning strategy in* $\mathrm{EF}_\omega^{\aleph_1}(\mathcal{M}, \mathcal{M}')$ *whenever* $\mathcal{M}$ *and* $\mathcal{M}'$ *are* $\aleph_1$*-like dense linear orders without first element.*

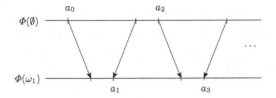

Figure 9.2

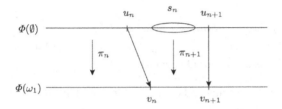

Figure 9.3

Proof During the game player **II** maintains an element $u_n \in M$, an element $v_n \in M'$, and an isomorphism π_n between the initial segment of $\mathcal{M}$ determined by u_n and the initial segment of $\mathcal{M}'$ determined by v_n. Player **II** takes care that the partial isomorphism of played elements is a subfunction of π_n. Suppose **I** then plays a countable sequence s_n in one of the models. Now **II** moves her u_n and v_n to new positions u_{n+1}, v_{n+1} above the elements in s_n (see Figure 9.3).
The intervals (u_n, u_{n+1}) of $\Phi(\emptyset)$ and (v_n, v_{n+1}) of $\Phi(\omega_1)$ are countable dense linear orderings without endpoints, hence isomorphic. So **II** can maintain her strategy. □

We obtain an important corollary due to M. Morley:

Proposition 9.11 *There are non-isomorphic models $\mathcal{M}$ and $\mathcal{M}'$ of cardinality $\aleph_1$ such that player **II** has a winning strategy in $\mathrm{EF}_\omega^{\aleph_1}(\mathcal{M}, \mathcal{M}')$.*

This result should be compared with Proposition 5.16: If $\mathcal{M} \simeq_p \mathcal{M}'$, where $\mathcal{M}$ and $\mathcal{M}'$ are countable, then $\mathcal{M} \cong \mathcal{M}'$.

Definition 9.12 Suppose L is a vocabulary and λ is a cardinal. The class of L-formulas of $L_{\infty\lambda}$ is defined as follows:

(1) Variables $x_\alpha, \alpha < \lambda$, and constant symbols $c \in L$ are L-terms. Function symbols yield new L-terms as usual.

(2) If t and t' are L-terms, then $(0, t, t')$ is an L-formula denoted by $\approx tt'$.

(3) If $t_1, \ldots, t_n$ are L-terms and $R \in L$, then $(1, R, t_1, \ldots, t_n)$ is an L-formula denoted by $Rt_1 \ldots t_n$.

(4) If φ is an L-formula, so is $(2, \varphi)$, and we denote it by $\neg\varphi$.

(5) If Φ is a set of L-formulas with a fixed set V of free variables and $|V| < \lambda$, then $(3, \Phi)$ is an L-formula denoted by $\wedge\Phi$.

(6) If Φ is a set of L-formulas with a fixed set V of free variables and $|V| < \lambda$, then $(4, \Phi)$ is an L-formula denoted by $\vee\Phi$.

(7) If φ is an L-formula and $V \subseteq \{x_\alpha : \alpha < \lambda\}$ has cardinality $< \lambda$, then $(5, \varphi, V)$ is an L-formula denoted by $\forall V \varphi$.

(8) If φ is an L-formula and $V \subseteq \{x_\alpha : \alpha < \lambda\}$ has cardinality $< \lambda$, then $(6, \varphi, V)$ is an L-formula denoted by $\exists V \varphi$.

The concept of subformula and quantifier rank are defined in a natural way. The set $\mathrm{Sub}(\varphi)$ of subformulas $\varphi \in L_{\infty\lambda}$ is defined as follows:

$$\mathrm{Sub}(\varphi) = \{\varphi\} \text{ if } \varphi \text{ is atomic}$$
$$\mathrm{Sub}(\neg\varphi) = \mathrm{Sub}(\varphi) \cup \{\neg\varphi\}$$
$$\mathrm{Sub}(\wedge\Phi) = \bigcup_{\varphi\in\Phi} \mathrm{Sub}(\varphi) \cup \{\wedge\Phi\}$$
$$\mathrm{Sub}(\vee\Phi) = \bigcup_{\varphi\in\Phi} \mathrm{Sub}(\varphi) \cup \{\vee\Phi\}$$
$$\mathrm{Sub}(\forall V\varphi) = \mathrm{Sub}(\varphi) \cup \{\forall V\varphi\}$$
$$\mathrm{Sub}(\exists V\varphi) = \mathrm{Sub}(\varphi) \cup \{\exists V\varphi\}.$$

The quantifier rank $\mathrm{QR}(\varphi)$ of $\varphi \in L_{\infty\lambda}$ is defined analogously:

$$\mathrm{QR}(\varphi) = 0 \text{ if } \varphi \text{ is atomic}$$
$$\mathrm{QR}(\neg\varphi) = \mathrm{QR}(\varphi)$$
$$\mathrm{QR}(\wedge\Phi) = \sup\{\mathrm{QR}(\varphi) : \varphi \in \Phi\}$$
$$\mathrm{QR}(\vee\Phi) = \sup\{\mathrm{QR}(\varphi) : \varphi \in \Phi\}$$
$$\mathrm{QR}(\forall V\varphi) = \mathrm{QR}(\varphi) + 1$$
$$\mathrm{QR}(\exists V\varphi) = \mathrm{QR}(\varphi) + 1.$$

Definition 9.13 We use $L_{\infty\infty}$ to denote the union $\bigcup_\lambda L_{\infty\lambda}$, and $L_{\kappa\lambda}$ to denote the set of formulas φ of $L_{\infty\lambda}$ which satisfy the condition: If $\wedge\Phi$ or $\vee\Phi$ is a subformula of φ, then $|\Phi| < \kappa$.

Usually we consider $L_{\kappa\lambda}$ only when κ is regular. The logic $L_{\omega_1\omega_1}$ is a particularly interesting one. It is a perfect analogue of first-order logic $L_{\omega\omega}$ obtained by stepping up from finite to countable. However, the properties of $L_{\omega_1\omega_1}$ are very different from those of $L_{\omega\omega}$, mainly because of the expressive power of the former.

Example 9.14 Let $\varphi \in L_{\omega_1\omega_1}$ be the sentence

$$\forall\{x_n : n < \omega\}\exists x_\omega \bigwedge_{n<\omega} \neg\approx x_n x_\omega.$$

The subformulas of φ are φ itself,

$$\exists x_\omega \bigwedge_{n<\omega} \neg \approx x_n x_\omega$$

$$\bigwedge_{n<\omega} \neg \approx x_n x_\omega$$

$$\neg \approx x_n x_\omega$$

$$\approx x_n x_\omega,$$

and

$$QR(\varphi) = QR(\exists x_\omega \bigwedge_{n<\omega} \neg\approx x_n x_\omega) + 1$$

$$= QR\left(\bigwedge_{n<\omega} \neg\approx x_n x_\omega\right) + 2$$

$$= \sup_n QR(\neg\approx x_n x_\omega) + 2$$

$$= 2.$$

Example 9.15 Let $\varphi \in L_{\omega_1\omega_1}$ be the sentence

$$\forall\{x_n : n < \omega\} \bigvee_{n<\omega} \neg x_{n+1} < x_n.$$

Then the subformulas of φ are φ itself,

$$\bigvee_{n<\omega} \neg x_{n+1} < x_n$$

$$\neg x_{n+1} < x_n$$

and

$$QR(\varphi) = \sup_n QR(\neg x_{n+1} < x_n) + 1 = 1.$$

Functions $s : \{x_\alpha : \alpha < \lambda\} \to M$ are called assignments of M. If in addition $V \subseteq \{x_\alpha : \alpha < \lambda\}$ and $a : V \to M$, we define

$$s[a/V](x_\alpha) = \begin{cases} a(x_\alpha) & \text{if} \quad x_\alpha \in V \\ s(x_\alpha) & \text{if} \quad x_\alpha \notin V. \end{cases}$$

$t^{\mathcal{M}}(s)$ is defined as usual.

Definition 9.16 The concept of an assignment $s : \{x_\alpha : \alpha < \lambda\} \to M$ *satisfying* a formula $\varphi \in L_{\infty\lambda}$ in a model $\mathcal{M}$, $\mathcal{M} \models_s \varphi$ is defined as follows:

$$\begin{array}{lll} \mathcal{M} \models_s \approx t_1 t_2 & \text{iff} & t_1^{\mathcal{M}}(s) = t_2^{\mathcal{M}}(s) \\ \mathcal{M} \models_s R t_1 \dots t_n & \text{iff} & (t_1^{\mathcal{M}}(s), \dots, t_n^{\mathcal{M}}(s)) \in \mathrm{Val}_{\mathcal{M}}(R) \\ \mathcal{M} \models_s \neg\varphi & \text{iff} & \mathcal{M} \not\models_s \varphi \\ \mathcal{M} \models_s \wedge\Phi & \text{iff} & \mathcal{M} \models_s \varphi \text{ for all } \varphi \in \Phi \\ \mathcal{M} \models_s \vee\Phi & \text{iff} & \mathcal{M} \models_s \varphi \text{ for some } \varphi \in \Phi \\ \mathcal{M} \models_s \forall V \varphi & \text{iff} & \mathcal{M} \models_{s[a/V]} \varphi \text{ for all } a : V \to M \\ \mathcal{M} \models_s \exists V \varphi & \text{iff} & \mathcal{M} \models_{s[a/V]} \varphi \text{ for some } a : V \to M. \end{array}$$

Example 9.17 Let φ be the $L_{\omega_1\omega_1}$-sentence:

$$\forall\{x_n : x < \omega\} \bigvee_{n<\omega} \neg x_{n+1} < x_n.$$

A linear order satisfies φ if and only if it is a well-order. So this shows that $L_{\infty\omega_1}$ extends $L_{\infty\omega}$ properly.

Example 9.18 Let φ be the $L_{\omega_1\omega_1}$-sentence:

$$\exists\{x_n : x < \omega\}\forall\{x_\omega, x_{\omega+1}\} \bigvee_n (x_\omega < x_n \wedge x_n < x_{\omega+1}).$$

A linear order satisfies φ if and only if it is separable (i.e. has a countable dense subset).

Example 9.19 Let $\varphi \in L_{\omega_1\omega_1}$ be the conjunction of the axioms of ordered fields plus firstly the *Archimedian axiom*

$$\forall x_0 \bigvee_{n<\omega} x_0 < \underbrace{1 + \cdots + 1}_{n}$$

and secondly the *completeness axiom*

$$\forall\{x_n : n < \omega\}(\ (\exists x_\omega \wedge_n x_n < x_\omega) \to \\ \exists x_\omega \ (\wedge_n x_n < x_\omega \wedge \forall x_{\omega+1} \\ ((\wedge_n x_n < x_{\omega+1}) \to x_\omega \le x_{\omega+1}))).$$

An ordered field satisfies φ if and only if it is isomorphic to the ordered field of reals.

Example 9.20 Let φ be the $L_{\lambda^+\lambda^+}$-sentence

$$\exists\{x_\alpha : \alpha < \lambda\} \bigwedge_{\alpha<\beta<\lambda} x_\alpha E x_\beta.$$

A graph satisfies φ if and only if it has a clique of size λ.

Example 9.21 Let φ be the $L_{\lambda^+\lambda^+}$-sentence

$$\exists\{x_\alpha : \alpha < \lambda\}\forall x_\lambda \bigvee_{\alpha<\lambda} \approx x_\alpha x_\lambda.$$

A model $\mathcal{M}$ satisfies φ iff $|M| \le \lambda$. Let ψ be the $L_{\lambda^+\lambda^+}$-sentence

$$\exists\{x_\alpha : \alpha < \lambda\} \bigwedge_{\alpha<\beta<\lambda} \neg\approx x_\alpha x_\beta.$$

A model $\mathcal{M}$ satisfies ψ iff $|M| \ge \lambda$.

We use $[A]^{<\lambda}$ to denote the set $\{a \subseteq A : |a| < \lambda\}$.

Definition 9.22 Suppose $\mathcal{A}$ and $\mathcal{B}$ are L-structures. A λ-*back-and-forth set* for $\mathcal{A}$ and $\mathcal{B}$ is any non-empty set $P \subseteq \mathrm{Part}(\mathcal{A}, \mathcal{B})$ such that

(1) $\forall f \in P \forall a \in [A]^{<\lambda} \exists g \in P(f \subseteq g$ and $a \subseteq \mathrm{dom}(g))$;
(2) $\forall f \in P \forall b \in [B]^{<\lambda} \exists g \in P(f \subseteq g$ and $a \subseteq \mathrm{rng}(g))$.

The structures $\mathcal{A}$ and $\mathcal{B}$ are said to be λ-*partially isomorphic*, in symbols

$$\mathcal{A} \simeq_\lambda \mathcal{B},$$

if there is a λ-back-and-forth set for them.

It is easy to see that $\simeq_\lambda$ is an equivalence relation on $\mathrm{Str}(L)$ for any L. The big drawback of $\simeq_\lambda$ in comparison to $\simeq_p$ (i.e. $\simeq_2$) is that for $\lambda > \omega$ there is no guarantee that $\mathcal{A} \simeq_\lambda \mathcal{B}, |A| \le \lambda, |B| \le \lambda$ implies $\mathcal{A} \cong \mathcal{B}$.

Example 9.23 If $\mathcal{A}$ and $\mathcal{B}$ are $\aleph_1$-like dense linear orders without first elements, then $\mathcal{A} \simeq_{\aleph_1} \mathcal{B}$. We can let P consist of such $f \in \mathrm{Part}(\mathcal{A}, \mathcal{B})$ that $\mathrm{dom}(f)$ and $\mathrm{ran}(f)$ both are initial segments and both have a last element (see Figure 9.4).

Proposition 9.24 *There are linear orders $\mathcal{A}$ and $\mathcal{B}$ such that*

(1) $\mathcal{A} \not\cong \mathcal{B}$.
(2) $\mathcal{A} \simeq_{\aleph_1} \mathcal{B}$.
(3) $|A| = |B| = \aleph_1$.

Proof We can choose $\mathcal{A} = \Phi(\emptyset)$ and $\mathcal{B} = \Phi(\omega_1)$. $\qquad\square$

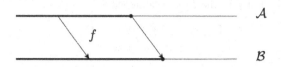

Figure 9.4

The following theorem of Benda (1969), the proof of which we omit, illuminates how reduced products can improve the chances of player **II** in Ehrenfeucht–Fraïssé Games:

Theorem 9.25 *Suppose D is a (κ, λ)-regular[2] countably incomplete filter on I. If* **II** *has a winning strategy in $EF_n^\kappa(\mathcal{M}_i, \mathcal{N}_i)$ for each $n < \omega$ and each $i \in I$, then she has a winning strategy in $EF_\omega^{\lambda^+}(\prod \mathcal{M}_i/D, \prod \mathcal{N}_i/D)$.*

We now establish the marriage of truth and separation for $L_{\infty\lambda}$, manifesting the Strategic Balance of Logic in the case of the infinite quantifier logics:

Theorem 9.26 *The following are equivalent:*

(1) $\mathcal{M} \equiv_{\infty\lambda} \mathcal{N}$ (i.e. $\mathcal{M}$ and $\mathcal{N}$ satisfy the same sentences of $L_{\infty\lambda}$).
(2) Player **II** *has a winning strategy in $\mathrm{EF}_\omega^\lambda(\mathcal{M}, \mathcal{N})$.*
(3) $\mathcal{M} \simeq_\lambda \mathcal{N}$.

Proof As in Proposition 5.21 and Proposition 7.48. □

For working with winning strategies of **II** in $\mathrm{EFD}_\alpha^\lambda(\mathcal{A}, \mathcal{B})$ we have the appropriate version of the back-and-forth sequence: A λ-*back-and-forth sequence* $(P_\beta : \beta \le \alpha)$ is defined by the conditions

$$\emptyset \ne P_\alpha \subseteq \cdots \subseteq P_0 \subseteq \mathrm{Part}(\mathcal{A}, \mathcal{B})$$
$$\forall f \in P_{\beta+1} \forall a \in [A]^{<\lambda} \exists g \in P_\beta (a \subseteq \mathrm{dom}(g) \text{ and } f \subseteq g)$$
$$\forall f \in P_{\beta+1} \forall b \in [B]^{<\lambda} \exists g \in P_\beta (b \subseteq \mathrm{ran}(g) \text{ and } f \subseteq g).$$

We write

$$\mathcal{A} \simeq_\lambda^\alpha \mathcal{B}$$

if there is a λ-back-and-forth-sequence of length α for $\mathcal{A}$ and $\mathcal{B}$.

Theorem 9.27 *The following are equivalent:*

---

[2] A filter on I is (κ, λ)-regular if it contains a subset E such that $|E| = \lambda$ and $|\{e \in E : i \in e\}| < \kappa$ for each $i \in I$.

(1) $\mathcal{A} \equiv^{\alpha}_{\infty\lambda} \mathcal{B}$ *(i.e. $\mathcal{A}$ and $\mathcal{B}$ satisfy the same sentences of $L_{\infty\lambda}$ of quantifier rank $\leq \alpha$).*

(2) Player **II** *has a winning strategy in* $\mathrm{EFD}^{\lambda}_{\alpha}(\mathcal{A}, \mathcal{B})$.

(3) $\mathcal{A} \simeq^{\alpha}_{\lambda} \mathcal{B}$.

Proof (1)→(2). For $\beta \leq \alpha$ let P_{β} consist of $f : A \to B$ of cardinality $< \lambda$ such that if $\mathrm{dom}(f) = \{a_{\xi} : \xi < \mu\}$ then

$$(\mathcal{A}, \ldots, a_{\xi}, \ldots) \equiv^{\beta}_{\infty\lambda} (\mathcal{B}, \ldots, fa_{\xi}, \ldots).$$

Clearly, $\emptyset \in P_{\beta}$ for all $\beta < \alpha$. For an inductive argument, suppose $f \in P_{\beta+1}$, $\beta < \alpha$, and $(a'_{\zeta} : \zeta < \gamma) \in {}^{\gamma}A$. We need a $(b'_{\zeta} : \zeta < \gamma) \in {}^{\gamma}B$ such that

$$(\mathcal{A}, \ldots, a_{\xi}, \ldots, a'_{\zeta}, \ldots) \equiv^{\beta}_{\infty\lambda} (\mathcal{B}, \ldots, fa_{\xi}, \ldots, b'_{\zeta}, \ldots).$$

If such a $(b'_{\zeta} : \zeta < \gamma)$ exists, we are done. Otherwise $\overline{b} = (b'_{\zeta} : \zeta < \gamma) \in {}^{\gamma}B$ implies

$$(\mathcal{A}, \ldots, a_{\xi}, \ldots, a'_{\zeta}, \ldots) \not\equiv^{\beta}_{\infty\lambda} (\mathcal{B}, \ldots, fa_{\xi}, \ldots, b'_{\zeta}, \ldots).$$

Let $\varphi_{\overline{b}}(\ldots, x_{\xi}, \ldots, x'_{\zeta}, \ldots) \in L_{\infty\lambda}$ with quantifier rank $\leq \beta$ so that

$$\mathcal{A} \models \varphi_{\overline{b}}(\ldots, a_{\xi}, \ldots, a'_{\zeta}, \ldots)$$

and

$$\mathcal{B} \not\models \varphi_{\overline{b}}(\ldots, fa_{\xi}, \ldots, b'_{\zeta}, \ldots).$$

Let

$$\psi(\ldots, x_{\xi}, \ldots, x'_{\zeta}, \ldots) = \bigwedge_{\overline{b} \in B} \varphi_{\overline{b}}(\ldots, x_{\xi}, \ldots, x'_{\zeta}, \ldots).$$

Now

$$\mathcal{A} \models \exists x'_0 \ldots x'_{\zeta} \ldots \psi(\ldots, a_{\xi}, \ldots, x'_{\zeta}, \ldots)$$

and $\exists x'_0 \ldots x'_{\zeta} \ldots \psi(\ldots, x_{\xi}, \ldots, x'_{\zeta}, \ldots)$ has quantifier rank $\leq \beta + 1$, so

$$\mathcal{B} \models \exists x'_0 \ldots x'_{\zeta} \ldots \psi(\ldots, fa_{\xi}, \ldots, x'_{\zeta}, \ldots),$$

a contradiction. Hence there is $\overline{b}$ such that $f \cup \{(a'_{\xi}, b'_{\xi}) : \xi < \gamma\} \in P_{\beta}$. The other condition is proved similarly.

(2)→(1). Suppose $(P_{\beta} : \beta \leq \alpha)$ is a λ-back-and-forth sequence for $\mathcal{A}$ and $\mathcal{B}$. As in the proof of Proposition 7.47, it is easy to use induction on $\beta \leq \alpha$ to show: If $f \in P_{\beta}$ and $a_0, \ldots, a_{\zeta}, \ldots \in \mathrm{dom}(f)$ for $\zeta < \mu$, where $\mu < \lambda$, then

$$(\mathcal{A}, \ldots, a_{\zeta}, \ldots) \equiv^{\beta}_{\infty\lambda} (\mathcal{B}, \ldots, fa_{\zeta}, \ldots).$$

Then (i) follows from the assumption that $P_{\beta} \neq \emptyset$ for all $\beta \leq \alpha$.

The equivalence of (2) and (3) is proved as Proposition 7.17. $\square$

Theorem 9.28 *Suppose* $\kappa = |A|^{<\lambda} + |B|^{<\lambda}$ *and* **II** *has a winning strategy in* $\mathrm{EFD}_\alpha^\lambda(A, B)$ *for each* $\alpha < (\kappa^{<\lambda})^+$. *Then* **II** *has a winning strategy in* $\mathrm{EF}_\omega^\lambda(A, B)$.

Proof Let P consist of such $f \in \mathrm{Part}_\lambda(A, B)$ that if $\mathrm{dom}(f) = \{a_\zeta : \zeta < \mu\}$, then

$$\forall \alpha < (\kappa^{<\lambda})^+((A, \ldots, a_\zeta, \ldots) \equiv_{\infty\lambda}^\alpha (B, \ldots, f a_\zeta, \ldots)).$$

The set P is non-empty, for Theorem 9.27 implies $\emptyset \in P$. Suppose $f \in P$ and $(a_\zeta' : \zeta < \gamma) \in {}^\gamma A$. We need a $\overline{b} = (b_\zeta' : \zeta < \gamma) \in {}^\gamma B$ so that

$$\forall \alpha < (\kappa^{<\lambda})^+((A, \ldots, a_\xi, \ldots, a_\zeta', \ldots) \equiv_{\infty\lambda}^\alpha (B, \ldots, f a_\xi, \ldots, b_\zeta', \ldots)).$$

If no such $\overline{b}$ exists, then for all $\overline{b}$ there is some $\alpha_{\overline{b}} < (\kappa^{<\lambda})^+$ so that

$$(A, \ldots, a_\xi, \ldots, a_\zeta', \ldots) \not\equiv_p^{\alpha_b} (B, \ldots, f a_\xi, \ldots, b_\zeta', \ldots). \tag{9.1}$$

Let $\alpha = \sup\{\alpha_{\overline{b}} : \overline{b} \in {}^\gamma B\} < (\kappa^{<\lambda})^+$. Since $f \in P$,

$$(A, \ldots, a_\xi, \ldots) \equiv_{\infty\lambda}^{\alpha+1} (B, \ldots, f a_\xi, \ldots).$$

Let $(P_\beta : \beta \le \alpha + 1)$ be a back-and-forth sequence for $(A, \ldots, a_\xi, \ldots)$ and $(B, \ldots, f a_\xi, \ldots)$. Let $g \in P_{\alpha+1}$ and $\overline{b} \in {}^\gamma B$ so that $g \cup \{(a_\zeta', b_\zeta') : \zeta < \gamma\} \in P_\alpha$. Let $P_\beta', \beta \le \alpha$, consist of

$$h \in \mathrm{Part}_\lambda((A, \ldots, a_\xi, \ldots, a_\zeta', \ldots), (B, \ldots, f a_\xi, \ldots, b_\zeta', \ldots))$$

such that $h \subseteq h'$ for some $h' \in P_\beta$ such that $g \cup \{(a_\zeta', b_\zeta') : \zeta < \gamma\} \subseteq h'$. It is clear that $(P_\beta' : \beta \le \alpha)$ is a back-and-forth sequence which contradicts (9.1). $\qquad\square$

We can define the λ-*Scott watershed* of $\mathcal{M}$ and $\mathcal{M}'$ as the least ordinal α such that **II** has a winning strategy in $\mathrm{EFD}_\alpha^\lambda(\mathcal{M}, \mathcal{M}')$ and **I** has a winning strategy in $\mathrm{EFD}_{\alpha+1}^\lambda(\mathcal{M}, \mathcal{M}')$.

In the case $\lambda = 2$ the Scott watershed of countable models was countable and every pair of countable non-isomorphic models had a Scott watershed. For $\lambda > \omega$ the situation is less satisfactory. For non-isomorphic models of size λ the λ-Scott watershed need not exist, as it is possible that $\mathcal{M} \simeq_\lambda \mathcal{N}$ and $\mathcal{M} \not\cong \mathcal{N}$ (Proposition 9.11). Also, if e.g. $\lambda = \aleph_1$, even if the λ-Scott watershed exists, we only know it is less than the cardinal $(2^\omega)^+$, and this cardinal may be very big.

The λ-*Scott height* λ-$\mathrm{SH}(\mathcal{M})$ of a model $\mathcal{M}$ is the least α such that if $a_1, \ldots, a_n, b_1, \ldots, b_n \in [M]^{<\lambda}$ and

$$\mathrm{len}(a_i) = \mathrm{len}(b_i) \text{ for } 1 \le i \le n$$

and

$$(\mathcal{M}, a_1, \ldots, a_n) \simeq_\lambda^\alpha (\mathcal{M}, b_1, \ldots, b_n)$$

then

$$(\mathcal{M}, a_1, \ldots, a_n) \simeq_\lambda^{\alpha+1} (\mathcal{M}, b_1, \ldots, b_n).$$

(Here $(\mathcal{M}, a_1, \ldots, a_n)$ means $(\mathcal{M}, (a_1(\xi))_{\xi<\mathrm{len}(a_1)}, \ldots, (a_n(\xi))_{\xi<\mathrm{len}(a_n)}).$)
If none exist, then $\lambda\text{-SH}(\mathcal{M}) = 0$.

Theorem 9.29 *If* $\mathcal{M} \simeq_\lambda^{\lambda\text{-SH}(\mathcal{M})+2} \mathcal{M}'$, *then* $\mathcal{M} \simeq_\lambda \mathcal{M}'$.

Proof Exercise 9.20. □

Note that $\lambda\text{-SH}(\mathcal{M}) < (|M|^{<\lambda})^+$. While we have for countable models

$$\mathcal{M} \simeq_p^{\mathrm{SH}(\mathcal{M})+\omega} \mathcal{M}' \iff \mathcal{M} \cong \mathcal{M}',$$

for models of size λ we only obtain

$$\mathcal{M} \simeq_\lambda^{\mathrm{SH}(\mathcal{M})+\omega} \mathcal{M}' \iff \mathcal{M} \equiv_{\infty\lambda} \mathcal{M}'.$$

This shortcoming takes away some of the interest of $\lambda\text{-SH}(\mathcal{M})$.

There is a special case where we can salvage the equivalence between $\mathcal{M} \equiv_{\infty\lambda} \mathcal{M}'$ and $\mathcal{M} \cong \mathcal{M}'$:

Theorem 9.30 *Suppose λ has cofinality ω. Then for all $\mathcal{M}$ and $\mathcal{M}'$ of cardinality λ:*

$$\mathcal{M} \equiv_{\infty\lambda} \mathcal{M}' \iff \mathcal{M} \cong \mathcal{M}'.$$

Proof Suppose $\lambda = \sup_n \lambda_n$. Suppose P is a λ-back-and-forth set for $\mathcal{M}$ and $\mathcal{M}'$. Let $f_0 \in P$. Let $M = \{a_\alpha : \alpha < \lambda\}$ and $M' = \{b_\alpha : \alpha < \lambda\}$. We can construct $f_0 \subseteq f_a \subseteq \cdots$ in P in such a way that $a_\alpha \in \mathrm{dom}(f_n)$ and $b_\alpha \in \mathrm{rng}(f_n)$ whenever $\alpha < \lambda_n$. Then $\bigcup_n f_n : \mathcal{M} \cong \mathcal{M}'$. □

The following deep result of Shelah gives some idea of what is it about uncountable models that makes $\mathcal{M} \equiv_{\infty\lambda} \mathcal{M}'$ fail to imply $\mathcal{M} \cong \mathcal{M}'$ for models of cardinality λ:

Theorem 9.31 *Suppose T is a countable complete first-order theory. Then the following are equivalent:*

(1) T is superstable and has NDOP and NOTOP.[3]
(2) All models $\mathcal{M}$ and $\mathcal{M}'$ of T of any cardinality $\lambda > 2^\omega$ satisfy

$$\mathcal{M} \equiv_{\infty\lambda} \mathcal{M}' \iff \mathcal{M} \cong \mathcal{M}'.$$

[3] For the concepts of superstable, NDOP, and NOTOP see Shelah (1990).

Thus for models $\mathcal{M}$ and $\mathcal{M}'$ of cardinality $\lambda > 2^\omega$ satisfying $\mathcal{M} \equiv_{\infty\lambda} \mathcal{M}'$ it is the model-theoretic property of the first-order theory of $\mathcal{M}$ (which is the same as the first-order theory of $\mathcal{M}'$) of lacking stability that bars us from concluding $\mathcal{M} \cong \mathcal{M}'$. In the proof of Theorem 9.31 in the case that (1) fails two non-isomorphic models of T are constructed from indiscernibles in such a way that they satisfy $\mathcal{M} \equiv_{\infty\lambda} \mathcal{M}'$. There is some resemblance to the proof of Lemma 9.10, although the situation is much more complicated. On the other hand, when (1) holds, a "decomposition theorem" holds for models of T of cardinality λ, and this helps to conclude $\mathcal{M} \cong \mathcal{M}'$ from $\mathcal{M} \equiv_{\infty\lambda} \mathcal{M}'$.

Chang proved the following facts about elementary equivalence of ordinals:

Proposition 9.32   *For any β and ξ*

(1) $(\lambda^{\beta+1}, <) \simeq_\lambda^\beta (\lambda^{\beta+1} \cdot (\xi+1), <)$.

(2) $(\lambda^{\beta+1}, <) \simeq_\lambda^\beta (\lambda^{\beta+1} \cdot \xi, <)$ *if ξ is a limit ordinal of cofinality $\geq \lambda$.*

(3) $(\lambda^\beta, <) \simeq_\lambda^\beta (\lambda^\beta \cdot (\xi+1), <)$ *if β is a limit ordinal limit of cofinality $\geq \lambda$.*

(4) $(\lambda^\beta, <) \simeq_\lambda^\beta (\lambda^\beta \cdot \xi, <)$ *if β, ξ are limit ordinals of cofinality $\geq \lambda$.*

Proof   Exercise 9.21.   □

It follows that $\lambda\text{-SH}((\lambda^\beta, <)) \geq \beta$ for all limit β and all cardinals λ. Compare this with Theorem 7.26 which gives $\text{SH}(\mathcal{M})((\alpha, <)) = \alpha$ if $\alpha = \sup_{\beta<\alpha} \omega^\beta$.

Below we use $\bar{x}, \bar{y}$, etc. to denote sequences $\{x_\alpha : \alpha < \beta\}, \{y_\alpha : \alpha < \gamma\}$, etc. of variables of length $< \lambda$.

Definition 9.33   Let L be a vocabulary. An L-*fragment* of $L_{\infty\lambda}$ is any set $\mathcal{T}$ of formulas of $L_{\infty\lambda}$ in the vocabulary L such that

(1) $\mathcal{T}$ contains the atomic L-formulas.
(2) $\varphi \in \mathcal{T}$ if and only if $\neg\varphi \in \mathcal{T}$.
(3) $\wedge\Phi \in \mathcal{T}$ implies $\Phi \subseteq \mathcal{T}$.
(4) $\Phi \subseteq \mathcal{T}$ implies $\wedge\Phi \in \mathcal{T}$ for finite Φ.
(5) $\vee\Phi \in \mathcal{T}$ implies $\Phi \subseteq \mathcal{T}$.
(6) $\Phi \subseteq \mathcal{T}$ implies $\vee\Phi \in \mathcal{T}$ for finite Φ.
(7) $\forall\bar{y}\varphi \in \mathcal{T}$ if and only if $\varphi \in \mathcal{T}$.
(8) $\exists\bar{y}\varphi \in \mathcal{T}$ if and only if $\varphi \in \mathcal{T}$.

Note that a fragment is necessarily closed under subformulas. The applicability of fragments arises from the following fact:

Lemma 9.34 *Suppose $\lambda \leq \kappa^+$, λ regular, and $\mu = \kappa^{<\lambda}$. Suppose $\varphi \in L_{\kappa^+\lambda}$ with a vocabulary L, $|L| \leq \mu$. Then there is a fragment $\mathcal{T} \subseteq L_{\kappa^+\lambda}$ such that $\varphi \in \mathcal{T}$ and $|\mathcal{T}| \leq \mu$.*

Proof Let $\mathcal{T}_0$ consist of atomic L-formulas and the formula φ. For successor ordinals we let

$$
\begin{aligned}
\mathcal{T}_{\alpha+1} = \mathcal{T}_\alpha \quad &\cup \{\psi : \neg\psi \in \mathcal{T}_\alpha\} \cup \{\neg\psi : \psi \in \mathcal{T}_\alpha\} \\
&\cup \{\psi : \psi \in \Phi, \wedge\Phi \in \mathcal{T}_\alpha \text{ or } \vee\Phi \in \mathcal{T}_\alpha\} \\
&\cup \{\wedge\Phi : \Phi \subseteq \mathcal{T}_\alpha \text{ finite}\} \cup \{\vee\Phi : \Phi \subseteq \mathcal{T}_\alpha \text{ finite}\} \\
&\cup \{\psi : \forall\bar{y}\psi \in \mathcal{T}_\alpha \text{ or } \exists\bar{y}\psi \in \mathcal{T}_\alpha\} \\
&\cup \{\forall\bar{y}\psi : \psi \in \mathcal{T}_\alpha, \operatorname{len}(\bar{y}) < \lambda\} \cup \{\exists\bar{y}\psi : \psi \in \mathcal{T}_\alpha, \operatorname{len}(\bar{y}) < \lambda\}
\end{aligned}
$$

and $\mathcal{T}_\nu = \bigcup_{\alpha<\nu} \mathcal{T}_\alpha$ for limit ν. Finally, let $\mathcal{T} = \bigcup_{\alpha<\mu} \mathcal{T}_\alpha$. Now $|\mathcal{T}_\alpha| \leq \mu$ for all $\alpha < \mu$, as $\mu^{<\lambda} = \mu$ by the regularity of λ. $\qquad\square$

Definition 9.35 Suppose L is a vocabulary and $\mathcal{M}$ is an L-structure. A *Skolem function* for a formula $\varphi(\bar{x}, \bar{y}) \in L_{\infty\lambda}$ is any function $f_\varphi : M^{<\lambda} \to M^{<\lambda}$ such that for all $\bar{a} \in M^{<\lambda}$

$$
\operatorname{len}(f(\bar{a})) = \operatorname{len}(\bar{a})
$$

and

$$
\mathcal{M} \models \exists\bar{x}\varphi(\bar{a}, \bar{x}) \to \varphi(\bar{a}, f(\bar{a})).
$$

By the Axiom of Choice it is clear that Skolem functions exist in any model for any formula of $L_{\infty\lambda}$.

Proposition 9.36 *Suppose $\mathcal{T}$ is an L-fragment of $L_{\kappa^+\lambda}$, $\mathcal{M}$ is an L-structure, and $\mathcal{F}$ is a family of functions such that every formula in $\mathcal{T}$ has a Skolem function in $\mathcal{F}$. Suppose $M_0 \subseteq M$ is closed under all functions in $\mathcal{F}$. Then M_0 is the universe of a substructure $\mathcal{M}_0$ of $\mathcal{M}$ such that for all $\varphi \in \mathcal{T}$ and s:*

$$
\mathcal{M}_0 \models_s \varphi \quad\Longleftrightarrow\quad \mathcal{M} \models_s \varphi.
$$

Proof See the proof of Proposition 8.27. $\qquad\square$

Theorem 9.37 (Löwenheim–Skolem Theorem) *Suppose $\lambda \leq \kappa^+$, λ is regular, and $\mu = \kappa^{<\lambda}$. Suppose L is a vocabulary of cardinality $\leq \mu$, φ is a sentence of $L_{\kappa^+\lambda}$ in the vocabulary L, and $\mathcal{M}$ is an L-structure such that $\mathcal{M} \models \varphi$. Then there is $\mathcal{M}_0 \subseteq \mathcal{M}$ such that $\mathcal{M}_0 \models \varphi$ and $|M_0| \leq \mu$.*

Proof Let $\mathcal{T}$ be an L-fragment of cardinality $\leq \mu$ such that $\varphi \in \mathcal{T}$. Let $\mathcal{F}$ be a set of cardinality $\leq \mu$ of functions $M^{<\lambda} \to M^{<\lambda}$ containing a Skolem function for each $\psi \in \mathcal{T}$. Let M_0 be a subset of M of cardinality $\leq \mu$ closed under the functions of $\mathcal{F}$. By Proposition 8.27, M_0 is the universe of a substructure $\mathcal{M}_0$ of $\mathcal{M}$ such that $\mathcal{M}_0 \models \varphi$.   □

For more powerful Löwenheim–Skolem Theorems in infinitary logic see Dickmann (1975).

It might seem that $L_{\infty\infty}$ is a very powerful logic. In fact any model class which is closed under isomorphisms and has only models of cardinality $\leq \lambda$ for some λ is definable in $L_{\infty\infty}$ (Exercise 9.9). Nevertheless, some very innocent looking properties of models are non-expressible in $L_{\infty\infty}$:

Proposition 9.38   *The class of well-ordered models of the form $(\gamma + \gamma, <)$ is definable in $L_{\omega_1 V}$ but not in $L_{\infty\infty}$.*

Proof   We shall first present a sentence in $L_{\omega_1 V}$, which characterizes the given class of models. In the sentence that follows, the variable x_0 picks the mid-point of the order $(\gamma + \gamma, <)$. Let Φ be the conjunction of the axiom which says that $<$ well-orders the universe, and the following closed game sentence:

$$\exists x_0 \forall x_1 \exists y_1 \forall y_2 \exists x_2 \forall x_3 \exists y_3 \ldots \bigwedge_{i<\omega} \varphi_i(x_0, x_{i-1}, y_{i-1}, x_i, y_i),$$

where $\varphi_0(x_0, x_1, y_1)$ is the conjunction of:

$x_1 > x_0 \rightarrow y_1 < x_0$
$x_1 < x_0 \rightarrow y_1 > x_0,$

and $\varphi_i(x_0, x_{i-1}, y_{i-1}, x_i, y_i)$ for $i > 0$ is the conjunction of:

$(x_0 < x_i \wedge x_i < x_{i-1}) \rightarrow y_i < y_{i-1}$
$x_i < y_{i-1} \rightarrow (x_0 < y_i \wedge y_i < x_{i-1}).$

Suppose $\mathcal{A} = (\gamma + \gamma, <)$. The strategy of **II** in $G_S(\mathcal{A}, \Phi)$ is to play first $x_0 = \gamma$. Then always, when **I** plays x_i, **II** plays either $y_i = \gamma + x_i$ or $\gamma + y_i = x_i$, according to whether $x_i < \gamma$ or $x_i \geq \gamma$. This shows that $\mathcal{A} \models \Phi$.

Conversely, suppose $\mathcal{A} = (\gamma, <) \models \Phi$. Let a be the choice for x_0 that the winning strategy of **II** gives. Let us divide γ into a sum $\gamma + \beta$ of two ordinals so that γ is the set of predecessors of a. If $\gamma = \beta$, we are done. Otherwise, say, $\beta < \gamma$. Now **I** plays $x_1 = \beta$. Whenever **II** has played y_i, **I** plays $x_i = \gamma + y_i$ or $\gamma + x_i = y_i$, according to whether $y_i < \gamma$ or $y_i \geq \gamma$. This leads to an infinite descending chain in $\mathcal{A}$, a contradiction. This ends the first part of the proof.

We shall now prove that the given class of models is not definable in $L_{\infty\infty}$. Suppose it were. W.l.o.g. it is definable in $L_{\kappa^+\kappa^+}$ for some κ with a sentence ψ of quantifier rank δ. Let μ be the cardinality of the set of all (up to logical equivalence) $L_{\infty\kappa^+}$-formulas in the vocabulary $\{<\}$ of quantifier rank $\leq \delta$ (see Exercise 9.10). Let $\gamma = 2^\mu$. Let φ be the conjunction of all sentences of

$L_{\infty\kappa^+}$ of quantifier rank $\leq \delta$ that are true in $(\gamma, <)$. Thus $\varphi \in L_{\mu^+\kappa^+}$. By Theorem 9.37 there is $\mathcal{B} \subseteq (\gamma, <)$ such that $\mathcal{B} \models \varphi$ and $|B| \leq 2^\mu$. Let $\beta < \gamma$ so that $\mathcal{B} \cong (\beta, <)$. Now $(\gamma, <) \equiv^\delta_{\infty\lambda} (\beta, <)$. By the preservation of $\equiv^\delta_{\infty\lambda}$ under sums of models (see Exercise 9.19),

$$(\gamma + \beta, <) \equiv^\delta_{\infty\lambda} (\gamma + \gamma, <).$$

This contradicts the fact that $(\gamma + \beta, <) \not\models \psi$ and $(\gamma + \gamma, <) \models \psi$. □

Theorem 9.39   *There is a sentence in $L_{\omega_1\omega_1}$ which defines implicitly a relation which is not explicitly definable in $L_{\infty\infty}$. Hence the Craig Interpolation Theorem and the Beth Definability Theorem fail for all $L_{\kappa\lambda}$, where $\omega < \lambda \leq \kappa$.*

Proof   Let $\mathcal{K}$ be the class of well-ordered models $(\gamma + \gamma, <, R)$, where $R = \{(\alpha, \gamma + \alpha) : \alpha < \gamma\}$. Since the isomorphism between two well-ordered structures is always unique, the class $\mathcal{K}$ is definable in $L_{\omega_1\omega_1}$. Moreover, for any γ there is a unique R so that $(\gamma + \gamma, <, R) \in \mathcal{K}$. On the other hand, suppose R were explictly definable in $L_{\kappa\lambda}$, where κ is regular and $\omega < \lambda \leq \kappa$. Then the class of well-ordered models $(\gamma + \gamma, <)$ would be definable in $L_{\kappa\lambda}$, contrary to Proposition 9.38. □

Since the class of models $(\gamma + \gamma, <)$ used in the above proof of the failure of interpolation is definable in $L_{\omega_1 V}$, the question arises, whether $L_{\omega_1 V}$ itself satisfies the Craig Interpolation Theorem or the Beth Definability Theorem. The answer to both questions is negative, as was proved in Gostanian and Hrbáček (1976).

To overcome the failure of $\simeq_{\aleph_1}$ to characterize isomorphism in models of size $\aleph_1$ we now introduce a stronger form of $\simeq_{\aleph_1}$:

Definition 9.40   Suppose $\mathcal{A}$ and $\mathcal{B}$ are L-structures. A *strong λ-back-and-forth set* for $\mathcal{A}$ and $\mathcal{B}$ is a λ-back-and-forth set P which satisfies the following condition:

(3)   If $\{f_\alpha : \alpha < \beta\}, \beta < \lambda$, is an $\subseteq$-increasing sequence in P, then there is $f \in P$ such that $f_\alpha \subseteq f$ for all $\alpha < \beta$.

If there is a strong λ-back-and-forth set for $\mathcal{A}$ and $\mathcal{B}$, we write

$$\mathcal{A} \simeq^s_\lambda \mathcal{B}.$$

Theorem 9.41   *If $|A| \leq \lambda$ and $|B| \leq \lambda$, then the following conditions are equivalent:*

(1) $\mathcal{A} \cong \mathcal{B}$.
(2) $\mathcal{A} \simeq^s_\lambda \mathcal{B}$.

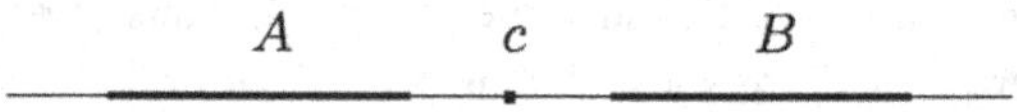

$$A \qquad c \qquad B$$

Figure 9.5

Proof   Suppose P is a strong λ-back-and-forth set for $\mathcal{A}$ and $\mathcal{B}$. Let

$$A = \{a_\alpha : \alpha < \lambda\}$$
$$B = \{b_\alpha : \alpha < \lambda\}.$$

Let $f_0 \in P$. If f_α has been defined for $\alpha < \gamma$ and

$$\alpha < \beta < \gamma \quad \Longrightarrow \quad f_\alpha \subseteq f_\beta$$

then let $g \in P$ be such that $f_\alpha \subseteq g$ for all $\alpha < \gamma$. Then, let $f_\gamma \in P$ be such that $g \subseteq f_\gamma, a_\gamma \in \mathrm{dom}(f_\gamma)$ and $b_\gamma \in \mathrm{ran}(f_\gamma)$. Then $\bigcup_{\alpha < \lambda} f_\alpha$ is an isomorphism $\mathcal{A} \to \mathcal{B}$. $\qquad \square$

Example 9.42   A linear order $(M, <)$ is an η_α-set if for all sets $A \subseteq M$ and $B \subseteq M$ such that $|A| < \aleph_\alpha, |B| < \aleph_\alpha$ and $a < b$ for all $a \in A$ and $b \in B$, there is $c \in M$ such that

$$a < c < b$$

for all $a \in A$ and $b \in B$ (See Figure 9.5).

Proposition 9.43   *Suppose $\mathcal{A}$ and $\mathcal{B}$ are η_α-sets and $\aleph_\alpha$ is regular. Then $\mathcal{A} \simeq^s_{\aleph_\alpha} \mathcal{B}$. Hence if $|A| = |B| \le \aleph_\alpha$, then $\mathcal{A} \cong \mathcal{B}$.*

Proof   Let P be the set $\mathrm{Part}_{\aleph_\alpha}(\mathcal{A}, \mathcal{B})$. It is easy to see that P is an $\aleph_\alpha$-back-and-forth set. $\qquad \square$

A model $\mathcal{M}$ is λ-*homogeneous* if for all $\{a_\alpha : \alpha \le \beta\}$ and $\{b_\alpha : \alpha < \beta\}$ such that $\beta < \lambda$ and

$$(\mathcal{M}, a_\alpha)_{\alpha < \beta} \equiv (\mathcal{M}, b_\alpha)_{\alpha < \beta}$$

there is $b_\beta \in M$ such that

$$(\mathcal{M}, a_\alpha)_{\alpha \le \beta} \equiv (\mathcal{M}, b_\alpha)_{\alpha \le \beta}.$$

The ordered sets $(\mathbb{Q}, <)$ and $(\mathbb{R}, <)$ are $\aleph_0$-homogeneous. Every countable consistent first-order theory has for every λ a λ-homogeneous model (see e.g. Chang and Keisler (1990)).

Proposition 9.44 *If $\mathcal{A}$ and $\mathcal{B}$ are λ-homogeneous, λ is regular, and $\mathcal{A} \simeq_\lambda \mathcal{B}$, then $\mathcal{A} \simeq_\lambda^s \mathcal{B}$.*

Proof We let P consist of

$$f = \{(a_\alpha, b_\alpha) : \alpha < \beta\}$$

such that $\beta < \lambda$ and

$$(\mathcal{A}, a_\alpha)_{\alpha < \beta} \equiv (\mathcal{B}, b_\alpha)_{\alpha < \beta}. \tag{9.2}$$

P is non-empty, as $\emptyset \in P$. Let us prove that P is a λ-back-and-forth set. Suppose (9.2) holds and $a_\beta \in A$ (we treat this case first and deal with the case $a_\beta \in [A]^{<\lambda}$ later). By $\mathcal{A} \simeq_\lambda \mathcal{B}$ there are $b'_\alpha, \alpha \le \beta$ such that

$$(\mathcal{A}, a_\alpha)_{\alpha \le \beta} \equiv (\mathcal{B}, b'_\alpha)_{\alpha \le \beta}.$$

Thus

$$(\mathcal{B}, b_\alpha)_{\alpha < \beta} \equiv (\mathcal{B}, b'_\alpha)_{\alpha < \beta}.$$

By λ-homogeneity there is b_β such that

$$(\mathcal{B}, b_\alpha)_{\alpha \le \beta} \equiv (\mathcal{B}, b'_\alpha)_{\alpha \le \beta}.$$

Thus

$$(\mathcal{A}, a_\alpha)_{\alpha \le \beta} \equiv (\mathcal{B}, b_\alpha)_{\alpha \le \beta}.$$

The property (9.2) is clearly preserved under unions of chains, i.e. if $f_\alpha \in P$ for $\alpha < \gamma$, where $\gamma < \lambda$, and

$$\alpha < \beta < \gamma \quad \Longrightarrow \quad f_\alpha \subseteq f_\beta$$

then $\bigcup_{\alpha < \gamma} f_\alpha \in P$ (here we use regularity of λ). Thus the above back-and-forth argument extends to $a_\beta \in [A]^{<\lambda}$. $\square$

A model $\mathcal{A}$ is λ-*saturated* if for all $a_\alpha \in A, \alpha < \beta$, where $\beta < \lambda$, every type of the structure

$$(\mathcal{A}, a_\alpha)_{\alpha < \beta}$$

is realized in $(\mathcal{A}, a_\alpha)_{\alpha < \beta}$. λ-saturated models are always λ-homogeneous. Every countable consistent first-order theory has for every λ a λ-saturated model.

Proposition 9.45 *Suppose $\mathcal{A}$ and $\mathcal{B}$ are λ-saturated structures and $\mathcal{A} \equiv \mathcal{B}$. Then $\mathcal{A} \simeq_\lambda \mathcal{B}$. If moreover λ is regular, then $\mathcal{A} \simeq_\lambda^s \mathcal{B}$.*

Proof Let P consist of

$$f = \{(a_\alpha, b_\alpha) : \alpha < \beta\}$$

such that $\beta < \lambda$ and

$$(\mathcal{A}, a_\alpha)_{\alpha < \beta} \equiv (\mathcal{B}, b_\alpha)_{\alpha < \beta}. \tag{9.3}$$

By assumption, $\emptyset \in P$. Suppose then (9.3) holds and $a_\beta \in A$ (the case $a_\beta \in [A]^{<\lambda}$ follows from this). Let Σ be the set of first-order sentences, $\varphi(x_{\alpha_1}, \ldots, x_{\alpha_n}, x_\beta)$, $\alpha_1, \ldots, \alpha_n < \beta$, satisfying

$$\mathcal{A} \models \varphi(a_{\alpha_1}, \ldots, a_{\alpha_n}, a_\beta).$$

Condition (9.3) guarantees that for any finite $\Sigma_0 \subseteq \Sigma$ there is $b_\beta \in B$ such that

$$\mathcal{B} \models \varphi(b_{\alpha_1}, \ldots, b_{\alpha_n}, b_\beta) \tag{9.4}$$

for all $\varphi \in \Sigma_0$. By λ-saturation, there is $b_\beta \in B$ such that (9.4) holds for all $\varphi \in \Sigma$. We have proved one half of the λ-back-and-forth criterion for P. The other half is similar. $\qquad\square$

A theory is κ-categorical if all its models of cardinality κ are isomorphic. The first-order theory of $(\mathbb{Q}, <)$ is $\aleph_0$-categorical. The first-order theory of $(\mathbb{C}, +, \cdot, 0, 1)$ is $\aleph_1$-categorical. A countable first-order theory without finite models which is κ-categorical for some $\kappa \geq \omega$ is necessarily complete. By a deep theorem of Morley, if a countable first-order theory is κ-categorical for some uncountable κ, it is κ-categorical for all uncountable κ.

Corollary *Let λ be a regular cardinal $> \omega$. Suppose T is a countable complete first-order theory which is κ-categorical for some (hence all) uncountable κ. Then for any $\mathcal{M} \models T$ and $\mathcal{N} \models T$ of cardinality $\geq \lambda$ we have*

$$\mathcal{M} \simeq^s_\lambda \mathcal{N}.$$

Proof For such T and λ we know that $\mathcal{M}$ and $\mathcal{N}$ have to be λ-saturated (see e.g. Chang and Keisler (1990)). $\qquad\square$

Open Problem: Is the relation $\mathcal{M} \simeq^s_\lambda \mathcal{N}$ transitive?

There are many partial results[4] about the transitivity of $\mathcal{M} \simeq^s_\lambda \mathcal{N}$ but the general question is open. Thus it is not known whether there is some version L^* of infinitary logic such that $\mathcal{M} \simeq^s_\lambda \mathcal{N}$ is equivalent to $\mathcal{M} \models \varphi \iff \mathcal{N} \models \varphi$ for all $\varphi \in L^*$.

---

[4] See Väänänen and Veličković (2004).

9.3 The Transfinite Ehrenfeucht–Fraïssé Game

All our games up to now have had at most ω rounds. There is no difficulty in imagining what a game of, say, length $\omega + \omega$ would look like: it would be like playing two games of length ω one after the other. For example, it is by now well known to the reader that the second player has a winning strategy in the Ehrenfeucht–Fraïssé Game of length ω on $(\mathbb{R}, <)$ and $(\mathbb{R} \setminus \{0\}, <)$. But if player **I** is allowed one more move after the ω moves, he wins.

For a more enlightening example, suppose $\mathcal{M}$ and $\mathcal{N}$ are equivalence relations such that $\mathcal{M}$ has $\aleph_1$ countable classes and $\aleph_0$ uncountable classes while $\mathcal{N}$ has $\aleph_1$ countable classes and $\aleph_1$ uncountable classes. Does **II** have a winning strategy in EF_ω? Yes! She just keeps matching different equivalence classes with different equivalence classes. But she can actually win the game of length $\omega + \omega$, too! During the first ω moves she matches countable equivalence classes with countable ones and uncountable equivalence classes with uncountable ones. After the first ω moves she may have to match a countable equivalence class with an uncountable class, but **I** will not be able to call **II**'s bluff. It is only when **I** has $\omega + \omega + 1$ moves that he has a winning strategy: During the first ω moves **I** plays one element from each uncountable class of $\mathcal{M}$. Then **I** plays one element b from an unused uncountable equivalence class of $\mathcal{N}$. Player **II** will match this element with an element c from a countable equivalence class of $\mathcal{M}$. During the next ω rounds player **I** enumerates the countable equivalence class of c. Finally he plays an unplayed element equivalence to b. Player **II** loses as all elements equivalent to c have been played already.

Let L be a vocabulary and $\mathcal{A}_0$ and $\mathcal{A}_1$ two L-structures. We give a rigorous definition of a transfinite version of the Ehrenfeucht–Fraïssé Game on the two models $\mathcal{A}_0$ and $\mathcal{A}_1$. We let the number of rounds of this game be an arbitrary ordinal δ.

In the sequel we allow the domains of $\mathcal{M}_0$ and $\mathcal{M}_1$ to intersect and incorporate a mechanism to account for this.

We shall all the time refer to sequences

$$\bar{z} = \langle z_\alpha : \alpha < \delta \rangle, \text{ where } z_\alpha = (c_\alpha, x_\alpha),$$

of elements of $\{0, 1\} \times (A_0 \cup A_1)$. If $\bar{y} = \langle y_\alpha : \alpha < \delta \rangle$ is a sequence of elements of $A_0 \cup A_1$, the relation

$$p_{\bar{z}, \bar{y}} \subseteq (A_0 \cup A_1)^2$$

is defined as follows:

$$p_{\bar{z}, \bar{y}} = \{(a_\alpha, b_\alpha) : \alpha < \delta\}$$

I	z_0	z_1	$\cdots$	z_α	$\cdots$	$(\alpha < \delta)$
II	y_0	y_1	$\cdots$	y_α	$\cdots$	$(\alpha < \delta)$

Figure 9.6 The Ehrenfeucht–Fraïssé Game.

where

$$a_\alpha = \left\{ \begin{array}{ll} x_\alpha & \text{if } c_\alpha = 0 \\ y_\alpha & \text{if } c_\alpha = 1 \end{array} \right. \quad b_\alpha = \left\{ \begin{array}{ll} y_\alpha & \text{if } c_\alpha = 0 \\ x_\alpha & \text{if } c_\alpha = 1. \end{array} \right.$$

Remark We shall often use the fact that $p_{\bar{z},\bar{y}} = \cup_{\sigma < \delta} p_{\bar{z}\restriction_\sigma, \bar{y}\restriction_\sigma}$ if δ is a limit ordinal.

We are interested in the question whether

$$p_{\bar{z},\bar{y}} \in \text{Part}(\mathcal{A}_0, \mathcal{A}_1) \tag{9.5}$$

or not. In the Ehrenfeucht–Fraïssé Game one player chooses $\bar{z}$ trying to make (9.5) false, and the other player chooses $\bar{y}$ trying to make (9.5) true. Let

$$\text{Seq}_\delta(A_0, A_1)$$

be the set of all sequences $\langle (c_\alpha, x_\alpha) : \alpha < \delta \rangle$ where $c_\alpha \in \{0, 1\}$ and $x_\alpha \in A_{c_\alpha}$.

Definition 9.46 Let $\delta \in On$. The Ehrenfeucht–Fraïssé Game of length δ on $\mathcal{A}_0$ and $\mathcal{A}_1$, in symbols

$$\text{EF}_\delta(\mathcal{A}_0, \mathcal{A}_1),$$

is defined as follows. There are two players **I** and **II** . During one round of the game player **I** chooses an element c_α of $\{0, 1\}$ and an element x_α of A_{c_α}, and then player **II** chooses an element y_α of A_{1-c_α}. Let $z_\alpha = (c_\alpha, x_\alpha)$. There are δ rounds and in the end we have $\bar{z} = \langle z_\alpha : \alpha < \delta \rangle$ and $\bar{y} = \langle y_\alpha : \alpha < \delta \rangle$. We say that player **II** *wins* this sequence of rounds, if $p_{\bar{z},\bar{y}} \in \text{Part}(\mathcal{A}_0, \mathcal{B}_1)$. Otherwise player **I** wins this sequence of rounds.

The above definition is useful as an intuitive model of the game. However, it is not mathematically precise because we have not defined what choosing an element means. The idea is that a player is free to choose any element. Also, we are really interested in the existence of a winning strategy for a player by means of which he can win every sequence of rounds. The following exact definition of a winning strategy is our mathematical model for the intuitive concept of a game.

Definition 9.47 A *strategy* of **II** in $\mathrm{EF}_\delta(\mathcal{A}_0, \mathcal{A}_1)$ is a sequence

$$\tau = \langle \tau_\alpha : \alpha < \delta \rangle$$

of functions such that

$$\mathrm{dom}(\tau_\alpha) = \mathrm{Seq}_{\alpha+1}(\mathcal{A}_0, \mathcal{A}_1)$$

and

$$\mathrm{rng}(\tau_\alpha) \subseteq \mathcal{A}_0 \cup \mathcal{A}_1$$

for each $\alpha < \delta$. If $\bar{z} \in \mathrm{Seq}_\delta(\mathcal{A}_0, \mathcal{A}_1)$ and $\bar{y} = \langle y_\alpha : \alpha < \delta \rangle$, where

$$y_\alpha = \tau_\alpha(\bar{z}{\upharpoonright}_{\alpha+1})$$

for all $\alpha < \delta$, then we denote $p_{\bar{z}, \bar{y}}$ by $p_{\bar{z}, \tau}$. The strategy τ of **II** is a *winning strategy* if $p_{\bar{z}, \tau} \in \mathrm{Part}(\mathcal{A}_0, \mathcal{A}_1)$ for all $\bar{z} \in \mathrm{Seq}_\delta(\mathcal{A}_0, \mathcal{A}_1)$. A *strategy* of **I** in $\mathrm{EF}_\delta(\mathcal{A}_0, \mathcal{A}_1)$ is a sequence

$$\rho = \langle \rho_\alpha : \alpha < \delta \rangle$$

of functions such that

$$\mathrm{rng}(\rho_\alpha) \subseteq \{0, 1\} \times (\mathcal{A}_0 \cup \mathcal{A}_1)$$

and $\mathrm{dom}(\rho_\alpha)$ is defined inductively as follows: $\mathrm{dom}(\rho_\alpha)$ is the set of sequences $\bar{y} = \langle y_\beta : \beta < \alpha \rangle$ such that for all $\beta < \alpha$,

$$y_\beta \in \begin{cases} \mathcal{A}_1, & \text{if } \rho_\beta(\bar{y}{\upharpoonright}_\beta) = (0, x) \text{ for some } x \\ \mathcal{A}_0, & \text{if } \rho_\beta(\bar{y}{\upharpoonright}_\beta) = (1, x) \text{ for some } x. \end{cases}$$

If $\bar{y} = \langle y_\alpha : \alpha < \delta \rangle \in \mathrm{dom}(\rho)$ (i.e $\langle y_\alpha : \alpha < \beta \rangle \in \mathrm{dom}(\rho_\beta)$ for all $\beta < \delta$) and $\bar{z} = \langle z_\alpha : \alpha < \delta \rangle$ satisfy

$$\rho_\alpha(\bar{y}{\upharpoonright}_\alpha) = z_\alpha,$$

then $p_{\bar{z}, \bar{y}}$ is denoted by $p_{\rho, \bar{y}}$. The strategy ρ is a *winning strategy* of **I** if there is no $\bar{y} \in \mathrm{dom}(\rho)$ such that $p_{\rho, \bar{y}} \in \mathrm{Part}(\mathcal{A}, \mathcal{B})$. We say that a player *wins* $\mathrm{EF}_\delta(\mathcal{A}_0, \mathcal{A}_1)$ if he has a winning strategy in it.

Remark If $\tau = \langle \tau_\alpha : \alpha < \delta \rangle$ is a strategy of **II** in $\mathrm{EF}_\delta(\mathcal{A}_0, \mathcal{A}_1)$ and $\sigma < \delta$, then $\tau{\upharpoonright}_\sigma = \langle \tau_\alpha : \alpha < \sigma \rangle$ is a strategy of **II** in $\mathrm{EF}_\sigma(\mathcal{A}_0, \mathcal{A}_1)$. If τ is winning, then so is $\tau{\upharpoonright}_\sigma$. Moreover, if $\bar{z} \in \mathrm{Seq}_\delta(\mathcal{A}_0 \cup \mathcal{A}_1)$, then $p_{\bar{z}{\upharpoonright}_\sigma, \tau{\upharpoonright}_\sigma} \subseteq p_{\bar{z}, \tau}$. If δ is a limit ordinal, we have $p_{\bar{z}, \tau} = \cup_{\sigma < \delta} p_{\bar{z}{\upharpoonright}_\sigma, \tau{\upharpoonright}_\sigma}$ and τ is winning if and only if $\tau{\upharpoonright}_\sigma$ is winning for all $\sigma < \delta$.

Example 9.48 Let $L = \emptyset$. Let $\mathcal{A}_0$ and $\mathcal{A}_1$ be two L-structures of cardinalities κ and λ respectively. Let us first assume $\delta \leq \kappa \leq \lambda$. Then **II** wins $\mathrm{EF}_\delta(\mathcal{A}_0, \mathcal{A}_1)$. Her winning strategy $\tau = \langle \tau_\alpha : \alpha < \delta \rangle$ is defined as follows. Let $F(X) \in X$ for every non-empty $X \subseteq \mathcal{A}_0 \cup \mathcal{A}_1$. Suppose τ_α is defined for $\alpha < \sigma$, where $\sigma < \delta$. We define τ_σ. For any $\bar{z} = \langle (c_\alpha, x_\alpha) : \alpha < \sigma \rangle$, let $Y_{\bar{z}} = \{ x_\zeta : \zeta < \sigma \} \cup \{ \tau_\zeta(\bar{z} {\restriction}_{\zeta+1}) : \zeta + 1 < \sigma \}$. Let now

$$
\tau_\sigma(\bar{z}{\restriction}_{\sigma+1}) = \begin{cases} \tau_\zeta(\bar{z}{\restriction}_{\zeta+1}) & \text{if } x_\sigma = x_\zeta, \zeta < \sigma \\ x_\zeta & \text{if } x_\sigma = \tau_\zeta(\bar{z}{\restriction}_{\zeta+1}), \zeta < \sigma \\ F(\mathcal{A}_{1-c_i} \setminus Y_{\bar{z}{\restriction}_\sigma}) & \text{otherwise.} \end{cases}
$$

It is clear that for all $\bar{z}$ the relation $p_{\bar{z},\tau}$ is a partial isomorphism. In fact it suffices that $p_{\bar{z},\tau}$ is a one-one function, since $L = \emptyset$.

Let us then assume $\kappa < \lambda$ and $\kappa < \delta$. Then **I** wins $\mathrm{EF}_\delta(\mathcal{A}_0, \mathcal{B}_1)$. His winning strategy $\rho = \langle \rho_\alpha : \alpha < \delta \rangle$ is defined as follows. Let $A_0 = \{ u_\eta : \eta < \kappa \}$.

$$
\rho_\alpha(\langle y_\eta : \eta < \alpha \rangle) = \begin{cases} u_\alpha & \text{if } \alpha < \kappa \\ F(A_1 \setminus \{ \rho_\zeta(\langle y_\eta : \eta < \zeta \rangle) : \zeta < \alpha \}) & \text{if } \kappa \leq \alpha < \delta. \end{cases}
$$

The intuitive argument behind Example 9.48 based on Definition 9.46 can be described very succinctly: If $\delta \leq \kappa \leq \lambda$, the strategy of **II** is to copy the old moves if **I** plays an old element and choose some new element if **I** plays a new element. The assumption $\delta \leq \kappa < \lambda$ guarantees that there are enough elements to choose from. If $\kappa < \delta$, the strategy of **I** is to first enumerate A_0 during the first κ rounds of the game and then pick an element $x_\kappa \in A_1$, which has not been played yet by **II**. Then **II** has no elements in A_0 left to play and he loses the game.

Lemma 9.49 (i) *If* **II** *wins the game* $\mathrm{EF}_\alpha(\mathcal{A}_0, \mathcal{A}_1)$ *and* $\beta < \alpha$, *then* **II** *wins the game* $\mathrm{EF}_\beta(\mathcal{A}_0, \mathcal{A}_1)$.

(ii) *If* **I** *wins the game* $\mathrm{EF}_\alpha(\mathcal{A}_0, \mathcal{A}_1)$ *and* $\alpha < \beta$, *then* **I** *wins the game* $\mathrm{EF}_\beta(\mathcal{A}_0, \mathcal{A}_1)$.

(iii) *There is no* α *such that both* **II** *and* **I** *win* $\mathrm{EF}_\alpha(\mathcal{A}_0, \mathcal{A}_1)$.

Proof (i) If $\langle \tau_\xi : \xi < \alpha \rangle$ is a winning strategy of **II** in $\mathrm{EF}_\alpha(\mathcal{A}_0, \mathcal{A}_1)$, then $\langle \tau_\xi : \xi < \beta \rangle$ is a winning strategy of **II** in $\mathrm{EF}_\beta(\mathcal{A}_0, \mathcal{A}_1)$.

(ii) If $\langle \rho_\xi : \xi < \alpha \rangle$ is a winning strategy of **I** in $\mathrm{EF}_\alpha(\mathcal{A}_0, \mathcal{A}_1)$, then $\langle \rho_\xi : \xi < \beta \rangle$ is a winning strategy of **I** in $\mathrm{EF}_\beta(\mathcal{A}_0, \mathcal{A}_1)$, where

$$
\rho_\xi(\langle y_\eta : \eta < \xi \rangle) = \rho_\alpha(\langle y_\eta : \eta < \alpha \rangle)
$$

for $\alpha \leq \xi < \beta$.

(iii) Suppose $\langle \tau_\xi : \xi < \alpha \rangle$ is a winning strategy of **II** and $\langle \rho_\xi : \xi < \alpha \rangle$ a winning strategy of **I** in $\mathrm{EF}_\alpha(\mathcal{A}_0, \mathcal{A}_1)$. Define inductively

$$
\begin{aligned}
x_\xi &= \rho_\xi(\langle y_\eta : \eta < \xi \rangle) \\
y_\xi &= \tau_\xi(\langle x_\eta : \eta \le \xi \rangle).
\end{aligned}
$$

If $\bar{z} = \langle x_\xi : \xi < \alpha \rangle$ and $\bar{y} = \langle y_\xi : \xi < \alpha \rangle$, then $p_{\bar{z},\bar{y}}$ is a partial isomorphism because **II** wins, and not a partial isomorphism because **I** wins, a contradiction. $\qquad \square$

Lemma 9.50 *(i) If $\mathcal{A}_0 \cong \mathcal{A}_1$, then **II** wins $\mathrm{EF}_\alpha(\mathcal{A}_0, \mathcal{A}_1)$ for all α.*
*(ii) If $\mathcal{A}_0 \not\cong \mathcal{A}_1$, then **I** wins $\mathrm{EF}_\alpha(\mathcal{A}_0, \mathcal{A}_1)$ for all $\alpha \ge |A_0| + |A_1|$.*

Proof (i) Suppose $f : \mathcal{A}_0 \cong \mathcal{A}_1$. Let

$$
\tau_\xi(\langle (c_\eta, x_\eta) : \eta \le \xi \rangle) = \begin{cases} f(x_\xi) & \text{if } c_\xi = 0 \\ f^{-1}(x_\xi) & \text{if } c_\xi = 1. \end{cases}
$$

Then $\langle \tau_\xi : \xi < \alpha \rangle$ is a winning strategy of **II** in $\mathrm{EF}_\alpha(\mathcal{A}_0, \mathcal{A}_1)$.
(ii) Let $\{0, 1\} \times (A_0 \cup A_1) = \{z_\xi : \xi < \alpha\}$ and $\rho = \langle \rho_\xi : \xi < \alpha \rangle$, where

$$
\rho_\xi(\langle y_\eta : \eta < \xi \rangle) = z_\xi
$$

for $\xi < \alpha$. For any $\bar{y} = \langle y_\xi : \xi < \alpha \rangle$ the relation $p_{\rho, \bar{y}}$ is a partial isomorphism between $\mathcal{A}_0$ and $\mathcal{A}_1$. Since no isomorphism exists, ρ is a winning strategy of **I**. $\qquad \square$

Corollary *(i) If **I** wins $\mathrm{EF}_\alpha(\mathcal{A}_0, \mathcal{A}_1)$, then $\mathcal{A}_0 \not\cong \mathcal{A}_1$.*
*(ii) If **II** wins $\mathrm{EF}_\alpha(\mathcal{A}_0, \mathcal{A}_1)$, where $\alpha \ge |A_0| + |A_1|$, then $\mathcal{A}_0 \cong \mathcal{A}_1$.*

There is always at least one α for which **II** wins $\mathrm{EF}_\alpha(\mathcal{A}_0, \mathcal{A}_1)$, namely $\alpha = 0$. If $\mathcal{A}_0 \cong \mathcal{A}_1$, then by Lemmas 9.49 and 9.50 there cannot be any α for which **I** wins $\mathrm{EF}_\alpha(\mathcal{A}_0, \mathcal{A}_1)$. But if $\mathcal{A}_0 \not\cong \mathcal{A}_1$, then **I** wins $\mathrm{EF}_\alpha(\mathcal{A}_0, \mathcal{A}_1)$ from some α onwards.

There may be ordinals α for which neither player has a winning strategy (Exercises 9.29 and 9.30 below). Then the game is non-determined. The game of length ω_1 may also be non-determined, see Mekler et al. (1993). There may also be a limit ordinal α such that **II** wins $\mathrm{EF}_\beta(\mathcal{A}_0, \mathcal{A}_1)$ for each $\beta < \alpha$ but not $\mathrm{EF}_\alpha(\mathcal{A}_0, \mathcal{A}_1)$. We already know that this can happen if $\alpha = \omega$.

Lemma 9.51 *Let L be a vocabulary and α an ordinal. The relation*

$$
\mathcal{A}_0 \sim_\alpha \mathcal{A}_1 \Leftrightarrow \exists \text{ wins } \mathrm{EF}_\alpha(\mathcal{A}_0, \mathcal{A}_1)
$$

is an equivalence relation on $\mathrm{Str}(L)$.

Proof Reflexivity of $\sim_\alpha$ follows from Lemma 9.50(i). In fact, **II** wins the game $\mathrm{EF}_\alpha(\mathcal{A}_0, \mathcal{A}_0)$ with the trivial strategy $\tau_\xi(\langle (c_\eta, x_\eta) : \eta \leq \xi \rangle) = x_\xi$. Symmetry is also trivial: Suppose **II** wins $\mathrm{EF}_\alpha(\mathcal{A}_0, \mathcal{A}_1)$ with $\tau = \langle \tau_\xi : \xi < \alpha \rangle$. The following strategy $\tau' = \langle \tau'_\xi : \xi < \alpha \rangle$ is winning for **II** in $\mathrm{EF}_\alpha(\mathcal{A}_1, \mathcal{A}_0)$:

$$\tau'(\langle (c_\eta, x_\eta) : \eta \leq \xi \rangle) = \tau(\langle (1 - c_\eta, x_\eta) : \eta \leq \xi \rangle).$$

To see this, suppose $\bar{z} = \langle z_\xi : \xi < \alpha \rangle$ is given. Then $p_{\bar{z}, \tau}$ is a partial isomorphism between $\mathcal{A}_0$ and $\mathcal{A}_1$, and the relation

$$p'_{\bar{z}, \tau} = \{ (b, a) : (a, b) \in p_{\bar{z}, \tau} \}$$

is a partial isomorphism between $\mathcal{A}_1$ and $\mathcal{A}_0$, witnessing the victory of **II** in $\mathrm{EF}_\alpha(\mathcal{A}_1, \mathcal{A}_0)$. To prove transitivity of $\sim_\alpha$, suppose $\tau = \langle \tau_\alpha : \xi < \alpha \rangle$ is a winning strategy of **II** in $\mathrm{EF}_\alpha(\mathcal{A}_0, \mathcal{A}_1)$ and $\tau' = \langle \tau'_\xi : \xi < \alpha \rangle$ is a winning strategy of **II** in $\mathrm{EF}_\alpha(\mathcal{A}_1, \mathcal{A}_2)$. We describe a winning strategy $\tau'' = \langle \tau''_\xi : \xi < \alpha \rangle$ of **II** in $\mathrm{EF}_\alpha(\mathcal{A}_0, \mathcal{A}_2)$. The idea is that **II** plays $\mathrm{EF}_\alpha(\mathcal{A}_0, \mathcal{A}_1)$ and $\mathrm{EF}_\alpha(\mathcal{A}_1, \mathcal{A}_2)$ simultaneously. Suppose $\bar{z}'' = \langle (c''_\eta, x''_\eta) : \eta \leq \xi \rangle \in \mathrm{Seq}_{\xi+1}(\mathcal{A}_0, \mathcal{A}_1)$. We define by induction over $\eta \leq \xi$ the sequences $\bar{z} = \langle (c_\eta, x_\eta) : \eta \leq \xi \rangle$, $\bar{z}' = \langle (c'_\eta, x'_\eta) : \eta \leq \xi \rangle$, and $\tau'' = \langle \tau''_\xi : \xi < \alpha \rangle$ as follows:

If	c''_η	$=$	0	1
Then	(c_η, x_η)	$=$	$(0, x''_\eta)$	$(1, \tau'_\eta(\bar{z}' \restriction_\eta))$
	(c'_η, x'_η)	$=$	$(0, \tau_\eta(\bar{z} \restriction_\eta))$	$(1, x''_\eta)$
	$\tau''_\eta(\bar{z}'' \restriction_\eta)$	$=$	$\tau'_\eta(\bar{z}' \restriction_\eta)$	$\tau_\eta(\bar{z} \restriction_\eta)$

Now $\langle \tau''_\xi : \xi < \alpha \rangle$ is a winning strategy of **II** in $\mathrm{EF}_\alpha(\mathcal{A}_0, \mathcal{A}_2)$. $\square$

The relations $\sim_\alpha$ form a sequence of finer and finer partitions of $\mathrm{Str}(L)$, starting from the one-class partition $\sim_0$ and eventually approaching the ultimate refinement $\cong$ of every $\sim_\alpha$.

9.4 A Quasi-Order of Partially Ordered Sets

Before we define the dynamic version of the transfinite game EF_α we develop some useful theory of po-sets.

Definition 9.52 Suppose $\mathcal{P}$ and $\mathcal{P}'$ are po-sets. We define

$$\mathcal{P} \leq \mathcal{P}'$$

if there is a mapping $f : P \to P'$ such that for all $x, y \in P$:

$$x <_\mathcal{P} y \to f(x) <_{\mathcal{P}'} f(y).$$

We write $\mathcal{P} < \mathcal{P}'$, if $\mathcal{P} \leq \mathcal{P}'$ and $\mathcal{P}' \not\leq \mathcal{P}$, and we write $\mathcal{P} \equiv \mathcal{P}'$, if $\mathcal{P} \leq \mathcal{P}'$ and $\mathcal{P}' \leq \mathcal{P}$.

Note that $\leq$ is a transitive relation among po-sets. The $\equiv$-classes of $\leq$ form a quasi-ordered class. This quasi-order is the topic of this section. It is not a total order, for there are incomparable po-sets, for example $(\omega, <)$ and its inverse ordering $(\omega, >)$. For simplicity, we call $\leq$ itself the quasi-order of po-sets, without recourse to the $\equiv$-classes.

Definition 9.53 Suppose $\mathcal{P}$ is a po-set. The tree $\sigma\mathcal{P}$ is defined as follows. Its domain is the set of functions s with $\text{dom}(s) \in \text{On}$ such that for all $\alpha, \beta \in \text{dom}(s)$

$$\alpha < \beta \to s(\alpha) <_{\mathcal{P}} s(\beta).$$

The order is

$$s \leq s' \leftrightarrow s = s' \restriction_{\text{dom}(s)}.$$

$\sigma'\mathcal{P}$ is the suborder of $\sigma\mathcal{P}$ consisting of sequences $s \in \sigma\mathcal{P}$ of successor length.

The σ-operation was introduced by Kurepa (1956) and studied further, e.g. in Hyttinen and Väänänen (1990) and Todorčević and Väänänen (1999).

Example 9.54 For any ordinal α let B_α be the tree of descending sequences $\beta_0 > \ldots > \beta_n$ of elements of α ordered by end-extension. Show that $\alpha \leq \beta$ (as ordinals) if and only if $B_\alpha \leq B_\beta$ as po-sets. Every well-founded tree is $\equiv$-equivalent to some B_α. (See Exercise 9.36.)

Lemma 9.55 *(i)* $\sigma'\mathcal{P} \leq \mathcal{P}$.
(ii) $\sigma\mathcal{P} \not\leq \mathcal{P}$.
(iii) $\sigma'\mathcal{P} < \sigma\mathcal{P}$.
(iv) *If T is a tree, then $T \equiv \sigma'T$.*

Proof (i) If $s \in \sigma'\mathcal{P}$, let $f(s) = s(\text{dom}(s) - 1)$. Then $f : \sigma'\mathcal{P} \to \mathcal{P}$ is order-preserving.

(ii) Suppose $f : \sigma\mathcal{P} \to \mathcal{P}$ were order-preserving. Define inductively $s : \text{On} \to \mathcal{P}$ by $s(\alpha) = f(s\restriction_\alpha)$. Since $\alpha < \beta$ implies $s(\alpha) <_{\mathcal{P}} s(\beta)$, we get the result that $\mathcal{P}$ is a proper class, a contradiction.

(iii) $\sigma'\mathcal{P} \leq \sigma\mathcal{P}$ trivially. On the other hand, if $\sigma\mathcal{P} \leq \sigma'\mathcal{P}$, then $\sigma\mathcal{P} \leq \mathcal{P}$ contrary to (ii),

(iv) We already know $\sigma'T \leq T$. Suppose $t \in T$ and $\langle t_\alpha : \alpha \leq \beta \rangle$ is the set of $t' \in T$ with $t' \leq_T t$ in ascending order. Let $\text{dom}(s) = \beta + 1$ and $s_t(\alpha) = t_\alpha$. Then $s_t \in \sigma'T$ and $t \mapsto s_t$ is order-preserving. $\square$

Example 9.56 $Q \not\leq \sigma Q$ since σQ is well-founded while Q is not. In particular $Q \not\leq \sigma' Q$. Hence $\sigma' Q < Q$. Note that $\sigma' Q$ is a special tree while σQ is non-special. (See Exercise 9.40.)

Lemma 9.57   *There is no sequence $\mathcal{P}_0, \mathcal{P}_1, \ldots$ so that $\sigma \mathcal{P}_{n+1} \leq \mathcal{P}_n$ for all $n < \omega$.*

Proof   Suppose $f_n : \sigma \mathcal{P}_{n+1} \to \mathcal{P}_n$ is order-preserving. For each fixed α, let $s_\alpha^n \in \mathcal{P}_n$ so that

$$f_n(\langle s_\beta^{n+1} : \beta < \alpha \rangle) = s_\alpha^n.$$

Then each $\mathcal{P}_n$ is a proper class, a contradiction.   □

Definition 9.58   Suppose $\mathcal{P}$ and $\mathcal{P}'$ are po-sets. The game $G(\mathcal{P}, \mathcal{P}')$ is defined as follows. Player **I** plays $p_0 \in \mathcal{P}$, then player **II** plays $p_0' \in \mathcal{P}'$. After this **I** plays $p_1 \in \mathcal{P}$ with $p_0 <_\mathcal{P} p_1$, and then player **II** plays $p_1' \in \mathcal{P}'$ with $p_0' <_{\mathcal{P}'} p_1'$, and so on. At limits player **I** moves first $p_\nu \in \mathcal{P}$ with $p_\alpha <_\mathcal{P} p_\nu$ for all $\alpha < \nu$. Then **II** moves $p_\nu' \in \mathcal{P}'$ with $p_\alpha' <_\mathcal{P} p_\nu'$ for all $\alpha < \nu$. If a player cannot move, he loses and the other player wins. Since $\mathcal{P}$ and $\mathcal{P}'$ are sets, one of the players eventually wins.

Lemma 9.59   (i) $\sigma' \mathcal{P} \leq \mathcal{P}'$ *if and only if **II** wins $G(\mathcal{P}, \mathcal{P}')$.*
(ii) *If $\mathcal{P}$ is a tree, then $\mathcal{P} \leq \mathcal{P}'$ if and only if **II** wins $G(\mathcal{P}, \mathcal{P}')$.*

Proof   (i) Suppose $f : \sigma' \mathcal{P} \to \mathcal{P}'$ is order-preserving. If **I** has played $p_0 < \ldots < p_\alpha$ in $G(\mathcal{P}, \mathcal{P}')$, **II** plays $p_\alpha' = f((p_0, \ldots, p_\alpha))$. In this way she ends up the winner. Conversely, suppose **II** wins $G(\mathcal{P}, \mathcal{P}')$ and $s \in \sigma' \mathcal{P}$ with $\text{dom}(s) = \alpha + 1$. Let us play $G(\mathcal{P}, \mathcal{P}')$ so that **I** plays $p_\beta = s(\beta)$ for $\beta \leq \alpha$ and **II** uses her winning strategy. After **I** plays p_α, **II** plays p_α'. If we define $f(s) = p_\alpha'$, we get an order-preserving mapping $\sigma' \mathcal{P} \to \mathcal{P}$. This ends the proof of (i). (ii) follows from (i) and Lemma 9.55 (iv).   □

Lemma 9.60 $\sigma \mathcal{P}' \leq \mathcal{P}$ *if and only if **I** wins $G(\mathcal{P}, \mathcal{P}')$.*

Proof   Suppose $f : \sigma \mathcal{P}' \to \mathcal{P}$ is order-preserving. If **II** has played

$$p_0' < \ldots < p_\beta' < \ldots \quad (\beta < \alpha) \tag{9.6}$$

in $G(\mathcal{P}, \mathcal{P}')$, **I** plays $p_\alpha = f((p_0', \ldots, p_\beta', \ldots))$ in $\mathcal{P}'$. In this way **I** wins $G(\mathcal{P}, \mathcal{P}')$. On the other hand, if **I** wins $G(\mathcal{P}, \mathcal{P}')$ and (9.6) is an ascending chain in $\mathcal{P}'$, we can let **I** play against the moves $p_0', \ldots, p_\beta', \ldots$ of **II** in $G(\mathcal{P}, \mathcal{P}')$. Finally **I** plays p_α according to his winning strategy. We let

$$f((p_0', \ldots, p_\beta', \ldots)) = p_\alpha.$$

Now $f : \sigma \mathcal{P}' \to \mathcal{P}$ is order-preserving.   □

Example 9.61 Suppose $S \subseteq \omega_1$. Let $T(S)$ be the tree of closed ascending sequences of elements of S. Choose disjoint stationary sets S_1 and S_2. Then $T(S_1) \not\leq T(S_2)$ and $T(S_2) \not\leq T(S_1)$ (Exercise 9.35). Thus the game $G(T(S_1), T(S_2))$ is non-determined.

Definition 9.62 We use $\mathcal{T}_{\lambda,\kappa}$ to denote the class of trees of cardinality $\leq \lambda$ without branches of length κ.

The simplest uncountable tree in $\mathcal{T}_{\kappa,\kappa}$ is the κ-*fan* which consists of branches of all lengths $< \kappa$ joined at the root, or in symbols,

$$F_\kappa = \{s_\alpha : 0 < \alpha < \kappa\}, s_\alpha = \langle a_\beta^\alpha : \beta < \alpha \rangle,$$

$$a_0^\alpha = 0, a_\beta^\alpha = (\alpha, \beta) \text{ for } \beta > 0,$$

ordered by end-extension. Aronszajn trees are in $\mathcal{T}_{\aleph_1, \aleph_1}$. The trees $T(S)$ of Example 9.61 are in $\mathcal{T}_{2^\omega, \aleph_1}$.

Definition 9.63 A tree T is a *persistent* if for all $t \in T$ and all $\alpha < \text{ht}(T)$ there is $t' \in T$ such that $t <_T t'$ and $\text{ht}(t') \geq \alpha$.

Persistency is a kind of non-triviality assumption for a tree. It means that from any node you can go as high as you like. The κ-fan is certainly non-persistent. On the other hand, the tree

$$T_p^\kappa = (F_\kappa)^{<\omega}, (s_{\alpha_0}, \ldots, s_{\alpha_{n-1}}) \leq (s_{\beta_0}, \ldots, s_{\beta_{m-1}}) \iff$$

$$n \leq m \text{ and } \alpha_i = \beta_i \text{ for } i < n$$

is persistent and indeed the $\leq$-smallest persistent tree in $\mathcal{T}_{\kappa,\kappa}$ (Exercise 9.41).

Definition 9.64 A po-set $\mathcal{P}$ is a *bottleneck* in a class K of po-sets if $\mathcal{P}' \leq \mathcal{P}$ or $\mathcal{P} \leq \mathcal{P}'$ for all $\mathcal{P}'$ in the class K. A tree T is a *strong bottleneck* for a class K if the game $G(T, \mathcal{P})$ is determined for all $\mathcal{P} \in K$.

Every well-founded tree is a strong bottleneck in the class of all trees. If $S \subseteq \omega_1$ is bistationary, then $T(S)$ is by Example 9.61 not a bottleneck in the class of all trees. The smallest persistent tree T_p^κ is a strong bottleneck in the class $\mathcal{T}_{\kappa,\kappa}$ (Exercise 9.42). It is an interesting problem whether there are bottlenecks in the class $\mathcal{T}_{\kappa,\kappa}$ above T_p^κ. The following partial result is known:

Theorem 9.65 *Suppose κ is a regular cardinal and $\mathcal{P}$ is the forcing notion for adding κ^+ Cohen subsets to κ. Then $\mathcal{P}$ forces that there are no bottlenecks in the class $\mathcal{T}_{\kappa,\kappa}$ above T_p^κ.*

Proof Suppose T is a bottleneck. Let $\alpha < \kappa^+$ such that $T \in V[G_\alpha]$. Let A_α be the Cohen subset of κ added at stage α. Note that A_α is a bistationary subset of κ. We first show that $\Vdash T \not\le T(A_\alpha)$. Suppose

$$p \Vdash \hat{f} : T(A_\alpha) \to T \text{ is strictly increasing.}$$

When we force with A_α, calling the forcing notion $\mathcal{P}'$, an uncountable branch appears in $T(A_\alpha)$, hence also in T. The product forcing $\mathcal{P}_\alpha \star \mathcal{P}'$ contains a κ-closed dense set (Exercise 9.45). Hence it cannot add a branch of length κ to T. We have shown that $T(A_\alpha) \not\le T$ in $V[G]$. Since T is a bottleneck, $T \le T(A_\alpha)$. By repeating the same with $-A_\alpha$ we get $T \le T(-A_\alpha)$. In sum, $T \le T(A_\alpha) \otimes T(-A_\alpha)$ (see Exercise 9.44 for the definition of $\otimes$). But $T(A_\alpha) \otimes T(-A_\alpha) \le T_p^\kappa$ (Exercise 9.46). Hence $T \le T_p^\kappa$. □

It is also known (Todorčević and Väänänen (1999)) that if $V = L$, then there are no bottlenecks in the class $\mathcal{T}_{\aleph_1, \aleph_1}$ above $T_p^{\aleph_1}$.

9.5 The Transfinite Dynamic Ehrenfeucht–Fraïssé Game

In this section we introduce a more general form of the Ehrenfeucht–Fraïssé Game. The new game generalizes both the usual Ehrenfeucht–Fraïssé Game and the dynamic version of it. In this game player **I** makes moves not only in the models in question but also moves up a po-set, move by move. The game goes on as long as **I** can move. This game generalizes at the same time the games $EF_\alpha(\mathcal{A}_0, \mathcal{A}_1)$ and $EFD_\delta(\mathcal{A}_0, \mathcal{A}_1)$. Therefore we denote it by $EF_\mathcal{P}$ rather than by $EFD_\mathcal{P}$.

If $\mathcal{P}$ is a po-set, let $b(\mathcal{P})$ denote the least ordinal δ so that $\mathcal{P}$ does not have an ascending chain of length δ.

Definition 9.66 Suppose $\mathcal{A}_0$ and $\mathcal{A}_1$ are L-structures and $\mathcal{P}$ is a po-set. The *Transfinite Dynamic Ehrenfeucht–Fraïssé Game* $EF_\mathcal{P}(\mathcal{A}_0, \mathcal{A}_1)$ is like the game $EF_\delta(\mathcal{A}_0, \mathcal{A}_1)$ except that on each round **I** chooses an element $c_\alpha \in \{0, 1\}$, an element $x_\alpha \in A_{c_\alpha}$, and an element $p_\alpha \in \mathcal{P}$. It is required that

$$p_0 <_\mathcal{P} \ldots <_\mathcal{P} p_\alpha <_\mathcal{P} \ldots.$$

Finally **I** cannot play a new p_α anymore because $\mathcal{P}$ is a set. Suppose **I** has played $\bar{z} = \langle (c_\beta, x_\beta) : \beta < \alpha \rangle$ and **II** has played $\bar{y} = \langle y_\beta : \beta < \alpha \rangle$. If $p_{\bar{z}, \bar{y}}$ is a partial isomorphism between $\mathcal{A}_0$ and $\mathcal{A}_1$, **II** has won the game, otherwise **I** has won.

Thus a winning strategy of **I** in $\mathrm{EF}_{\mathcal{P}}(\mathcal{A}_0, \mathcal{A}_1)$ is a sequence $\rho = \langle \rho_\alpha : \alpha < \mathrm{b}(\mathcal{P}) \rangle$ and a strategy of **II** is a sequence $\tau = \langle \tau_\alpha : \alpha < \mathrm{b}(\mathcal{P}) \rangle$. Note that

$$\mathrm{EF}_\alpha(\mathcal{A}_0, \mathcal{A}_1) \text{ is the same game as } \mathrm{EF}_{(\alpha, <)}(\mathcal{A}_0, \mathcal{A}_1),$$

and

$$\mathrm{EFD}_\alpha(\mathcal{A}_0, \mathcal{A}_1) \text{ is the same game as } \mathrm{EF}_{(\alpha, >)}(\mathcal{A}_0, \mathcal{A}_1).$$

Naturally, if α is finite, the games $\mathrm{EF}_{(\alpha, <)}(\mathcal{A}_0, \mathcal{A}_1)$ and $\mathrm{EF}_{(\alpha, >)}(\mathcal{A}_0, \mathcal{A}_1)$ are one and the same game. But if α happens to be infinite, there is a big difference: The first is a transfinite game while the second can only go on for a finite number of moves.

The ordering $\mathcal{P} \leq \mathcal{P}'$ of po-sets has a close connection to the question who wins the game $\mathrm{EF}_{\mathcal{P}}(\mathcal{A}_0, \mathcal{A}_1)$, as the following two results manifest:

Lemma 9.67 *If* **II** *wins the game* $\mathrm{EF}_{\mathcal{P}'}(\mathcal{A}_0, \mathcal{A}_1)$ *and* $\mathcal{P} \leq \mathcal{P}'$, *then* **II** *wins the game* $\mathrm{EF}_{\mathcal{P}}(\mathcal{A}_0, \mathcal{A}_1)$. *If* **I** *wins the game* $\mathrm{EF}_{\mathcal{P}}(\mathcal{A}_0, \mathcal{A}_1)$ *and* $\mathcal{P} \leq \mathcal{P}'$, *then* **I** *wins the game* $\mathrm{EF}_{\mathcal{P}'}(\mathcal{A}_0, \mathcal{A}_1)$.

Proof Exercise 9.50. □

Proposition 9.68 *Suppose* **II** *wins* $\mathrm{EF}_{\mathcal{P}}(\mathcal{A}_0, \mathcal{A}_1)$ *and* **I** *wins* $\mathrm{EF}_{\mathcal{P}'}(\mathcal{A}_0, \mathcal{A}_1)$. *Then* $\sigma\mathcal{P} \leq \mathcal{P}'$.

Proof Suppose **II** wins $\mathrm{EF}_{\mathcal{P}}(\mathcal{A}_0, \mathcal{A}_1)$ with τ and **I** wins $\mathrm{EF}_{\mathcal{P}'}(\mathcal{A}_0, \mathcal{A}_1)$ with ρ. We describe a winning strategy of **I** in $G(\mathcal{P}', \mathcal{P})$, and then the claim follows from Lemma 9.60. Suppose $\rho_0(\emptyset) = (c_0, x_0, p'_0)$. The element p'_0 is the first move of **I** in $G(\mathcal{P}', \mathcal{P})$. Suppose **II** plays $p_0 \in \mathcal{P}$. Let

$$y_0 = \tau_0(((c_0, x_0, p_0))),$$
$$(c_1, x_1, p'_1) = \rho_1((y_0)).$$

The element p'_1 is the second move of **I** in $G(\mathcal{P}', \mathcal{P})$. More generally the equations

$$y_\beta = \tau_\beta(\langle (c_\gamma, x_\gamma, p_\gamma) : \gamma \leq \beta \rangle)$$
$$(c_\alpha, x_\alpha, p'_\alpha) = \rho_\alpha(\langle y_\beta : \beta < \alpha \rangle)$$

define the move p'_α of **I** in $G(\mathcal{P}', \mathcal{P})$ after **II** has played $\langle p_\beta : \beta < \alpha \rangle$. The game can only end if **II** cannot move p_α at some point, so **I** wins. □

Suppose $\mathcal{A}_0 \not\cong \mathcal{A}_1$. Then there is a least ordinal

$$\delta \leq \mathrm{Card}(A_0) + \mathrm{Card}(A_1)$$

such that **II** does not win $\mathrm{EF}_\delta(\mathcal{A}_0, \mathcal{A}_1)$. Thus for all $\alpha + 1 < \delta$ there is a winning strategy for **II** in $\mathrm{EF}_{\alpha+1}(\mathcal{A}_0, \mathcal{A}_1)$. Let $K(= K(\mathcal{A}_0, \mathcal{A}_1))$ be the set of all winning strategies of **II** in $\mathrm{EF}_{\alpha+1}(\mathcal{A}_0, \mathcal{A}_1)$ for $\alpha + 1 < \delta$. We can make K a tree by letting

$$\langle \tau_\xi : \xi \leq \alpha \rangle \leq \langle \tau'_\xi : \xi \leq \alpha' \rangle$$

if and only if $\alpha \leq \alpha'$ and $\forall \xi \leq \alpha (\tau_\xi = \tau'_\xi)$.

Definition 9.69   We call K, as defined above, the *canonical Karp tree* of the pair $(\mathcal{A}_0, \mathcal{A}_1)$.

Note that even when δ is a limit ordinal K does not have a branch of length δ, for otherwise **II** would win $\mathrm{EF}_\delta(\mathcal{A}_0, \mathcal{A}_1)$.

Lemma 9.70   *Suppose $\mathcal{P}$ is a po-set. Then*

$$\exists \ wins \ \mathrm{EF}_{\mathcal{P}}(\mathcal{A}_0, \mathcal{A}_1) \iff \sigma'\mathcal{P} \leq K.$$

Proof $\Rightarrow$ Suppose **II** wins $\mathrm{EF}_{\mathcal{P}}(\mathcal{A}_0, \mathcal{A}_1)$ with τ. If $s = \langle s_\xi : \xi \leq \alpha \rangle \in \sigma'\mathcal{P}$, we can define a strategy τ' of **II** in $\mathrm{EF}_{\alpha+1}(\mathcal{A}_0, \mathcal{A}_1)$ as follows

$$\tau'_\xi(\langle (c_\eta, x_\eta) : \eta \leq \xi \rangle) = \tau_\xi(\langle (c_\eta, x_\eta, s_\eta) : \eta \leq \xi \rangle).$$

Since K does not have a branch of length δ, $\alpha < \delta$, and hence $\tau' \in K$. The mapping $s \mapsto \tau'$ is an order-preserving mapping $\sigma'\mathcal{P} \to K$.

$\Leftarrow$ Suppose $f : \sigma'\mathcal{P} \to K$ is order-preserving. We can define a winning strategy of **II** in $\mathrm{EF}_{\mathcal{P}}(\mathcal{A}_0, \mathcal{A}_1)$ by the equation

$$\tau_\alpha(\langle (c_\xi, x_\xi, s_\xi) : \eta \leq \xi \rangle) = f(\langle s_\xi : \xi \leq \alpha \rangle)(\langle (c_\xi, x_\xi, s_\eta) : \xi \leq \alpha \rangle).$$

$\square$

Proposition 9.71   *Suppose δ is a limit ordinal and **II** wins $\mathrm{EF}_\alpha(\mathcal{A}_0, \mathcal{A}_1)$ for all $\alpha < \delta$. The following are equivalent:*

(i) **II** *wins* $\mathrm{EF}_\delta(\mathcal{A}_0, \mathcal{A}_1)$.
(ii) **II** *wins* $\mathrm{EF}_{\mathcal{P}}(\mathcal{A}_0, \mathcal{A}_1)$ *for every po-set $\mathcal{P}$ with no branches of length δ.*

Proof   To prove (ii)→(i), suppose **II** does not win $\mathrm{EF}_\delta(\mathcal{A}_0, \mathcal{A}_1)$. Let $\mathcal{P} = K(\mathcal{A}_0, \mathcal{A}_1)$. Then $\sigma\mathcal{P}$ does not have branches of length δ, hence by (ii) **II** wins $\mathrm{EF}_{\sigma\mathcal{P}}(\mathcal{A}_0, \mathcal{A}_1)$ and we get $\sigma\mathcal{P} \leq \mathcal{P}$ from Lemma 9.70, a contradiction with Lemma 9.55. The other direction (i)→(ii) is trivial. $\square$

Note   Suppose $\kappa = \mathrm{Card}(A_0) + \mathrm{Card}(A_1)$. Then we can compute $\mathrm{Card}(K) \leq \sup_{\alpha<\delta}(\kappa^{\kappa^\alpha})^\alpha = \sup_{\alpha<\delta} \kappa^{\kappa^\alpha}$. If GCH and κ is regular, then $\mathrm{Card}(K) \leq \kappa^+$. Furthermore, if we assume GCH, we can assume $\mathrm{Card}(\mathcal{P}) \leq \kappa$ in (ii) above (Hyttinen). For $\delta = \omega$ this does not depend on GCH. **II** wins $\mathrm{EF}_{\mathcal{P}}(\mathcal{A}_0, \mathcal{A}_1)$ if and only if **II** wins $\mathrm{EF}_{\sigma'\mathcal{P}}(\mathcal{A}_0, \mathcal{A}_1)$. So from the point of view of the existence of a winning strategy for **II** we could always assume that $\mathcal{P}$ is a tree.

Corollary **II** *never wins* $\mathrm{EF}_{\sigma K}(\mathcal{A}_0, \mathcal{A}_1)$.

Definition 9.72 A po-set $\mathcal{P}$ is a *Karp po-set* of the pair $(\mathcal{A}_0, \mathcal{A}_1)$ if **II** wins $\mathrm{EF}_{\mathcal{P}}(\mathcal{A}_0, \mathcal{A}_1)$ but not $\mathrm{EF}_{\sigma\mathcal{P}}(\mathcal{A}_0, \mathcal{A}_1)$. If a Karp po-set is a tree, we call it a *Karp tree*.

By Lemma 9.70 and the above corollary, there are always Karp trees for every pair of non-isomorphic structures.

Suppose **I** wins $\mathrm{EF}_{\mathcal{P}}(\mathcal{A}_0, \mathcal{A}_1)$ with the strategy ρ. Let S_ρ be the set of sequences $\bar{y} = \langle y_\xi : \xi \leq \alpha \rangle \in \mathrm{dom}(\rho)$ such that

$$p_{\rho \upharpoonright \alpha+1, y} \in \mathrm{Part}(\mathcal{A}_0, \mathcal{A}_1).$$

Thus S_ρ is the set of sequences of moves of **II** before she loses $\mathrm{EF}_{\mathcal{P}}(\mathcal{A}_0, \mathcal{A}_1)$, when **I** plays ρ. We can make S_ρ a tree by ordering it as follows

$$\langle y_\xi : \xi \leq \alpha \rangle \leq \langle y'_\xi : \xi \leq \alpha' \rangle$$

if and only if $\alpha \leq \alpha'$ and $\forall \xi \leq \alpha(y_\xi = y'_\xi)$.

Lemma 9.73 **I** *wins* $\mathrm{EF}_{\sigma S_\rho}(\mathcal{A}_0, \mathcal{A}_1)$.

Proof The following equation defines a winning strategy ρ' of **I** in the game $\mathrm{EF}_{\sigma S_\rho}(\mathcal{A}_0, \mathcal{A}_1)$:

$$\rho'_\alpha \langle y_\xi : \xi < \alpha \rangle) = \langle c_\alpha, x_\alpha, \langle (y_\xi : \xi \leq \beta) : \beta < \alpha \rangle,$$

where

$$\rho_\alpha(\langle y_\xi : \xi < \alpha \rangle) = (c_\alpha, x_\alpha, p_\alpha).$$

$\square$

Lemma 9.74 $\sigma S_\rho \leq \mathcal{P}$.

Proof Suppose $s = \langle \langle y_\xi : \xi \leq \beta \rangle : \beta < \alpha \rangle \in \sigma S_\rho$, where

$$\beta_0 < \beta_1 < \ldots < \beta_\eta < \ldots (\eta < \alpha).$$

Let $\delta = \sup_{\eta < \alpha} \beta_\eta$ and

$$\rho_\delta(\langle y_\xi : \xi < \delta \rangle) = (c_\delta, x_\delta, p_\delta).$$

We define $f(s) = p_\delta$. Then $f : \sigma S_\rho \to \mathcal{P}$ is order-preserving. $\square$

Note that Lemma 9.74 implies $\mathcal{P} \not\leq S_\rho$. In particular, if **I** wins $\mathrm{EF}_\delta(\mathcal{A}_0, \mathcal{A}_1)$ with ρ, then S_ρ is a tree with no branches of length δ.

Suppose $\mathcal{P}_0$ is such that $\sigma\mathcal{P}_0 \leq \mathcal{P}$ and **I** wins $\mathrm{EF}_{\sigma\mathcal{P}_0}$. So $\mathcal{P}_0$ could be S_ρ. Suppose furthermore that there is no $\mathcal{P}_1$ such that $\sigma\mathcal{P}_1 \leq \mathcal{P}_0$ and **I** wins $\mathrm{EF}_{\sigma\mathcal{P}_1}$. Lemma 9.57 implies that this assumption can always be satisfied.

Lemma 9.75   **I** *does not win* $\mathrm{EF}_{\mathcal{P}_0}(\mathcal{A}_0, \mathcal{A}_1)$.

Proof   Suppose **I** wins $\mathrm{EF}_{\mathcal{P}_0}(\mathcal{A}_0, \mathcal{A}_1)$ with ρ'. Then **I** wins $\mathrm{EF}_{\sigma S_{\rho'}}(\mathcal{A}_0, \mathcal{A}_1)$ and $\sigma S_{\rho'} \leq \mathcal{P}_0$, contrary to the choice of $\mathcal{P}_0$.   □

Definition 9.76   A po-set $\mathcal{P}$ is a *Scott po-set* of $(\mathcal{A}_0, \mathcal{A}_1)$ if **I** wins the game $\mathrm{EF}_{\sigma\mathcal{P}}(\mathcal{A}_0, \mathcal{A}_1)$ but not the game $\mathrm{EF}_{\mathcal{P}}(\mathcal{A}_0, \mathcal{A}_1)$. If a Scott po-set is a tree, we call is a *Scott tree*. If $\mathcal{P}$ is both a Scott and a Karp po-set, it is called a *determined* Scott po-set.

By Lemma 9.73 and Lemma 9.75, S_ρ is always a Scott tree of $(\mathcal{A}_0, \mathcal{A}_1)$, so Scott trees always exist. Note that

$$\mathrm{Card}(S_\rho) \leq \sup_{\alpha < \mathrm{b}(\mathcal{P})} (\mathrm{Card}(\mathcal{A}_0) + \mathrm{Card}(\mathcal{A}_1))^\alpha.$$

Lemma 9.77   *Suppose* **I** *wins* $\mathrm{EF}_{\mathcal{P}}(\mathcal{A}_0, \mathcal{A}_1)$ *with* ρ *and* $K = K(\mathcal{A}_0, \mathcal{A}_1)$. *Then* $K \leq S_\rho$.

Proof   Suppose $\tau \in K$. Let **II** play τ against ρ in $\mathrm{EF}_{\mathcal{P}}(\mathcal{A}_0, \mathcal{A}_1)$. The resulting sequence $\bar{y}$ of moves of **II** is an element of S_ρ. The mapping $\tau \mapsto \bar{y}$ is order-preserving.   □

Suppose **II** wins $\mathrm{EF}_{\mathcal{P}_0}(\mathcal{A}_0, \mathcal{A}_1)$ and **I** wins $\mathrm{EF}_{\mathcal{P}_1}(\mathcal{A}_0, \mathcal{A}_1)$ with ρ. Figure 9.7 shows the resulting picture.

In summary, we have proved:

Theorem 9.78   *Suppose* **II** *wins* $\mathrm{EF}_{\mathcal{P}_0}(\mathcal{A}_0, \mathcal{A}_1)$ *and* **I** *wins* $\mathrm{EF}_{\mathcal{P}_1}(\mathcal{A}_0, \mathcal{A}_1)$. *Then there are trees* T_0 *and* T_1 *such that*

(i) $\sigma'\mathcal{P}_0 \leq T_0 \leq T_1 \leq \mathcal{P}_1$.
(ii) **II** *wins* $\mathrm{EF}_{T_0}(\mathcal{A}_0, \mathcal{A}_1)$ *but not* $\mathrm{EF}_{\sigma T_0}(\mathcal{A}_0, \mathcal{A}_1)$.
(iii) **I** *wins* $\mathrm{EF}_{\sigma T_1}(\mathcal{A}_0, \mathcal{A}_1)$ *but not* $\mathrm{EF}_{T_1}(\mathcal{A}_0, \mathcal{A}_1)$.

Example 9.79   Suppose **I** wins $\mathrm{EF}_\omega(\mathcal{A}_0, \mathcal{A}_1)$. By Proposition 7.19 there is a unique $\delta = \delta(\mathcal{A}_0, \mathcal{A}_1)$ such that **II** wins $\mathrm{EF}_{(\delta,>)}(\mathcal{A}_0, \mathcal{A}_1)$ and **I** wins $\mathrm{EF}_{(\delta+1,>)}(\mathcal{A}_0, \mathcal{A}_1)$. Then $(\delta, >)$ is both a Karp and a Scott po-set for $\mathcal{A}_0$ and $\mathcal{A}_1$.

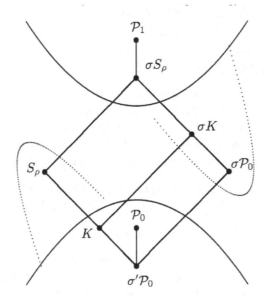

Figure 9.7 The boundary between **II** winning and **I** winning.

Example 9.80 Suppose **II** wins $EF_\alpha(\mathcal{A}_0, \mathcal{A}_1)$ but not $EF_{\alpha+1}(\mathcal{A}_0, \mathcal{A}_1)$. Then $(\alpha, <)$ is a Karp tree (in fact a Karp well-order) of $\mathcal{A}_0$ and $\mathcal{A}_1$. This follows from the fact that $\sigma(\alpha, <) \equiv (\alpha + 1, <)$.

Example 9.81 Suppose **I** wins $EF_{\alpha+1}(\mathcal{A}_0, \mathcal{A}_1)$ but not $EF_\alpha(\mathcal{A}_0, \mathcal{A}_1)$. Then $(\alpha, <)$ is a Scott tree (in fact a Scott well-order) of $\mathcal{A}_0$ and $\mathcal{A}_1$.

If T is a tree, $T + 1$ is the tree which is obtained from T by adding a new element at the end of every maximal branch of T. Note that $T + 1$ may be uncountable even if T is countable.

Lemma 9.82 *Suppose $S \subseteq \omega_1$ is bistationary, $\mathcal{A}_0 = \Phi(S)$, $\mathcal{A}_1 = \Phi(\emptyset)$, and $\mathcal{P} = T(\omega_1 \setminus S) + 1$. Then* **I** *wins* $EF_{\sigma\mathcal{P}}(\mathcal{A}_0, \mathcal{A}_1)$.

Proof Suppose **I** has already played $(c_\beta, x_\beta, p_\beta)$ and **II** has played y_β for $\beta < \alpha$. Suppose **I** now has to decide how to play $(c_\alpha, x_\alpha, p_\alpha)$ in $EF_\mathcal{P}(\mathcal{A}_0, \mathcal{A}_1)$. We assume that **I** has played in such a way that

1. $p_\beta = \langle\langle \delta_\delta : \delta \leq \gamma \rangle : \gamma < \beta \rangle \, (\in \sigma(T(\omega_1 \setminus S) + 1)$.
2. $x_{\nu+2n} < y_{\nu+2n+1}$ in $\mathcal{A}_0$.
3. $x_{\nu+2n+1} < y_{\nu+2n+2}$ in $\mathcal{A}_1$.

4. $\rho(y_{\nu+2n+1}) < \delta_{\nu+2n+1} < \rho(x_{\nu+2n+2})$.
5. $\rho(y_{\nu+2n}) < \delta_{\nu+2n} < \rho(x_{\nu+2n+1})$.
6. $\delta_\beta < \delta_\gamma$ in S for $\beta < \gamma$.
7. $\delta_\nu = \sup_{\beta<\nu} \delta_\beta$ for limit ν.

It is clear that **I** can continue playing so that the above conditions hold until $\delta_\nu \in S$ for some limit ν. At this point **I** plays the supremum x_ν of the previous moves in A_0. Suppose **II** moves $y_\nu \in A_1$. Now y_ν cannot be the supremum of the previous moves in A_1 (as $\delta_\nu \in S$), so **I** wins with his next move. The "+1" part of $\mathcal{P}$ guarantees that **I** does indeed have a next move. $\square$

Lemma 9.83 *Suppose $S \subseteq \omega_1 \setminus \{0\}$ is bistationary, $A_0 = \Phi(S)$, $A_1 = \Phi(\emptyset)$, and $\mathcal{P} = T(\omega_1 \setminus S) + 1$. Then* **I** *does not win* $\mathrm{EF}_{\mathcal{P}}(A_0, A_1)$.

Proof Exercise 9.51. See also Exercise 9.29. $\square$

Example 9.84 Let $S \subseteq \omega_1 \setminus \{0\}$ be a bistationary set. Then $(\omega + 1, <)$ is a Karp tree of $\mathcal{A}(S)$ and $\mathcal{A}(\emptyset)$ (see Lemma 9.10 and the proof of Lemma 9.9), and $T(\omega_1 \setminus S) + 1$ is a Scott tree of $\mathcal{A}(S)$ and $\mathcal{A}(\emptyset)$ (Lemmas 9.82 and 9.83, see also Exercise 9.30). In this example all Karp po-sets $\mathcal{P}_0$ must satisfy $\mathrm{b}(\mathcal{P}_0) \le \omega + 1$ and all Scott po-sets must satisfy $\mathrm{b}(\mathcal{P}_1) \ge \omega_1$, so the Karp po-sets and Scott po-sets are far apart.

In the above example all Karp po-sets and Scott po-sets were far apart. In contrast we now consider two examples where some Karp po-sets and some Scott po-sets are very close to each other, indeed they can be the same po-set.

Proposition 9.85 *There are trees $\mathcal{A}_0$ and $\mathcal{A}_1$ such that $(\mathbb{Q}, <)$ is both a Karp and a Scott po-set of $\mathcal{A}_0$ and $\mathcal{A}_1$.*

Proof Let $\mathcal{A}_0$ be the tree of sequences $s = (\alpha_0, q_0, \ldots, \alpha_\gamma, q_\gamma)$, where $q_0 < \cdots < q_\gamma$ in $\mathbb{Q}$ and $\alpha_\xi \in \omega_1$, ordered by end-extension. Let $\mathcal{A}_1$ be the tree of sequences $s = (\alpha_0, q_0, \ldots, \alpha_\gamma)$ where $q_0 < q_1 < \cdots$ in $\mathbb{Q}$, $\sup q_\xi < \infty$, and $\alpha_\xi \in \omega_1$, ordered by end-extension. If $r \in \mathbb{Q}$, let $\lfloor r \rfloor$ be the integer part of r. If $s \in A_0 \cup A_1$, $\lfloor s \rfloor$ is the maximum value of $\lfloor q_\xi \rfloor$ in s. Similarly, $\sup(s)$ and $\max(s)$ are defined as $\sup(q_\xi)$ and $\max(q_\xi)$.

We need a new po-set for the proof. This po-set, which we denote by $\mathbb{P}_Q$, consists of triples (q, α, β), where $q \in \mathbb{Q}$ and $\beta < \alpha < \omega_1$. The order is as follows:

$$(q, \alpha, \beta) \le (q', \alpha', \beta') \text{ iff } q < q' \text{ or } q = q', \alpha = \alpha' \text{ and } \beta \le \beta'.$$

Thus $\mathbb{P}_Q$ can be obtained from $(\mathbb{Q}, <)$ by replacing rationals by well-ordered sets $< \omega_1$, in all possible ways.

Claim (1) $\mathbb{P}_Q \leq (\mathbb{Q}, <)$.

Proof If $\alpha < \omega_1$, let $\delta_\alpha^n, n < \omega$, enumerate the elements of α. Let $\mathbb{Q} = \{q_n : n < \omega\}$. Let $\pi : \mathbb{N} \times \mathbb{N} \to \mathbb{N}$ be one-one and onto. If $s = (q, \alpha, \beta) \in \mathbb{P}_Q$, let

$$g(s) = \pi(n, m)$$

where $\beta = \delta_\alpha^n$ and $q = q_m$. Clearly, if $s < s'$ in T, then $g(s) \neq g(s')$. Thus $\mathbb{P}_Q$ is a union of $\aleph_0$ antichains. It is well-known that this implies $\mathbb{P}_Q \leq (\mathbb{Q}, <)$ (Exercise 9.38). $\qquad\square$

Let $\mathrm{EF}'_{\mathcal{P}}(\mathcal{A}_0, \mathcal{A}_1)$ be the variant of $\mathrm{EF}_{\mathcal{P}}(\mathcal{A}_0, \mathcal{A}_1)$ in which **I** can only play $(c_\alpha, x_\alpha, p_\alpha)$ if for all predecessors z of x_α in A_{c_α} there is $\beta < \alpha$ such that $c_\beta = c_\alpha$ and $x_\beta = z$.

Claim (2) If **II** wins $\mathrm{EF}'_{\mathbb{Q}}(\mathcal{A}_0, \mathcal{A}_1)$, she wins $\mathrm{EF}_{\mathbb{Q}}(\mathcal{A}_0, \mathcal{A}_1)$.

Proof Suppose **II** wins $\mathrm{EF}'_{\mathbb{Q}}(\mathcal{A}_0, \mathcal{A}_1)$. By Claim 1 she wins $\mathrm{EF}'_{\mathbb{P}_Q}(\mathcal{A}_0, \mathcal{A}_1)$. Suppose τ' is a winning strategy of **II** in $\mathrm{EF}'_{\mathbb{P}_Q}(\mathcal{A}_0, \mathcal{A}_1)$. Suppose the first move of **I** in $\mathrm{EF}_{\mathbb{Q}}(\mathcal{A}_0, \mathcal{A}_1)$ is (c_0, x_0, r_0). Let

$$\langle a_{c_0}^\beta : \beta \leq \alpha_0 \rangle$$

be the sequence of predecessors of x_0 in $\mathcal{A}_{c_0}$ in ascending order with $a_{c_0}^{\alpha_0} = x_0$. Let $y_0 = \tau'_{\alpha_0}(\langle (c_0, a_{c_0}^\beta, (r_0, \alpha_0, \beta)) : \beta \leq \alpha_0 \rangle)$. This is the first move of **II** in $\mathrm{EF}_{\mathbb{Q}}(\mathcal{A}_0, \mathcal{A}_1)$. Suppose (c_1, x_1, r_1) is the second move of **I** in $\mathrm{EF}_{\mathbb{Q}}(\mathcal{A}_0, \mathcal{A}_1)$. Let

$$\langle a_{c_1}^\beta : \beta \leq \alpha_1 \rangle$$

be the sequence of predecessors of x_1 in $\mathcal{A}_{c_1}$ in ascending order with $a_{c_1}^{\alpha_1} = x_1$. The second move of **II** in $\mathrm{EF}_{\mathbb{Q}}(\mathcal{A}_0, \mathcal{A}_1)$ is, with the above notation,

$$y_1 = \tau'_{\alpha_0 + \alpha_1}(\langle (c_0, a_{c_0}^\beta, (r_0, \alpha_0, \beta)) : \beta \leq \alpha_0 \rangle ^\frown$$
$$\langle (c_1, a_{c_1}^\beta, (r_1, \alpha_1, \beta)) : \beta \leq \alpha_1 \rangle).$$

This indicates how **II** wins $\mathrm{EF}_{\mathbb{Q}}(\mathcal{A}_0, \mathcal{A}_1)$. $\qquad\square$

Claim (3) **II** wins $\mathrm{EF}'_{\mathbb{Q}}(\mathcal{A}_0, \mathcal{A}_1)$.

Proof Suppose the moves $(c_\beta, x_\beta, r_\beta), \beta \leq \alpha$, have already been made by **I** and the moves $y_\beta, \beta < \alpha$, by **II**. Now **II** should decide what her next move y_α is. During the game **II** has maintained a certain strategy. To describe it, let

$$g_n : \mathbb{Q} \to \mathbb{Q} \cap [n, n+1)$$

be order-preserving. The conditions are

(1) $\lfloor y_\beta \rfloor = \lfloor x_\beta \rfloor + 1$.

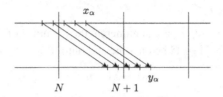

Figure 9.8 Subcase 1.1.

(2) $\sup(y_\beta) \leq g_{\lfloor y_\beta \rfloor}(r_\beta)$.

Case 1 The rank of x_α in $\mathcal{A}_{c_\alpha}$ is a limit ordinal. Thus there is an ascending sequence a_{α_β}, $\beta < \nu$, of predecessors of x_α in $\mathcal{A}_{c_\alpha}$ and they have been played already. Here $\nu = \cup\nu$. Thus $a_{\alpha_\beta} = x_{\alpha_\beta}$ or $a_{\alpha_\beta} = y_{\alpha_\beta}$ for each $\beta < \nu$. Because of (1) there is $\beta_0 < \nu$ such that $a_{\alpha_\beta} = x_{\alpha_\beta}$ for $\beta_0 \leq \beta < \nu$ or $a_{\alpha_\beta} = y_{\alpha_\beta}$ for $\beta \leq \beta < \nu$. We may also assume $\lfloor a_{\alpha_\beta} \rfloor = N$ for some constant $N \in \mathbb{N}$ for $\beta_0 \leq \beta < \nu$.

Subcase 1.1 $a_{\alpha_\beta} = x_{\alpha_\beta}$ for $\beta_0 \leq \beta < \nu$. Let $b_{\alpha_\beta} = y_{\alpha_\beta}$. By (1), $\lfloor b_{\alpha_\beta} \rfloor = N + 1$. By (2),

$$\sup_{\beta < \nu}(\sup(b_{\alpha_\beta})) \leq \sup_{\beta < \nu}(g_{N+1}(r_\beta))$$
$$\leq g_{N+1}(r_\alpha).$$

Thus we can find y_α such that (see Figure 9.8)

(3) $\forall\beta < \nu(y_{\alpha_\beta} < y_\alpha)$.
(4) $\sup(y_\alpha) \leq g_{N+1}(r_\alpha)$.
(5) $\lfloor y_\alpha \rfloor = N + 1$.
(6) $\forall\beta < \alpha(y_\alpha = y_\beta \leftrightarrow x_\alpha = x_\beta)$.
(7) $\forall\beta < \alpha(y_\alpha = x_\beta \leftrightarrow x_\alpha = y_\beta)$.

Subcase 1.2 $a_{\alpha_\beta} = y_{\alpha_\beta}$ for $\beta_0 \leq \beta < \nu$. Let $b_{\alpha_\beta} = x_{\alpha_\beta}$. By (1), $\lfloor b_{\alpha_\beta} \rfloor = N - 1$. Let y_α be such that

$$\forall\beta < \nu(x_{\alpha_\beta} < y_\alpha)$$

and (4)–(7) above hold (see Figure 9.9).

Case 2 x_α has an immediate predecessor x_α^- in $\mathcal{A}_{c_\alpha}$. The predecessor x_α^- is x_β or y_β for some $\beta < \alpha$.

Subcase 2.1 $x_\alpha^- = x_\beta$. Let $N = \lfloor x_\alpha^- \rfloor$. Player **II** chooses an immediate successor y_α of y_β in such a way that (5)–(7) hold. Note that

$$\sup(y_\beta) \leq g_{N+1}(r_\beta)$$

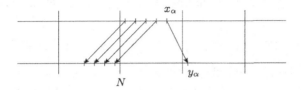

Figure 9.9 Subcase 1.2.

$$< g_{N+1}(r_\alpha)$$

so (4) can be satisfied as well.

Subcase 2.2 $x_\alpha^- = y_\beta$. Similar to 2.1.   □

Claim (4)   **I** wins $\mathrm{EF}_{\sigma\mathbb{Q}}(\mathcal{A}_0, \mathcal{A}_1)$.

Proof   Let $f : (\mathbb{Q}, <) \to (\mathbb{Q} \cap (0,1), <)$ be order-preserving. The strategy of **I** is to play an increasing sequence in $\mathcal{A}_1$, imitating the moves of **II** . The idea is that player **I** uses f to translate the moves of **II** into his own moves in $\mathbb{Q}\cap(0,1)$. Then eventually **II** cannot move anymore because $\sup_{\beta<\alpha}(\max y_\beta) = \infty$. At this point **I** jumps to $\mathbb{Q} \cap [1, \infty)$ and wins. More exactly, suppose **I** has played $(c_\beta, x_\beta, s_\beta), \beta < \alpha$, so far and **II** has responded with $y_\beta, \beta < \alpha$. The idea of **I** is to maintain the conditions:

(1) $c_\beta = 1$.
(2) $x_\beta = (0, \max(y_0), 0, \max(y_1), \ldots, 0, \max(y_\gamma), \ldots, 0)$ $(\gamma < \beta)$.
(3) $s_\beta = \langle \max(y_\xi) : \xi < \beta \rangle$ $(\in \sigma\mathbb{Q})$.

It is evident that **I** can always choose $(c_\alpha, x_\alpha, s_\alpha)$ so that (1)–(3) continue to hold. Since the game cannot go on for ω_1 moves, a point is reached where **II** cannot choose y_α anymore, because

$$\sup_{\beta<\alpha}(\max(y_\beta)) = \infty.$$

□

Proposition 9.85 is proved.   □

Theorem 9.86   *There are structures* $(\mathcal{A}_0, \mathcal{A}_1)$ *of cardinality* 2^ω *so that the following conditions hold for any tree* T:

(i) T *is* $\mathbb{R}$*-embeddable if and only if* **II** *wins* $\mathrm{EF}_T(\mathcal{A}_0, \mathcal{A}_1)$.
(ii) $\sigma\mathbb{R} \leq T$ *if and only if* **I** *wins* $\mathrm{EF}_T(\mathcal{A}_0, \mathcal{A}_1)$.

Proof Let P_0 be the set of functions $f : \mathbb{R} \to 3$. If $f \in P_0$, let $\mathrm{Supp}(f) = \{r \in \mathbb{R} : f(r) \neq 0\}$. Let

$$P = \{f \in P_0 : \mathrm{Supp}(f) \text{ is a well-ordered subset of } \mathbb{R}\}.$$

If $d \in 3$, let $\#(d)$ be defined by $\#(d) = 0$ if $d = 0$ and $\#(d) = 1$ otherwise. If $f \in P$ let $\#(f)$ be the function $\#(f)(r) = \#(f(r))$. Let F be the set of functions f with $\mathrm{dom}(f)$ an open initial segment of $\mathbb{R}$ and $\mathrm{rng}(f) \subseteq \{0, 1\}$. For $f \in F$ we define $(1 - f)(r) = 1 - f(r)$, when $r \in \mathrm{dom}(f)$. The construction that follows depends on certain decisions at limit stages and these decisions are not canonically determined by earlier stages of the construction. For this reason we introduce a decision-making function m. Let $m : F \to 2$ be a function so that $m(f)$ is arbitrarily chosen, subject to the conditions that:

1. If $f \in F$ is eventually constant and equals d, then $m(f) = d$.
2. For all f, $m(1 - f) = 1 - m(f)$.
3. If f and g eventually agree, then $m(f) = m(g)$.

Our models will have $P \cup \mathbb{R}$ as the universe, an auxiliary predicate E, and a unary predicate U the interpretation of which makes the models different. The predicate U is interpreted by defining two "smashed" versions $f^{\mathcal{A}}$ and $f^{\mathcal{B}}$ of every $f \in P$. These are defined by induction along $\mathrm{Supp}(f)$. At successor stages the idea is to use the smash function $\#$ to let $f(r)$ generate a value $\#(f(r))$ or $1 - \#(f(r))$, according to what decisions have been made before, for $f^{\mathcal{A}}(r')$ and $f^{\mathcal{B}}(r')$, $r' < r$. At limit stages we use the function m.

Suppose $f \in P$ and $\mathrm{Supp}(f) = \langle x_\alpha : \alpha < \beta \rangle$ in increasing order. Suppose $\mathcal{C} \in \{\mathcal{A}, \mathcal{B}\}$. We define $f^{\mathcal{C}}(x)$ by cases. If g is a function with $\mathrm{dom}(g) \subseteq \mathbb{R}$, denote the restriction of g to $(-\infty, x)$ by $g\restriction_x$ and the restriction to $(-\infty, x]$ by $g\restriction_{\leq x}$.

If $x \in \mathbb{R}$ and $\mathrm{Supp}(f) = \emptyset$ or $x < x_0$, we let $f^{\mathcal{C}}(x) = 0$ if $\mathcal{C} = \mathcal{A}$ and $f^{\mathcal{C}}(x) = 1$ if $\mathcal{C} = \mathcal{B}$. For other x we let

$$
f^{\mathcal{C}}(x) = \begin{cases}
\#(f(x)), & \text{if } x = x_\alpha, \text{ and } m(f^{\mathcal{C}}\restriction_x) = 0 \\
1 - \#(f(x)), & \text{if } x = x_\alpha, \text{ and } m(f^{\mathcal{C}}\restriction_x) = 1 \\
f^{\mathcal{C}}(x_\alpha), & \text{if } x_\alpha < x < x_{\alpha+1} \\
m(f^{\mathcal{C}}\restriction_x), & \text{if } x = \sup_{\alpha < \nu} x_\alpha < x_\nu, \nu = \cup\nu \\
f^{\mathcal{C}}(x_\alpha), & \text{if } x > \sup_{\alpha < \beta} x_\alpha \text{ and } \beta = \alpha + 1 \\
m(f^{\mathcal{C}}\restriction_{x^*}), & \text{if } x \geq x^* = \sup_{\alpha < \beta} x_\alpha \text{ and } \beta = \cup\beta.
\end{cases}
$$

Now we are ready to define the models needed in the theorem:

$$\mathcal{A} = (P \cup \mathbb{R}, E, \{f \in P : f^{\mathcal{A}} \text{ is eventually } 0\})$$
$$\mathcal{B} = (P \cup \mathbb{R}, E, \{f \in P : f^{\mathcal{B}} \text{ is eventually } 0\}),$$

where $E = \{(f, g, r) : f, g \in P, r \in \mathbb{R} \text{ and } f\!\restriction_r = g\!\restriction_r\}$.

Claim: **I** wins $\mathrm{EF}_{\sigma\mathbb{R}}(\mathcal{A}, \mathcal{B})$.

The strategy of **I** is the following. Player **I** will play elements h_α^0 and h_α^1 together with s_α^0 and $s_\alpha^1 = s_\alpha^0 \frown (r_\alpha)$ in $\sigma\mathbb{R}$. We construe round α of the game as consisting of actually two rounds. First **I** plays h_α^0 and s_α^0, and after this h_α^1 and s_α^1. The responses of **II** are respectively, k_α^0 and k_α^1. As part of his strategy **I** then chooses one of h_α^0 and h_α^1, say $h_\alpha^{d_\alpha}$, to be denoted by a_α, if the move $h_\alpha^{d_\alpha}$ was made in $\mathcal{A}$ and the corresponding $k_\alpha^{d_\alpha}$ in $\mathcal{B}$ will then be denoted by b_α. If the move $h_\alpha^{d_\alpha}$ was made in $\mathcal{B}$, then $h_\alpha^{d_\alpha}$ will be denoted by b_α and the corresponding $k_\alpha^{d_\alpha}$ by a_α. Eventually a limit ordinal δ will emerge such that $\sup_{\alpha<\delta} r_\alpha = \infty$ and then the unions $f = \cup_\alpha a_\alpha$ and $f' = \cup_\alpha b_\alpha$ will be elements of P. The point is that **I** takes care that $f^{\mathcal{A}}$ and $f'^{\mathcal{B}}$ will not be both eventually 0, so when he finally plays f forcing **II** (because of the predicate E) to play f', he has won the game.

Suppose a_α, r_α, and b_α for $\alpha < \beta$ have been played as above and we have $\sup_{\alpha<\beta}(r_\alpha) < \infty$. Part of the strategy of **I** is to maintain the condition $a_\alpha^{\mathcal{A}}\!\restriction_{\leq r_\alpha} = 1 - b_\alpha^{\mathcal{B}}\!\restriction_{\leq r_\alpha}$.

Case 1: β **is a successor** $\alpha + 1$. Since $a_\alpha^{\mathcal{A}}\!\restriction_{\leq r_\alpha} = 1 - b_\alpha^{\mathcal{B}}\!\restriction_{\leq r_\alpha}$ there is a smallest $r > r_\alpha$ such that $a_\alpha(r) \neq 0$ or $b_\alpha(r) \neq 0$, for otherwise **II** has lost the game already. Now **I** plays h_β^0 and h_β^1, choosing carefully $h_\beta^0(r)$ and $h_\beta^1(r)$, together with s_β^0 and $s_\beta^1 = s_\beta^0 \frown (r_\beta)$, $r_\beta = r$, in such a way that, to avoid an immediate loss, **II** has to play k_β^0 and k_β^1 so that $a_\beta^{\mathcal{A}}\!\restriction_{\leq r_\beta} = 1 - b_\beta^{\mathcal{B}}\!\restriction_{\leq r_\beta}$.

Case 2: β **is limit.** Let $r_\beta = \sup_{\alpha<\beta} r_\alpha$. Now **I** plays h_β^0 and h_β^1, choosing carefully $h_\beta^0(r_\beta)$ and $h_\beta^1(r_\beta)$, together with $s_\beta^0 = \cup_{\alpha<\beta} s_\alpha^1$ and $s_\beta^1 = s_\beta^0 \frown (r_\beta)$ in such a way that, to avoid an immediate loss, **II** has to play k_β^0 and k_β^1 so that $a_\beta^{\mathcal{A}}\!\restriction_{\leq r_\beta} = 1 - b_\beta^{\mathcal{B}}\!\restriction_{\leq r_\beta}$

Claim: **II** has a winning strategy in $\mathrm{EF}_{\sigma'\mathbb{R}}(\mathcal{A}, \mathcal{B})$.

Suppose h_α and $r_\alpha \in \mathbb{R}$ have been played by **I** and k_α by **II** for $\alpha < \beta$, and $r_\beta = \sup_{\alpha<\beta} \sup(s_\alpha) < \infty$. Let $a_\alpha = h_\alpha$ if h_α was played in $\mathcal{A}$ and $a_\alpha = k_\alpha$ otherwise. Similarly, let $b_\alpha = h_\alpha$ if h_α was played in $\mathcal{B}$ and $b_\alpha = k_\alpha$ otherwise. The strategy of **II** is to keep $a_\alpha^{\mathcal{A}}(r_\alpha) = b_\alpha^{\mathcal{B}}(r_\alpha)$ and $a_\alpha(x) = b_\alpha(x)$ for $x > r_\alpha$. If **II** can keep playing like this he wins since then $a_\alpha^{\mathcal{A}}$ is eventually 0 if and only if $b_\alpha^{\mathcal{B}}$ is. $\qquad\square$

Note In the above proof we showed that $\sigma\mathbb{Q}$ is both a Karp and a Scott tree of $(\mathcal{A}, \mathcal{B})$. Thus $\sigma\mathbb{Q}$ is a determined Scott tree of $(\mathcal{A}, \mathcal{B})$.

9.6 Topology of Uncountable Models

Countable models with countable vocabulary can be thought of as points in the Baire space ω^ω. Likewise, models $\mathcal{M}$ of cardinality κ with vocabulary of cardinality κ can be thought of as points $f_\mathcal{M}$ in the set κ^κ. We can make κ^κ a topological space by letting the sets

$$N(f, \alpha) = \{g \in \omega^\kappa : f \upharpoonright \alpha = g \upharpoonright \alpha\},$$

where $\alpha < \kappa$, form the basis of the topology. Let us denote this *generalized Baire space* κ^κ by $\mathcal{N}_\kappa$. Now properties of models of size κ correspond to subsets of $\mathcal{N}_\kappa$. In particular, modulo coding, isomorphism of structures of cardinality κ becomes an "analytic" property in this space.

One of the basic questions about models of size κ that we can try to attack with methods of logic is the question which of those models can be identified up to isomorphism by means of a set of invariants. Shelah's Main Gap Theorem gives one answer: If $\mathcal{M}$ is any structure of cardinality $\kappa \geq \omega_1$ in a countable vocabulary, then the first-order theory of $\mathcal{M}$ is either of the two types:

Structure Case All uncountable models elementary equivalent to $\mathcal{M}$ can be characterized in terms of dimension-like invariants.

Non-structure Case In every uncountable cardinality there are non-isomorphic models elementary equivalent to $\mathcal{M}$ that are extremely difficult to distinguish from each other by means of invariants.

The game-theoretic methods we have developed in this book help us to analyze further the non-structure case. For this we need to develop some basic topology of $\mathcal{N}_\kappa$. A set $A \subseteq \mathcal{N}_\kappa$ is *dense* if A meets every non-empty open set. The space $\mathcal{N}_\kappa$ has a dense subset of size $\kappa^{<\kappa}$ consisting of all eventually constant functions. If the *Generalized Continuum Hypothesis GCH* is assumed, then $\kappa^{<\kappa} = \kappa$ for all regular κ and $\kappa^{<\kappa} = \kappa^+$ for singular κ.

Theorem 9.87 (Baire Category Theorem) *Suppose A_α, $\alpha < \kappa$, are dense open subsets of $\mathcal{N}_\kappa$. Then $\bigcap_\alpha A_\alpha$ is dense.*

Proof Let $f_0 \in \mathcal{N}_\kappa$ and $\alpha_0 < \kappa$ be arbitrary. If f_ξ and α_ξ for $\xi < \beta$ have been defined so that

$$\alpha_\zeta < \alpha_\xi \text{ and } f_\xi \in N(f_\zeta, \alpha_\zeta)$$

for $\zeta < \xi < \beta$, then we define f_β and α_β as follows: Choose some $g \in \mathcal{N}_\kappa$ such that $g \in N(f_\xi, \alpha_\xi)$ for all $\xi < \beta$ and let $\alpha_\beta = \sup_{\xi < \beta} \alpha_\xi$. Since A_β is dense, there is $f_\beta \in A_\beta \cap N(g, \alpha_\beta)$. When all f_ξ and α_ξ for $\xi < \kappa$ have been defined, we let f be such that $f \in N(f_\xi, \alpha_\xi)$ for all $\xi < \kappa$. Then $f \in \bigcap_\alpha A_\alpha \cap N(f_0, \alpha_0)$. □

Definition 9.88 A subset A of $\mathcal{N}_\kappa$ is said to be Σ^1_1 (or *analytic*) if it is a projection of a closed subset of $\mathcal{N}_\kappa \times \mathcal{N}_\kappa$. A set is Π^1_1 (or *co-analytic*) if its complement is analytic. Finally, a set is Δ^1_1 if it is both Σ^1_1 and Π^1_1.

Example 9.89 Examples of analytic sets relevant if κ is a regular cardinal $> \omega$, are

$$\mathrm{CUB}_\kappa = \{f \in \mathcal{N}_\kappa : \{\alpha < \kappa : f(\alpha) = 0\} \text{ contains a club}\}$$

and

$$\mathrm{NS}_\kappa = \{f \in \mathcal{N}_\kappa : \{\alpha < \kappa : f(\alpha) \neq 0\} \text{ contains a club}\}.$$

The set of α-sequences of elements of κ for various $\alpha < \kappa$ form a tree $\mathcal{N}_{<\kappa}$ under the subsequence relation. Any subset T of $\mathcal{N}_{<\kappa}$ which is closed under subsequences is called a *tree* in this section. A κ-branch of such a tree is any linear subtree (branch) of height κ. Let us denote $\langle g(\beta) : \beta < \alpha \rangle$ by $\bar{g}(\alpha)$.

Lemma 9.90 *A set $A \subseteq \mathcal{N}_\kappa$ is analytic iff there is a tree $T \subseteq \mathcal{N}_{<\kappa} \times \mathcal{N}_{<\kappa}$ such that for all f:*

$$f \in A \iff T(f) \text{ has a } \kappa\text{-branch}, \tag{9.7}$$

where $T(f) = \{\bar{g}(\alpha) : (\bar{g}(\alpha), \bar{f}(\alpha)) \in T\}$. Such a tree is called a tree representation *of A.*

Proof Suppose first A is analytic and $B \subseteq \kappa^\kappa \times \kappa^\kappa$ is a closed set such that

$$f \in A \iff \exists g((f, g) \in B).$$

Let

$$T = \{(\bar{f}(\alpha), \bar{g}(\alpha)) : (f, g) \in B, \alpha < \kappa\}.$$

Clearly now $f \in A$ if and only if $T(f)$ has a κ-branch. Conversely, suppose such a T exists. Let B be the set of (f, g) such that $(\bar{f}(\alpha), \bar{g}(\alpha)) \in T$ for all $\alpha < \kappa$. The set B is closed and its projection is A. □

Respectively, a set is co-analytic if and only if there is a tree $T \subseteq \mathcal{N}_{<\kappa} \times \mathcal{N}_{<\kappa}$ such that for all f:

$$f \in A \iff T(f) \text{ has no } \kappa\text{-branches}. \tag{9.8}$$

Let $\mathcal{T}_\kappa$ denote the class of all trees without κ-branches. Let $\mathcal{T}_{\lambda,\kappa}$ denote the set of subtrees of $\lambda^{<\kappa}$ of cardinality $\leq \lambda$ without any κ-branches.

Proposition 9.91   *Suppose B is a co-analytic subset of $\mathcal{N}_\kappa$ and T is as in (9.8). For any tree $S \in \mathcal{T}_\kappa$ let*

$$B_S = \{f \in B : T(f) \leq S\}.$$

Then

$$B = \bigcup_{S \in \mathcal{T}_{\lambda,\kappa}} B_S,$$

where $\lambda = \kappa^{<\kappa}$.

Proof   Clearly $B_S \subseteq B$ if $S \in \mathcal{T}_\kappa$. Conversely, suppose $f \in B$. Then of course $f \in B_{T(f)}$. It remains to observe that $|T(f)| \leq \kappa^{<\kappa}$.   □

Suppose $A \subseteq B$ is analytic and S is a tree as in (9.7). Let

$$T' = \{(\bar{f}(\alpha), \bar{g}(\alpha), \bar{h}(\alpha)) : \bar{g}(\alpha) \in T(f), \bar{h}(\alpha) \in S(f)\}. \tag{9.9}$$

Note that $|T'| \leq \kappa^{<\kappa}$ and T' has no κ-branches, for such a branch would give rise to a triple (f, g, h) which would satisfy $f \in A \setminus B$. Note also that if $f \in A$, then there is a κ-branch $\{\bar{h}(\alpha) : \alpha < \kappa\}$ in $S(f)$, and hence the mapping

$$\bar{g}(\alpha) \mapsto (\bar{f}(\alpha), \bar{g}(\alpha), \bar{h}(\alpha))$$

witnesses

$$T(f) \leq T'.$$

We have proved:

Proposition 9.92 (Covering Theorem for $\mathcal{N}_\kappa$)   *Suppose B is a co-analytic subset of $\mathcal{N}_\kappa$ and S is as in (9.8). Suppose $A \subseteq B$ is analytic. Then*

$$A \subseteq B_T$$

for some $T \in \mathcal{T}_{\lambda,\kappa}$, where $\lambda = \kappa^{<\kappa}$.

The idea is that the sets B_T, $T \in \mathcal{T}_{\lambda,\kappa}$, cover the co-analytic set B completely, and moreover any analytic subset of B can be already covered by a single B_T. Especially if B happens to be $\mathbf{\Delta}_1^1$, then there is $T \in \mathcal{T}_{\lambda,\kappa}$ such that $B = B_T$.

Corollary (Souslin–Kleene Theorem for $\mathcal{N}_\kappa$)   *Suppose B is a $\mathbf{\Delta}_1^1$ subset of $\mathcal{N}_\kappa$. Then*

$$B = B_T$$

for some $T \in \mathcal{T}_{\lambda,\kappa}$, where $\lambda = \kappa^{<\kappa}$.

Corollary (Luzin Separation Theorem for $\mathcal{N}_\kappa$) *Suppose A and B are disjoint analytic subsets of $\mathcal{N}_\kappa$. Then there is a set of the form C_T for some co-analytic set C and some $T \in \mathcal{T}_{\lambda,\kappa}$, where $\lambda = \kappa^{<\kappa}$, that separates A and B, i.e. $A \subseteq C$ and $C \cap B = \emptyset$.*

In the case of classical descriptive set theory, which corresponds to assuming $\kappa = \omega$, the sets B_T are Borel sets. If we assume CH, then CUB and NS cannot be separated by a Borel set.

Proposition 9.93 *If $\kappa^{<\kappa} = \kappa$, then the sets B_T are analytic. If in addition T is a strong bottleneck, then B_T is Δ_1^1.*

Let us call a family $\mathcal{B}$ of elements of $\mathcal{T}_{\lambda,\kappa}$ *universal* if for every $T \in \mathcal{T}$ there is some $S \in \mathcal{B}$ such that $T \le S$. If $\mathcal{T}_{\lambda,\kappa}$ has a universal family of size μ, and $\kappa^{<\kappa} = \kappa$, then by the above results every co-analytic set in $\mathcal{N}_\kappa$ is the union of μ analytic sets. By results in Mekler and Väänänen (1993) it is consistent relative to the consistency of ZFC that $\mathcal{T}_{\kappa^+}$, $2^\kappa = \kappa^+$, has a universal family of size κ^{++} while $2^{\kappa^+} = \kappa^{+++}$.

Definition 9.94 The class of *Borel* subsets of $\mathcal{N}_\kappa$ is the smallest class containing the open sets and the closed sets which is closed under unions and intersections of length κ.

Note that every closed set in $\mathcal{N}_\kappa$ is the union of $\kappa^{<\kappa}$ open sets (Exercise 9.57). So if $\kappa^{<\kappa} = \kappa$, then the definition of Borelness can be simplified.

Theorem 9.95 *Assume $\kappa^{<\kappa} = \kappa > \omega$. Then $\mathcal{N}_\kappa$ has two disjoint analytic sets that cannot be separated by Borel sets.*

Proof Note that κ is a regular cardinal. Every Borel set A has a "Borel code" c such that $A = B_c$. Let us suppose $A = B_c$ separates the disjoint analytic sets $\mathcal{CUB}_\kappa$ and NS_κ defined in Example 9.89. For example, $\mathcal{CUB} \subseteq A$ and $A \cap \mathrm{NS}_\kappa = \emptyset$. Let $\mathcal{P} = (2^{<\kappa}, \le)$ be the Cohen forcing for adding a generic subset for κ. Let G be $\mathcal{P}$-generic and $g = \bigcup \mathcal{P}$. Now either $g \in A$ or $g \notin A$. Let us assume, w.l.o.g., that $g \in A$. Let $p \Vdash \check{g} \in B_{\check{c}}$. Let $M \prec (H(\mu), \in, <^*)$ for a large μ such that $\kappa, p, \mathcal{P}, TC(c) \in M$, $M^{<\kappa} \subseteq M$, and $<^*$ is a well-order of $H(\mu)$. Since $\kappa^{<\kappa} = \kappa > \omega$, we may also assume $|M| = \kappa$. Since $\mathcal{P}$ is $< \kappa$-closed, it is easy to construct a $\mathcal{P}$-generic G' over M in V such that

$$\{\alpha < \kappa : M \models "(\check{g})_{G'}(\alpha) \ne 0"\} \text{ contains a club.} \tag{9.10}$$

It is easy to show that $B_c = (B_{\check{c}})_{G'}$. Since

$$M \models "p \Vdash \check{g} \in B_{\check{c}}",$$

whence $(\check{g})_{G'} \in B_c$ and therefore $(\check{g})_{G'} \notin \mathrm{NS}_\kappa$. This contradicts (9.10). □

Example 9.96 Suppose $\mathcal{M}$ is a structure with $M = \kappa$. We call the analytic set

$$\{\mathcal{N} : N = \kappa \text{ and } \mathcal{N} \cong \mathcal{M}\}$$

the *orbit* of $\mathcal{M}$. Let $\mathcal{N} \not\cong \mathcal{M}$. Now player **I** has an obvious winning strategy ρ in $\mathrm{EF}_\kappa(\mathcal{M}, \mathcal{N})$: he simply makes sure that all elements of both models are played. Obviously there are many ways to play all the elements but any of them will do. Let us consider the co-anaytic set $B = \{f_\mathcal{N} : N = \kappa \text{ and } \mathcal{N} \not\cong \mathcal{M}\}$. Let $S(\mathcal{N})$ be the Scott tree S_ρ of the pair $(\mathcal{M}, \mathcal{N})$. Let us choose a tree representation T of B in such a way that for all $\mathcal{N}$ with $N = \kappa$, $T(f_\mathcal{N}) = S(\mathcal{N})$. If now $f_\mathcal{N} \in B_{T'}$, then player **I** wins $\mathrm{EF}_{T'}(\mathcal{M}, \mathcal{N})$.

Recall that if $\mathcal{M}$ is a countable structure and α is the Scott height of $\mathcal{M}$, then **I** wins $\mathrm{EFD}_{\alpha+\omega}(\mathcal{M}, \mathcal{N})$ whenever $\mathcal{M} \not\cong \mathcal{N}$ and N is countable. Equivalently, using the notation of Example 9.54, player **I** wins $\mathrm{EF}_{B_{\alpha+\omega}}(\mathcal{M}, \mathcal{N})$ whenever $\mathcal{M} \not\cong \mathcal{N}$ and N is countable. We now generalize this property of $B_{\alpha+\omega}$ to uncountable structures.

Definition 9.97 Suppose κ is an infinite cardinal and $\mathcal{M}$ is a structure of cardinality κ. A tree T is a *universal Scott tree* of a structure $\mathcal{M}$ if T has no branches of length κ and player **I** wins $\mathrm{EF}_{\sigma T}(\mathcal{M}, \mathcal{N})$ whenever $\mathcal{M} \not\cong \mathcal{N}$ and $|N| = |M|$.

The idea of the universal Scott tree is that the tree T alone suffices as a clock for player **I** to win all the 2^κ different games $\mathrm{EF}_T(\mathcal{M}, \mathcal{N})$ where $\mathcal{M} \not\cong \mathcal{N}$ and $|N| = |M|$. Universal Scott trees exist: there is always a universal Scott tree of cardinality $\le 2^\kappa$ as we can put the various Scott trees of the pairs $(\mathcal{M}, \mathcal{N})$, $\mathcal{M} \not\cong \mathcal{N}$, $|M| = |N|$, each of them of the size $\le \kappa^{<\kappa}$, together into one tree. So the question is: How small universal Scott trees does a given structure have?

If $\kappa^{<\kappa} = \lambda$ and $\mathcal{T}_{\lambda,\kappa}$ has a universal family of size μ, then every structure of size κ has a universal Scott tree of size μ.

If we allowed T to have a branch of length κ, any such tree would be a universal Scott tree of any structure of cardinality κ.

We ask whether **I** wins $\mathrm{EF}_{\sigma T}(\mathcal{M}, \mathcal{N})$ rather than in $\mathrm{EF}_T(\mathcal{M}, \mathcal{N})$ in order to preserve the analogy with the concept of a Scott tree. A universal Scott tree T in our sense would give rise to a universal Scott tree σT in the latter sense. Note that $|\sigma T| = |T|^{<\kappa}$, so this is the order of magnitude of a difference in the size of universal Scott trees in the two possible definitions.

Proposition 9.98 *Suppose $\kappa^{<\kappa} = \kappa$ and $\mathcal{M}$ is a structure with $M = \kappa$. The following are equivalent:*

(1) The orbit of $\mathcal{M}$ is $\mathbf{\Delta}_1^1$.

(2) $\mathcal{M}$ *has a universal Scott tree of cardinality* κ.

Proof Suppose first (2) is true. Then

$$\mathcal{M} \not\cong \mathcal{N} \iff \text{player } \mathbf{I} \text{ wins } \text{EF}_{\sigma T}(\mathcal{M}, \mathcal{N}).$$

The existence of a winning strategy of $\mathbf{I}$ can be written in Π_1^1 form since we assume $\kappa^{<\kappa} = \kappa$. Assume then (1). Let ρ be a strategy of player $\mathbf{I}$ in $\text{EF}_\kappa(\mathcal{M}, \mathcal{N})$ in which he simply enumerates the universes. Note that this is independent of $\mathcal{N}$. Let $S(\mathcal{N})$ be the Scott tree S_ρ of the pair $(\mathcal{M}, \mathcal{N})$. Let us consider the co-anaytic set $B = \{ f_\mathcal{N} : N = \kappa \text{ and } \mathcal{N} \not\cong \mathcal{M} \}$. Let us choose a tree representation T of B as in Example 9.96. If now $f_\mathcal{N} \in B_{T'}$, then player $\mathbf{I}$ wins $\text{EF}_{T'}(\mathcal{M}, \mathcal{N})$. By the above Souslin–Kleene Theorem, (1) implies the existence of a tree T' such that $B = B'_T$. Thus for any $\mathcal{N}$ with $N = \kappa$, $\mathcal{M} \not\cong \mathcal{N}$ implies that player $\mathbf{I}$ wins $\text{EF}_{T'}(\mathcal{M}, \mathcal{N})$. Thus T' is a universal Scott tree of $\mathcal{M}$. Moreover, $|T'| = \kappa^{<\kappa} = \kappa$. $\square$

The question whether the orbit of $\mathcal{M}$ is Δ_1^1 is actually highly connected to stability-theoretic properties of the first-order theory of $\mathcal{M}$, see Hyttinen and Tuuri (1991) for more on this.

9.7 Historical Remarks and References

Excellent sources for stronger infinitary languages are the textbook Dickmann (1975), the handbook chapter Dickmann (1985), and the book chapter Kueker (1975). The Ehrenfucht-Fraïssé Game for the logics $L_{\infty\lambda}$ appeared in Benda (1969) and Calais (1972). Proposition 9.32, Proposition 9.45, and the corollary of Proposition 9.45 are due to Chang (1968). The concept of Definition 9.40 and its basic properties were isolated independently by Dickmann (1975) and Kueker (1975). Theorem 9.31 is from Shelah (1990).

Looking at the origins of the transfinite Ehrenfeucht–Fraïssé Game, one can observe that the game plays a role in Shelah (1990), and is then systematically studied, first in the framework of back-and-forth sets in Karttunen (1984), and then explicitly as a game in Hyttinen (1987), Hyttinen (1990), Hyttinen and Väänänen (1990) and Oikkonen (1990).

The importance of trees in the study of the transfinite Ehrenfeucht–Fraïssé Game was first recognized in Karttunen (1984) and Hyttinen (1987). The crucial property of trees, or more generally partial orders, is Lemma 9.55 part (ii), which goes back to Kurepa (1956). A more systematic study of the quasi-order $\mathcal{P} \leq \mathcal{P}'$ of partial orders, with applications to games in mind, was started in Hyttinen and Väänänen (1990), where Lemma 9.57, Definition 9.58,

Lemma 9.59, and Lemma 9.60 originate. The important role of the concept of persistency (Definition 9.63) gradually emerged and was explicitly isolated and exploited in Huuskonen (1995). Once it became clear that trees may be incomparable by $\leq$, the concept of bottleneck arose quite naturally. Definition 9.64 is from Todorčević and Väänänen (1999). The relative consistency of the non-existence of non-trivial bottlenecks (Theorem 9.65) was proved in Mekler and Väänänen (1993). For more on the structure of trees see Todorčević and Väänänen (1999) and Džamonja and Väänänen (2004).

The point of studying trees in connection with the transfinite Ehrenfeucht–Fraïssé Game is that there are two very natural tree structures behind the game. The first tree that arises from the game is the tree of sequences of moves, as in Lemma 9.73. This tree originates in Karttunen (1984). The second, and in a sense more powerful tree is the tree of strategies of a player, as in Definition 9.69 and the subsequent Proposition 9.71. This idea originates from Hyttinen (1987).

The "transfinite" analogues of Scott ranks are the Scott and Karp trees, introduced in Hyttinen and Väänänen (1990). Because of problems of incomparability of some trees, the picture of the "Scott watershed" is much more complicated than in the case of games of length ω, as one can see by comparing Figure 7.4 and Figure 9.7. Proposition 9.85 and Theorem 9.86 are from Tuuri (1990).

There is a form of infinitary logic the elementary equivalence of which corresponds exactly to the existence of a winning strategy for **II** in EF_α, in the spirit of the Strategic Balance of Logic. These infinitary logics are called *infinitely deep languages*. Their formulas are like formulas of $L_{\kappa\lambda}$ but there are infinite descending chains of subformulas. Thus, if we think of the syntax of a formula as a tree, the tree may have transfinite rank. These languages were introduced in Hintikka and Rantala (1976) and studied in Karttunen (1979), Rantala (1981), Karttunen (1984), Hyttinen (1990), and Tuuri (1992). See Väänänen (1995) for a survey on the topic.

There is also a transfinite version of the Model Existence Game, the other leg of the Strategic Balance of Logic, with applications to undefinability of (generalized) well-order and Separation Theorems, see Tuuri (1992) and Oikkonen (1997).

It was recognized already in Shelah (1990) that the roots of the problem of extending the Scott Isomorphism Theorem to uncountable cardinalities lie in stability theoretic properties of the models in question. This was made explicit in the context of transfinite Ehrenfeucht–Fraïssé Games in Hyttinen and Tuuri (1991). It turns out that there is indeed a close connection between the structure of Scott and Karp trees of elementary equivalent uncountable models and the stability theoretic properties such as superstability, DOP, and OTOP, of the (common) first-order theory. For more on this, see Hyttinen (1992), Hyttinen et al. (1993), and Hyttinen and Shelah (1999).

A good testing field for the power of long Ehrenfeucht–Fraïssé Games turned out to be the area of almost free groups, where it seemed that the applicability of the infinitary languages $L_{\kappa\lambda}$ had been exhausted. For results in this direction, see Mekler and Oikkonen (1993), Eklof et al. (1995), Shelah and Väisänen (2002), and Väisänen (2003).

An alternative to considering transfinite Ehrenfeucht–Fraïssé Games is to study isomorphism in a forcing extension. Isomorphism in a forcing extension is called potential isomorphism. The basic reference is Nadel and Stavi (1978). See also Huuskonen et al. (2004).

Early on it was recognized that the trees $T(S)$ (see Example 9.61) are very useful and in some sense fundamental in the area of transfinite Ehrenfeucht–Fraïssé Games. The question arose, whether there is a largest such tree for $S \subseteq \omega_1$ bistationary. Quite unexpectedly the existence of a largest such tree turned out to be consistent relative to the consistency of ZF. The name "Canary trees" was coined for them, because such a tree would indicate whether some stationary set was killed. See Mekler and Shelah (1993) and Hyttinen and Rautila (2001) for results on the Canary tree.

While the Ehrenfeucht–Fraïssé Game of length ω is almost trivially determined, the Ehrenfeucht–Fraïssé Game of length ω_1 (and also of length $\omega + 1$) can be non-determined, see Hyttinen (1992), Mekler et al. (1993), and Hyttinen et al. (2002). This has devastating consequences for attempts to use transfinite Ehrenfeucht–Fraïssé Games to classify uncountable models. It is a phenomenon closely related to the incomparability of non-well-founded trees by the relation $\leq$. This non-determinism is ultimately also the reason why the simple picture in Figure 7.4 becomes Figure 9.7.

Some of the complexities of uncountable models can be located already on the topological level, as is revealed by the study of the spaces $\mathcal{N}_\kappa$. These spaces were studied under the name of κ-metric spaces in Sikorski (1950), Juhász and Weiss (1978), and Todorčević (1981b). Their role as spaces of models, in the spirit of Vaught (1973), was emphasized in Mekler and Väänänen (1993). For more on the topology of uncountable models, see Väänänen (1991), Väänänen (1995), and Shelah and Väänänen (2000). See Väänänen (2008) for an informal exposition of some basic ideas. Theorem 9.95 is from Shelah and Väänänen (2000).

Exercise 9.22 is from Nadel and Stavi (1978). Exercises 9.29 and 9.30 are from Hyttinen (1987). Exercise 9.35 is from Hyttinen and Väänänen (1990).

Exercise 9.40 is from Kurepa (1956). Exercise 9.41 is from Huuskonen (1995). Exercise 9.47 is from Todorčević (1981a). Exercise 9.56 is due to Lauri Hella.

Exercises

9.1 Show that player **II** wins $\text{EF}^{\aleph_0}_\omega(\mathcal{M}, \mathcal{M}')$ if and only if she has a winning strategy in $\text{EF}_\omega(\mathcal{M}, \mathcal{M}')$.

9.2 Show that **I** wins $\text{EFD}^{\omega_1}_2(\mathcal{M}, \mathcal{N})$ if $\mathcal{M} = (\mathbb{Q}, <)$ and $\mathcal{N} = (\mathbb{R}, <)$.

9.3 Show that in Example 9.2 player **I** has a winning strategy already in $\text{EFD}^{\omega_1}_2(\mathcal{M}, \mathcal{M}')$.

9.4 Show that $\mathcal{M} \simeq_p \mathcal{N}$, where $\mathcal{M}$ and $\mathcal{N}$ are as in Example 9.4.

9.5 Prove the claim of Example 9.5.

9.6 Prove the claim of Example 9.19.

9.7 Give necessary and sufficient conditions for player **I** to have a winning strategy in $\text{EFD}^\kappa_\alpha(\mathcal{M}, \mathcal{M}')$, when $\mathcal{M}$ and $\mathcal{M}'$ are L-structures for a unary vocabulary L.

9.8 Show that if $\mathcal{M} = (M, d, \mathbb{R}, <_\mathbb{R})$ and $\mathcal{M}' = (M', d', \mathbb{R}, <_\mathbb{R})$ are separable metric spaces so that **II** has a winning strategy in $\text{EFD}^{\omega_1}_3(\mathcal{M}, \mathcal{M}')$, then $\mathcal{M}$ is complete if and only if $\mathcal{M}'$ is.

9.9 Prove that any model class which is closed under isomorphisms and has only models of cardinality $\leq \lambda$ for some λ is definable in $L_{\infty\infty}$.

9.10 Fix λ and a vocabulary L. Prove that for every α there is only a set of logically non-equivalent formulas of $L_{\infty\lambda}$ of the vocabulary L and of quantifier rank $\leq \alpha$.

9.11 Prove that $\simeq_\lambda$ is an equivalence relation on $\text{Str}(L)$ for any vocabulary L.

9.12 Suppose $\text{cf}(\kappa) = \omega$ (i.e. $\kappa = \sup_n \kappa_n$, where $\kappa_0 < \kappa_1 < \cdots$). Show that $\mathcal{A} \simeq_\kappa \mathcal{B}$ implies $\mathcal{A} \cong \mathcal{B}$ if $|A| = |B| = \kappa$.

9.13 Suppose $\{\mathcal{A}_i : i \in I\}$ is a family of L-structures for a relational vocabulary L. Suppose furthermore $A_i \cap A_j = \emptyset$ for $i \neq j$. The *disjoint sum* of the family $\{\mathcal{A}_i : i \in I\}$ is the L-structure:

$$\biguplus_{i \in I} \mathcal{A}_i = \left(\bigcup_{i \in I} A_i, \left(\bigcup_{i \in I} R^{\mathcal{A}_i} \right)_{R \in L} \right).$$

Show that if $\{\mathcal{A}_i : i \in I\}$ and $\{\mathcal{B}_i : i \in I\}$ are families of L-structures for a relational vocabulary L and for each i

$$\mathcal{A}_i \simeq_\lambda \mathcal{B}_i,$$

then

$$\biguplus_{i \in I} \mathcal{A}_i \simeq_\lambda \biguplus_{i \in I} \mathcal{B}_i.$$

9.14 Suppose $\{\mathcal{A}_i : i \in I\}$ is a family of L-structures. The *direct product* of the family $\{\mathcal{A}_i : i \in I\}$ is the L-structure

$$\prod_{i \in I} \mathcal{A}_i = \left(\prod_{i \in I} A_i, \left(\prod_{i \in I} R^{\mathcal{A}_i} \right)_{R \in L}, (\mathrm{prod}_{i \in I} f^{\mathcal{A}_i})_{f \in L}, ((c : i \in I))_{c \in L} \right)$$

where

$$(\mathrm{prod}_{i \in I} f^{\mathcal{A}_i})((a : i \in I)) = (f_i^{\mathcal{A}_i}(a_i) : i \in I).$$

Show that if $\{\mathcal{A}_i : i \in I\}$ and $\{\mathcal{B}_i : i \in I\}$ are families of L-structures and for each $i \in I$

$$\mathcal{A}_i \simeq_\lambda \mathcal{B}_i,$$

then

$$\prod_{i \in I} \mathcal{A}_i \simeq_\lambda \prod_{i \in I} \mathcal{B}_i.$$

9.15 Suppose $\{\mathcal{A}_i : i \in I\}$ is a family of L-structures in a vocabulary L containing a distinguished constant symbol 0_L. The direct sum $\bigoplus_{i \in I} \mathcal{A}_i$ of the family $\{\mathcal{A}_i : i \in I\}$ is the substructure of $\prod_{i \in I} \mathcal{A}_i$ consisting of $(a_i : i \in I)$ such that $a_i = 0_L^{\mathcal{A}_i}$ for all but finitely many $i \in I$. Show that if $\{\mathcal{A}_i : i \in I\}$ and $\{\mathcal{B}_i : i \in I\}$ are such families and for all $i \in I$

$$\mathcal{A}_i \simeq_\lambda \mathcal{B}_i,$$

then

$$\bigoplus_{i \in I} \mathcal{A}_i \simeq_\lambda \bigoplus_{i \in I} \mathcal{B}_i.$$

9.16 Suppose $\mathcal{M}$ is an L-structure for a relational vocabulary L. Let $I \subseteq J$ be sets of size $\geq \lambda$. Show that

$$\bigoplus_{i \in I} \mathcal{M} \simeq_\lambda \bigoplus_{i \in J} \mathcal{M}.$$

9.17 Consider $\mathbb{Z}$ as an abelian group. Show that for any set I:

$$\bigoplus_{i \in I} \mathbb{Z} \simeq_p \prod_{i \in I} \mathbb{Z}.$$

9.18 Show that "has a clique of size λ" is not definable in $L_{\infty\lambda}$.

9.19 Prove Exercise 9.13 for $\equiv_{\infty\omega}^\alpha$.

9.20 Prove Theorem 9.29.

9.21 Prove Proposition 9.32.

9.22 Let us write $\mathcal{M}(\lambda - \text{PI})\mathcal{N}$ if there is a forcing notion which does not add new sets of cardinality $< \lambda$ (such forcing is called $< \lambda$-distributive) which forces $\mathcal{M}$ and $\mathcal{N}$ to be isomorphic. This is a form of "potential isomorphism", i.e. isomorphism in a forcing extension. Show that $\mathcal{M}(\lambda - \text{PI})\mathcal{N}$ is not a transitive relation among structures and thereby does not correspond to elementary equivalence relative to any logic. (Hint: Use the models $\Phi(A)$ of Definition 9.8.)

9.23 Show that if $\mathcal{M}$ is λ-homogeneous, then for any sequences $\vec{a}$ and $\vec{b}$ of the same length from M:

$$(\mathcal{M}, \vec{a}) \equiv (\mathcal{M}, \vec{b}) \Rightarrow (\mathcal{M}, \vec{a}) \simeq^s_\lambda (\mathcal{M}, \vec{b}).$$

9.24 Let H_α be the lexicographically ordered set of sequences $s \in {}^{\omega_\alpha}\{0,1\}$ (i.e. $s <_{H_\alpha} s'$ if $s(\xi) < s'(\xi)$ for the least ξ such that $s(\xi) \neq s'(\xi)$) for which there is a $\beta < \omega_\alpha$ such that $s(\beta) = 1$ and $s(\gamma) = 0$ for $\beta \leq \gamma < \omega_\alpha$. Show that $H_{\alpha+1}$ is an $\eta_{\alpha+1}$-set.

9.25 Show that H_α is an η_α-set if and only if $\aleph_\alpha$ is regular.

9.26 Show that any η_α-set for singular $\aleph_\alpha$ is also an $\eta_{\alpha+1}$-set.

9.27 Prove that if $\mathcal{A}$ and $\mathcal{B}$ are η_α-sets, then $\mathcal{A} \simeq_{\aleph_\alpha} \mathcal{B}$, and if moreover $\aleph_\alpha$ is regular, then $\mathcal{A} \simeq^s_{\aleph_\alpha} \mathcal{B}$.

9.28 Suppose $\mathcal{M}$ and $\mathcal{M}'$ are real-closed fields whose underlying orders are η_α sets. Show that $\mathcal{M} \simeq_{\aleph_\alpha} \mathcal{M}'$ and if moreover $\aleph_\alpha$ is regular, then $\mathcal{M} \simeq^s_{\aleph_\alpha} \mathcal{M}'$.

9.29 Suppose $S \subseteq \omega_1$. Show that S contains a cub if and only if **I** wins the game $\text{EF}_{\omega+2}(\Phi(S), \Phi(\emptyset))$. (Hint: It is a good idea to consider for a given strategy the set of ordinals $< \omega_1$ which are in some appropriate sense "closed under the first ω moves of the strategy".)

9.30 Show that $S \subseteq \omega_1$ is disjoint from a cub if and only if **II** wins the game $\text{EF}_{\omega+2}(\Phi(S), \Phi(\emptyset))$. (Hint: It is a good idea to consider for a given strategy the set of ordinals $< \omega_1$ which are in some appropriate sense "closed under the first ω moves of the strategy".)

9.31 Show that if $\mathcal{M} \simeq^s_{\aleph_1} \mathcal{N}$, then **II** wins the game $\text{EF}_{\omega_1}(\mathcal{M}, \mathcal{N})$.

9.32 Show that **II** wins the game $\text{EF}_{\omega_1}(\mathcal{M}, \mathcal{N})$ if and only if $\mathcal{M}$ and $\mathcal{N}$ are potentially isomorphic in the following sense: there is a countably closed[5] forcing notion $\mathcal{P}$ such that $\mathcal{P}$ forces $\mathcal{M} \cong \mathcal{N}$. (Hint: Note that the forcing which collapses $|M \cup N|$ to $\aleph_1$ is countably closed.)

---

[5] I.e. every countable descending chain of conditions has a lower bound.

9.33 Show that $(\omega_1, <)$ and $(\mathbb{R}, <)$ are incomparable by the quasi-order $\leq$ of po-sets.

9.34 Prove $\sigma\mathcal{P} \leq \sigma\mathcal{P}' \iff \sigma'\mathcal{P} \leq \mathcal{P}'$.

9.35 Suppose $S \subseteq \omega_1$. Let $T(S)$ be the tree of closed ascending sequences of elements of S. Choose disjoint bistationary sets S_1 and S_2. Then $T(S_1) \not\leq T(S_2)$ and $T(S_2) \not\leq T(S_1)$ (see Example 9.61).

9.36 Prove the claims of Example 9.54.

9.37 Suppose T is a tree. Show that T has no infinite branches if and only if there is an ordinal α so that $T \equiv (\alpha, >)$.

9.38 Prove that if a po-set $\mathcal{P}$ is a union of countably many antichains, it satisfies $\mathcal{P} \leq (\mathbb{Q}, <)$.

9.39 Show that $F_{\aleph_1}$ and $T_p^{\aleph_1}$ are special trees.

9.40 Prove the claim in Example 9.56 that $\sigma'\mathbb{Q}$ is special but $\sigma\mathbb{Q}$ non-special.

9.41 Show that T_p^κ is the $\leq$-smallest persistent tree in $\mathcal{T}_\kappa$.

9.42 Prove that T_p^κ is a strong bottleneck in the class $\mathcal{T}_\kappa$

9.43 If T_i, $i \in I$, is a family of trees, let $\bigoplus_{i \in I} T_i$ be the tree which consists of a union of disjoint copies of T_i, $i \in I$, identified at the root. Show that $\bigoplus_{i \in I} T_i$ is the supremum of $\{T_i : i \in I\}$ in the sense that $T_i \leq \bigoplus_{i \in I} T_i$ for all $i \in I$ and if $T_i \leq T$ for all $i \in I$, then $\bigoplus_{i \in I} T_i \leq T$.

9.44 If T_i, $i \in I$, is a family of trees, let $\prod_{i \in I} T_i$ be the product tree

$$\prod_{i \in I} T_i = \{s : \mathrm{dom}(s) = I, \forall i \in I(s(i) \in T_i)\}.$$

$$s \leq s' \iff \forall i \in I(s(i) \leq_{T_i} s'(i)).$$

Let $\bigotimes_{i \in I} T_i$ be the subtree

$$\bigotimes_{i \in I} T_i = \left\{ s \in \prod_{i \in I} T_i : \forall i \in I \forall j \in I(\mathrm{ht}_{T_i}(s(i)) = \mathrm{ht}_{T_j}(s(j))) \right\}.$$

We denote $\bigotimes_{i \in \{0,1\}} T_i$ by $T_0 \otimes T_1$. Prove that $\bigotimes_{i \in I} T_i$ is the infimum of $\{T_i' : i \in I\}$, that is, $\bigotimes_{i \in I} T_i \leq T_i$ for each $i \in I$, and if $T \leq T_i$ for all $i \in I$, then $T \leq \bigotimes_{i \in I} T_i$.

9.45 Show that $\mathcal{P}_\alpha \star \mathcal{P}'$ in the proof of Theorem 9.65 contains a κ-closed dense set. (Hint: Suppose $(s, s') \in \mathcal{P}_\alpha \star \mathcal{P}'$. Thus $s : \kappa \to \{0, 1\}$, $|s| < \kappa$, and s forces that s' is a closed sequence of length $< \kappa$ in A_α. Consider the sets of (s, s') for which $\sup\{\beta : s(\beta) = 1\} = \max(s')$.)

9.46 Suppose A and B are disjoint stationary subsets of a regular cardinal κ. Show that $T(A) \otimes T(B) \leq T_p^\kappa$. (Hint: Show that **II** has a winning strategy in the game $G(T(A) \otimes T(B), T_p^\kappa)$.)

9.47 Suppose $S \subseteq \omega_1$. Prove that $T(S) \leq \mathbb{Q}$ if and only if S is non-stationary.

9.48 Suppose $S \subseteq \omega_1$ is bistationary and T is Aronszajn. Show that $T(S) \not\leq T$ and $T \not\leq T(S)$.

9.49 Prove $b((\mathbb{Q}, <)) = b((\mathbb{R}, <)) = \omega_1$.

9.50 Prove Lemma 9.67.

9.51 Prove Lemma 9.83.

9.52 Show that if **I** wins $\mathrm{EF}_{T_i}(\mathcal{A}_0, \mathcal{A}_1)$ for all $i \in I$, then **II** does not win $\mathrm{EF}_{\bigotimes_{i \in I} T_i}(\mathcal{A}_0, \mathcal{A}_1)$.

9.53 Show that the family of Scott trees of $(\mathcal{A}_0, \mathcal{A}_1)$ is closed under suprema.

9.54 Show that the family of Karp trees of $(\mathcal{A}_0, \mathcal{A}_1)$ is closed under suprema.

9.55 Suppose $\mathcal{P}$ is a Scott po-set of $(\mathcal{A}_0, \mathcal{A}_1)$, where $\mathrm{Card}(A_0), \mathrm{Card}(A_1) \leq 2^{\aleph_0}$. Show that there is a Scott tree of $(\mathcal{A}_0, \mathcal{A}_1)$ such that $T \leq \mathcal{P}$ and $\mathrm{Card}(T) \leq 2^{\aleph_0}$.

9.56 Show that if $2^\kappa = 2^{\kappa^+}$, then $\mathcal{T}_{\kappa^+, \kappa^+}$ has an upper bound in $\mathcal{T}_{2^\kappa, \kappa^+}$.

9.57 Show that every closed set in $\mathcal{N}_\kappa$ is the union of $\kappa^{<\kappa}$ open sets.

9.58 Show that if $\mathrm{cf}(\kappa)\omega$, then the intersection of countably many open sets in $\mathcal{N}_\kappa$ is again open. Topological spaces with this property are called *σ-additive*.

9.59 Show $\mathcal{N}_\kappa$ has a basis consisting of clopen sets. Topological spaces with this property are called *zero-dimensional*.

10
Generalized Quantifiers

10.1 Introduction

First-order logic is not able to express "there exists infinitely many x such that ..." nor "there exists uncountably many x such that ...". Also, if we restrict ourselves to finite models, first-order logic is not able to express "there exists an even number of x such that ...". These are examples of new logical operations called *generalized quantifiers*. There are many others, such as the Magidor–Malitz quantifiers, cofinality quantifiers, stationary logic, and so on. We can extend first-order logic by adding such new quantifiers. In the case of "there exists infinitely many x such that ..." the resulting logic is not axiomatizable, but in the case of "there exists uncountably many x such that ..." the new logic is indeed axiomatizable. The proof of the Completeness Theorem for this quantifier is non-trivial going well beyond the Completeness Theorem of first-order logic.

10.2 Generalized Quantifiers

Generalized quantifiers occur everywhere in our language. Here are some examples:

> **Two thirds** voted for John
> **Exactly half** remains.
> **Most** wanted to leave.
> **Some but not all** liked it.
> **Between 10% and 20%** were students.
> **Hardly anybody** touched the cake.
> The number of white balls **is even**.
> **There are infinitely many** primes.
> **There are uncountably many** reals.

These are instances of generalized quantifiers in natural language.[1] The mathematical study of quantifiers provides an exact framework in which such quantifiers can be investigated. An overall goal is to find *invariants* for such objects, that is, to classify them and find the characteristic properties of quantifiers in each class. Typical questions that we study are: which quantifier is "definable" in terms of another given quantifier, which quantifiers can be axiomatized, which satisfy the Compactness Theorem, etc. We start with a very general concept of a quantifier and then later we impose restrictions. Usually in the literature the generalized quantifiers are assumed to be what we call bijection closed (see Definition 10.16).

Definition 10.1 A *weak (generalized) quantifier* is a mapping Q which maps every non-empty set A to a subset of $\mathcal{P}(A)$. A *weak (generalized) quantifier on* a domain A is any subset of $\mathcal{P}(A)$.

Virtually all quantifiers we consider are quantifiers in the first sense, i.e. mappings $A \longmapsto Q(A)$. However, most actual results and examples are about a fixed given domain A, whence the concept of a quantifier *on* a domain. The domain is assumed to be a set.

The set-theoretic nature of a quantifier (as a mapping) is somewhat problematic. We cannot call a quantifier a function in the set-theoretical sense since its domain consists of all possible non-empty sets. However, this problem does not arise in practice. Our quantifiers are in general definable so we can treat them as classes. If we have to talk about all quantifiers, definable or not, we have to restrict ourselves to considering domains A contained in one sufficiently

---

[1] Quantifiers occurring in natural language are usually of a slightly more complex form, such as "Two thirds of the people voted for John", "Exactly half of the cake remains", "Most students wanted to leave", "Some but not all viewers liked it".

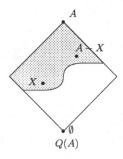

$Q(A)$

Figure 10.1 Generalized quantifier.

big "monster domain". There are no considerations here that would make this necessary.

Example 10.2 1. The *existential quantifier* $\exists$ is the mapping

$$\exists(A) = \{X \subseteq A : X \neq \emptyset\}.$$

2. The *universal quantifier* $\forall$ is the mapping

$$\forall(A) = \{X \subseteq A : X = A\} = \{A\}.$$

3. The *counting quantifier* $\exists^{\geq n}$ is the mapping

$$\exists^{\geq n}(A) = \{X \subseteq A : |X| \geq n\},$$

where we assume n is a natural number.

4. The *infinity quantifier* $\exists^{\geq \omega}$ is the mapping

$$\exists^{\geq \omega}(A) = \{X \subseteq A : X \text{ is infinite}\}.$$

5. The *finiteness quantifier* $\exists^{<\omega}$ is the mapping

$$\exists^{<\omega}(A) = \{X \subseteq A : X \text{ is finite}\}.$$

6. The following subsets of $\mathcal{P}(\mathbb{N})$ are weak quantifiers on $\mathbb{N}$:

$$[\{5\}] = \{X \subseteq \mathbb{N} : 5 \in X\}$$
$$[X_0] = \{X \subseteq \mathbb{N} : X_0 \subseteq X\}, \text{ where } X_0 \subseteq \mathbb{N} \text{ is fixed}$$
$$[X_0]^* = \{X \subseteq \mathbb{N} : X_0 \cap X \neq \emptyset\}, \text{ where } X_0 \subseteq \mathbb{N} \text{ is fixed}.$$

We can draw pictures of quantifiers on a domain A by thinking of $\mathcal{P}(A)$ as a Boolean algebra under $\subseteq$, as in Figure 10.1.

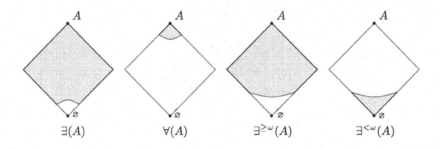

Figure 10.2 Some generalized quantifiers.

The reason we call $\{X \subseteq A : X \neq \emptyset\}$ the existential quantifier on A is the following:

$$\mathcal{A} \models \exists x \varphi(x) \iff \{a \in A : \mathcal{A} \models \varphi(a)\} \neq \emptyset$$
$$\iff \{a \in A : \mathcal{A} \models \varphi(a)\} \in \exists(A).$$

Respectively

$$\mathcal{A} \models \forall x \varphi(x) \iff \{a \in A : \mathcal{A} \models \varphi(a)\} = A$$
$$\iff \{a \in A : \mathcal{A} \models \varphi(a)\} \in \forall(A).$$

Later we will associate with every quantifier Q an extension of first-order logic based on the above idea.

Some quantifiers make sense only in a *finite context*. By this we mean that only finite domains A are considered. If we allow countable domains too we work in a *countable context*.

Example 10.3 (Finite context)   1. The *even-cardinality quantifier* Q^{even} is the mapping (see Figure 10.3)

$$Q^{\text{even}}(A) = \{X \subseteq A : |X| \text{ is even}\}.$$

Similarly

$$Q^D(A) = \{X \subseteq A : |X| \in D\} \text{ for any } D \subseteq \mathbb{N}.$$

2. The *at-least-one-half quantifier* $\exists^{\geq \frac{1}{2}}$ is the mapping (see Figure 10.4)

$$\exists^{\geq \frac{1}{2}}(A) = \{X \subseteq A : |X| \geq |A|/2\}.$$

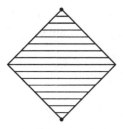

Figure 10.3 Even cardinality quantifier.

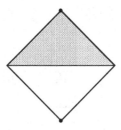

Figure 10.4 At-least-one-half quantifier.

The quantifier $\exists^{\geq r}$ is defined similarly

$$\exists^{\geq r} = \{X \subseteq A : |X| \geq r \cdot |A|\} \text{ for any real } r \in [0, 1].$$

Thus at-least-two-thirds would be the quantifier $\exists^{\geq \frac{2}{3}}$. It is obvious how to define the quantifier less-than-two-thirds, in symbols $\exists^{< \frac{2}{3}}$, and more generally $\exists^{<r}$, $\exists^{\leq r}$ and $\exists^{>r}$.

3. The *most quantifier* $\exists^{\text{most}}$ is the mapping

$$\exists^{\text{most}}(A) = \{X \subseteq A : |X| > |A - X|\}.$$

We can define Boolean operations for weak quantifiers in a natural way:

$$(Q \cap Q')(A) = Q(A) \cap Q'(A)$$
$$(Q \cup Q')(A) = Q(A) \cup Q'(A)$$
$$(-Q)(A) = \{X \subseteq A : X \notin Q(A)\}.$$

These operations obey the familiar laws of Boolean algebras, such as idempotency, commutativity, associativity, distributivity, and the de Morgan laws:

$$-(Q \cap Q') = -Q \cup -Q'$$
$$-(Q \cup Q') = -Q \cap -Q'.$$

The quantifier $-Q$ is called the *complement* of Q. There is also another kind of complement of a quantifier, the quantifier

$$(Q-)(A) = \{A \setminus X : X \in Q(A)\}$$

called the *postcomplement* of Q.

Example 10.4 The complement of "everybody" is "not everybody", while the postcomplement of "everybody" is "nobody". The complement of $\exists^{\geq \frac{2}{3}}$ is $\exists^{< \frac{2}{3}}$, while the postcomplement of $\exists^{\geq \frac{2}{3}}$ is $\exists^{< \frac{1}{3}}$.

The postcomplement satisfies $(Q-)- = Q$, but does not obey the de Morgan laws. Rather:

$$(Q \cap Q')- = (Q'-) \cap (Q-)$$
$$(Q \cup Q')- = (Q'-) \cup (Q-).$$

Note that complement and postcomplement obey the following associativity law:

$$(-Q)- = -(Q-).$$

Thus we may leave out parentheses and write simply $-Q-$. The existential and the universal quantifier have a special relationship called *duality*, exemplified by the equation

$$\exists = -\forall - \quad \text{and} \quad \forall = -\exists - .$$

Duality is an important phenomenon among quantifiers and gives rise to the following definition:

Definition 10.5 The *dual* of a weak quantifier Q is the quantifier

$$\check{Q} = -Q-,$$

that is, the mapping

$$\check{Q}(A) = \{X \subseteq A : A \setminus X \notin Q(A)\}.$$

The dual of a weak quantifier on a domain is defined in the same way. (See Figure 10.5.)

Example 10.6 1. The dual of $\exists$ is $\forall$ and vice versa: the dual of $\forall$ is $\exists$.
2. The dual of $\exists^{\geq \omega}$ is the quantifier *all-but-finite*

$$\forall^{< \omega}(A) = \{X : |A - X| \text{ is finite}\} = (\exists^{< \omega})-$$

and vice versa: the dual of $\forall^{< \omega}$ is the quantifier $\exists^{\geq \omega}$.

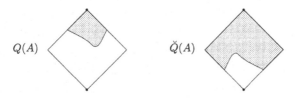

$Q(A)$ $\check{Q}(A)$

Figure 10.5 Dual of a quantifier.

3. The dual of $\exists^{<\omega}$ is the quantifier *infinite-failure*

$$\forall^{\geq\omega}(A) = \{X : |A - X| \text{ is infinite}\} = (\exists^{\geq\omega})-$$

and vice versa: the dual of $\forall^{\geq\omega}$ is the quantifier $\exists^{<\omega}$.

Lemma 10.7 *1. $\check{\check{Q}} = Q$.*
2. $(Q_1 \cup Q_2)\check{} = \check{Q}_1 \cap \check{Q}_2$.
3. $(Q_1 \cap Q_2)\check{} = \check{Q}_1 \cup \check{Q}_2$.
4. $(-Q)\check{} = \check{Q}-, (Q-)\check{} = -\check{Q}$.
5. If $Q_1 \subseteq Q_2$, then $\check{Q}_2 \subseteq \check{Q}_1$.

Proof Exercise 10.3. □

One possible intuitive meaning of $X \in Q(A)$ is that X is "large". This is by no means the only relevant meaning but it is the meaning which is most commonly used. Largeness should be closed under further extensions. Therefore it is no exaggeration to say that the following concept is extremely important in the study of generalized quantifiers:

Definition 10.8 A weak quantifier Q is (upwards) *monotone* if $X \in Q(A)$ and $X \subseteq Y \subseteq A$ imply $Y \in Q(A)$.

Lemma 10.9 *If Q is monotone, then so is $\check{Q}$.*

Proof Suppose $X \in \check{Q}$ and $X \subseteq Y \subseteq A$. Then $A\backslash X \notin Q$ and $A\backslash X \supseteq A\backslash Y$. Since Q is monotone, $A - Y \notin Q$. Hence $Y \in \check{Q}$. □

Example 10.10 The weak quantifiers $\exists, \forall, \exists^{\geq\omega}, \forall^{<\omega}, \exists^{\text{most}}, [X_0], [X_0]^*, \exists^{\geq r}$ are all monotone, while the weak quantifiers $Q^{\text{even}}, \exists^{<\omega}, \forall^{\geq\omega}$ are obviously non-monotone.

Definition 10.11 A *basis* of a monotone weak quantifier Q on a domain A is any set $\mathcal{C} \subseteq \mathcal{P}(A)$ such that

Figure 10.6 Basis of a quantifier.

(i) $C \subseteq Q(A)$.

(ii) If $X \in Q(A)$, then $Y \subseteq X$ for some $Y \in C$.

(iii) C forms an *antichain*, i.e. if X_0 and X_1 are different elements of C, then $X_0 \nsubseteq X_1$ and $X_1 \nsubseteq X_0$ (see Figure 10.6).

Two weak monotone quantifiers with the same basis clearly coincide.

Definition 10.12 An *atom* of a monotone weak quantifier Q on a domain A is any set $X \subseteq A$ such that

(i) $X \in Q(A)$.

(ii) If $Y \subsetneq X$, then $Y \notin Q(A)$.

The atoms of $\exists$ are the singletons. The one and only atom of $\forall$ on a domain A is the set A itself. The atoms of $[X_0]^*$ are the singletons $\{a\}$, $a \in X_0$. The only atom of $[X_0]$ is X_0. Note that the atoms of $Q(A)$ always form an antichain. However, there need not be any atoms. For example, $\exists^{\geq \omega}$ has no atoms on any infinite domain.

Proposition 10.13 *Suppose Q is a weak monotone quantifier on A. If Q has a basis, it consists of atoms. If A is finite, then the atoms of Q on A form a basis of Q on A.*

Proof For the second claim, suppose $X \in Q$. Let $n \in \mathbb{N}$ be the least n such that there is some $Y \subseteq X$ in Q of size n. Let Y be one such set. If $Z \subsetneq Y$, then $Z \notin Q$ by the minimality of n. Thus Y is an atom. □

Thus on finite domains weak monotone quantifiers are completely determined by their sets of atoms. If C is an arbitrary antichain on a domain A, we get a monotone quantifier on A:

$$Q_C^{\text{mon}} = \{X \subseteq A : \text{there is some } Y \in C \text{ with } Y \subseteq X\},$$

and then C is the set of atoms of Q_C^{mon} on A.

A weak quantifier is, as its name reveals, one that quantifies rather than qualifies. Thus whether X is in Q or not should not depend on anything but the "quantity of objects" in X. This idea is captured by the following concept:

Definition 10.14 A weak quantifier Q on a domain A is *permutation closed* if $X \in Q$ implies $\pi''X \in Q$ for every permutation π of A. If G is a group of permutations of A and $X \in Q$ implies $\pi''X \in Q$ for every $\pi \in G$, we call Q *G-closed*.

Note that the dual, complement, and postcomplement of a permutation closed (G-closed) quantifier are permutation closed (G-closed).

Example 10.15 $[\{5\}]$ is G-closed for the group of permutations π of $\mathbb{N}$ that satisfy $\pi(5) = 5$. $[X_0]$ and $[X_0]^*$ are G-closed for the group of permutations π of $\mathbb{N}$ that map X_0 to X_0, i.e. $X_0 = \pi''X_0$.

Note that we require not only that the permutation π is defined on X but that it is a permutation of all of the domain A. This is because quantity is not just about how many objects there are in X, but also about how many are *not* in X.

Definition 10.16 A weak quantifier Q is *bijection closed* if $X \in Q(A)$ implies $\pi''X \in Q(B)$ for all bijections $\pi : A \to B$.

Example 10.17 All the weak quantifiers defined above

$$\exists, \ \forall, \ \exists^{\geq\omega}, \ \forall^{<\omega}, \ \exists^{<\omega}, \ \forall^{\geq\omega}$$

are bijection closed.

Note again that the dual, complement, and postcomplement of a bijection closed quantifier are bijection closed.

Definition 10.18 If a weak (generalized) quantifier on a domain is permutation closed we drop "weak" and call it just a (generalized) quantifier. If a weak (generalized) quantifier is bijection closed we drop "weak" and call it just a (generalized) quantifier.

This concept was introduced in Mostowski (1957). The idea of Mostowski was that in order for a quantifier to be really about quantity, as the name says, it ought to be bijection closed.

Note that an equivalent condition for a weak quantifier Q to be bijection closed is that

$$(A, X) \cong (B, Y) \text{ and } X \in Q(A) \text{ imply } Y \in Q(B).$$

Lemma 10.19 *The following are equivalent for any $A, B, X,$ and Y:*

(1) $(A, X) \cong (B, Y)$.

(2) $X \sim Y$ *and* $A \setminus X \sim B \setminus Y$.

If A and B are finite then (1) and (2) are equivalent to

(3) $X \sim Y$ *and* $A \sim B$.

Proof If $f : (A, X) \cong (B, Y)$, then $f \upharpoonright X$ demonstrates $X \sim Y$ and $f \upharpoonright (A \setminus X)$ demonstrates $A \setminus X \sim B \setminus Y$. Conversely, if $f : X \to Y$ and $g : A \setminus X \to B \setminus Y$ are bijections, then $f \cup g$ demonstrates $(A, X) \cong (B, Y)$. If A and B are finite, then $A \sim B$ and $A \setminus X \sim B \setminus Y$ are equivalent under the assumption $X \sim Y$.   □

Definition 10.20 (Finite context)   Suppose Q is a generalized quantifier. We define

$$M_Q$$

to be the set of pairs (m, n) such that for some A and $X \subseteq A$ we have

$$X \in Q(A), |X| = m, |A \setminus X| = n.$$

If we put all pairs (m, n) into a "number triangle"

$$
\begin{array}{ccccccccc}
 & & & & (0,1) & & (1,0) & & \\
 & & & (0,2) & & (1,1) & & (2,0) & \\
 & & (0,3) & & (1,2) & & (2,1) & & (3,0) \\
 & (0,4) & & (1,3) & & (2,2) & & (3,1) & & (4,0) \\
 & \vdots & & \vdots & & \vdots & & \vdots & & \vdots
\end{array}
$$

we can picture a generalized quantifier Q by putting "+" at a place which is in the set M_Q and "−" at other places. For example for $\exists^{\geq \frac{1}{2}}$ we get the triangle:

$$
\begin{array}{ccccccccc}
 & & & & - & & + & & \\
 & & & - & & + & & + & \\
 & & - & & - & & + & & + \\
 & - & & - & & + & & + & & + \\
 - & & - & & - & & + & & + & & + \\
 - & & - & & - & & + & & + & & + & & + \\
 \vdots & & \vdots & & \vdots & & \vdots & & \vdots & & \vdots & & \vdots
\end{array}
$$

For Q^{even} we get the triangle:

```
                    +       −
               +       −       +
          +       −       +       −
     +       −       +       −       +
+       −       +       −       +       −
+       −       +       −       +       −       +
:       :       :       :       :       :       :
```

Definition 10.21 (Countable context)   Suppose Q is a generalized quantifier. We define N_Q as the union of M_Q and the set of the following pairs

(ω, n) if there are A and $X \subseteq A$ such that
$X \in Q(A), X \sim \mathbb{N}$ and $|A \setminus X| = n$.

(n, ω) if there are A and $X \subseteq A$ such that
$X \in Q(A), |X| = n$ and $A \sim \mathbb{N}$.

(ω, ω) if there are A and $X \subseteq A$ such that
$X \in Q(A), X \sim \mathbb{N}$ and $A \setminus X \sim \mathbb{N}$.

```
                    (0,1)           (1,0)
              (0,2)         (1,1)         (2,0)
    (0,3)           (1,2)         (2,1)         (3,0)
      :               :             :             :
(0,ω)         (1,ω)    ···   (ω,ω)    ···   (ω,1)         (ω,0)
```

For $\exists^{\geq \omega}$ we get the triangle:

```
                    −       −
               −       −       −
          −       −       −       −
     −       −       −       −       −       −
     :       :       :       :       :       :       :
   −       −    ···    +    ···    +       +
```

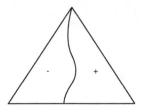

Figure 10.7 Monotone quantifier.

For $\forall^{<\omega}$ we get the triangle:

$$
\begin{array}{ccccccccc}
 & & & & + & & + & & \\
 & & & + & & + & & + & \\
 & & + & & + & & + & & + \\
 & + & & + & & + & & + & & + \\
+ & & + & & + & & + & & + & & + \\
\vdots & & \vdots & & \vdots & & \vdots & & \vdots & & \vdots \\
- & & - & & \cdots & & - & & \cdots & + & & +
\end{array}
$$

Note that in the finite context

$$(m, n) \in M_Q \text{ if and only if } (n, m) \notin M_{\check{Q}}.$$

Thus there is an easy algorithm for getting the triangle of the dual from the triangle of the quantifier itself. The same is true in the countable context:

$$(\omega, n) \in M_Q \text{ if and only if } (n, \omega) \notin M_{\check{Q}}$$

$$(m, \omega) \in M_Q \text{ if and only if } (\omega, m) \notin M_{\check{Q}}$$

$$(\omega, \omega) \in M_Q \text{ if and only if } (\omega, \omega) \notin M_{\check{Q}}.$$

As a consequence, there cannot be *self-dual* permutation closed quantifiers (i.e. $Q = \check{Q}$) on a domain of even or countably infinite size.

The number triangle of a monotone bijection closed quantifier Q has all plus signs after all minus signs (Figure 10.7). The line which separates the signs is an important invariant of Q:

Definition 10.22 (Finite context) Suppose Q is a monotone bijection closed quantifier. The *threshold function* of Q is the function $f_Q : \mathbb{N} \to \mathbb{N}$ defined by

$$f_Q(n) = \begin{cases} \text{the least } m \text{ such that } (m, n - m) \in M_Q & \text{if any exist} \\ n + 1 & \text{otherwise.} \end{cases}$$

Example 10.23

$$f_\exists(n) = 1$$
$$f_\forall(n) = n$$
$$f_{\exists \geq \frac{1}{2}}(n) = \lceil \tfrac{n}{2} \rceil \quad (\lceil r \rceil \text{ is the smallest integer } \geq r).$$

We can take any function $f : \mathbb{N} \to \mathbb{N}$ which satisfies $f(n) \leq n + 1$ for all $n \in \mathbb{N}$ and define in the finite context the quantifier:

$$Q^{\geq f}(A) = \{X \subseteq A : |X| \geq f(|A|)\}.$$

Then

$$f_1 = f_2 \iff Q^{\geq f_1} = Q^{\geq f_2}$$

and

$$f_{Q^{\geq g}} = g.$$

Thus the study of monotone bijection closed quantifiers reduces in the finite context to a study of functions $f : \mathbb{N} \to \mathbb{N}$ with $f(n) \leq n + 1$ for all $n \in \mathbb{N}$.

Definition 10.24 (Finite context)   A monotone bijection closed quantifier Q is *eventually counting* if there are $k, m \in \mathbb{N}$ such that

(1) $f_Q(n) = k$ for all $n \geq m$, or
(2) $f_{\check{Q}}(n) = k$ for all $n \geq m$.

Q is *bounded* if there is a finite set $S \subseteq \mathbb{N}$ such that for all $n \in \mathbb{N}$

(3) $f_Q(n) \in S$, or
(4) $f_{\check{Q}}(n) \in S$.

Otherwise Q is *unbounded*.

The classification of monotone quantifiers given in Definition 10.24 is very important. Eventually counting quantifiers as well as unbounded quantifiers occur abundantly in natural language. The bounded case is more technical.

Example 10.25 $\exists$ and $\forall$ are eventually counting. If

$$f(n) = \begin{cases} 0 & \text{for even } n \\ 1 & \text{for odd } n \end{cases}$$

then $Q^{\geq f}$ is bounded but not eventually counting. $Q^{\geq \frac{1}{2}}$ is unbounded.

Note that Q is eventually counting (bounded) if and only if $\check{Q}$ is.

10.3 The Ehrenfeucht–Fraïssé Game of Q

It was realized early on that whatever the generalized quantifier Q is, and whether we work in the finite context or not, the method of Ehrenfeucht–Fraïssé Games works perfectly. It has become one of the main tools in the study of generalized quantifiers. The definition of the Ehrenfeucht–Fraïssé Game is slightly easier in the case of monotone quantifiers, so we concentrate on them.

Definition 10.26 Suppose Q is a monotone generalized quantifier, L is a vocabulary, $\mathcal{M}$ and $\mathcal{M}'$ are L-structures, $M \cap M' = \emptyset$, and n is a natural number. The *Ehrenfeucht–Fraïssé Game of Q* of n moves on $\mathcal{M}$ and $\mathcal{M}'$, $\mathrm{EF}_n^Q(\mathcal{M}, \mathcal{M}')$, is defined as follows: There are n moves during which player **I** plays x_i and player **II** plays y_i. Each move is of one of two kinds:

First-order move: **I** chooses $x_i \in M \cup M'$ and then **II** chooses $y_i \in M \cup M'$ such that

$$x_i \in M \iff y_i \in M'.$$

Q-move: **I** first chooses $Y \in Q(M) \cup Q(M')$. Then **II** chooses $X \in Q(M) \cup Q(M')$ such that

$$Y \in Q(M) \iff X \in Q(M').$$

Now **I** chooses x_i and finally **II** chooses y_i such that $x_i \in X \Rightarrow y_i \in Y$ (see Figure 10.8).

After $\{(x_0, y_0), \ldots, (x_{n-1}, y_{n-1})\}$ has been played, player **II** wins if, denoting

$$v_i = \begin{cases} x_i & \text{if } x_i \in M \\ y_i & \text{if } y_i \in M \end{cases} \quad \text{and} \quad v_i' = \begin{cases} x_i & \text{if } x_i \in M' \\ y_i & \text{if } y_i \in M', \end{cases}$$

the corresponding relation $\{(v_0, v_0'), \ldots, (v_{n-1}, v_{n-1}')\} \subseteq M \times M'$ is a partial isomorphism $\mathcal{M} \to \mathcal{M}'$.

Recall that the intuition behind the first-order moves is that **I** suspects there is a difference between $\mathcal{M}$ and $\mathcal{M}'$, and he tries to locate it. The intuition behind the Q-move is that **I** again suspects there is a difference between $\mathcal{M}$ and $\mathcal{M}'$, but this time of a different kind, for example, if P is a unary predicate symbol in L, the difference could be of the kind

$$P^{\mathcal{M}} \in Q(M) \text{ but } P^{\mathcal{M}'} \notin Q(M')$$

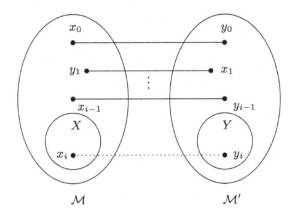

Figure 10.8 The Q-move.

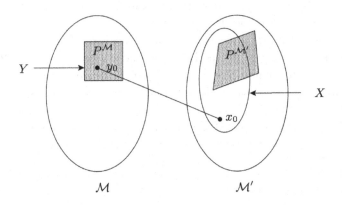

Figure 10.9 **I** wins.

or, in other words,

$$P^{\mathcal{M}} \text{ is "large" but } P^{\mathcal{M}'} \text{ is not "large".}$$

Player **I** would exploit this difference by playing $Y = P^{\mathcal{M}} \in Q(M)$ after which **II** plays $X \in Q(M')$. As **I** suspects $P^{\mathcal{M}'} \notin Q(M')$ and Q is monotone he goes on to suspect that $X \not\subseteq P^{\mathcal{M}'}$. Thus he plays $x_0 \in X$ trying to make sure, if possible, that $x_0 \notin P^{\mathcal{M}'}$. Now **II** has to play $y_0 \in Y = P^{\mathcal{M}}$ (see

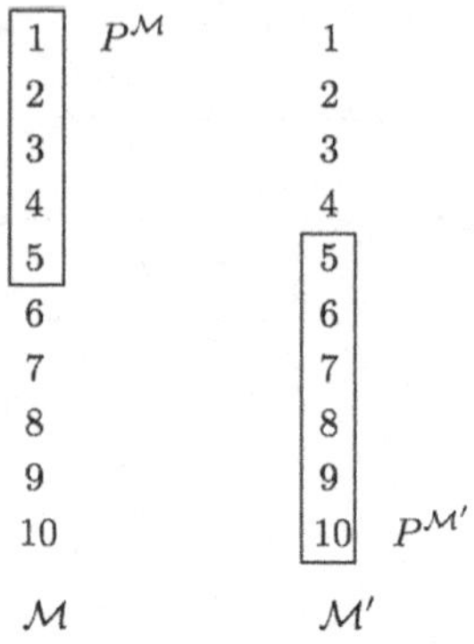

$$
\begin{array}{cccc}
1 & P^{\mathcal{M}} & 1 & \\
2 & & 2 & \\
3 & & 3 & \\
4 & & 4 & \\
5 & & 5 & \\
6 & & 6 & \\
7 & & 7 & \\
8 & & 8 & \\
9 & & 9 & \\
10 & & 10 & P^{\mathcal{M}'} \\
\mathcal{M} & & \mathcal{M}' &
\end{array}
$$

Figure 10.10 The models $\mathcal{M}$ and $\mathcal{M}'$.

Figure 10.9). If **I** was successful in his attempt to achieve $x_0 \in X \setminus P^{\mathcal{M}'}$, then **I** has won: $\{(x_0, y_0)\}$ is not a partial isomorphism.

Since $X \in Q(M)$ for a monotone Q can be read as "X is large", we can think that **I** plays a large set Y and **II** has to respond with a large set X. When **II** decides what to put into X (to make it large), she has to be careful to avoid letting any x into X for which she is not prepared to find a corresponding y from the set Y. Thus if $P^{\mathcal{M}}$ is large and **I** plays $Y = P^{\mathcal{M}}$, **II** is out of luck unless $P^{\mathcal{M}'}$ is large, too. For if she lets an element from the complement of $P^{\mathcal{M}'}$ slip into X, **I** will immediately play that "wrong" element.

Example 10.27 We have two unary structures $\mathcal{M}$ and $\mathcal{M}'$ (see Figure 10.10). More exactly, suppose $L = \{P\}, \#(P) = 1, \mathcal{M} = (\{1, \ldots, 10\}, P^{\mathcal{M}})$, where $P^{\mathcal{M}} = \{1, \ldots, 5\}, \mathcal{M}' = (\{1, \ldots, 10\}, P^{\mathcal{M}'})$, where $P^{\mathcal{M}'} = \{5, \ldots, 10\}$. Clearly, **II** has a winning strategy in $\mathrm{EF}_n(\mathcal{M}, \mathcal{M}')$ if and only if $n \leq 4$. Player **I** has a winning strategy in $\mathrm{EF}_1^{\exists^{\geq \frac{1}{2}}}(\mathcal{M}, \mathcal{M}')$ as he can play $Y = M \setminus P^{\mathcal{M}}$. If **II** plays $X \in \exists^{\geq \frac{1}{2}}(M)$, then X will contain an element b from $P^{\mathcal{M}'}$. We let **I** play $x_0 = b$. Whatever $y_0 \in Y$ **II** plays, she loses as $y_0 \notin P^{\mathcal{M}}$ while $x_0 \in P^{\mathcal{M}'}$ (see Figure 10.11). But **II** has a winning strategy in $\mathrm{EF}_4^{\exists^{\geq \frac{1}{3}}}(\mathcal{M}, \mathcal{M}')$. Her strategy for each Q-move is the same: If Y meets $P^{\mathcal{M}}$ (or $P^{\mathcal{M}'}$), she plays $X = P^{\mathcal{M}'}$ (or $X = P^{\mathcal{M}}$). If Y meets $M \setminus P^{\mathcal{M}}$ (or $M' \setminus P^{\mathcal{M}'}$), she correspondingly plays $X = M' \setminus P^{\mathcal{M}'}$ (or $X = M \setminus P^{\mathcal{M}}$). The point is that both $P^{\mathcal{M}}$ and $M \setminus P^{\mathcal{M}}$ have at least $|M|/3$ elements and the same with $P^{\mathcal{M}'}$ and $M' \setminus P^{\mathcal{M}'}$. Likewise **II** has a winning strategy in $\mathrm{EF}_4^{\exists^{\geq \frac{2}{3}}}(\mathcal{M}, \mathcal{M}')$.

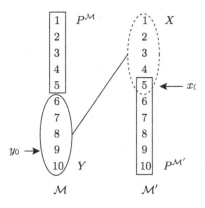

Figure 10.11 **I** wins $\mathrm{EF}_1^{\exists \geq \frac{1}{2}}(\mathcal{M}, \mathcal{M}')$.

Example 10.28 Let us consider two linear orders (see Figure 10.12). Let $L = \{<\}, \#(<) = 2, \mathcal{M} = (\mathbb{Q}, <)$, and $\mathcal{M}' = (\mathbb{R}, <)$. Then **II** has a winning strategy in $\mathrm{EF}_n^{\exists \geq \omega}(\mathcal{M}, \mathcal{M}')$ for all $n \in \mathbb{N}$. The strategy of **II** is the same for all Q-moves: Suppose **I** plays an infinite $Y \subseteq \mathbb{Q}$. Suppose $v_0 < \cdots < v_{i-1}$ have been played in $\mathcal{M}$ and $v_0' < \cdots < v_{i-1}'$ in $\mathcal{M}'$. Suppose first $Y \subseteq M$. Y must meet one of the open intervals

$$(\leftarrow, v_0), (v_0, v_1), \ldots, (v_{i-1}, \rightarrow),$$

say (v_j, v_{j+1}). Player **II** will let $X = (v_j', v_{j+1}')$, which is indeed infinite. Whatever $x \in X$ **I** now plays, **II** can choose $y \in Y$ maintaining the property that the induced relation $\{(v_0, v_0'), \ldots, (v_{i-1}, v_{i-1}'), (y, x)\}$ is a partial isomorphism (see Figure 10.12). The argument is the same if Y meets $(\leftarrow, v_0)$ or $(v_{i-1}, \rightarrow)$, and also if **I** plays $Y \subseteq M'$.

Definition 10.29 We define for L-structures $\mathcal{M}$ and $\mathcal{M}'$

$$\mathcal{M} \simeq_Q^n \mathcal{M}'$$

if player **II** has a winning strategy in the game $\mathrm{EF}_n^Q(\mathcal{M}, \mathcal{M}')$.

Lemma 10.30 $\simeq_Q^n$ *is an equivalence relation among L-structures.*

Proof It is clear that $\simeq_Q^n$ is reflexive and symmetric, so we only prove transitivity. Suppose $\mathcal{M} \simeq_Q^n \mathcal{M}'$ and $\mathcal{M}' \simeq_Q^n \mathcal{M}''$. In order to prove $\mathcal{M} \simeq_Q^n \mathcal{M}''$, suppose $v_0, \ldots, v_{i-1}$ has been played in M and $v_0'', \ldots, v_{i-1}''$ has been played

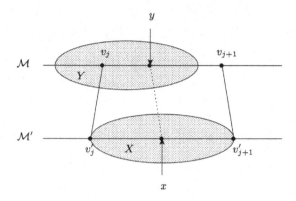

Figure 10.12 **II** wins $\mathrm{EF}^{\geq \omega}(\mathcal{M}, \mathcal{M}')$.

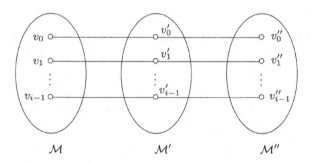

Figure 10.13 Transitivity of $\simeq_Q^n$.

in M'' in such a way that

$$\{(v_0, v_0''), \ldots, (v_{i-1}, v_{i-1}'')\}$$

is a partial isomorphism $\mathcal{M} \to \mathcal{M}''$. Let us assume that in addition, elements $v_0', \ldots, v_{i-1}'$ have been played in M' so that the position

$$p_0 = \{(v_0, v_0'), \ldots, (v_{i-1}, v_{i-1}')\}$$

is a position in $\mathrm{EF}_n^Q(\mathcal{M}, \mathcal{M}')$ while **II** has played her winning strategy. Also we assume that $p_1 = \{(v_0', v_0''), \ldots, (v_{i-1}', v_{i-1}'')\}$ is a position in the game $\mathrm{EF}_n^Q(\mathcal{M}', \mathcal{M}'')$ arising when **II** plays her winning strategy (see Figure 10.13). Now **I** plays $Y \in Q(M)$ in $\mathrm{EF}_n^Q(\mathcal{M}, \mathcal{M}'')$. We continue $\mathrm{EF}_n^Q(\mathcal{M}, \mathcal{M}')$ from position p_0 with this move of **I**.

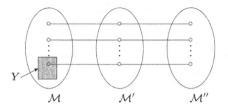

The winning strategy of **II** in $EF_n^Q(\mathcal{M}, \mathcal{M}')$ instructs her to choose $Z \in Q(M')$.

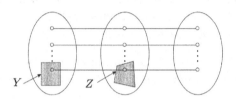

We submit this Z as a move of **I** in the game $EF_n^Q(\mathcal{M}', \mathcal{M}'')$ in position p_1. The winning strategy of **II** in $EF_n^Q(\mathcal{M}', \mathcal{M}'')$ instructs her to play $X \in Q(M'')$.

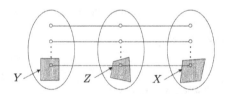

This set X is the response of **II** to Y in the game $EF_n^Q(\mathcal{M}, \mathcal{M}'')$. Next **I** picks $v_i'' \in X$. **II** uses her strategy in $EF_n^Q(\mathcal{M}', \mathcal{M}'')$ to get $v_i' \in Z$ and then her strategy in $EF_n^Q(\mathcal{M}, \mathcal{M}')$ to get $v_i \in Y$ (see Figure 10.14).

Now $\{(v_0, v_0''), \ldots, (v_i, v_i'')\}$ is a partial isomorphism $\mathcal{M} \to \mathcal{M}''$ and **II** has been able to maintain her winning strategy. $\qquad\square$

Note that $EF_n^Q(\mathcal{M}, \mathcal{M}')$ is always determined. There are some special cases in which the game $EF_n^Q(\mathcal{M}, \mathcal{M}')$ is particularly simple. First of all, if $Q(M) = Q(M') = \emptyset$ then **I** can never play the Q-move and the game reduces to the ordinary $EF_n(\mathcal{M}, \mathcal{M}')$. On the other hand, if **I** has a winning strategy in $EF_n(\mathcal{M}, \mathcal{M}')$, he never has to appeal to Q-moves as he wins with the first-

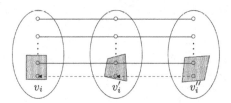

Figure 10.14

order moves alone. If $Q = \exists$, then $\mathrm{EF}_n^Q(\mathcal{M}, \mathcal{M}')$ again reduces to $\mathrm{EF}_n(\mathcal{M}, \mathcal{M}')$ (Exercise 10.36). Of course, the same applies to $Q = \forall$.

An important extreme case is the case that $\mathcal{M} \cong \mathcal{M}'$:

Proposition 10.31 *Suppose Q is a monotone bijection closed generalized quantifier. If $\mathcal{M}$ and $\mathcal{M}'$ are isomorphic L-structures, then Player **II** has a winning strategy in $\mathrm{EF}_n^Q(\mathcal{M}, \mathcal{M}')$.*

Proof The strategy that we will describe can be called, for good reason, the *isomorphism-strategy*. To commence, suppose π is an isomorphism $\mathcal{M} \to \mathcal{M}'$. The strategy of Player **II** is to respond to all moves of Player **I** by applying π. In particular, she makes sure that if a position

$$p = \{(x_0, y_0), \dots, (x_{i-1}, y_{i-1})\}$$

is reached during the game, and

$$f_p = \{(v_0, v_0'), \dots, (v_{i-1}, v_{i-1}')\}$$

is the mapping associated with p, then

$$v_j' = \pi(v_j)$$

for all $j < i$. Suppose now Player **I** makes a Q-move, say $Y \in Q(M)$. Player **II** chooses $X = \pi''Y$ noting that $X \in Q(\mathcal{M}')$ as Q is bijection closed. Next Player **I** chooses $v_i' \in X$. Let $v_i' = \pi(y)$, where $y \in Y$. The strategy of **II** is to choose $v_i = y$. She has been able to maintain the isomorphism strategy. □

Example 10.32 **II** has a winning strategy in

$$\mathrm{EF}_n^{\exists^{\geq \omega}}((\mathbb{R}, <), (\mathbb{R}, >))$$
$$\mathrm{EF}_n^{\exists^{\geq \frac{1}{2}}}(\mathcal{M} \times \mathcal{M}', \mathcal{M}' \times \mathcal{M}).$$

A far more interesting strategy than the above isomorphism strategy is the *invariance* strategy that we now define. First we discuss the topic of invariant subsets in a structure.

Definition 10.33 Suppose $\mathcal{M}$ is an L-structure. A subset X of $\mathcal{M}$ is said to be $\mathcal{M}$-*invariant* if X is closed under all automorphisms of $\mathcal{M}$.

Note that X is not required to be pointwise fixed[2] by automorphisms π of $\mathcal{M}$, merely that if $a \in X$, then also $\pi(a) \in X$. Intuitively, an invariant subset is one which has a description and has no arbitrariness in itself.

Example 10.34 Suppose $\mathcal{M}$ is an L-structure. If $L = \emptyset$, the only invariant subsets of $\mathcal{M}$ are $\emptyset$ and M, which are always invariant anyway. If $\mathcal{M}$ is a finite linear order, every subset of M is invariant (such structures are called *rigid*). If $L = \{c_1, \ldots, c_n\}$, where $c_1, \ldots, c_n$ are constant symbols, the only invariant subsets are the subsets of $\{c_1^{\mathcal{M}}, \ldots, c_n^{\mathcal{M}}\}$ and their complements (Exercise 10.40). If L consists of a unary predicate P and constant symbols $c_1, \ldots, c_n$, the only $\mathcal{M}$-invariant subsets of M are sets of the form

$$(P^{\mathcal{M}} \setminus \{c_i^{\mathcal{M}} : i \in I\}) \cup \{c_i^{\mathcal{M}} : i \notin I\}$$

and their complements, and, of course also subsets of $\{c_1^{\mathcal{M}}, \ldots, c_n^{\mathcal{M}}\}$ and their complements (Exercise 10.41).

Invariant sets can be decomposed into a kind of atoms or building blocks. We call them *orbits*.

Definition 10.35 Suppose $\mathcal{M}$ is an L-structure. The equivalence classes of the equivalence relation

$$xEy \iff \text{there is an automorphism } \pi \text{ of } \mathcal{M} \text{ such that } \pi(x) = y$$

are called *orbits* of $\mathcal{M}$.

It is easy to see that the above relation xEy is indeed an equivalence relation.

Example 10.36 Orbits of a finite linear order are mere singletons. Elements of a structure of the empty vocabulary are all in the same orbit. If $\mathcal{A}$ is a structure of the vocabulary with one unary predicate p, the orbits are $P^{\mathcal{A}}$ (if non-empty) and $A \setminus P^{\mathcal{A}}$ (if non-empty). If $\mathcal{A}$ is a structure of the vocabulary with two unary predicates p_1 and p_2, the possible orbits are $P_1^{\mathcal{A}} \cap P_2^{\mathcal{A}}$, $P_1^{\mathcal{A}} \setminus P_2^{\mathcal{A}}$, $P_2^{\mathcal{A}} \setminus P_1^{\mathcal{A}}$, and $A \setminus (P_1^{\mathcal{A}} \cup P_2^{\mathcal{A}})$ but only the non-empty ones count. The interpretation of any constant symbol is fixed by all automorphisms so they form a

---

[2] A set X is *pointwise fixed* by π if every $a \in X$ satisfies $\pi(a) = a$.

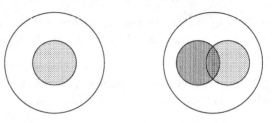

Figure 10.15 Orbits of unary structures.

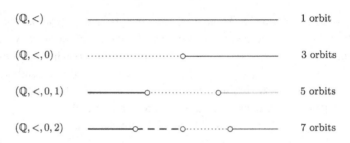

Figure 10.16 Orbits of linear orders with constants.

singleton orbit each. In an equivalence relation all equivalence classes of the same size (cardinality) together form an orbit.

Lemma 10.37 *Suppose $\mathcal{M}$ is an L-structure. A subset of M is $\mathcal{M}$-invariant if and only if it is a union of orbits.*

Proof Every orbit is invariant, hence every union of orbits is invariant. On the other hand if an orbit meets an invariant set, it is included in it. □

Corollary *Suppose $\mathcal{M}$ is an L-structure and $X \subseteq M$. The union of all orbits that meet X is $\mathcal{M}$-invariant. We call it the $\mathcal{M}$-invariant cover of X.*

Definition 10.38 The *invariant Q-Ehrenfeucht–Fraïssé* Game on $\mathcal{M}$ and $\mathcal{M}'$ is like $\mathrm{EF}_n^Q(\mathcal{M}, \mathcal{M}')$ except that when **I** plays a Q-move, he is allowed to choose only sets which are invariant. More exactly, if

$$\{(v_0, v_0'), \ldots, (v_{i-1}, v_{i-1}')\}$$

has been played during the game and **I** chooses $Y \in Q(A)$, it is required that Y is $(\mathcal{M}, v_0, \ldots, v_{i-1})$-invariant. If **I** chooses $Y \in Q(B)$, it is required that Y is $(\mathcal{M}', v_0', \ldots, v_{i-1}')$-invariant.

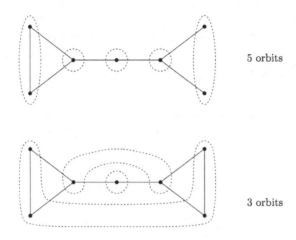

5 orbits

3 orbits

Figure 10.17 Orbits in graphs.

The invariant game is far easier for Player **II** than the original $\text{EF}_n^Q(\mathcal{A}, \mathcal{B})$, as the possible moves of Player **I** are severely restricted. However, for bijection closed generalized quantifiers there is surprisingly no difference:

Theorem 10.39 *Suppose Q is a monotone bijection closed generalized quantifier. Then the following are equivalent:*

*(1) Player **II** has a winning strategy in* $\text{EF}_n^Q(\mathcal{M}, \mathcal{M}')$.
*(2) Player **II** has a winning strategy in the invariant* $\text{EF}_n^Q(\mathcal{M}, \mathcal{M}')$.

Proof The implication from (1) to (2) is trivial. So we assume (2) and prove (1). Suppose τ is a winning strategy of **II** in the invariant game. We describe a winning strategy τ' of **II** in the ordinary, non-invariant game. The idea of **II** is to play the invariant game and the ordinary game simultaneously. She uses invariant covers to overcome the mismatch that in one game **I** plays invariant sets and in the other he plays arbitrary sets. Due to this mismatch we cannot play quite the same elements in the two games.

A partial mapping

$$f = \{(v_0, v_0'), \ldots, (v_{i-1}, v_{i-1}')\}$$

and a partial mapping

$$g = \{(w_0, w_0'), \ldots, (w_{i-1}, w_{i-1}')\}$$

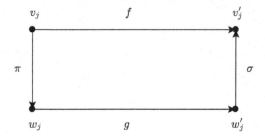

Figure 10.18 f and g are conjugate.

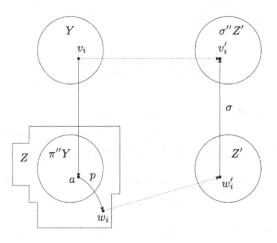

Figure 10.19 The strategy of **II**.

are called *conjugate* if there are an automorphism π of $\mathcal{M}$ and an automorphism σ of $\mathcal{M}'$ such that

$$\pi v_j = w_j$$
$$\sigma w'_j = v'_j$$

for all $j < i$.

The strategy of **II** is to play both games so that the mappings associated to the position of the two games are always conjugate. Suppose she has played in this way, a partial mapping f as above has been played in the non-invariant game and a conjugate partial mapping g as above has been played in the invariant game, and now **I** plays $Y \in Q(M)$ in the non-invariant game. Let π

and σ be as in the definition of conjugation. Since Q is permutation closed, $\pi''Y \in Q(M)$. We would now like to continue the invariant game by letting Player **I** play the set $\pi''Y$. However, this set may not be invariant. So we let Z be the $(M, w_0, \ldots, w_{i-1})$-invariant cover of $\pi''Y$. Since Q is monotone we continue to have $Z \in Q(M)$. Thus this set is a legal move for **I**. The winning strategy τ of **II** gives her a set $Z' \in Q(M')$ in the invariant game. Now comes the moment when we have to decide what is the response of **II** to the move Y by **I** in the non-invariant game. Her response is $X = \sigma''Z'$. Suppose then **I** plays $v_i' \in X$ in the non-invariant game. We know that $w_i' = \sigma^{-1}(v_i')$ is in Z'. Thus it is a legal move for **I** in the invariant game. The winning strategy τ of **II** gives $w_i \in Z$. It would be nice if we had $w_i \in \pi''Y$, but we only know w_i is on an orbit that meets $\pi''Y$. However, it follows from this that there is an automorphism p of $(M, w_0, \ldots, w_{i-1})$ and an element a of $\pi''Y$ such that $p_a = w_i$. Let $v_i \in Y$ be such that $\pi v_i = a$. The element v_i of M is the response of **II** to v_i' in the non-invariant game. Now we should check that the partial mappings

$$f' = \{(v_0, v_0'), \ldots, (v_i, v_i')\}$$

and

$$g' = \{(w_0, w_0'), \ldots, (w_i, w_i')\}$$

are conjugate. Let $\pi' = p \circ \pi$. Then

$$\pi'(v_j) = p\pi(v_i) = pw_j = w_j \text{ for } j < i$$
$$\pi'(v_i) = p\pi(v_i) = pa = w_i$$
$$\sigma(w_j') = v_j' \text{ for } j < i$$
$$\sigma(w_i') = v_i'.$$

Thus π' and σ witness the conjugacy of f' and g'. $\square$

10.4 First-Order Logic with a Generalized Quantifier

We have already defined the concept of a first-order L-formula for any vocabulary L. Now we add to the rules for forming L-formulas

$$\approx tt'$$
$$Rt_1 \ldots t_n$$
$$\neg\varphi$$
$$(\varphi \wedge \psi), (\varphi \vee \psi)$$
$$\forall x_n \varphi, \exists x_n \varphi$$

a new one, namely

$$Qx_n\varphi.$$

Here Q is a new quantifier symbol. If Q is a weak quantifier on M, then we can define

$$\mathcal{M} \models_s Qx_n\varphi \text{ iff } \{a \in M : \mathcal{M} \models_{s[a/x_n]} \varphi\} \in Q.$$

In this way we can associate with every (weak) generalized quantifier Q a logic that we denote by

$$L_{\omega\omega}(Q)$$

and call it the extension of first-order logic by the quantifier Q. When we have fixed the quantifier Q we abuse notation a bit and write

$$Qx_n\varphi \quad \text{for} \quad \mathcal{Q}x_n\varphi.$$

Example 10.40 $\exists^{\geq\frac{1}{2}}x_0 Px_0$ is true in a finite structure $\mathcal{M}$ if and only if $|P^{\mathcal{M}}| \geq |M|/2$. $\exists^{\geq\frac{1}{3}}x_0\exists^{\geq\frac{1}{2}}x_1(x_0 Ex_1)$ is true in a graph if and only if at least a third of the vertices are connected by an edge to at least half of the vertices. $\exists^{\geq\omega}x_0\neg\exists^{\geq\omega}x_1(x_0 \sim x_1)$ is true in an equivalence relation if and only if there are infinitely many finite equivalence classes. $Q^{\mathrm{even}}x_0 Px_0$ is true in a finite model $\mathcal{M}$ if and only if $|P^{\mathcal{M}}|$ is even. $\forall x_0 Q^{\mathrm{even}}x_1(x_0 Ex_1)$ is true in a finite graph if and only if the degree of every vertex is even.

Example 10.41 Suppose L consists of a unary predicate symbol P. Let $\mathcal{M}$ and $\mathcal{M}'$ be L-structures such that $\mathcal{M} \equiv_1^Q \mathcal{M}'$, where Q is a monotone quantifier. Then

$$\mathcal{M} \models Qx_0 Px_0 \quad \text{implies} \quad \mathcal{M}' \models Qx_0 Px_0.$$

For if $\mathcal{M} \models Qx_0 Px_0$, then we can play $\mathrm{EF}_1^Q(\mathcal{M}, \mathcal{M}')$ by letting **I** play $Y = P^{\mathcal{M}}$. Then **II** plays according to her strategy and submits $X \in Q(M')$. We wish to conclude $\mathcal{M}' \models_s Qx_0 Px_0$ so it suffices to prove $X \subseteq P^{\mathcal{M}'}$. So let $a \in X$. Let **I** play a. By the rules of the game, the winning strategy of **II** gives $b \in Y$ such that $\{(b, a)\}$ is a partial isomorphism $\mathcal{M} \to \mathcal{M}'$. As $b \in P^{\mathcal{M}}$, necessarily $a \in P^{\mathcal{M}'}$, as desired.

Example 10.42 Suppose $\mathcal{M} \equiv_1^Q \mathcal{M}'$ and $\mathcal{M} \models Qx_0(x_0 = x_0)$. Then $\mathcal{M}' \models Qx_0(x_0 = x_0)$. Note that $\mathcal{M} \models Qx_0(x_0 = x_0)$ means $M \in Q(M)$. If **I** plays $Y = M$ in $\mathrm{EF}_1^Q(\mathcal{M}, \mathcal{M}')$, the winning strategy of **II** gives $X \in Q(M')$. Since Q is monotone, $M' \in Q(M')$ follows. Thus $\mathcal{M}' \models Qx_0(x_0 = x_0)$.

Example 10.43 Suppose $\mathcal{M}$ and $\mathcal{M}'$ are finite graphs such that $\mathcal{M} \equiv_2^{\exists \geq \frac{1}{3}} \mathcal{M}'$. Then

$$\mathcal{M} \models \forall x_0 \exists^{\geq \frac{1}{3}} x_1 (x_0 E x_1) \quad \Longrightarrow \quad \mathcal{M}' \models \forall x_0 \exists^{\geq \frac{1}{3}} x_1 (x_0 E x_1)$$

that is, if every vertex of $\mathcal{M}$ is connected to at least a third of the vertices of $\mathcal{M}$, then the same is true of $\mathcal{M}'$. Let a be an arbitrary element of M'. Let us play $\text{EF}_2^{\exists \geq \frac{1}{3}}(\mathcal{M}, \mathcal{M}')$. First **I** makes a first-order move: $v_0' = a$. The winning strategy of **II** gives $v_0 \in M$. Now **I** plays the set Y of all neighbors of v_0. Since $\mathcal{M} \models \forall x_0 \exists^{\geq \frac{1}{3}} x_1 (x_0 E x_1)$, $|Y| \geq |M|/3$. The winning strategy of **II** gives $X \subseteq M'$ with $|X| \geq |M'|/3$. The set X must consist of neighbors of a for otherwise **I** could play a non-neighbor $b \in X$ and the winning strategy of **II** would give $c \in Y$ such that

$$cE^{\mathcal{M}}a \quad \Longleftrightarrow \quad bE^{\mathcal{M}'}v_0,$$

a contradiction, as all elements of Y are neighbors of v_0 but c is not a neighbor of a.

Definition 10.44 The *quantifier rank* of a formula φ of $L_{\omega\omega}(Q)$, denoted $\text{QR}(\varphi)$, is defined in the following way (just as for first-order logic): $\text{QR}(\approx tt') = \text{QR}(Rt_1 \ldots t_n) = 0$, $\text{QR}(\neg\varphi) = \text{QR}(\varphi)$, $\text{QR}(\exists x\varphi) = \text{QR}(\forall x\varphi) = \text{QR}(\varphi) + 1$, $\text{QR}((\varphi \wedge \psi)) = \text{QR}((\varphi \vee \psi)) = \max\{\text{QR}(\varphi), \text{QR}(\psi)\}$, $\text{QR}(Qx\varphi) = \text{QR}(\varphi) + 1$. A formula φ is *quantifier free* if $\text{QR}(\varphi) = 0$.

Proposition 10.45 *Suppose L is a finite vocabulary without function symbols. For every n and for every set $\{x_1, \ldots, x_n\}$ of variables, there are only finitely many logically non-equivalent L-formulas in $L_{\omega\omega}(Q)$ of quantifier rank $< n$ with the free variables $\{x_1, \ldots, x_n\}$.*

Proof The proof is exactly like that of Proposition 4.15. $\square$

We can now prove one leg of the Strategic Balance of Logic for generalized quantifiers, namely the marriage of truth and separation:

Theorem 10.46 *Suppose Q is a monotone bijection closed quantifier. The following are equivalent:*

1. $\mathcal{M} \simeq_Q^n \mathcal{M}'$.
2. *The models $\mathcal{M}$ and $\mathcal{M}'$ satisfy the same sentences of $L_{\omega\omega}(Q)$ of quantifier rank $\leq n$.*

Proof We leave (2)→(1) to the reader and prove only (1)→(2). In fact, we prove a slightly more general statement:

- If $(\mathcal{M}, a_1, \ldots, a_m) \simeq^n_Q (\mathcal{M}', b_1, \ldots, b_m)$, then $\mathcal{M} \models \varphi(a_1, \ldots, a_m) \iff \mathcal{M}' \models \varphi(b_1, \ldots, b_m)$, whenever $\varphi(x_1, \ldots, x_m)$ is a formula of $L_{\omega\omega}(Q)$ of quantifier rank $\leq n$.

For $n = 0$ this is trivial. The claim is also trivially preserved by Boolean operations. So let us then assume $\varphi(x_1, \ldots, x_m)$ is $\exists x_0 \psi(x_0, \ldots, x_m)$, where $\psi(x_0, \ldots, x_m)$ has quantifier rank $\leq n$. Suppose $\mathcal{M} \models \exists x_0 \psi(x_0, a_1, \ldots, a_m)$. Then there is $a_0 \in M$ such that

$$\mathcal{M} \models \psi(a_0, a_1, \ldots, a_m).$$

We let player **I** play this element a_0 as his first-order move. The winning strategy of **II** gives $b_0 \in M'$. Now $(\mathcal{M}, a_0, \ldots, a_m) \simeq^n_Q (\mathcal{M}', b_0, \ldots, b_m)$, whence by the induction hypothesis,

$$\mathcal{M}' \models \psi(b_0, b_1, \ldots, b_m).$$

Thus $\mathcal{M}' \models \exists x_0 \psi(x_0, b_1, \ldots, b_m)$. The converse is similar. Let us finally assume $\varphi(x_1, \ldots, x_m)$ is $Qx_0 \psi(x_0, \ldots, x_m)$, where $\psi(x_0, \ldots, x_m)$ has quantifier rank $\leq n$. Suppose $\mathcal{M} \models Qx_0 \psi(x_0, a_1, \ldots, a_m)$. By definition,

$$Y = \{a_0 : \mathcal{M} \models \psi(a_0, a_1, \ldots, a_m)\} \in Q(M).$$

We let player **I** play the set Y as his Q-move. The winning strategy of **II** gives $X \in Q(M')$. We claim that

$$X \subseteq \{b_0 : \mathcal{M}' \models \psi(b_0, b_1, \ldots, b_m)\}. \tag{10.1}$$

If not, then there is some $b_0 \in X$ such that

$$\mathcal{M}' \models \neg\psi(b_0, b_1, \ldots, b_m).$$

We let **I** play this element as the second part of his Q-move. The winning strategy of **II** gives $a_0 \in Y$. By the definition of Y we have

$$\mathcal{M} \models \psi(a_0, a_1, \ldots, a_m).$$

But this contradicts $(\mathcal{M}, a_0, \ldots, a_m) \simeq^n_Q (\mathcal{M}', b_0, \ldots, b_m)$, which follows from the induction hypothesis. Thus (10.1) holds, and

$$\mathcal{M}' \models Qx_0 \psi(x_0, b_1, \ldots, b_m)$$

follows. The converse is similar. $\qquad\square$

Definition 10.47 We say that a generalized quantifier Q is *definable* in terms of the generalized quantifier Q' if there is a sentence φ in $L_{\omega\omega}(Q')$ in the vocabulary $L = \{R\}$, $\#_L(R) = 1$, such that the following holds for all L-structures (M, X):

$$X \in Q(M) \iff (M, X) \models \varphi.$$

Equivalently, the property $R^{\mathcal{M}} \in Q(M)$ of an L-structure $\mathcal{M}$ is expressible in $L_{\omega\omega}(Q')$.

Example 10.48 $Q, -Q, Q-$, and $\check{Q}$ are trivially definable in terms of Q, for example:

$$(M, X) \in \check{Q}(M) \iff (M, X) \models \neg Qx \neg R(x).$$

The infinity quantifier $\exists^{\geq \omega}$ and the finiteness quantifier $\exists^{<\omega}$ are definable in terms of each other. In the finite context the most quantifier and the at-least-one-half quantifier $\exists^{\geq \frac{1}{2}}$ are definable from each other.

Example 10.49 Let $L = \{P\}$, $\#(P) = 1$. The following properties of finite L-structures $\mathcal{M}$ are expressible in $L_{\omega\omega}(\exists^{\geq \frac{1}{2}})$:

1. $|P^{\mathcal{M}}| = |M|/2$.
2. $|P^{\mathcal{M}}| = \lceil |M|/2 \rceil$.

In the first case we use the sentence

$$\exists^{\geq \frac{1}{2}} x_0 P x_0 \wedge \exists^{\geq \frac{1}{2}} x_0 \neg P x_0. \tag{10.2}$$

If $|P^{\mathcal{M}}| = |M|/2$, of course (10.2) holds in $\mathcal{M}$. Conversely, if $\mathcal{M}$ satisfies (10.2), then

$$|P^{\mathcal{M}}| \geq |M|/2, |M \setminus P^{\mathcal{M}}| \geq |M|/2.$$

Since we assume M is finite, we obtain

$$|M|/2 \leq |P^{\mathcal{M}}| \leq |M| - |M|/2 = |M|/2.$$

In the second case we use the sentence

$$\exists^{\geq \frac{1}{2}} x_0 P x_0 \wedge \exists x_1 (P x_1 \wedge \neg \exists^{\geq \frac{1}{2}} x_0 (P x_0 \wedge \neg \approx x_0 x_1)). \tag{10.3}$$

Example 10.50 Let $L = \{P\}$, $\#(P) = 1$. The property

$$|P^{\mathcal{M}}| = |M|/3 \tag{10.4}$$

is *not* expressible in $L_{\omega\omega}(\exists^{\geq \frac{1}{3}})$ (in the finite context). Equivalently, the quantifier $Q(M) = \{X \subseteq M : |X| = |M|/3\}$ is not definable in $L_{\omega\omega}(\exists^{\geq \frac{1}{3}})$. To see why, let $n \in \mathbb{N}$ be arbitrary, $n > 1$. Let $\mathcal{M}_n = (\{1, \ldots, 3n\}, \{1, \ldots, n\})$, so (10.4) holds in each $\mathcal{M}_n$. Let $\mathcal{N}_n = (\{1, \ldots, 3n + 1\}, \{1, \ldots, n + 1\})$. Now (10.4) is false in $\mathcal{N}_n$. So it suffices to prove $\mathcal{M}_n \equiv_n^{\exists^{\geq \frac{1}{3}}} \mathcal{N}_n$. The winning strategy of **II** is simply to play different elements if **I** plays different elements.

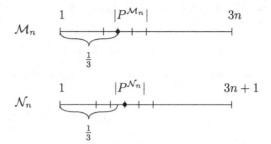

10.5 Ultraproducts and Generalized Quantifiers

In Section 6.10 we introduced the ultraproduct operation on models. This operation now allows us to prove the Compactness Theorem for the logic $L_{\omega\omega}(\exists^{>2^{\omega}})$, that is, the extension of first-order logic by the quantifier

$$Q(M) = \{X \subseteq M : |X| > |\mathbb{R}|\}$$

i.e. the quantifier

"there are more ... than there are reals".

Later we prove the Compactness Theorem also for the quantifier

$$Q(M) = \{X \subseteq M : |X| > |\mathbb{N}|\}$$

i.e. the quantifier

"there are more ... than there are natural numbers".

It is remarkable that compactness fails for the quantifier

"there are at least as many ... as there are natural numbers",

whereas the compactness of the quantifier

"there are at least as many ... as there are reals"

is a famous open problem.

We need some preliminary observations about ultraproducts and ultrafilters.

Lemma 10.51 *If F is an ultrafilter on I and $A \in F$, then*

$$F \restriction A = \{X \cap A : X \in F\}$$

is an ultrafilter on A.

Proof Exercise. □

Lemma 10.52 *If F is an ultrafilter on $\mathbb{N}$ and $A \in F$, then*

$$|\prod_{n \in \mathbb{N}} M_n/F| = |\prod_{n \in A} M_n/(F \restriction A)|.$$

Proof If $[f] \in \prod_n M_n/F$, let

$$\pi([f]) = [f \restriction A].$$

It is easy to see that π is well-defined and a bijection $\prod M_n/F \to \prod_n M_n/F \restriction A$. □

Lemma 10.52 tells us that the size of the ultraproduct is unaltered by restriction to a subset of the indices as long as the subset itself was large, i.e. an element of the ultrafilter. This stability of the cardinality of the ultraproduct is very beneficial for us.

Lemma 10.53 *If $\varphi \in L_{\omega\omega}(\exists^{>2^\omega})$, then*

$$\prod_n M_n/F \models_s \varphi \iff \{n \in \mathbb{N} : M_n \models_{s_n} \varphi\} \in F.$$

Proof We can concentrate on the case $\varphi = Qx_m\psi$. Let

$$X = \{[f] \in \prod M_n/F : \prod_n M_n/F \models_{s[[f]/x_m]} \psi\}$$

and

$$X_n = \{a \in M_n : M_n \models_{s_n[a/x_m]} \psi\}.$$

Then by the induction hypothesis

$$X = \prod_n X_n/F. \tag{10.5}$$

Let $A = \{n \in \mathbb{N} : M_n \models_{s_n} \varphi\}$. We want to prove

$$|X| > 2^\omega \iff A \in F. \tag{10.6}$$

Note that

$$n \in A \iff |X_n| > 2^\omega. \tag{10.7}$$

So if $A \in F$, then by (10.5), (10.7), and Lemma 10.52

$$|X| = |\prod_{n \in A} X_n/(F \restriction A)| > 2^\omega.$$

Conversely, if $A \notin F$, then again by (10.5), (10.7), and Lemma 10.52

$$|X| = |\prod_{n \notin A} X_n/(F \restriction -A)| \leq (2^\omega)^\omega = 2^\omega.$$

We have proved (10.6). □

Theorem 10.54 $L_{\omega\omega}(\exists^{>2^\omega})$ *satisfies the Compactness Theorem in countable vocabularies.*

Proof Suppose $T = \{\varphi_0, \varphi_1, \ldots\}$ is a set of sentences of $L_{\omega\omega}(\exists^{>2^\omega})$ such that each $\{\varphi_0, \ldots, \varphi_n\}$ has a model $\mathcal{M}_n$. Let $\mathcal{M} = \prod_n \mathcal{M}_n/F$. Since F does not contain finite sets,

$$\{n \in \mathbb{N} : \mathcal{M}_n \models \varphi_i\} \supseteq \{i, i+1, i+2, \ldots\} \in F.$$

Hence $\mathcal{M} \models \varphi_i$ for all $i \in \mathbb{N}$. □

If we want to prove compactness for vocabularies of size κ, we have to move to the logic $L_{\omega\omega}(\exists^{\geq\lambda})$, where $\lambda = (\kappa^\omega)^+$. Unlike with first order logic, there is no Compactness Theorem for any $L_{\omega\omega}(\exists^{\geq\lambda})$, which would be valid for all vocabularies.

10.6 Axioms for Generalized Quantifiers

A fundamental property of first-order logic is the fact that there are simple axioms and rules with respect to which the logic is complete, i.e. any set of sentences which is consistent with respect to these axioms and rules has a model. This even extended to countable theories in the logic $L_{\omega_1\omega}$. Mostowski in his original paper Mostowski (1957) asked whether $L_{\omega\omega}(\exists^{\geq\aleph_1})$ can be completely axiomatized. Vaught observed that the set of valid sentences of $L_{\omega\omega}(\exists^{\geq\aleph_1})$ is recursively enumerable, and subsequently Keisler (Keisler (1970)) found a beautifully simple complete axiomatization, which we now present.

We start by establishing criteria for sentences of $L_{\omega\omega}(Q)$ to have a model under various interpretations of Q. The main result is a criterion for a sentence of $L_{\omega\omega}(\exists^{\geq\aleph_1})$ to have a model. Earlier we defined the Model Existence Game $\mathrm{MEG}(T, L)$ and showed that a first-order theory T has a model if and only if **II** has a winning strategy in $\mathrm{MEG}(T, L)$. We define a similar game $\mathrm{MEG}^Q(T, L)$ for $L_{\omega\omega}(Q)$. This is not enough for $L_{\omega\omega}(\exists^{\geq\aleph_1})$ and we have to work harder to get a model in that case.

Every monotone quantifier Q satisfies the condition

(MON) $\qquad \models (\forall x(\varphi \to \psi) \wedge Qx\varphi) \to Qx\psi.$

Conversely, is every generalized quantifier that satisfies (MON) necessarily monotone? Certainly not as (MON) restricts to only definable sets in Q. However, we may reasonably expect that any sentence of $L_{\omega\omega}(Q)$ consistent with (MON) is satisfied in a model where Q is monotone.

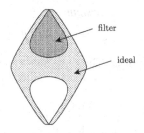

Suppose Q is a weak quantifier on the universe M of a model $\mathcal{M}$. Then we call the pair $(\mathcal{M}, Q)$ a *weak model*. Furthermore, $(\mathcal{M}, Q)$ is

monotone	if Q is monotone on $\mathbf{M}$.
ideal	if $\mathcal{P}(M) \setminus Q$ is an ideal on M.
filter	if Q is a filter on M.

Remark $(\mathcal{M}, Q)$ is a filter model iff $(\mathcal{M}, \check{Q})$ is an ideal model.

Example 10.55 $(\mathcal{M}, \exists^{\geq \omega})$ is an ideal model but not a filter model. $(\mathcal{M}, \forall^{<\omega})$ is a filter model but not an ideal model. Of course, $(\mathcal{M}, \exists)$ is an ideal model and $(\mathcal{M}, \forall)$ is a filter model.

Example 10.56   Every ideal model satisfies (MON) and

(IDE) $Qx(\varphi \vee \psi) \rightarrow (Qx\varphi \vee Qx\psi)$.

Every filter model satisfies (MON) and

(FIL) $(Qx\varphi \wedge Qx\psi) \rightarrow Qx(\varphi \wedge \psi)$.

If $(\mathcal{M}, Q)$ is a weak model, we can define a new weak quantifier as follows

$$\text{Def}(\mathcal{M}, Q) = \{\{a \in M : (\mathcal{M}, Q) \models_{s[a/x]} \varphi\} :$$
$$\varphi \in L_{\omega\omega}(Q), s \text{ assignment into } M\}.$$

Note that $\text{Def}(\mathcal{M}, Q) \cap Q \subseteq Q$ but $\text{Def}(\mathcal{M}, Q) \cap Q$ need not be monotone even if Q is. However, $(\mathcal{M}, Q)$ satisfies (MON), (IDE), or (FIL) if and only if $(\mathcal{M}, \text{Def}(\mathcal{M}, Q) \cap Q)$ does. So as to (MON), (IDE), and (FIL), $\text{Def}(\mathcal{M}, Q) \cap Q$ is the relevant part of Q.

Proposition 10.57   *Suppose $\mathcal{M}$ is an L-structure and Q and Q' are weak quantifiers on M such that*

$$\text{Def}(\mathcal{M}, Q) \cap Q = \text{Def}(\mathcal{M}, Q) \cap Q'. \tag{10.8}$$

Then for all $\varphi \in L_{\omega\omega}(\mathcal{Q})$ *and all assignments* s

$$(\mathcal{M}, Q) \models_s \varphi \iff (\mathcal{M}, Q') \models_s \varphi.$$

Proof We use induction on φ. The only non-trivial case is $\varphi = Qx\psi$. If $(\mathcal{M}, Q) \models_s \varphi$, then

$$X = \{a \in M : (\mathcal{M}, Q) \models_{s[a/x]} \psi\} \in Q.$$

By (10.8), $X \in Q'$. By the induction hypothesis

$$X = \{a \in M : (\mathcal{M}, Q') \models_{s[a/x]} \psi\}.$$

Thus $(\mathcal{M}, Q') \models_s \varphi$. Conversely, suppose $(\mathcal{M}, Q') \models_s \varphi$. Then

$$Y = \{a \in M : (\mathcal{M}, Q') \models_{s[a/x]} \psi\} \in Q'.$$

By (10.8) and the induction hypothesis

$$Y = \{a \in M : (\mathcal{M}, Q) \models_{s[a/x]} \psi\} \in Q.$$

Thus $(\mathcal{M}, Q) \models \varphi$. $\square$

Remark Another consequence of the assumptions of Proposition 10.57 is that

$$\text{Def}(\mathcal{M}, Q) = \text{Def}(\mathcal{M}, Q').$$

Lemma 10.58   *1. If* $(\mathcal{M}, Q)$ *satisfies (MON) and*

$$\overline{Q} = \{X \subseteq M : Y \subseteq X \text{ for some } Y \in \text{Def}(\mathcal{M}, Q) \cap Q\}$$

then $(\mathcal{M}, \overline{Q})$ *is a monotone model and*

$$(\mathcal{M}, Q) \models \varphi \iff (\mathcal{M}, \overline{Q}) \models \varphi$$

for all $\varphi \in L_{\omega\omega}(\mathcal{Q})$.

2. If $(\mathcal{M}, Q)$ *satisfies (MON) and (FIL), then* $(\mathcal{M}, \overline{Q})$ *is a filter model.*
3. If $(\mathcal{M}, Q)$ *satisfies (MON) and (IDE), then*

$$\overrightarrow{Q} = \{X \subseteq M : X \not\subseteq Y \text{ for all } Y \in \text{Def}(\mathcal{M}, Q) \setminus Q\}$$

is an ideal model and

$$(\mathcal{M}, Q) \models \varphi \iff (\mathcal{M}, \overrightarrow{Q}) \models \varphi$$

for all $\varphi \in L_{\omega\omega}(\mathcal{Q})$.

Proof Claims 1 and 2 are easy. To prove 3 it suffices to show that $\overrightarrow{Q}$ is closed under unions. Suppose $X_1, X_2 \notin \overrightarrow{Q}$. Thus there are $Y_1, Y_2 \in \text{Def}(\mathcal{M}, Q) \setminus Q$ such that $X_1 \subseteq Y_1$ and $X_2 \subseteq Y_2$. Since $(\mathcal{M}, Q)$ satisfies (IDE), $Y_1 \cup Y_2 \in \text{Def}(\mathcal{M}, Q) \setminus Q$. Thus $X_1 \cup X_2 \notin \overrightarrow{Q}$. $\square$

Note: Inside the Boolean algebra $\text{Def}(\mathcal{M}, Q)$:

$$\overline{\breve{Q}} = (\overline{Q})\breve{}, \quad \breve{\overline{Q}} = (\overline{Q})\breve{}.$$

Thus inside $\text{Def}(\mathcal{M}, Q)$

$$\overline{Q} = (\breve{\overline{Q}})\breve{}$$

and

$$\overline{Q} = (\overline{\breve{Q}})\breve{}.$$

Now we can conclude that $\varphi \in L_{\omega\omega}(\mathcal{Q})$ has a monotone, ideal, or filter model as long as φ is "consistent" respectively with (MON), (IDE), or (FIL).

Let us call a weak quantifier Q on M *trivial* if $\emptyset \in Q$, *plural* if it contains no singletons, *co-plural* if it contains the complements of singletons, and *bi-plural* if it is both plural and co-plural (i.e. both Q and $\breve{Q}$ are plural). The natural axioms for non-tiviality and plurality are

(NON-TRI) $\neg Qx \neg \approx xx.$

(PLU) $\forall x \neg Qy \approx yx.$

(CO-PLU) $\forall x Qy \neg \approx yx.$

(BI-PLU) $\forall x (\neg Qy \approx yx \wedge Qy \neg \approx yx).$

Thus a weak model is non-trivial, plural, co-plural, or bi-plural if and only if it satisfies respectively (NON-TRI), (PLU), (CO-PLU), or (BI-PLU).

In sum:

(1) φ has a monotone model $\iff$ $\varphi + (MON)$ has a model.
(2) φ has a filter model $\iff$ $\varphi + (MON) + (FIL)$ has a model.
(3) φ has an ideal model $\iff$ $\varphi + (MON) + (IDE)$ has a model.

We shall next introduce a criterion for consistency of sentences of $L_{\omega\omega}(\mathcal{Q})$. For this end we assume that we have a symbol

$$\breve{\mathcal{Q}}$$

for the dual of $\mathcal{Q}$. Thus the meaning of

$$\breve{\mathcal{Q}}x\varphi \quad \text{is} \quad \neg \mathcal{Q}x \neg \varphi.$$

Definition 10.59 The *(Monotone) Model Existence Game* $\text{MEG}^{\mathcal{Q}}(T, L)$ is obtained from the Model Existence Game $\text{MEG}(T, L)$ of $L_{\omega\omega}$ by adding the following rule:

I	II	Explanation
$Qx\varphi(x)$ $\check{Q}y\psi(y)$		I enquires about played sentences $Qx\varphi(x)$ and $\check{Q}y\psi(y)$
	$\varphi(c)$ $\psi(c)$	II chooses $c \in C$

The idea is that $Qx\varphi(x)$ and $\check{Q}x\psi(x)$ can simultaneously hold in a monotone model only if $\exists x(\varphi(x) \wedge \psi(x))$ also holds. Otherwise $\forall x(\varphi(x) \rightarrow \neg\psi(x))$ holds and then $Qx\varphi(x)$ implies $Qx\neg\psi(x)$ contrary to $\check{Q}x\psi(x)$.

Theorem 10.60 (Monotone Model Existence Theorem) *Suppose L is a countable vocabulary and T is a set of L-sentences of $L_{\omega\omega}(Q)$. The following are equivalent:*

1. *T has a monotone model $(\mathcal{M}, Q)$.*
2. *Player **II** has a winning strategy in $\text{MEG}^Q(T, L)$.*

Proof We follow the proof of Theorem 6.59. Let us first assume $(\mathcal{M}, Q) \models T$. The winning strategy of **II** is to keep adding interpretations to constants $c \in C$ while they appear in the game. Let us consider the different cases:

Case 1: $x_n = \varphi \in T$. $(\mathcal{M}, Q) \models \varphi$ by assumption.

Case 2: $x_n = {\approx}tt$. This is always true in $(\mathcal{M}, Q)$.

Case 3: $x_n = \varphi(t), \varphi(c)$ and ${\approx}ct$ have been played, $\varphi(c)$ is basic. Since $(\mathcal{M}, Q) \models \varphi(c) \wedge {\approx}ct$, we have $(\mathcal{M}, Q) \models \varphi(t)$.

Case 4: $x_n = \varphi_i, \varphi_0 \wedge \varphi_1$ has been played and $i \in \{0, 1\}$. Since $(\mathcal{M}, Q) \models \varphi_0 \wedge \varphi_1$, trivially $(\mathcal{M}, Q) \models \varphi_1$.

Case 5: $x_n = \varphi_0 \vee \varphi_1, \varphi_0 \vee \varphi_1$ has been played. Since $(\mathcal{M}, Q) \models \varphi_0 \vee \varphi_1$, player **II** can pick $i \in \{0, 1\}$ such that $(\mathcal{M}, Q) \models \varphi_i$.

Case 6: $x_n = \varphi(c), \forall x\varphi(x)$ has been played. Since $(\mathcal{M}, Q) \models \forall x\varphi(x)$ we can proceed as follows: If c is interpreted in $(\mathcal{M}, Q)$, then $(\mathcal{M}, Q) \models \varphi(c)$. Otherwise we expand $(\mathcal{M}, Q)$ by interpreting c in an arbitrary way. Again $(\mathcal{M}, Q) \models \varphi(c)$.

Case 7: $x_n = \exists x\varphi(x), \exists x\varphi(x)$ has been played. Since $(\mathcal{M}, Q) \models \exists x\varphi(x)$, player **II** can pick a new constant c and interpret it in $(\mathcal{M}, Q)$ in such a way that $(\mathcal{M}, Q) \models \varphi(c)$.

Case 8: $x_n = t$. Player **II** picks a new constant $c \in C$ and interprets it in $(\mathcal{M}, Q)$ so that $(\mathcal{M}, Q) \models \approx ct$.

Case 9: $x_n = Qx\varphi(x), x_{n+1} = \check{Q}y\psi(y)$, both $Qx\varphi(x)$ and $\check{Q}y\psi(y)$ have been played. Thus $(\mathcal{M}, Q) \models Qx\varphi(x) \wedge \check{Q}y\psi(y)$. Since $(\mathcal{M}, Q)$ is monotone, there is an interpretation for a new constant c in such a way that $(\mathcal{M}, Q) \models \varphi(c) \wedge \psi(c)$.

After the game is over, player **II** has won, as no atomic sentence can be both true and false in $(\mathcal{M}, Q)$.

For the converse, we use similar coding as in the proof of Theorem 6.59 in order to describe the enumeration strategy of player **I**. The new case is:

9. If $n = 256 \cdot 3^i \cdot 5^j$, y_i is $Qx\varphi(x)$ and y_j is $\check{Q}y\psi(y)$, then x_n is $Qx\varphi(x)$ and x_{n+1} is $\check{Q}y\psi(y)$.

Let H be the set of responses y_i of player **II** using her winning strategy. Clearly, H is a *Hintikka set* for $L_{\omega\omega}(Q)$, i.e.

1. $\approx tt \in H$ for every constant $L \cup C$-term t.
2. If $\varphi(c) \in H$ and $\approx ct \in H$, then $\varphi(t) \in H$.
3. If $\varphi \wedge \psi \in H$, then $\varphi \in H$ and $\psi \in H$.
4. If $\varphi \vee \psi \in H$, then $\varphi \in H$ or $\psi \in H$.
5. If $\forall x\varphi(x) \in H$, then $\varphi(c) \in H$ for all $c \in C$.
6. If $\exists x\varphi(x) \in H$, then $\varphi(c) \in H$ for some $c \in C$.
7. If $Qx\varphi(x) \in H$ and $\check{Q}y\psi(y) \in H$, then $\varphi(c) \in H$ and $\psi(c) \in H$ for some $c \in C$.
8. For every constant $L \cup C$-term t there is $c \in C$ such that $\approx ct \in H$.
9. There is no atomic sentence φ such that $\varphi \in H$ and $\neg\varphi \in H$.

Let us define in C:

$$c \sim d \iff \approx cd \in H.$$

This is an equivalence relation on H. Let $[c]$ be the equivalence class of c. Let $M = \{[c] : c \in C\}$. Note that

a If $Rc_1 \ldots c_n \in H$ and $c_1 \sim c_1', \ldots, c_n \sim c_n'$, then $Rc_1' \ldots c_n' \in H$.

b If $\approx cfc_1 \ldots c_n \in H$ and $c \sim c', c_1 \sim c_1', \ldots$ and $c_n \sim c_n'$, then $\approx c'fc_1' \ldots c_n' \in H$.

c There is for all $c_1 \ldots c_n$ some c such that $\approx cfc_1 \ldots c_n \in H$.

Thus we can define on M

- $R^{\mathcal{M}} = \{([c_1], \ldots, [c_n]) : Rc_1 \ldots c_n \in H\}$.

- $f^{\mathcal{M}}([c_1], \ldots, [c_n]) = [c]$ for the unique $[c]$ such that $\approx cfc_1 \ldots c_n \in H$.
- $c^{\mathcal{M}} = [c]$.

In order to define a weak quantifier Q on M, let

$$\varphi(x)^H = \{[c] : \varphi(c) \in H\}$$

and

$$Q = \overline{\{\varphi(x)^H : Qx\varphi(x) \in H\}}.$$

We prove by induction on $\varphi(x_1, \ldots, x_n)$ that for all $c_1, \ldots, c_n \in C$:

$$\varphi(c_1, \ldots, c_n) \in H \implies (\mathcal{M}, Q) \models \varphi(c_1, \ldots, c_n).$$

1. φ is basic. The claim is true by construction.
2. $\varphi = \psi \wedge \theta$.

$$
\begin{aligned}
\varphi(c_1, \ldots, c_n) \in H \implies & \psi(c_1, \ldots, c_n) \wedge \theta(c_1, \ldots, c_n) \in H \\
\implies & \psi(c_1, \ldots, c_n) \in H \text{ and } \theta(c_1, \ldots, c_n) \in H \\
\implies & (\mathcal{M}, Q) \models \psi(c_1, \ldots, c_n) \text{ and} \\
& (\mathcal{M}, Q) \models \theta(c_1, \ldots, c_n) \\
\implies & (\mathcal{M}, Q) \models \psi(c_1, \ldots, c_n) \wedge \theta(c_1, \ldots, c_n) \\
\implies & (\mathcal{M}, Q) \models \varphi(c_1, \ldots, c_n).
\end{aligned}
$$

3. $\varphi = \psi \vee \theta$.

$$
\begin{aligned}
\varphi(c_1, \ldots, c_n) \in H \implies & \psi(c_1, \ldots, c_n) \vee \theta(c_1, \ldots, c_n) \in H \\
\implies & \psi(c_1, \ldots, c_n) \in H \text{ or } \theta(c_1, \ldots, c_n) \in H \\
\implies & (\mathcal{M}, Q) \models \psi(c_1, \ldots, c_n) \text{ or} \\
& (\mathcal{M}, Q) \models \theta(c_1, \ldots, c_n) \\
\implies & (\mathcal{M}, Q) \models \psi(c_1, \ldots, c_n) \vee \theta(c_1, \ldots, c_n) \\
\implies & (\mathcal{M}, Q) \models \varphi(c_1, \ldots, c_n).
\end{aligned}
$$

4. $\varphi = \exists x \psi(x)$.

$$
\begin{aligned}
\varphi(c_1, \ldots, c_n) \in H \implies & \exists x \psi(x, c_1, \ldots, c_n) \in H \\
\implies & \psi(c, c_1, \ldots, c_n) \in H \text{ for some } c \in C \\
\implies & (\mathcal{M}, Q) \models \psi(c, c_1, \ldots, c_n) \in H \\
& \text{for some } c \in C \\
\implies & (\mathcal{M}, Q) \models \exists x \psi(x, c_1, \ldots, c_n) \\
\implies & (\mathcal{M}, Q) \models \varphi(c_1, \ldots, c_n).
\end{aligned}
$$

5. $\varphi = \forall x \psi(x)$.

$$
\begin{aligned}
\varphi(c_1, \ldots, c_n) \in H \implies\ & \forall x \psi(x, c_1, \ldots, c_n) \in H \\
\implies\ & \psi(c, c_1, \ldots, c_n) \in H \text{ for all } c \in C \\
\implies\ & (\mathcal{M}, Q) \models \psi(c, c_1, \ldots, c_n) \in H \\
& \text{for all } c \in C \\
\implies\ & (\mathcal{M}, Q) \models \forall x \psi(x, c_1, \ldots, c_n) \\
\implies\ & (\mathcal{M}, Q) \models \varphi(c_1, \ldots, c_n).
\end{aligned}
$$

6. $\varphi = Qx\psi(x)$

$$
\begin{aligned}
\varphi(c_1, \ldots, c_n) \in H \implies\ & Qx\psi(x, c_1, \ldots, c_n) \in H \\
\implies\ & \{[c] : \psi(c, c_1, \ldots, c_n) \in H\} \in Q \\
\implies\ & \{[c] : (\mathcal{M}, Q) \models \psi(c, c_1, \ldots, c_n)\} \in Q \\
\implies\ & (\mathcal{M}, Q) \models Qx\psi(x, c_1, \ldots, c_n) \\
\implies\ & (\mathcal{M}, Q) \models \varphi(c_1, \ldots, c_n).
\end{aligned}
$$

7. $\varphi = \check{Q}x\psi(x)$. If $\varphi(c_1, \ldots, c_n) \in H$ but $(\mathcal{M}, Q) \not\models \varphi(c_1, \ldots, c_n)$, then $(\mathcal{M}, Q) \models Qx\neg\psi(xc_1, \ldots, c_n)$. Thus there is $\theta(y, d_1, \ldots, d_n)$ such that

$$
\{c : (\mathcal{M}, Q) \models \neg\psi(c, c_1, \ldots, c_n)\} \supseteq \{c : \theta(c, d_1, \ldots, d_n)\} \\
\text{and } Qy\theta(y, d_1, \ldots, d_n) \in H. \tag{10.9}
$$

Since H is a Hintikka set, there is $c \in C$ such that

$$
\psi(c, c_1, \ldots, c_n) \in H, \ \theta(c, d_1, \ldots, d_n) \in H.
$$

But this contradicts (10.9).

We conclude the proof by pointing out that $T \subseteq H$, whence $(\mathcal{M}, Q) \models T$. $\qquad\square$

Corollary *Suppose φ and ψ are L-sentences of $L_{\omega\omega}(Q)$. Then the following are equivalent.*

(1) Every monotone model of φ is a model of ψ.
*(2) Player **I** has a winning strategy in $\mathrm{MEG}^Q(\{\varphi, \neg\psi\}, L)$.*

Corollary *Suppose φ is an L-sentence of $L_{\omega\omega}(Q)$. Then φ is valid in every monotone model if and only if player **I** has a winning strategy in the game $\mathrm{MEG}^Q(\{\neg\varphi\}, L)$.*

We can think of winning strategies of **I** in $\mathrm{MEG}^Q(\{\neg\varphi\}, L)$ as *proofs* of φ.

Example 10.61 $Qx\exists yRxy$ follows from $Qx\forall yRxy$ in monotone models.
Here is player **I**'s winning strategy in $\mathrm{MEG}^{\mathcal{Q}}(\{Qx\forall yRxy, \check{Q}x\forall y\neg Rxy\}, \{R\})$:

I	**II**
$Qx\forall yRxy$	
	$Qx\forall yRxy$
$\check{Q}x\forall y\neg Rxy$	
	$\check{Q}x\forall y\neg Rxy$
$Qx\forall yRxy$	
$\check{Q}x\forall y\neg Rxy$	
	$\forall yRcy$
	$\forall y\neg Rcy$
Rcc	
	Rcc
$\neg Rcc$	
	$\neg Rcc$

Example 10.62 $\{Qx\exists yRxy, \check{Q}x\exists y\neg Rxy\}$ has a monotone model. Here is
the winning strategy of player **II**:

I	**II**
$Qx\exists yRxy$	
	$Qx\exists yRxy$
$\check{Q}x\exists y\neg Rxy$	
	$\check{Q}x\exists y\neg Rxy$
$Qx\exists yRxy$	
$\check{Q}x\exists y\neg Rxy$	
	$\exists yRcy$
	$\exists y\neg Rcy$
$\exists yRcy$	
	Rcd
$\exists y\neg Rcy$	
	$\neg Rce$

Theorem 10.63 (Compactness of Monotone Logic)   *Suppose L is a count-able vocabulary and T is a set of L-sentences such that every finite subset of $L_{\omega\omega}(\mathcal{Q})$ of T has a monotone model. Then T has a monotone model.*

Proof   We use a countable set c of new constants and show that **II** has a win-ning strategy in $\mathrm{MEG}^{\mathcal{Q}}(T, L)$. As in the proof of Theorem 6.36 the strategy of player **II** is to maintain the condition that T, added with the sentences she has played so far, forms a "finitely consistent" set. In this case "finitely consistent" means every finite subset has a monotone model. Let us see how she maintains this condition when **I** plays $Qx\varphi(x)$ and $\check{Q}x\psi(x)$. Let $c \in C$ be a new constant. If **II** cannot play $\varphi(c)$ and $\psi(c)$ according to this strategy, there is a finite conjunction θ of played sentences and sentences of T such that $\{\theta, \varphi(c), \psi(c)\}$ has no monotone models. Thus

$$\theta \models \forall x(\varphi(x) \rightarrow \neg\psi(x))$$

in monotone models. But then $\{\theta, Qx\varphi(x), \check{Q}x\psi(x)\}$ cannot have a monotone model, contrary to our assumption. The other cases are exactly as in the proof of Theorem 6.36.   □

Example 10.64   Let $\mathcal{M} = (\mathbb{N}, +, \cdot, 0, 1, <)$ and $Q = \exists^{\geq\omega}(\mathbb{N})$, i.e. $Q = \{X \subseteq \mathbb{N} : X \text{ infinite}\}$. There is $(\mathcal{M}', Q')$ such that $\mathcal{M}' \not\cong \mathcal{M}$ but

$$(\mathcal{M}, \exists^{\geq\omega}) \models \varphi \iff (\mathcal{M}', Q') \models \varphi$$

for all $\varphi \in L_{\omega\omega}(\mathcal{Q})$. So a weak monotone quantifier cannot guarantee that every initial segment is finite. Such an $\mathcal{M}'$ cannot exist if we insist that $Q' = \exists^{\geq\omega}$, too.

Theorem 10.65 (Omitting Types for Monotone Logic)   *Suppose L is a countable vocabulary, T a consistent set of $L_{\omega\omega}(\mathcal{Q})$-sentences (i.e. T has a countable model), and $p_0, p_1, \ldots$ a sequence of non-principal types of T. Then there is a countable model of T which omits each p_n.*

Proof   We proceed as in the proof of Theorem 6.38. Let

$$p_n = \{\varphi_m^n : m \in \mathbb{N}\}.$$

We show that **II** can win against the enumeration strategy of **I** (see the proof of Theorem 10.60) even if player **II** is allowed the following additional move: if the round is $n = 512 \cdot 3^i \cdot 5^j$ player **I** can require player **II** to pick $f_j(i)$ and play $\varphi_{f_j(i)}^j(c_i)$. Her strategy is to keep the set of played sentences all the time consistent. Maintaining this condition throughout the game is accomplished as in the proof of Theorem 10.63. The choice of $f_j(i)$ is exactly as in the proof of Theorem 6.38.   □

Example 10.66   Let $(\mathcal{M}, Q)$ be a monotone model of $\varphi \in L_{\omega\omega}(\mathcal{Q})$. We assume $<^{\mathcal{M}}$ is a linear order, $c_0^{\mathcal{M}} < c_1^{\mathcal{M}} < \ldots$ are cofinal in $\mathcal{M}$, and $c_0^{\mathcal{M}}$ has infinitely many predecessors in $\mathcal{M}$. There is a monotone $(\mathcal{M}', Q') \models \varphi$ such that $c_0^{\mathcal{M}'} < c_1^{\mathcal{M}'} < \ldots$ are cofinal in $\mathcal{M}'$ and some element a below $c_0^{\mathcal{M}'}$ has infinitely many predecessors in $\mathcal{M}'$ (see Figure 10.20).

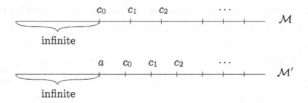

Figure 10.20 Adding elements, preserving cofinality.

Let T consist of all sentences true in $(\mathcal{M}, Q)$ and the axioms

$$d_i < d_{i+1}, \text{ for } i \in \mathbb{N},$$
$$d_i < c < c_0, \text{ for } i \in \mathbb{N},$$
$$c_i < c_{i+1}, \text{ for } i \in \mathbb{N},$$

where $c, d_0, d_1, \ldots$ are new constants. Let p be the following type of T:

$$c_0 < x, c_1 < x, c_2 < x, \ldots.$$

To see that p is a non-principal, suppose $T \cup \{\exists x \theta(x)\}$ is consistent (i.e. T has a monotone model). Thus $(\mathcal{M}, Q) \models \exists x \theta(x)$. Since the elements $c_n^{\mathcal{M}}$ are cofinal in $\mathcal{M}$, we can choose a large enough n such that $(\mathcal{M}, Q) \models \exists x (\theta(x) \wedge \neg c_n < x)$. This ends the proof that p is a non-principal. Clearly any model $(\mathcal{M}', Q')$ satisfying T and omitting p is as desired.

Example 10.66 is a demonstration of an important application of omitting types to extending one definable set while maintaining the cofinal set.

Definition 10.67 A weak model $(\mathcal{M}, Q)$ is an *elementary submodel* of a weak model $(\mathcal{M}', Q')$ if

(1) $M \subseteq M'$.
(2) $(\mathcal{M}, Q) \models_s \varphi \iff (\mathcal{M}', Q') \models_s \varphi$ for all assignments s into M and all $\varphi \in L_{\omega\omega}(\mathcal{Q})$.

Then $(\mathcal{M}', Q')$ is an *elementary extension* of $(\mathcal{M}, Q)$.

Definition 10.68 Weak models $(\mathcal{M}, Q)$ and $(\mathcal{M}', Q')$ are *isomorphic* if there is a bijection $\pi : M \to M'$ such that

(1) $\pi : \mathcal{M} \cong \mathcal{M}'$.
(2) For all $X \subseteq M : X \in Q \iff \pi``X \in Q'$.

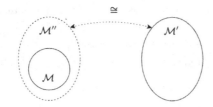

Definition 10.69 Weak models $(\mathcal{M}, Q)$ and $(\mathcal{M}', Q')$ are *elementary equivalent* if for all sentences $\varphi \in L_{\omega\omega}(Q)$

$$(\mathcal{M}, Q) \models \varphi \iff (\mathcal{M}', Q') \models \varphi.$$

Lemma 10.70 *Isomorphic weak models are elementary equivalent.*

Lemma 10.71 *Suppose $(\mathcal{M}, Q)$ is a countable weak model. Let $(\mathcal{M}^*, Q)$ be the expansion of $(\mathcal{M}, Q)$ obtained by taking a new constant symbol for each element of M. Then the following are equivalent for any $(\mathcal{M}', Q')$:*

(1) $(\mathcal{M}^*, Q) \equiv (\mathcal{M}', Q')$.
(2) There is $(\mathcal{M}'', Q'')$ such that $(\mathcal{M}, Q) \prec (\mathcal{M}'', Q'') \cong (\mathcal{M}', Q')$.

Definition 10.72 A sequence

$$(\mathcal{M}_0, Q_0), (\mathcal{M}_1, Q_1), \ldots$$

of weak models is an *elementary chain* if

$$(\mathcal{M}_n, Q_n) \prec (\mathcal{M}_{n+1}, Q_{n+1})$$

for all $n \in \mathbb{N}$. Its *union* is the weak model $(\mathcal{M}, Q)$, where

(1) $M = \bigcup_n M_n$.
(2) $R^{\mathcal{M}} = \bigcup\{R^{\mathcal{M}_n} : n \in \mathbb{N}\}, R \in L$.
(3) $f^{\mathcal{M}} = \bigcup\{f^{\mathcal{M}_n} : n \in \mathbb{N}\}, f \in L$.
(4) $c^{\mathcal{M}} = c^{\mathcal{M}_0}, c \in L$.
(5) $Q = \{X \subseteq M : \text{there is } n \in \mathbb{N} \text{ such that } X \cap M_m \in Q_m \text{ for all } m \geq n.\}$.

Lemma 10.73 (Union Lemma) *The union of an elementary chain is an elementary extension of each element of the chain.*

Proof Let $(\mathcal{M}_n, Q_n)$ and $(\mathcal{M}, Q)$ be as in Definition 10.72. We use induction on φ to prove

$$(\mathcal{M}_n, Q_n) \models_s \varphi \iff (\mathcal{M}, Q) \models_s \varphi$$

whenever $n \in \mathbb{N}$ and s is an assignment into M_n.

$$(10.10)$$

1. φ atomic. Now (10.10) is clear as $\mathcal{M}_n \subseteq \mathcal{M}$ by the definition of $\mathcal{M}$.
2. $\varphi = \neg\psi$. Clearly (10.10) is closed under negation.
3. $\varphi = \psi \wedge \theta$. Clearly (10.10) is closed under conjunction.
4. $\varphi = \exists x\psi$. If $(\mathcal{M}_n, Q_n) \models_s \exists x\psi$, then there is $a \in M_n$ such that

$$(\mathcal{M}_n, Q_n) \models_{s[a/x]} \psi.$$

By the induction hypothesis, $(\mathcal{M}, Q) \models_{s[a/x]} \psi$, whence $(\mathcal{M}, Q) \models_s \exists x\psi$.
Conversely, suppose $(\mathcal{M}, Q) \models_s \exists x\psi$. Choose $a \in M$ with

$$(\mathcal{M}, Q) \models_{s[a/x]} \psi.$$

There is $m \geq n$ such that $a \in M_m$. Now $s[a/x]$ is an assignment into M_m.
By the induction hypothesis, $(\mathcal{M}_m, Q_m) \models_{s[a/x]} \psi$. Thus $(\mathcal{M}_m, Q_m) \models_s$
$\exists x\psi$. But

$$(\mathcal{M}_n, Q_n) \prec (\mathcal{M}_m, Q_m), \tag{10.11}$$

so $(\mathcal{M}_n, Q_n) \models \exists x\psi$.

5. $\varphi = Qx\psi$. Suppose s is an assignment into M_n. Suppose $(\mathcal{M}, Q) \models_s$
$Qx\psi$. Then there is $m \geq n$ such that

$$\{a \in M_k : (\mathcal{M}, Q) \models_{s[a/x]} \psi\} \in Q_k$$

for all $k \geq m$. By the induction hypothesis

$$\{a \in M_k : (\mathcal{M}_k, Q_k) \models_{s[a/x]} \psi\} \in Q_k$$

i.e.

$$(\mathcal{M}_k, Q_k) \models_s Qx\psi$$

for all $k \geq m$. By (10.11)

$$(\mathcal{M}_n, Q_n) \models_s Qx\psi.$$

Conversely, suppose $(\mathcal{M}, Q) \not\models_s Qx\psi$. Then there is $m \geq n$ such that

$$\{a \in M_m : (\mathcal{M}, Q) \models_{s[a/x]} \psi\} \notin Q_m.$$

By the induction hypothesis

$$\{a \in M_m : (\mathcal{M}_m, Q_m) \models_{s[a/x]} \psi\} \notin Q_m$$

i.e.

$$(\mathcal{M}_m, Q_m) \not\models Qx\psi.$$

By (10.11), $(\mathcal{M}_n, Q_n) \not\models Qx\psi$.

$\square$

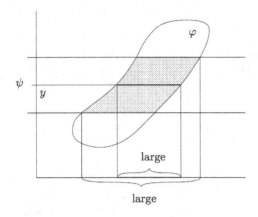

Figure 10.21 Countable-like formula.

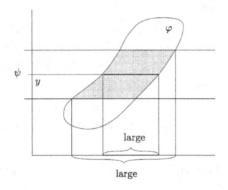

Figure 10.22 Pigeon-Hole Principle.

Definition 10.74 Suppose $(\mathcal{M}, Q)$ is a weak model. A formula $\psi(x)$ is *countable-like* in $(\mathcal{M}, Q)$ if

$$(\mathcal{M}, Q) \models \mathcal{Q}x\exists y(\varphi(x, y) \wedge \psi(y)) \rightarrow \exists y \mathcal{Q}x(\varphi(x, y) \wedge \psi(y)) \qquad (10.12)$$

for all formulas $\varphi(x, y)$. (See Figure 10.21.)

The assumption $\mathcal{Q}x\exists y(\varphi(x, y) \wedge \psi(y))$ says that many values of x are connected by $\varphi(x, y)$ to values of y. By the Pigeon-Hole Principle either there are many values of y or one y is related to many values of x (Figure 10.22).

Lemma 10.75   *If $\psi(x)$ is countable-like in a plural non-trivial $(\mathcal{M}, Q)$, then $(\mathcal{M}, Q) \models \neg Qx\psi(x)$.*

Proof   Suppose $(\mathcal{M}, Q) \models Qx\psi(x)$. Then $(\mathcal{M}, Q) \models Qx\exists y(\approx xy \wedge \psi(y))$. Since $\psi(x)$ is countable-like, $(\mathcal{M}, Q) \models \exists y Qx(\approx xy \wedge \psi(y))$. Thus Q contains $\emptyset$ or a singleton, contrary to assumption.   □

Definition 10.76   *Keisler's Axiom* is the schema

(KA) $Qx\exists y\varphi(x, y) \rightarrow (\exists y Qx\varphi(x, y) \vee Qy\exists x\varphi(x, y)).$

The intuition behind (KA) is that if there are many x which are connected to some y by $\varphi(x, y)$, then either some y is connected to many x or there are many y that are connected to some x (Figure 10.22). This is again an instance of the Pigeon-Hole Principle.

Lemma 10.77   *If $(\mathcal{M}, Q)$ satisfies (KA) and is monotone, then every $\psi(x)$ such that $(\mathcal{M}, Q) \models \neg Qx\psi(x)$ is countable-like.*

Proof   Suppose $(\mathcal{M}, Q) \models Qx\exists y(\varphi(x, y) \wedge \psi(y))$. By (KA), $(\mathcal{M}, Q) \models \exists y Qx(\varphi(x, y) \wedge \psi(y))$ or $(\mathcal{M}, Q) \models Qy\exists x(\varphi(x, y) \wedge \psi(y))$. The latter is impossible by monotonicity, as $(\mathcal{M}, Q) \models \neg Qy\psi(y)$. Thus the former holds.
                                                                               □

In conclusion, for plural non-trivial monotone weak models satisfying (KA), the countable-like formulas are exactly the formulas $\psi(x)$ that satisfy the sentence $\neg Qx\psi(x)$.

Lemma 10.78 (Main Lemma)   *Suppose $(\mathcal{M}, Q)$ is a countable monotone ideal plural non-trivial weak model satisfying (KA) and $\varphi(x)$ is a formula of $L_{\omega\omega}(Q)$ such that*[3]

$$(\mathcal{M}^*, Q) \models Qx\varphi(x).$$

Then there is a weak model $(\mathcal{N}, Q')$ such that

(1) $(\mathcal{M}, Q) \prec (\mathcal{N}, Q')$.
(2) For some $b \in N \setminus M$ we have $(\mathcal{N}, Q') \models \varphi(b)$.
(3) For every countable-like formula with constants from M $\psi(x)$ we have
$$(\mathcal{N}^*, Q) \models \psi(a) \Longrightarrow a \in M.$$

Proof   Let c be a new constant symbol and T the theory

$$\{\theta : (\mathcal{M}^*, Q) \models \theta\} \cup \{\varphi(c)\} \cup \{\neg\psi(c) : (\mathcal{M}^*, Q) \models \neg Qx\psi(x)\}.$$

---

[3] $\mathcal{M}^*$ is a structure obtained from $\mathcal{M}$ by introducing a name for each element of M.

We now prove a very useful criterion:

$T \cup \{\theta(c)\}$ is consistent if and only if $(\mathcal{M}^*, Q) \models Qx(\theta(x) \wedge \varphi(x))$. (10.13)

The meaning of (10.13) is that while $\varphi(x)$ is satisfied by many elements in $(\mathcal{M}^*, Q)$, any definable large subset of $\varphi(\cdot)$ can be consistently extended to a model of T but even more, *only* such definable subsets of φ are extended in any extension of $(\mathcal{M}^*, Q)$ that are large already in $(\mathcal{M}^*, Q)$.

The proof of (10.13) is not difficult. Suppose first $(\mathcal{M}^*, Q) \models Qx(\theta(x) \wedge \varphi(x))$ and $T \models \neg\theta(c)$. Then by the Compactness Theorem there is a finite set T_0 of sentences true in $(\mathcal{M}^*, Q)$ and countable-like $\psi_1(x), \ldots, \psi_n(x)$ such that

$$T_0 \models (\varphi(c) \wedge \theta(c)) \rightarrow (\psi_1(c) \vee \ldots \vee \psi_n(c)).$$

So

$$(\mathcal{M}^*, Q) \models \forall u((\varphi(u) \wedge \theta(u)) \rightarrow (\psi_1(u) \vee \ldots \vee \psi_n(u))).$$

By monotonicity,

$$(\mathcal{M}^*, Q) \models Qu(\varphi(u) \wedge \theta(u)) \rightarrow Qu(\psi_1(u) \vee \ldots \vee \psi_n(u)).$$

Since $(\mathcal{M}^*, Q)$ is ideal, and we assume $(\mathcal{M}^*, Q) \models Qu(\varphi(u) \wedge \theta(u))$, there is an i such that

$$(\mathcal{M}^*, Q) \models Qu\psi_i(u)$$

contrary to the assumption that $(\mathcal{M}^*, Q) \models \neg Qx\psi_i(x)$.

Conversely, suppose $(\mathcal{M}^*, Q) \models \neg Qx(\theta(x) \wedge \varphi(x))$. In this case $\neg(\theta(c) \wedge \varphi(c))$ is in T and therefore trivially $T \models \neg(\theta(c) \wedge \varphi(c))$.

Criterion (10.13) is proved. Now we continue the proof of the Main Lemma. Let us apply the criterion to the sentence $\approx cc$. By assumption, $(\mathcal{M}^*, Q) \models Qx\varphi(x)$, so $(\mathcal{M}^*, Q) \models Qx(\varphi(x) \wedge \approx xx)$. Thus by (10.13) $T \cup \{\approx cc\}$ is consistent. In particular, T itself is consistent.

Let $\psi_1(x), \psi_2(x), \ldots$ be a complete list of all countable-like formulas. Let, for each n, Γ_n be the type

$$\Gamma_n = \{\psi_n(y_n)\} \cup \{\neg \approx y_n c_a : a \in M\}.$$

To prove that each Γ_n is a non-principal type of T, suppose $\exists y_n \theta(y_n, c)$ is consistent with T. By (10.13)

$$(\mathcal{M}^*, Q) \models Qu(\varphi(u) \wedge \exists y_n \theta(y_n, u)).$$

By monotonicity and the ideal property we have

$$(\mathcal{M}^*, Q) \models Qu(\varphi(u) \wedge \exists y_n(\theta(y_n, u) \wedge \psi_n(y_n)))$$

or

$$(\mathcal{M}^*, Q) \models Qu(\varphi(u) \wedge \exists y_n(\theta(y_n, u) \wedge \neg\psi_n(y_n))).$$

In the latter case (10.13) implies that $\exists y_n(\theta(y_n, c) \wedge \neg\psi_n(y_n))$ is consistent with T, and we are done. In the former case

$$(\mathcal{M}^*, Q) \models Qu\exists y_n(\varphi(u) \wedge \theta(y_n, u) \wedge \psi_n(y_n)).$$

Since $\psi_n(y_n)$ is countable-like,

$$(\mathcal{M}^*, Q) \models \exists y_n Qu(\varphi(u) \wedge \theta(y_n, u) \wedge \psi_n(y_n)).$$

Let $a \in M$ be such that

$$(\mathcal{M}^*, Q) \models Qu(\varphi(u) \wedge \theta(c_a, u) \wedge \psi_n(c_a)).$$

By (10.13) $\theta(c_a, u) \wedge \psi_n(c_a)$ is consistent with T. Hence $\exists y_n(\theta(y_n, u) \wedge \approx y_n c_a)$ is consistent with T, and we are done.

By the Omitting Types Theorem there is a countable weak model $(\mathcal{N}, Q')$ of T which omits each Γ_n. W.l.o.g., $(\mathcal{M}, Q) \prec (\mathcal{N}, Q')$. Let b be the value of c in $(\mathcal{N}, Q')$. Since $(\mathcal{M}^*, Q) \models \neg Qx(\approx x c_a)$ for all $a \in M$, it follows that $\neg\approx c c_a \in T$ and so $b \in N \setminus M$. $\square$

Lemma 10.79 (Precise Extensions) *Suppose $(\mathcal{M}, Q)$ is a countable monotone ideal plural non-trivial weak model satisfying (KA). There is a countable elementary extension $(\mathcal{N}, R)$ of $(\mathcal{M}, Q)$ such that for all formulas $\varphi(x)$ of $L_{\omega\omega}(\mathcal{Q})$ of the vocabulary of $\mathcal{M}^*$ the following conditions are equivalent:*

(1) $(\mathcal{M}^*, Q) \models Qx\varphi(x).$
(2) *There is $b \in N \setminus M$ such that $(\mathcal{N}^*, R) \models \varphi(b)$.*

Such a model $(\mathcal{N}, R)$ is called a precise extension *of $(\mathcal{M}, Q)$.*

Proof Let $\varphi_0(x), \varphi_1(x), \ldots$ be a complete list of all $\varphi_i(x)$ in the vocabulary of $(\mathcal{M}^*, Q)$ such that

$$(\mathcal{M}^*, Q) \models Qx\varphi_i(x).$$

We can use Lemma 10.78 (the Main Lemma) to construct an elementary chain

$$(\mathcal{M}_0, Q_0) \prec (\mathcal{M}_1, Q_1) \prec \cdots$$

so that

(3) $(\mathcal{M}_0, Q_0) = (\mathcal{M}, Q).$
(4) *There is $a_n \in M_{n+1} \setminus M_n$ such that $(\mathcal{M}^*_{n+1}, Q_{n+1}) \models \varphi_n(a_n)$.*

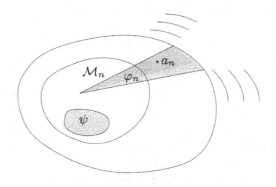

Figure 10.23

(5) If $\psi(x)$ is a countable-like formula of $(\mathcal{M}_n^*, Q_n)$, then

$$(\mathcal{M}_{n+1}^*, Q_{n+1}) \models \psi(a) \implies a \in M_n.$$

Let $(\mathcal{N}, R)$ be the union of this chain. Then by the Chain Lemma, $(\mathcal{M}, Q) \prec (\mathcal{N}, R)$.

Suppose now (1) holds. Then $\varphi(x)$ is $\varphi_n(x)$ for some n. By (4) there is $b \in N \setminus M$ with $(\mathcal{N}^*, R) \models \varphi(b)$. Conversely, if $(\mathcal{M}^*, Q) \models \neg Qx\varphi(x)$, then Lemma 10.78 implies that $\varphi(x)$ is a countable-like formula of each $(\mathcal{M}_n, Q_n)$ and by (5) no new elements satisfy $\varphi(x)$. So (2) fails. $\qquad \square$

Theorem 10.80 (Completeness Theorem for $\exists^{\geq \omega_1}$) *Suppose T is a countable theory in $L_{\omega\omega}(\mathcal{Q})$. Then the following conditions are equivalent:*

(1) *T has a monotone ideal bi-plural non-trivial weak model $(\mathcal{M}, Q)$ which satisfies (KA).*
(2) *T has a model $(\mathcal{M}, \exists^{\geq \omega_1})$.*

Proof (2) $\to$ (1) trivially as $(\mathcal{M}, \exists^{\geq \omega_1})$ satisfies the conditions of (1).

(1) $\to$ (2). We iterate the lemma on precise extensions ω_1 times as follows: Let

$$(\mathcal{M}_0, Q_0) = (\mathcal{M}, Q).$$

If $(\mathcal{M}_\alpha, Q_\alpha)$ is defined, let

$$(\mathcal{M}_\alpha, Q_\alpha) \prec (\mathcal{M}_{\alpha+1}, Q_{\alpha+1})$$

so that $(\mathcal{M}_{\alpha+1}, Q_{\alpha+1})$ is a precise extension of $(\mathcal{M}_\alpha, Q_\alpha)$. If $(\mathcal{M}_\alpha, Q_\alpha)$ is defined for all $\alpha < \beta$ where β is a limit ordinal, then

$$(\mathcal{M}_\beta, Q_\beta) = \bigcup_{\alpha < \beta} (\mathcal{M}_\alpha, Q_\alpha).$$

Our definition for the union of a chain was for a chain $\{(\mathcal{M}_n, Q_n) : n \in \mathbb{N}\}$ only. For an arbitrary chain $(\mathcal{M}_\alpha, Q_\alpha) \prec (\mathcal{M}_\beta, Q_\beta)$ $(\alpha < \beta < \gamma)$ we use the same definition: $\bigcup_{\alpha < \gamma} (\mathcal{M}_\alpha, Q_\alpha) = (\mathcal{M}, Q)$ where $\mathcal{M} = \bigcup_{\alpha < \gamma} \mathcal{M}_\alpha, Q = \{X \subseteq M : \text{ there is } \alpha < \gamma \text{ such that } X \cap M_\beta \in Q_\beta \text{ for } \alpha \le \beta < \gamma\}$. The Union Lemma still holds (Exercise 10.106).

Let $(\mathcal{M}, Q)$ be the union of the elementary chain $(\mathcal{M}_\alpha, Q_\alpha), \alpha < \omega_1$. We prove

$$(\mathcal{M}, Q) \models_s \varphi \iff (\mathcal{M}, \exists^{\ge \omega_1}) \models_s \varphi$$

for all $\varphi \in L_{\omega\omega}(Q)$ and all assignments s. The induction step for Q is the only one to worry about.

Suppose $(\mathcal{M}, Q) \models_s Qx\varphi(x)$. Choose $\alpha < \omega_1$ such that s is an assignment for $(\mathcal{M}_\alpha, Q_\alpha)$. By the Union Lemma, $(\mathcal{M}_\beta, Q_\beta) \models_s Qx\varphi(x)$ for all $\alpha \le \beta < \omega_1$. By the preciseness of the chain there are uncountably many $a \in M$ such that $(\mathcal{M}, Q) \models_{s[a/x]} \varphi(x)$. Thus there are (by the induction hypothesis) uncountably many $a \in M$ such that $(\mathcal{M}, \exists^{\ge \omega_1}) \models_{s[a/x]} \varphi(x)$, whence $(\mathcal{M}, \exists^{\ge \omega_1}) \models Qx\varphi(x)$.

Conversely, suppose $(\mathcal{M}, Q) \not\models_s Qx\varphi(x)$. Assume again s is an assignment for $(\mathcal{M}_\alpha, Q_\alpha)$. By the preciseness of the chain, $(\mathcal{M}_\beta, Q_\beta)$ has no new elements satisfying $\varphi(x)$ over and above those in M_α. Thus neither has $(\mathcal{M}, Q)$. By the induction hypothesis, $(\mathcal{M}, \exists^{\ge \omega_1})$ has no new elements satisfying $\varphi(x)$ over and above those in M_α. Thus $(\mathcal{M}, \exists^{\ge \omega_1}) \not\models Qx\varphi(x)$. $\qquad \square$

Corollary *If T is a countable theory in $L_{\omega\omega}(Q)$, then the following are equivalent:*

(1) T has a model $(\mathcal{M}, \exists^{\ge \omega_1})$.

(2) II has a winning strategy in $\mathrm{MEG}^Q(T^*)$, where T^* consists of T, (IDE), (NON-TRI), (PLU) and (KA).

Corollary (Compactness Theorem for $L_{\omega\omega}(\exists^{\ge \omega_1})$) *Suppose T is a countable theory in $L_{\omega\omega}(\exists^{\ge \omega_1})$ so that every finite subset of T has a model. Then T itself has a model.*

10.7 The Cofinality Quantifier

We prove a Completeness Theorem and a Compactness Theorem for the quantifier

$$\text{"linear order } \varphi \text{ has cofinality } \omega\text{"} \tag{10.14}$$

in a vocabulary of arbitrary cardinality. This is in sharp contrast to quantifiers like $\exists^{\geq \omega_1}$ where we get a Completeness Theorem and a Compactness Theorem in a countable vocabulary only. A linear order $(L, <_L)$ is said to have cofinality ω if there are $a_0 <_L a_1 <_L \cdots$ in L such that

$$\forall b \in L \exists n \in \mathbb{N}(b <_L a_n)$$

i.e. the increasing sequence $a_0, a_1, \ldots$ is *cofinal* in $(L, <_L)$. We can make some immediate observations about ω-cofinal linear orders:

(1) $(L, <_L)$ has no last element.
(2) Every strictly increasing sequence $a_\alpha, \alpha < \omega_1$, in $(L, <_L)$ has an upper bound. (Exercise)
(3) Every linear order without last element has an ω-cofinal sub-order.

The quantifier (10.14) is a new kind of generalized quantifier. Until now our quantifiers have been sets of subsets of the domain. Now we have a set of subsets of the Cartesian product:

$$Q_\omega^{\text{cf}}(M) = \{R \subseteq M \times M : R \text{ is an } \omega\text{-cofinal linear order of } M\}.$$

Note that $Q_\omega^{\text{cf}}(M)$ is by no means monotone, so it is a quite different object from what we are used to.

A *weak cofinality model* is a pair $(\mathcal{M}, Q)$, where $\mathcal{M}$ is an ordinary model and $Q \subseteq \mathcal{P}(M \times M)$. Likewise, we can add a new quantifier symbol $\mathcal{Q}$ to $L_{\omega\omega}$ and define

$$(\mathcal{M}, Q) \models_s \mathcal{Q}xy\varphi(x,y) \iff \{(a,b) : (\mathcal{M}, Q) \models_{s[a/x, b/y]} \varphi\} \in Q.$$

What kind of axioms should $\varphi \in L_{\omega\omega}(\mathcal{Q})$ be consistent with in order to have a model of the form $(\mathcal{M}, Q_\omega^{\text{cf}})$? We have some obvious candidates such as

(LO) $\qquad \mathcal{Q}xy\varphi(x,y) \to \mathcal{Q}^{\text{LO}}xy\varphi(x,y)$, where

$$\mathcal{Q}^{\text{LO}}xy\varphi(x,y) \quad = \quad \forall x \neg \varphi(x,x) \wedge$$
$$\forall x \forall y \forall z((\varphi(x,y) \wedge \varphi(y,z)) \to \varphi(x,z)) \wedge$$
$$\forall x \forall y(\varphi(x,y) \vee \varphi(y,x) \vee \approx xy)$$

and

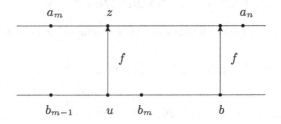

Figure 10.24

(NLE) $Qxy\varphi(x,y) \to \forall x \exists y \varphi(x,y).$

Let us define

$$Q^*xy\varphi(x,y) = Q^{\mathrm{LO}}xy\varphi(x,y) \wedge \forall x \exists y \varphi(x,y) \wedge \neg Qxy\varphi(x,y).$$

Thus $Q^*xy\varphi(x,y)$ "says" that φ is a linear order without last element but the cofinality is not ω. So it is a formalization of

$$Q^{\mathrm{cf}}_{>\omega}(M) = \{R \subseteq M \times M : R \text{ is a linear order of } M \text{ with cofinality} > \omega\}.$$

Let us make some observations about the case

$$R \in Q^{\mathrm{cf}}_{\omega}(M) \quad \& \quad S \in Q^{\mathrm{cf}}_{>\omega}(M). \tag{10.15}$$

First of all we may observe that there is no order-preserving mapping

$$f : (M, <_S) \to (M, <_R)$$

$$x <_S y \to f(x) <_R f(y)$$

whose range is cofinal in $<_S$. Why? Suppose f is one such. Let $a_0 <_R a_1 <_R \cdots$ be cofinal in $<_R$. We define $b_0 <_S b_1 <_S \cdots$ as follows. If $n = 0$ or b_{n-1} is defined let $a_n <_R z = f(n)$. Let $b_{n-1} <_S b_n$ be such that also $n <_S b_n$. Now b_n is defined. Let b be such that $b_n <_S b$ for all n (remember that $S \in Q^{\mathrm{cf}}_{>\omega}(M)$). Let n be such that $f(b) <_R a_n$. Then

$$a_n <_R f(b_n) <_R f(b) <_R a_n,$$

a contradiction.

We now use a similar inference but with a relation instead of a function:

Lemma 10.81   *If (10.15) holds, then there is* no *relation* $T \subseteq M \times M$ *such that*

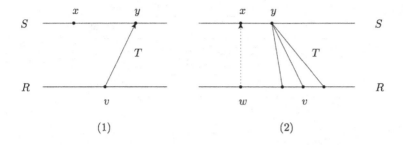

Figure 10.25

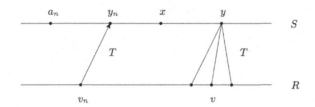

Figure 10.26

(1) $\forall x \exists y >_S x \exists v (vTy)$.

(2) $\forall w \exists x \forall y >_S x \forall v (vTy \to w <_R v)$.

Proof Let $a_0 <_R a_1 <_R \cdots$ be cofinal in $<_R$. We define $b_0 <_S b_1 <_S \cdots$ as follows. If $n = 0$, b_0 is arbitrary. If b_{n-1} is defined choose (by (1)) some $y_n >_S b_n$ and v_n such that $v_n T y_n$. Use (2) to find x such that for all $y >_S x$ and all v, vTy implies $\max(a_n, v_n) <_R v$. Let $b_{n+1} >_S x$ and $v_{n+1} T b_{n+1}$. By (10.15) there is b such that $b_n <_S b$ for all $n \in \mathbb{N}$. By (1) there is $y >_S b$ and vTy. For some n $a_n <_R v$. This is a contradiction. $\qquad\square$

Shelah's Axiom is

(SA) $\quad \neg(\mathcal{Q}xy\psi(x,y) \wedge \mathcal{Q}^*xy\varphi(x,y) \wedge \forall x \exists y \exists v (\theta(v,y) \wedge \varphi(x,y))$
$\qquad\qquad \wedge \forall w \exists x \forall y \forall v ((\varphi(x,y) \wedge \theta(v,y)) \to \psi(w,v)))$.

This may look complicated, but the above lemma shows that it is indeed a very natural axiom for $L_{\omega\omega}(\mathcal{Q})$, being as it is merely a formal statement of the mathematical fact captured by the lemma.

I	II	Explanation
$Qxy\varphi(x,y)$		a played formula
	$Q^{\mathrm{LO}}xy\varphi(x,y)\wedge$ $\forall x\exists y\varphi(x,y)$	
$Q^{*}xy\varphi(x,y)$		a played formula
	$Q^{\mathrm{LO}}xy\varphi(x,y)\wedge$ $\forall x\exists y\varphi(x,y)$	
$Qxy\varphi(x,y)$ $Q^{*}xy\psi(x,y)$		played formulas
	$\varphi(c,d)$ $\neg\psi(c,d)$ $\quad$ or $\neg\varphi(c,d)$ $\psi(c,d)$	
$\varphi\vee\neg\varphi$		φ any sentence
	φ $\neg\varphi$	or

Figure 10.27 $\mathrm{MEG}_{\kappa}^{Q,\mathrm{cf}}(T,L)$.

Definition 10.82 The Model Existence Game $\mathrm{MEG}_{\kappa}^{Q,\mathrm{cf}}(T,L)$ is obtained from the Model Existence Game $\mathrm{MEG}(T,L)$ of $L_{\omega\omega}$ by adding the rules of Figure 10.27.

Theorem 10.83 (Model Existence Theorem for Cofinality Logic) *Suppose L is a vocabulary of cardinality $\leq\kappa$ and T is a set of L-sentences of $L_{\omega\omega}(Q)$. The following are equivalent.*

(1) T has a model $(\mathcal{M},Q)$ satisfying (LO)+(NLE).
(2) Player II has a winning strategy in $\mathrm{MEG}_{\kappa}^{Q,\mathrm{cf}}(T,L)$.

Proof If $(\mathcal{M},Q)\models$ (T) + (LO) + (NLE), then clearly (2) holds. Conversely, suppose (2) holds. We let Player **I** play the obvious enumeration strategy. Let H be the set of responses of **II**, using her winning strategy. By construction, H gives rise to a model of (T) + (LO) + (NLE). Now the details: Let H be the set of responses of **II**, using her winning strategy, to a maximal play of **I**. Let $\mathcal{M}$ be defined from H as before. We define a weak cofinality quantifier Q on $\mathcal{M}$ as follows:

$$Q = \{\{([c],[d]) : \varphi(c,d)\in H\} : Qxy\varphi(x,y)\in H\}.$$

Now we show $(\mathcal{M}, Q) \models T$ by proving the following claim. By our previous work we have

1. $\approx tt \in H$.
2. If $\varphi(c) \in H$ and $\approx ct \in H$ then $\varphi(t) \in H$.
3. If $\varphi \wedge \psi \in H$, then $\varphi \in H$ and $\psi \in H$.
4. If $\varphi \vee \psi \in H$, then $\varphi \in H$ and $\psi \in H$.
5. If $\forall x \varphi(x) \in H$, then $\varphi(c) \in H$ for all $c \in C$.
6. If $\exists x \varphi(x) \in H$, then $\varphi(c) \in H$ for some $c \in C$.

Now we can note further

7. If $\varphi \notin H$, then $\neg \varphi \in H$ ($\neg \varphi$ has to be written in NNF).

The reason for 7 is simply that **I** can play $\varphi \vee \neg \varphi$ whenever he wants.

Claim

$$\varphi \in H \iff M \models \varphi.$$

Proof Note that:

- If $\varphi \in H$ and $\psi \in H$, then $\varphi \wedge \psi \in H$ for otherwise $\neg(\varphi \wedge \psi) \in H$ whence $\neg \varphi \in H$ or $\neg \psi \in H$. This is not possible as then $M \models \varphi \wedge \neg \varphi$ or $M \models \psi \wedge \neg \psi$.
- If $\varphi \in H$ or $\psi \in H$, then $\varphi \vee \psi \in H$ for otherwise $\neg \varphi \in H$ and $\neg \psi \in H$.
- If $\varphi(c) \in H$ for all $c \in C$, then $\forall x \varphi(x) \in H$ for otherwise $\neg \forall x \varphi(x)$, which in NNF is $\exists x \neg \varphi(x)$ is in H, leading to the conclusion that $M \models \varphi(c) \wedge \neg \varphi(c)$ for some c.
- If $\varphi(c) \in H$ for some $c \in C$, then $\exists x \varphi(x) \in H$ for otherwise $\neg \exists x \varphi(x) \in H$, leading to a contradiction.
- If $Qxy\varphi(x, y) \in H$, then $M \models Qxy\varphi(x, y)$, for let $R = \{([c], [d]) : M \models \varphi(c, d)\}$. By the induction hypothesis

$$R = \{([c], [d]) : \varphi(c, d) \in H\}.$$

By construction, $R \in Q(M)$.
- If $Q^*xy\varphi(x, y) \in H$, then $M \models Q^*xy\varphi(x, y)$, for let $R = \{([c], [d]) : M \models \varphi(c, d)\}$. As above, $R = \{([c], [d]) : \varphi(c, d) \in H\}$. By construction, R is a linear order without last element. If $M \models Qxy\varphi(x, y)$, then $R = \{([c], [d]) : \psi(c, d) \in H\}$ for some ψ such that $Qxy\psi(x, y) \in H$. By the rules of the game, there are c and d such that $\varphi(c, d) \in H \leftrightarrow \psi(c, d) \in H$, contrary to the choice of ψ.

- If $M \models \mathcal{Q}xy\varphi(x,y)$ then $\mathcal{Q}xy\varphi(x,y) \in H$, for otherwise $\neg\mathcal{Q}xy\varphi(x,y) \in H$. By induction hypothesis, $M \models \mathcal{Q}xy\varphi(x,y)$ implies

$$\begin{aligned} R &= \{([c],[d]) : \varphi(c,d) \in H\} \\ &= \{([c],[d]) : M \models \varphi(c,d)\} \end{aligned}$$

  is a linear order without last element and $\mathcal{Q}^{\mathrm{LO}}xy\varphi(x,y) \wedge \forall x \exists y \varphi(x,y) \in H$. If $\mathcal{Q}^*xy\varphi(x,y) \in H$, 9 leads to a contradiction. Hence $\neg\mathcal{Q}^*xy\varphi(x,y) \in H$, whence $\mathcal{Q}xy\varphi(x,y) \in H$.
- If $M \models \mathcal{Q}^*xy\varphi(x,y)$, then $\mathcal{Q}^*xy\varphi(x,y) \in H$, for otherwise we have $\neg\mathcal{Q}^*xy\varphi(x,y) \in H$. Since $\mathcal{Q}^{\mathrm{LO}}xy\varphi(x,y) \wedge \forall x \exists y \varphi(x,y) \in H$, we have $\mathcal{Q}xy\varphi(x,y) \in H$. By 8, $M \models \mathcal{Q}xy\varphi(x,y)$, a contradiction.

$\square$

Theorem 10.84 (Weak Compactness of Cofinality Logic) *If T is a set of sentences of $L_{\omega\omega}(\mathcal{Q})$ and every finite subset has a weak cofinality model satisfying (LO) + (NLE), then so does the whole T.*

Proof As in Theorem 10.63. $\square$

Theorem 10.85 (Weak Omitting Types Theorem of Cofinality Logic) *Assume κ is an infinite cardinal. Let L be a vocabulary of cardinality $\leq \kappa$, T an $L_{\omega\omega}(\mathcal{Q})$-theory and for each $\xi < \kappa$, Γ_ξ is a set $\{\varphi_\alpha^\xi(x) : \alpha < \kappa\}$ of $L_{\omega\omega}(\mathcal{Q})$-formulas in the vocabulary L. Assume that*

1. *If $\alpha \leq \beta < \kappa$, then $T \vdash \varphi_\beta^\xi(x) \to \varphi_\alpha^\xi(x)$.*
2. *For every $L_{\omega\omega}(\mathcal{Q})$-formula $\psi(x)$, for which $T \cup \{\psi(x)\}$ is consistent, and for every $\xi < \kappa$, there is an $\alpha < \kappa$ such that $T \cup \{\psi(x)\} \cup \{\neg\varphi_\alpha^\xi(x)\}$ is consistent.*

Then T has a weak cofinality model which omits Γ.

Proof As in Theorem 6.62. $\square$

Definition 10.86 The *union* of an elementary chain $(\mathcal{M}_\alpha, Q_\alpha)$ of weak cofinality models of (LO) + (NLE) is $(\mathcal{M}, Q)$, where $\mathcal{M} = \bigcup_\alpha \mathcal{M}_\alpha$ and

$$Q = \{R \subseteq M \times M : R \text{ is a linear order without last element and}$$
$$\text{there is } \alpha < \kappa \text{ such that } R \cap (M_\beta \times M_\beta) \in Q_\beta \text{ for all } \beta \geq \alpha\}.$$

Lemma 10.87 (Union Lemma) *The union of an elementary chain is an elementary extension of each member of the chain.*

Proof We will do only the case of $\mathcal{Q}xy\varphi(x,y)$. Suppose first $(\mathcal{M}, Q) \models_s$

$Qxy\psi(x, y)$, where s is an assignment into M_α. Then $R \in Q$ where for $a, b \in M$

$$aRb \longleftrightarrow (\mathcal{M}, Q) \models_{s[a/x, b/y]} \psi(x, y).$$

By definition there is $\beta \geq \alpha$ such that $R \cap (M_\gamma \times M_\gamma) \in Q_\gamma$ for $\gamma \geq \beta$. By the induction hypothesis, for $a, b \in M_\gamma$

$$aRb \longleftrightarrow (\mathcal{M}_\gamma, Q_\gamma) \models_{s[a/x, b/y]} \psi(x, y)$$

i.e.

$$(\mathcal{M}_\gamma, Q_\gamma) \models_s Qxy\psi(x, y).$$

By assumption $(\mathcal{M}_\alpha, Q_\alpha) \models_s Qxy\psi(x, y)$. Conversely, suppose $(\mathcal{M}, Q) \not\models_s Qxy\psi(x, y)$. Then for all $\beta \geq \alpha$: $R \cap (M_\beta \times M_\beta) \notin Q_\beta$ where for $a, b \in M$

$$aRb \iff (\mathcal{M}, Q) \models_{s[a/x, b/y]} \psi(x, y).$$

By the induction hypothesis for $a, b \in M_\beta$

$$aRb \iff (\mathcal{M}_\beta, Q_\beta) \models_{s[a/x, b/y]} \psi(x, y)$$

i.e.

$$(\mathcal{M}_\beta, Q_\beta) \not\models_s Qxy\varphi(x, y)$$

and hence $(\mathcal{M}_\alpha, Q_\alpha) \not\models_s Qxy\varphi(x, y)$. $\square$

Lemma 10.88 *For every infinite weak cofinality model $(\mathcal{M}, Q)$ and every $\kappa \geq |M|$ there is $(\mathcal{M}', Q')$ such that $(\mathcal{M}, Q) \prec (\mathcal{M}', Q')$ and every linear order on $\mathcal{M}$, which has no last element and which is $L_{\omega\omega}(\mathcal{Q})$-definable on $\mathcal{M}$ with parameters, has cofinality κ.*

Proof See Exercise 10.108. $\square$

Lemma 10.89 (Main Lemma) *Suppose $(\mathcal{M}, Q)$ is an infinite weak cofinality model of (SA) and $\varphi(x, y)$ is a formula of $L_{\omega\omega}(\mathcal{Q})$ such that $(\mathcal{M}^*, Q) \models Qxy\varphi(x, y)$. Then there is a weak cofinality model $(\mathcal{M}', Q')$ such that*

(1) $(\mathcal{M}, Q) \prec (\mathcal{M}', Q')$.
(2) *For some $b \in M' \setminus M$ we have $(\mathcal{M}', Q') \models \varphi(a, b)$ for all $a \in M$.*
(3) *For every $\psi(x, y)$ such that $(\mathcal{M}, Q) \models Q^* xy\psi(x, y)$ and every $d \in M'$ we have $(\mathcal{M}', Q') \models \psi(d, a)$ for some $a \in M$.*

Proof In the light of Lemma 10.88 may assume, without loss of generality, that the model $\mathcal{M}$ and the vocabulary L have an infinite cardinality κ, and every linear order on $\mathcal{M}$ which has no last element and which is $L_{\omega\omega}(\mathcal{Q})$-definable

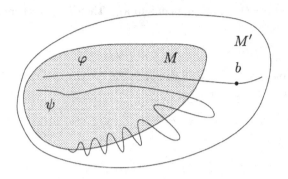

Figure 10.28

on $\mathcal{M}$ with parameters, has cofinality κ. Let c be a new constant symbol and T the theory

$$\{\theta : (\mathcal{M}^*, Q) \models \theta\} \cup \{\varphi(a, c) : a \in M\}.$$

The useful criterion, familiar from the proof of Lemma 10.78, is in this case very simple:

$$T \cup \{\theta(c)\} \text{ is consistent iff } (\mathcal{M}^*, Q) \models \forall x \exists y (\varphi(x, y) \wedge \theta(y)). \quad (10.16)$$

Proof of (10.16). Suppose $(\mathcal{M}^*, Q) \models \forall x \exists y (\varphi(x, y) \wedge \theta(y))$. Let $T_0 \subseteq T$ be finite. Let $a_0, \ldots, a_n$ be the constants occurring in T_0. Let b be φ-above every a_i in $\mathcal{M}$. By assumption there is d such that $(\mathcal{M}^*, Q) \models \varphi(b, d) \wedge \theta(d)$. Thus $T_0 \cup \{\theta(c)\}$ is consistent. Hence $T \cup \{\theta(c)\}$ is consistent. Conversely, suppose $(\mathcal{N}^*, Q') \models T \cup \{\theta(c)\}$. If $a \in M$, then $(\mathcal{N}^*, Q') \models \varphi(a, c) \wedge \theta(c)$, so $(\mathcal{M}^*, Q) \models \exists y (\varphi(a, y) \wedge \theta(y))$. $\square$

Since $(\mathcal{M}^*, Q) \models \forall x \exists y (\varphi(x, y) \wedge \approx yy)$, we conclude that T itself is consistent. Let $\psi_\xi(x, y)$, $\xi < \kappa$, be a complete list of all formulas such that

$$(\mathcal{M}^*, Q) \models Q^* xy \psi_\xi(x, y).$$

Let w_α^ξ, $\alpha < \kappa$, be a cofinal strictly ψ_ξ-increasing sequence in $\mathcal{M}$. Let, for each ξ, Γ_ξ be the type

$$\Gamma_\xi = \{\psi_n(w_\alpha^\xi, x) : \alpha < \kappa\}.$$

Thus Γ_n "says" that x is $<_{\psi_\xi}$-above every element of M. This is the situation we want to avoid, so we want to omit each type Γ_ξ. To prove using Theorem 10.85 that all the sets Γ_ξ can be simultaneously omitted suppose $\exists x \theta(x, c)$

is consistent with T. By (10.16)

$$(\mathcal{M}^*, Q) \models \forall x \exists y \exists v (\theta(v, y) \wedge \varphi(x, y)).$$

If there is no $\alpha < \kappa$ such that

$$\exists y (\theta(y, c) \wedge \neg \psi_\xi(w_\alpha^\xi, y))$$

is consistent with T, then for all $\alpha < \kappa$ (by (10.16))

$$(\mathcal{M}^*, Q) \models \exists x \forall y \forall v ((\varphi(x, y) \wedge \theta(v, y)) \rightarrow \psi_\xi(w_\alpha^\xi, v))$$

i.e.

$$(\mathcal{M}^*, Q) \models \forall w \exists x \forall y \forall v ((\varphi(x, y) \wedge \theta(v, y)) \rightarrow \psi_\xi(w, v))$$

contrary to $(\mathcal{M}^*, Q) \models$ (SA).

By the Omitting Types Theorem there is a countable weak cofinality model $(\mathcal{M}', Q')$ of T which omits each Γ_ξ. This is clearly as required. □

Lemma 10.90 (Precise Extension Lemma) *Suppose $(\mathcal{M}, Q)$ is an infinite weak cofinality model satisfying (SA). There is an elementary extension $(\mathcal{N}, R)$ of $(\mathcal{M}, Q)$ such that for all formulas $\varphi(x, y)$ of $L_{\omega\omega}(Q)$ of the vocabulary of $\mathcal{M}^*$ the following are equivalent:*

(1) $(\mathcal{M}^*, Q) \models Qxy\varphi(x, y)$.
(2) $(\mathcal{M}^*, Q) \models Q^{\text{LO}} xy\varphi(x, y) \wedge \forall x \exists y \varphi(x, y)$ *and there is $b \in N \setminus M$ such that $(\mathcal{N}^*, R) \models \varphi(a, b)$ for all $a \in M$.*

Such $(\mathcal{N}, R)$ is called a precise extension *of $(\mathcal{M}, Q)$.*

Proof Let $\varphi_0(x, y), \varphi_1(x, y)$ list all $\varphi(x, y)$ with $(\mathcal{M}^*, Q) \models Qxy\varphi(x, y)$. By the Main Lemma there is an elementary chain

$$(\mathcal{M}_0, Q_0) \prec (\mathcal{M}_1, Q_1) \prec \cdots$$

such that

(3) $(\mathcal{M}_0, Q_0) = (\mathcal{M}, Q)$.
(4) There is $b_n \in M_{n+1} \setminus M_n$ such that $(\mathcal{M}_{n+1}^*, Q_{n+1}) \models \varphi_n(a, b_n)$ for all $a \in M_n$
(5) If $(\mathcal{M}_n^*, Q_n) \models Q^* xy\varphi(x, y)$, then for all $b \in M_{n+1}$ there is $a \in M_n$ such that $(\mathcal{M}_{n+1}^*, Q_{n+1}) \models \psi(b, a)$.

Let $(\mathcal{N}, R)$ be the union of this chain. Then by the Union Lemma $(\mathcal{M}, Q) \prec (\mathcal{N}, R)$. Conditions (1) and (2) clearly hold. □

Theorem 10.91 (Completeness Theorem for Cofinality Logic) *Suppose T is a theory in $L_{\omega\omega}(Q)$. Then the following conditions are equivalent:*

(1) T *has a model* $(\mathcal{M}, Q_\omega^{\text{cf}})$.

(2) T *has a weak cofinality model satisfying (SA).*

(3) $T \cup \{(LO)\} \cup \{(NLE)\} \cup \{(SA)\}$ *is consistent.*

Proof To prove (3) → (1) we start with an $\aleph_1$-saturated model $(\mathcal{M}, Q)$ of $T \cup \{(\text{LO})\} \cup \{(\text{NLE})\} \cup \{(\text{SA})\}$. Thus in $(\mathcal{M}, Q)$ every definable linear order without a last element has uncountable cofinality. We then iterate the lemma on precise extensions ω times:

$$(\mathcal{M}, Q) = (\mathcal{M}_0, Q_0) \prec \cdots \prec (\mathcal{M}_n, Q_n) \prec \cdots \quad (n < \omega).$$

Let $(\mathcal{N}, R)$ be the union of this chain. We prove

$$(\mathcal{N}, R) \models_s \varphi \iff (\mathcal{N}, Q_\omega^{\text{cf}}) \models_s \varphi$$

for all φ and s. Suppose first $(\mathcal{N}, R) \not\models_s Qxy\varphi(x, y)$. Suppose s is into M_n. W.l.o.g., $(\mathcal{N}, R) \models_s Q^{\text{LO}}xy\varphi(x, y) \wedge \forall x \exists \varphi(x, y)$. Then $(\mathcal{N}, R) \models_s Q^* xy\varphi(x, y)$, whence $(\mathcal{M}_n, Q_n) \models_s Q^* xy\varphi(x, y)$. Now $\varphi^{(\mathcal{N}, R)}$ is cofinal with $\varphi^{(\mathcal{M}_n, Q_n)}$. Thus $(\mathcal{N}, Q_\omega^{\text{cf}}) \models_s Q^* xy\varphi(x, y)$.

Conversely, suppose $(\mathcal{N}, R) \models_s Qxy\varphi(x, y)$. Assume s is again into M_n. Thus $(\mathcal{N}, R) \models_s Qxy\varphi(x, y)$. By virtue of the Union Lemma $(\mathcal{M}_n, Q_n) \models_s Qxy\varphi(x, y)$. At every step $(\mathcal{M}_m, Q_m), n < m < \omega$, a new element is put into φ after the old ones. Thus the ordering φ has in $(\mathcal{N}, R)$ a cofinal ω-sequence. Hence $(\mathcal{N}, Q_\omega^{\text{cf}}) \models_s Qxy\varphi(x, y)$. □

Theorem 10.92 (Compactness Theorem for Cofinality Logic) *Suppose T is a theory in $L_{\omega\omega}(Q_\omega^{\text{cf}})$ so that every finite subset of T has a model. Then T itself has a model.*

We have proved the Compactness Theorem for $L_{\omega\omega}(Q_\omega^{\text{cf}})$. What is remarkable is that the Compactness Theorem holds also for uncountable languages, unlike in the case of the logic $L_{\omega\omega}(\exists^{\geq\omega_1})$. Therefore $L_{\omega\omega}(Q_\omega^{\text{cf}})$ is said to be *fully compact*. It is the best known fully compact logic.

The Compactness Theorem of cofinality logic opens the possibility to develop a theory of models which emphasizes the order type of the model $(A, <, \ldots)$ rather than the cardinality of the set A. A suggestion to study model theory from this angle is Open Problem 23 in the classic model theory book Chang and Keisler (1990).

10.8 Historical Remarks and References

Generalized quantifiers were introduced in Mostowski (1957), and in a more general sense in Lindström (1966). For more recent developments in the area of generalized quantifiers, see Barwise and Feferman (1985), Krynicki and Szczerba (1995), and Väänänen (Ed.) (1999). While generalized quantifiers were originally conceived as important in expressing mathematical concepts such as infinity and uncountability, they became subsequently popular in computer science and in linguistics. In computer science the relevant framework is the framework of finite models. An early study of generalized quantifiers on finite models is Hájek (1976). The topic was explicitly studied in van Benthem (1984), where a version of the number triangle first appears, and in Kolaitis and Väänänen (1995). Currently generalized quantifiers are a standard tool in computer science logic.

The study of interdefinability of generalized quantifiers leads quickly to difficult problems. It has turned out that Ramsey theory can help to solve some of these problems, see Kolaitis and Väänänen (1995), Luosto (2000), and Nešetřil and Väänänen (1996).

The Härtig quantifier (Exercise 10.91) is from Härtig (1965). See Herre et al. (1991) for a survey.

An early paper on the role of generalized quantifiers in linguistics is Barwise and Cooper (1981). A recent textbook on generalized quantifiers in natural language is Peters and Westerståhl (2008). The concept of a smooth quantifier is from Väänänen and Westerståhl (2002).

The Completeness Theorem for $\exists^{\geq \omega_1}$ in the current form is due to Keisler (1970). Previously it was just known that a complete axiomatization exists (Vaught (1964), using Fuhrken (1964)). The first known fully compact generalized quantifier, i.e. one that satisfies the Compactness Theorem in arbitrary vocabularies, was the cofinality quantifier Q_κ^{cf}. This quantifier and the above axiom (SA) were introduced by Shelah (1975). In Makowsky and Shelah (1981) a proof of Theorem 10.92 based on ultraproducts is given, see also (Barwise and Feferman, 1985, p. 48). For more on cofinality quantifiers, see Mekler and Shelah (1986), Mildenberger (1992), and Shelah (1985).

Stationary logic is an interesting axiomatizable logic in which both $\exists^{\geq \omega_1}$ and Q_ω^{cf} are definable, see Barwise et al. (1978).

Exercise 10.15 is due to E. Keenan. Exercise 10.105 is due to R. Robinson see Morley and Vaught (1962). Exercise 9.56 is from Hella (1989). Exercise 10.89 is from Väänänen (1977) and Caicedo (1980). Exercise 10.90 is from Hella and Sandu (1995).

Exercises

10.1 What are the duals of $[\{5\}]$, $[X_0]$, and $[X_0]^*$?

10.2 Prove $(-Q)- = -(Q-)$, $(-Q)\check{} = -\check{Q}$ and $(Q-)\check{} = \check{Q}-$.

10.3 Prove Lemma 10.7.

10.4 Find the dual, the complement,, and the postcomplement of the quantifier "some but not all".

10.5 Find the dual, the complement and the postcomplement of the quantifier "between 10% and 20%".

10.6 What are the duals of Q^{even} and $\exists^{\geq \frac{1}{2}}$?

10.7 What is the dual of $\exists^{\text{most}}$?

10.8 What is the dual of $\exists^{\geq r}$?

10.9 Show that $\exists^{\geq \omega}$ and $\forall^{<\omega}$ have no atoms on any infinite domains.

10.10 If we know the atoms of a weak monotone quantifier Q on a finite domain A, how can we find the atoms of $\check{Q}$?

10.11 Show that the set of atoms of Q_C^{mon} is C.

10.12 Show that there are at least $2^{\binom{n}{n/2}}$ weak monotone quantifiers on a domain of even size n.

10.13 Show that there are exactly $2^{2^{|A|}}$ weak monotone quantifiers on any infinite domain A.

10.14 Prove that the dual of a permutation closed quantifier is permutation closed.

10.15 Show that there are exactly 2^{n+1} self-dual permutation closed quantifiers on a domain of size $2n + 1$.

10.16 Show that there are exactly 2^{n+1} permutation closed quantifiers on a domain of size $n > 0$.

10.17 Show that there are exactly $n + 2$ permutation closed monotone quantifiers on a domain of size $n > 0$.

10.18 (Finite context) Prove that for monotone bijection closed Q always
$$f_Q(n) + f_{\check{Q}}(n) = n + 1.$$

10.19 (Finite context) Show that there are uncountably many monotone bijection closed quantifiers Q. Can they be chosen so that $f_Q(n) \leq 1$ for all $n \in \mathbb{N}$?

10.20 (Finite context) Draw the number triangles of the quantifiers $\{X \subseteq A : |X| \geq m \cdot |A|/n\}$, $\{X \subseteq A : |X| \geq \sqrt{|A|}\}$, and $\{X \subseteq A : |X| \geq \log |A|\}$ and show that they are all unbounded.

10.21 (Finite context) Show that if Q is bounded but not eventually counting, then there are finite sets S_0 and S_1 and $m \in \mathbb{N}$ such that

(1) For $n \geq m$, $f_Q(n) \in S_0$ or $f_{\check{Q}}(n) \in S_1$.
(2) If $a \in S_0$, then $f_Q(n) = a$ for infinitely many n.
(3) If $a \in S_1$, then $f_{\check{Q}}(n) = a$ for infinitely many n.

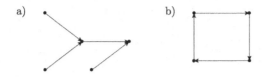

Figure 10.29

10.22 (Finite context) Show that a monotone bijection closed Q is unbounded if and only if $\overline{\lim}_{n\to\infty} f_Q(n) = \infty$ and $\overline{\lim}_{n\to\infty} f_{\breve{Q}}(n) = \infty$.

10.23 (Finite context) A monotone bijection closed quantifier Q is *smooth* if $f_Q(n) \le f_Q(n+1) \le f_Q(n)+1$ for all $n \in \mathbb{N}$. Show that the quantifier "for m out of every n" (i.e. $\exists^{\ge m/n}$) is smooth for all m and n such that $1 \le m < n$.

10.24 (Finite context) Show that a smooth quantifier is always eventually counting or unbounded.

10.25 (Finite context) Show that if Q is a smooth unbounded bijection closed quantifier, then $\forall k \exists m \forall n > m(k < f_Q(n) < n - k)$.

10.26 (Finite context) Show that a monotone bijection closed quantifier Q is smooth if and only if both f_Q and $f_{\breve{Q}}$ are non-decreasing.[4]

10.27 (Finite context) Show that there are unbounded monotone bijection closed quantifiers Q and Q' such that $Q \cup Q'$ is not unbounded.

10.28 Find the orbits of a complete graph K_n and a complete bi-partite graph $K_{n,m}$.

10.29 Find the orbits of $(\mathbb{N}, +, \cdot, 0, 1)$, $(\mathbb{Z}, <)$, and $(\mathbb{Z}, <, 0)$.

10.30 Find the orbits of $(\mathbb{Z} + \mathbb{Z}, <)$.

10.31 Give a 3-regular[5] graph with five orbits.

10.32 Find the orbits of the directed graphs of Figure 10.29.

10.33 Find the orbits of an arbitrary finite Boolean algebra.

10.34 What are the orbits of $(\mathbb{R}, <)$? What about $(\mathbb{R}, <, \mathbb{Q})$?

10.35 Find the orbits of a given finite chain.

10.36 Show that Player **II** has a winning strategy in $\mathrm{EF}_n^{\exists}(\mathcal{M}, \mathcal{M}')$ if and only if she has a winning strategy in $\mathrm{EF}_n(\mathcal{M}, \mathcal{M}')$.

10.37 Show that the $\mathcal{M}$-invariant subsets of M form a Boolean algebra under $\subseteq$.

---

[4] A function $f : \mathbb{N} \to \mathbb{N}$ is *non-decreasing* if $f(n) \le f(n+1)$ for all $n \in \mathbb{N}$.

[5] A graph is *n-regular* if every node has degree n.

Figure 10.30

10.38 Suppose Q is monotone. Prove: **II** (or **I**) has a winning strategy in the game $\text{EF}_n^Q(\mathcal{A}, \mathcal{B})$ if and only if **II** (respectively **I**) has a winning strategy in $\text{EF}_n^{\check{Q}}(\mathcal{A}, \mathcal{B})$.

10.39 Suppose L consists of a unary predicate and $\mathcal{A}$ is an L-structure. Show that the $\mathcal{A}$-invariant subsets of A are $\emptyset$, $P^{\mathcal{A}}$, $A \setminus P^{\mathcal{A}}$, and A.

10.40 Suppose L consists of constant symbols $c_1, \ldots, c_n$ and $\mathcal{A}$ is an L-structure. Show that the subsets of A that are invariant are the subsets of $\{c_1^{\mathcal{A}}, \ldots, c_n^{\mathcal{A}}\}$ and their complements.

10.41 Suppose L is a vocabulary consisting of a unary predicate P and constant symbols $c_1, \ldots, c_n$. Show that if $\mathcal{A}$ is an L-structure, the invariant subsets of A are of the form

$$(P^{\mathcal{A}} \setminus \{c_i^{\mathcal{A}} : i \in I\}) \cup \{c_i^{\mathcal{A}} : i \notin I\}$$

or

$$\{c_i^{\mathcal{A}} : i \in I\}$$

or their complements, where $I \subseteq \{1, \ldots, n\}$.

10.42 Suppose L consists of unary predicates $P_1, \ldots, P_n$ and constant symbols $c_1, \ldots, c_m$. Describe the invariant subsets of an arbitrary L-structure.

10.43 Suppose $\mathcal{M} = (M, \sim)$ is an equivalence relation and $a_1, \ldots, a_n \in M$. Describe the $(M, \sim, a_1, \ldots, a_n)$-invariant subsets of M.

10.44 Describe the orbits of the trees of Figure 10.30.

10.45 Describe the invariant subsets of the trees of Figure 10.31.

10.46 Suppose $\mathcal{M}$ is a well-order. Show that the orbits of $\mathcal{M}$ are singletons.

10.47 Suppose $\mathcal{M}$ is a well-founded tree. Show that all elements in the same orbit have the same height.

10.48 Describe the invariant subsets of an arbitrary successor structure.

10.49 Let $L = \{P\}$, $\#(P) = 1$. What would be a good first move for player **I** in $\text{EF}^{\geq \frac{1}{3}}(\mathcal{M}, \mathcal{M}')$ if $\mathcal{M}$ and $\mathcal{M}'$ are the L-structures of Figure 10.32?

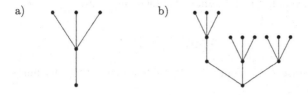

Figure 10.31

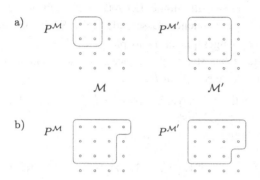

Figure 10.32

10.50 Suppose $L = \emptyset$. Let $\mathcal{M}$ and $\mathcal{M}'$ be L-structures such that $\mathcal{M}$ has 30 elements and $\mathcal{M}'$ has 20 elements. Show that $\mathcal{M} \equiv_{10}^{\exists \geq \frac{1}{2}} \mathcal{M}'$ but $\mathcal{M} \not\equiv_{11}^{\exists \geq \frac{1}{2}} \mathcal{M}'$.

10.51 Suppose L consists of one unary predicate symbol P. Let $\mathcal{M}$ and $\mathcal{M}'$ be L-structures such that $|M| = |M'| = 100$, $|P^{\mathcal{M}}| = 10$ and $|P^{\mathcal{M}'}| = 20$. Show that $\mathcal{M} \equiv_{10}^{\exists \geq \frac{1}{2}} \mathcal{M}'$ but $\mathcal{M} \not\equiv_{11}^{\exists \geq \frac{1}{2}} \mathcal{M}'$.

10.52 Let $P_n = \{0, \ldots, n\}$. Suppose $k \in \mathbb{N}$. For which m and n do we have $(\mathbb{N}, P_n) \equiv_k^{\geq \omega} (\mathbb{N}, P_m)$?

10.53 Suppose Q is a bounded monotone quantifier. Show that for all $m \in \mathbb{N}$ there is $n \in \mathbb{N}$ such that

$$(\{0, \ldots, 2n\}, \{0, \ldots, n\}) \equiv_k^Q (\{0, \ldots, 2n\}, \{0, \ldots, n+1\}).$$

10.54 Suppose Q is a bounded monotone quantifier and Q' is an unbounded monotone quantifier. Let L be a vocabulary consisting of one unary predicate. Let $k \in \mathbb{N}$. Show that there are L-structures $\mathcal{M}$ and $\mathcal{M}'$ such that

$$\mathcal{M} \equiv_k^Q \mathcal{M}' \text{ but } \mathcal{M} \not\equiv_k^{Q'} \mathcal{M}'.$$

10.55 Exercise 10.54 continued: Show that there are L-structures $\mathcal{A}$ and $\mathcal{A}'$ such that

$$\mathcal{A} \equiv_k^{Q'} \mathcal{A}' \text{ but } \mathcal{A} \not\equiv_k^{Q} \mathcal{A}'.$$

10.56 Let $k \in \mathbb{N} \setminus \{0\}$. Suppose $\mathcal{M}$ is an equivalence relation with exactly k equivalence classes, all infinite. Let $\mathcal{M}'$ be an equivalence relation with infinitely many equivalence classes, all infinite. Show that $\mathcal{M} \equiv_k^{\exists^{\geq \omega}} \mathcal{M}'$.

10.57 Player **I** challenges player **II** to play $\mathrm{EF}_1^{\geq \omega}((\mathbb{Q}, \mathbb{N}), (\mathbb{N}, P))$ and **II** agrees. After **I** has started the game, **II** already regrets she agreed to play. What can we say about P?

10.58 Player **II** challenges player **I** to play $\mathrm{EF}_{100}^{\exists^{\geq \frac{1}{2}}}((\mathbb{Q}, <, 0, 1), (\mathbb{Q}, <, 1, 0))$. **I** thinks **II** is crazy. Why?

10.59 Player **II** challenges player **I** to play

$$\mathrm{EF}_{100}^{\exists^{\geq \frac{1}{2}}}((\{0, \ldots, 1000\}, \{0, \ldots, 100\}), (\{0, \ldots, 1000\}, \{0, \ldots, 101\})).$$

Player **I** first refuses, then says "Wait a minute!", but finally refuses. Did he make the right decision?

10.60 Player **II** challenges player **I** to play

$$\mathrm{EF}_1^{\exists^{\geq \frac{1}{2}}}((\{0, \ldots, 1024\}, <, 512), (\{0, \ldots, 1024\}, <, 513)).$$

Player **I** says "This is a no-brainer!" Was he justified in being so confident?

10.61 Let $\mathcal{M}$ be a linear order of length 4 and $\mathcal{M}'$ a linear order of length 5. Show that **II** has a winning strategy in $\mathrm{EF}_n^{\exists^{\geq \frac{1}{2}}}(\mathcal{M}, \mathcal{M}')$ if and only if $n \leq 1$.

10.62 Let $\mathcal{M}$ be a linear order of length 5 and $\mathcal{M}'$ a linear order of length 6. Show that **II** has a winning strategy in $\mathrm{EF}_n^{\exists^{\geq \frac{1}{2}}}(\mathcal{M}, \mathcal{M}')$ if and only if $n \leq 1$.

10.63 Let $\mathcal{M}$ be a linear order of length 9 and $\mathcal{M}'$ a linear order of length 10. Show that **II** has a winning strategy in $\mathrm{EF}_n^{\exists^{\geq \frac{1}{3}}}(\mathcal{M}, \mathcal{M}')$ if and only if $n \leq 2$.

10.64 Let $\mathcal{M}$ and $\mathcal{M}'$ be equivalence relations such that $M = \mathbb{R}, M' = \mathbb{R}^2$,

$$x \sim^{\mathcal{M}} y \iff x - y \in \mathbb{Q}$$

and

$$(x, y) \sim^{\mathcal{M}'} (u, v) \qquad \Longleftrightarrow \qquad x = u.$$

Prove $\mathcal{M} \equiv_n^{\exists^{\geq \aleph_1}} \mathcal{M}'$ for all $n \in \mathbb{N}$.

10.65 Prove that $\mathcal{M} \simeq_p \mathcal{M}'$ implies $\mathcal{M} \equiv_n^{\exists^{\geq \omega}} \mathcal{M}'$ for all $n \in \mathbb{N}$.

10.66 Show that there are $\mathcal{M}$ and $\mathcal{M}'$ with $\mathcal{M} \simeq_p \mathcal{M}'$ but $\mathcal{M} \not\equiv_1^{\exists^{\geq \aleph_1}} \mathcal{M}'$.

10.67 Suppose $\mathcal{M}$ is a linear order such that $\mathcal{M} \equiv_2^{\exists^{\geq \omega}} (\mathbb{N}, <)$. Prove $\mathcal{M} \cong (\mathbb{N}, <)$.

10.68 Show that $\mathcal{M} \equiv_3^{\exists^{\geq \omega}} (\mathbb{Z}, <)$ implies $\mathcal{M} \cong (\mathbb{Z}, <)$.

10.69 Give a linear order $\mathcal{M}$ such that $\mathcal{M} \equiv_1^{\exists^{\geq \omega}} (\mathbb{N}, <)$, but $\mathcal{M} \not\cong (\mathbb{N}, <)$.

10.70 Give a linear order $\mathcal{M}$ such that $\mathcal{M} \equiv_2^{\exists^{\geq \omega}} (\mathbb{Z}, <)$, but $\mathcal{M} \not\cong (\mathbb{Z}, <)$.

10.71 Prove (2)→(1) in Theorem 10.46.

10.72 Let $L = \emptyset$. Write an L-sentence of $L_{\omega\omega}(\exists^{\geq \frac{1}{2}})$ of quantifier rank ≤ 4 which is true in a finite L-structure $\mathcal{M}$ if and only if

  (a) $|M| \leq 5$.

  (b) $|M| \geq 7$.

10.73 Let $L = \emptyset$. Write an L-sentence of $L_{\omega\omega}(\exists^{\geq \frac{2}{3}})$ of quantifier rank ≤ 4 which is true in a finite L-structure $\mathcal{M}$ if and only if

  (a) $|M| \leq 7$.

  (b) $|M| \geq 9$.

10.74 Let $L = \{P\}, \#(P) = 1$. Write an L-sentence of $L_{\omega\omega}(\exists^{\geq \frac{1}{2}})$ which is true in a finite L-structure $\mathcal{M}$ if and only if

  (a) $|P^{\mathcal{M}}| < |M| - 2$.

  (b) $|P^{\mathcal{M}}| < \lceil |M|/2 \rceil + 1$.

  (c) $|P^{\mathcal{M}}| = |M|/2 - 2$.

10.75 Let $L = \{P\}, \#(P) = 1$. Write an L-sentence of $L_{\omega\omega}(\exists^{\geq \frac{1}{3}})$ which is true in a finite L-structure $\mathcal{M}$ if and only if

  (a) $|M| \leq 3 \cdot |P^{\mathcal{M}}| - 3$.

  (b) $|M| > 3 \cdot |P^{\mathcal{M}}| + 6$.

10.76 Suppose Q is a monotone generalized quantifier with $f_Q(n) = \lceil \sqrt{n} \rceil$. Write a sentence of graph-theory in $L_{\omega\omega}(Q)$ which is true in a finite graph $\mathcal{M}$ if and only if

  (a) Every vertex has at least $\sqrt{|M|} + 2$ neighbors.

  (b) Some vertex has less than $\sqrt{|M|} + 3$ neighbors.

10.77 Show that a property of L-structures is expressible in $L_{\omega\omega}(Q)$ if and only if it is expressible in $L_{\omega\omega}(\check{Q})$.

10.78 Let $L = \{P\}, \#(P) = 1$. Which of the following properties of finite L-structures are expressible in $L_{\omega\omega}(\exists^{\geq\frac{1}{2}})$:

   (a) $P^{\mathcal{M}} \neq M$.

   (b) $|P^{\mathcal{M}}| = |M \setminus P^{\mathcal{M}}|$.

   (c) $|M| = |P^{\mathcal{M}}|^2$.

10.79 Let $L = \{P_1, P_2\}, \#(P_1) = \#(P_2) = 1$. Which of the following properties of L-structures are expressible in $L_{\omega\omega}(\exists^{\geq\omega})$:

   (a) $P_1^{\mathcal{M}} \cap P_2^{\mathcal{M}}$ is finite.

   (b) All but finitely many elements of $P_1^{\mathcal{M}}$ are in $P_2^{\mathcal{M}}$.

   (c) $|P_1^{\mathcal{M}}| = |P_2^{\mathcal{M}}|$.

10.80 Let $L = \emptyset$. Into how many classes does $\equiv_n^{\exists^{\geq\omega}}$ divide $\mathrm{Str}(L)$?

10.81 Let $L = \emptyset$. Into how many classes does $\equiv_n^{\exists^{\geq\frac{1}{2}}}$ divide $\mathrm{Str}(L)$?

10.82 Write a sentence of $L_{\omega\omega}(\exists^{\geq\omega})$ of quantifier rank 2 which holds in a graph if and only if infinitely many vertices are isolated. Show that there is no such sentence of quantifier rank 1.

10.83 Write a sentence of $L_{\omega\omega}(\exists^{\geq\omega})$ of quantifier rank 3 which holds in a graph if and only if at least two vertices have infinitely many neighbors. Show that there is no such sentence of quantifier rank 2.

10.84 Show that the property of a graph that it contains an infinite complete subgraph is not expressible in $L_{\omega\omega}(\exists^{\geq\omega})$.

10.85 Show that the property of a finite graph that it is 3-colorable is not expressible in $L_{\omega\omega}(\exists^{\geq\frac{1}{2}})$.

10.86 Show that connectedness of a finite graph is not expressible in $L_{\omega\omega}(\exists^{\geq r})$ for any real r with $0 \leq r \leq 1$.

10.87 Show that Hamiltonicity of a finite graph is not expressible in $L_{\omega\omega}(\exists^{\geq r})$ for any real r with $0 \leq r \leq 1$.

10.88 The *bijective Ehrenfeucht–Fraïssé* Game $\mathrm{BEF}_n(\mathcal{M}, \mathcal{N})$ is defined as follows: First player **II** picks a bijection $f_0 : M \to N$. Then player **I** picks an element a_0 from M. Then player **II** picks a bijection $f_1 : M \to N$ that maps a_0 to $f_0(a_0)$. Then player **I** picks an element a_1 from M. Next player **II** picks a bijection $f_2 : M \to N$ that maps a_0 to $f_0(a_0)$ and a_1 to $f_1(a_1)$. Then player **I** picks an element a_2 from M. This goes on until player **II** has picked f_{n-1} and player **I** has picked a_{n-1}. Then **II** wins if $f_{n-1} \upharpoonright \{a_0, \ldots, a_{n-1}\}$ extends to a partial isomorphism $\mathcal{M} \to \mathcal{N}$. Show that if player **II** has a winning strategy in $\mathrm{BEF}_n(\mathcal{M}, \mathcal{N})$, then $\mathcal{M} \simeq_Q^n \mathcal{N}$ for every monotone bijection closed quantifier Q.

10.89 Suppose $\mathcal{M}$ and $\mathcal{N}$ are equivalence relations with all equivalence classes infinite and both with infinitely many equivalence classes. Use the method of Exercise 10.88 to show that $\mathcal{M} \simeq_Q^n \mathcal{N}$ for every monotone bijection closed quantifier Q.

10.90 Suppose $n \in \mathbb{N}$. Construct a connected graph $\mathcal{M}$ and a non-connected graph $\mathcal{N}$ such that $\mathcal{M} \simeq_Q^n \mathcal{N}$ for all monotone unary quantifiers Q.

10.91 The *Härtig quantifier I* is defined as follows: $\mathcal{M} \models_s Ixy\varphi\psi$ if and only if $|\{a \in M : \mathcal{M} \models_{s(a/x)} \varphi\}| = |\{a \in M : \mathcal{M} \models_{s(a/x)} \psi\}|$. Show that the quantifier $\exists^{\geq\omega}$ is definable in terms of the Härtig quantifier, but the Härtig quantifier is not definable in terms of the quantifier $\exists^{\geq\omega}$.

10.92 Give an infinite set T of sentences of $L_{\omega\omega}(\exists^{\geq\omega})$ such that every proper subset of T has a model but T itself has no models.

10.93 Show that $L_{\omega\omega}(\exists^{\geq\aleph_1})$ does not satisfy the Compactness Theorem if uncountable vocabularies are allowed.

10.94 Prove Lemma 10.51.

10.95 Suppose $L = \{P\}$, P unary. Show that the L-structure $\mathcal{M} = (\mathbb{Q}, \mathbb{N})$ has a countable proper elementary extension $\mathcal{M}'$ with $P^{\mathcal{M}} = P^{\mathcal{M}'}$ and another countable elementary extension $\mathcal{M}''$ with $P^M \neq P^{\mathcal{M}''}$.

10.96 Suppose $\mathcal{M}_1 \subseteq \mathcal{M}_2$ and $\mathcal{M}_1$ and $\mathcal{M}_2$ have a common elementary extension. Show that $\mathcal{M}_1 \prec \mathcal{M}_2$.

10.97 Show that every proper elementary extension of $(\mathbb{R}, +, \cdot, 0, 1, <)$ has elements that are bigger than all $a \in \mathbb{R}$.

10.98 Suppose $(\mathcal{M}, Q)$ is a countable weak model, and $(\mathcal{M}', Q')$ is a proper elementary extension of $(\mathcal{M}, Q)$ such that for all $\varphi(x)$ in the vocabulary of $\mathcal{M}$ we have $(\mathcal{M}, Q) \models Qx\varphi(x)$ if and only if there is $a \in M' \setminus M$ with $\mathcal{M}' \models \varphi(a)$. Show that $(\mathcal{M}, Q)$ is

1. non-trivial
2. plural
3. a model of (MON)
4. a model of (IDE).

10.99 Suppose $(\mathcal{M}, Q)$ is as in the above exercise. Show that $(\mathcal{M}, Q)$ satisfies Keisler's Axioms (KA).

10.100 Show that (IDE) follows from (KA) under the stronger form of plurality $\neg Qx(\approx xy \vee \approx xz)$.

10.101 Suppose the vocabulary L contains a binary predicate symbol $<$. An L-structure $\mathcal{M}'$ is an *end extension* of another L-structure $\mathcal{M}$ if $\mathcal{M} \subseteq \mathcal{M}'$ and for every $a \in M$ and $a' \in M' \setminus M$, $\mathcal{M}' \models a < a'$. Suppose $\mathcal{M}$ is a countable L-structure with a proper elementary end extension. Let Q consist of the sets $X \subseteq M$ such that for every $a \in M$ there is $b \in X$ with $\mathcal{M} \models a < b$. Show the following:

1. $(\mathcal{M}, Q)$ has a precise extension.

2. $\mathcal{M}$ has an elementary extension $\mathcal{M}'$ in which $<^{\mathcal{M}'}$ is ω_1-like, i.e. uncountable but every initial segment is countable.

10.102 Suppose L is a countable vocabulary, $(\mathcal{M}, q)$ a countable non-trivial plural monotone ideal weak model satisfying (KA), and $p_0, p_1, \ldots$ a sequence of types omitted by $(\mathcal{M}, q)$ such that for each n there is some $\psi_n(x) \in p_n$ with $(\mathcal{M}, q) \models \neg \mathcal{Q}x\psi_n(x)$. Use the proof of the Completeness Theorem for $\exists^{\geq \aleph_1}$ to show that there is a model $(\mathcal{M}', \exists^{\geq \aleph_1})$ such that $(\mathcal{M}, q) \prec (\mathcal{M}', \exists^{\geq \aleph_1})$ and $(\mathcal{M}', \exists^{\geq \aleph_1})$ omits each p_n.

10.103 Suppose the countable vocabulary L contains a binary predicate symbol $<$. Show that if a countable L-model has a proper elementary end extension, it has an elementary end extension of cardinality $\aleph_1$.

10.104 Construct a first-order theory in a countable vocabulary which has a model $\mathcal{M}$ with $U^{\mathcal{M}}$ infinite and $|M| = |U^{\mathcal{M}}|^+$, and moreover in every model $\mathcal{M}$ of T we have $|M| \leq |U^{\mathcal{M}}|^+$.

10.105 Construct a first-order theory in a countable vocabulary which has a model $\mathcal{M}$ with $U^{\mathcal{M}}$ infinite and $|M| = 2^{|U^{\mathcal{M}}|}$, and moreover in every model $\mathcal{M}$ of T we have $|M| \leq 2^{|U^{\mathcal{M}}|}$.

10.106 Define the union of a chain $(\mathcal{M}_\alpha, Q_\alpha) \prec (\mathcal{M}_\beta, Q_\beta)$ $(\alpha < \beta < \gamma)$ by $\bigcup_{\alpha < \gamma}(\mathcal{M}_\alpha, Q_\alpha) = (\mathcal{M}, Q)$ where $\mathcal{M} = \bigcup_{\alpha < \gamma} \mathcal{M}_\alpha, Q = \{X \subseteq M : \text{there is } \alpha < \gamma \text{ such that } X \cap M_\beta \in Q_\beta \text{ for } \alpha \leq \beta < \gamma\}$. Show that the Union Lemma still holds.

10.107 Suppose L is a countable vocabulary and T is set of L-sentences of $L_{\omega\omega}(\mathcal{Q})$ such that every finite subset of T has, for some infinite κ, a model of the form $(\mathcal{M}, \exists^{\geq \kappa^+})$. Show that T has a model of the form $(\mathcal{M}, \exists^{\geq \aleph_1})$.

10.108 Use the Compactness Theorem and the Union Lemma to prove Lemma 10.88.

10.109 Prove Theorem 10.91 for the quantifier Q_κ^{cf}, where κ is an arbitrary regular cardinal.

10.110 Suppose κ and λ are regular cardinals $\geq \aleph_0$. Suppose a first-order theory T in an arbitrary vocabulary L has a model in which a binary predicate $R \in L$ is interpreted as a linear order of cofinality κ. Show that T has a model in which a binary predicate $R \in L$ is interpreted as a linear order of cofinality λ.

10.111 Show that every model in a vocabulary of size $\leq \kappa$, $\kappa > \aleph_1$, has an $L_{\omega\omega}(Q_\omega^{\text{cf}})$-elementary submodel of cardinality $\leq \kappa$.

References

Aczel, P. 1977. An introduction to inductive definitions. Pages 739–783 of: Barwise, Jon (ed), *Handbook of Mathematical Logic*. Amsterdam: North–Holland Publishing Co. Cited on page **125**.

Barwise, J. 1969. Remarks on universal sentences of $L_{\omega_1,\omega}$. *Duke Mathematical Journal*, **36**, 631–637. Cited on page **223**.

Barwise, J. 1975. *Admissible Sets and Structures*. Berlin: Springer-Verlag. Perspectives in Mathematical Logic. Cited on pages **71, 76, 171**, and **204**.

Barwise, J. 1976. Some applications of Henkin quantifiers. *Israel Journal of Mathematics*, **25**(1-2), 47–63. Cited on page **204**.

Barwise, J. and Cooper, R. 1981. Generalized quantifiers and natural language. *Linguistics and Philosophy*, 159–219. Cited on page **343**.

Barwise, J., and Feferman, S. (eds). 1985. *Model-Theoretic Logics*. Perspectives in Mathematical Logic. New York: Springer-Verlag. Cited on pages **118, 126**, and **343**.

Barwise, J., Kaufmann, M., and Makkai, M. 1978. Stationary logic. *Annals of Mathematical Logic*, **13**(2), 171–224. Cited on page **343**.

Bell, J. L., and Slomson, A. B. 1969. *Models and Ultraproducts: An Introduction*. Amsterdam: North-Holland Publishing Co. Cited on page **126**.

Benda, M. 1969. Reduced products and nonstandard logics. *Journal of Symbolic Logic*, **34**, 424–436. Cited on pages **125, 126, 238**, and **275**.

Beth, E. W. 1953. On Padoa's method in the theory of definition. *Nederl. Akad. Wetensch. Proc. Ser. A.* **56** = *Indagationes Mathematicae*, **15**, 330–339. Cited on page **126**.

Beth, E. W. 1955a. Remarks on natural deduction. *Nederl. Akad. Wetensch. Proc. Ser. A.* **58** = *Indagationes Mathematicae*, **17**, 322–325. Cited on page **125**.

Beth, E. W. 1955b. *Semantic Entailment and Formal Derivability*. Mededelingen der koninklijke Nederlandse Akademie van Wetenschappen, afd. Letterkunde. Nieuwe Reeks, Deel 18, No. 13. N. V. Noord-Hollandsche Uitgevers Maatschappij, Amsterdam. Cited on page **125**.

Bissell-Siders, R. 2007. Ehrenfeucht-Fraïssé games on linear orders. Pages 72–82 of: *Logic, Language, Information and Computation*. Lecture Notes in Computer Scence, vol. 4576. Berlin: Springer. Cited on page **126**.

Brown, J., and Hoshino, R. 2007. The Ehrenfeucht-Fraïssé game for paths and cycles. *Ars Combinatoria*, **83**, 193–212. Cited on page **126**.

Burgess, J. 1977. Descriptive set theory and infinitary languages. *Zbornik Radova Matematički Institut Beograd (Nova Serija)*, **2(10)**, 9–30. Set Theory, Foundations of Mathematics (Proc. Sympos., Belgrade, 1977). Cited on page **222**.

Burgess, J. 1978. On the Hanf number of Souslin logic. *Journal of Symbolic Logic*, **43**(3), 568–571. Cited on page **223**.

Caicedo, X. 1980. Back-and-forth systems for arbitrary quantifiers. Pages 83–102 of: *Mathematical logic in Latin America (Proc. IV Latin Amer. Sympos. Math. Logic, Santiago, 1978)*. Stud. Logic Foundations Math., vol. 99. Amsterdam: North-Holland. Cited on page **343**.

Calais, J.-P. 1972. Partial isomorphisms and infinitary languages. *Zeitschrift für mathematische Logik und Grundlagen der Mathematik*, **18**, 435–456. Cited on page **275**.

Cantor, G. 1895. Beiträge zur Begründung der transfiniten Mengenlehre, I. *Mathematische Annalen*, **46**, 481–512. Cited on page **65**.

Chang, C. C. 1968. Infinitary properties of models generated from indiscernibles. Pages 9–21 of: *Logic, Methodology and Philos. Sci. III (Proc. Third Internat. Congr., Amsterdam, 1967)*. Amsterdam: North-Holland. Cited on page **275**.

Chang, C. C., and Keisler, H. J. 1990. *Model theory*. Third edn. Studies in Logic and the Foundations of Mathematics, vol. 73. Amsterdam: North-Holland Publishing Co. Cited on pages **125**, **203**, **246**, **248**, and **342**.

Craig, W. 1957a. Linear reasoning. A new form of the Herbrand-Gentzen theorem. *Journal of Symbolic Logic*, **22**, 250–268. Cited on page **126**.

Craig, W. 1957b. Three uses of the Herbrand-Gentzen theorem in relating model theory and proof theory. *Journal of Symbolic Logic*, **22**, 269–285. Cited on page **126**.

Craig, W. 2008. The road to two theorems of logic. *Synthese*, **164**(3), 333–339. Cited on page **126**.

Devlin, K. 1993. *The Joy of Sets*. Second edn. New York: Springer-Verlag. Cited on page **11**.

Dickmann, M. A. 1975. *Large Infinitary Languages*. Amsterdam: North-Holland Publishing Co. Model theory, Studies in Logic and the Foundations of Mathematics, Vol. 83. Cited on pages **171**, **244**, and **275**.

Dickmann, M. A. 1985. Larger infinitary languages. Pages 317–363 of: *Model-Theoretic Logics*. Perspect. Math. Logic. New York: Springer. Cited on page **275**.

Džamonja, M., and Väänänen, J. 2004. A family of trees with no uncountable branches. *Topology Proceedings*, **28**(1), 113–132. Cited on page **276**.

Ehrenfeucht, A. 1957. Application of games to some problems of mathematical logic. *Bulletin de l'Acadámie Polonaise des Science, Série des Sciences Mathématiques, Astronomiques et Physiques Cl. III.*, **5**, 35–37, IV. Cited on pages **48** and **71**.

Ehrenfeucht, A. 1960/1961. An application of games to the completeness problem for formalized theories. *Fundamenta Mathematicae*, **49**, 129–141. Cited on pages **48** and **71**.

Eklof, P., Foreman, M., and Shelah, S. 1995. On invariants for ω_1-separable groups. *Transactions of the American Mathematical Society*, **347**(11), 4385–4402. Cited on page **277**.

Ellentuck, E. 1975. The foundations of Suslin logic. *Journal of Symbolic Logic*, **40**(4), 567–575. Cited on page **210**.

Ellentuck, E. 1976. Categoricity regained. *Journal of Symbolic Logic*, **41**(3), 639–643. Cited on page **71**.

Enderton, H. 1970. Finite partially-ordered quantifiers. *Zeitschrift für mathematische Logik und Grundlagen der Mathematik*, **16**, 393–397. Cited on page **207**.

Enderton, H. 1977. *Elements of Set Theory*. New York: Academic Press [Harcourt Brace Jovanovich Publishers]. Cited on page **11**.

Enderton, H. 2001. *A Mathematical Introduction to Logic*. Second edn. Harcourt/Academic Press, Burlington, MA. Cited on page **102**.

Fagin, R. 1976. Probabilities on finite models. *Journal of Symbolic Logic*, **41**(1), 50–58. Cited on page **48**.

Fraïssé, R. 1955. Sur quelques classifications des relations, basées sur des isomorphismes restreints. II. Application aux relations d'ordre, et construction d'exemples montrant que ces classifications sont distinctes. *Publ. Sci. Univ. Alger. Sér. A.*, **2**, 273–295 (1957). Cited on pages **71** and **125**.

Fuhrken, G. 1964. Skolem-type normal forms for first-order languages with a generalized quantifier. *Fundamenta Mathematicae*, **54**, 291–302. Cited on page **343**.

Gaifman, H. 1964. Concerning measures in first order calculi. *Israel Journal for Mathematics*, **2**, 1–18. Cited on page **48**.

Gale, David, and Stewart, F. M. 1953. Infinite games with perfect information. Pages 245–266 of: *Contributions to the Theory of Games, vol. 2*. Annals of Mathematics Studies, no. 28. Princeton, N. J.: Princeton University Press. Cited on page **28**.

Gentzen, G. 1934. Untersuchungen über das logische Schließen. I. *Mathematische Zeitschrift*, **39**, 176–210. Cited on page **125**.

Gentzen, G. 1969. *The Collected Papers of Gerhard Gentzen*. Edited by M. E. Szabo. Studies in Logic and the Foundations of Mathematics. Amsterdam: North-Holland Publishing Co. Cited on page **125**.

Gostanian, R., and Hrbáček, K. 1976. On the failure of the weak Beth property. *Proceedings of the American Mathematical Society*, **58**, 245–249. Cited on page **245**.

Grädel, E., Kolaitis, P. G., Libkin, L., Marx, M., Spencer, J., Vardi, M. Y., Venema, Y., and Weinstein, S. 2007. *Finite Model Theory and its Applications*. Texts in Theoretical Computer Science. An EATCS Series. Berlin: Springer. Cited on page **48**.

Green, J. 1975. Consistency properties for finite quantifier languages. Pages 73–123. Lecture Notes in Math., Vol. 492 of: *Infinitary Logic: in Memoriam Carol Karp*. Berlin: Springer. Cited on page **222**.

Green, J. 1978. κ-Suslin logic. *Journal of Symbolic Logic*, **43**(4), 659–666. Cited on page **223**.

Green, J. 1979. Some model theory for game logics. *Journal of Symbolic Logic*, **44**(2), 147–152. Cited on pages **210** and **223**.

Gurevich, Y. 1984. Toward logic tailored for computational complexity. Pages 175–216 of: *Computation and Proof Theory (Aachen, 1983)*. Lecture Notes in Math., vol. 1104. Berlin: Springer-Verlag. Cited on page **126**.

Hájek, P. 1976. Some remarks on observational model-theoretic languages. Pages 335–345 of: *Set Theory and Hierarchy Theory (Proc. Second Conf., Bierutowice,*

1975), Lecture Notes in Math., Vol. 537. Berlin: Springer. Cited on pages **126** and **343**.

Harnik, V., and Makkai, M. 1976. Applications of Vaught sentences and the covering theorem. *Journal of Symbolic Logic*, **41**(1), 171–187. Cited on page **222**.

Härtig, K. 1965. Über einen Quantifikator mit zwei Wirkungsbereichen. Pages 31–36 of: *Colloq. Found. Math., Math. Machines and Appl. (Tihany, 1962)*. Budapest: Akad. Kiadó. Cited on page **343**.

Hella, L. 1989. Definability hierarchies of generalized quantifiers. *Annals of Pure and Applied Logic*, **43**(3), 235–271. Cited on page **343**.

Hella, L., and Sandu, G. 1995. Partially ordered connectives and finite graphs. Pages 79–88 of: Krynicki, M., M., Mostowski, and L., Szczerba (eds), *Quantifiers: Logics, Models and Computation, Vol. II*. Kluwer. Cited on page **343**.

Henkin, L. 1961. Some remarks on infinitely long formulas. Pages 167–183 of: *Infinitistic Methods (Proc. Sympos. Foundations of Math., Warsaw, 1959)*. Oxford: Pergamon. Cited on pages **125** and **204**.

Herre, H., Krynicki, M., Pinus, A., and Väänänen, J. 1991. The Härtig quantifier: a survey. *Journal of Symbolic Logic*, **56**(4), 1153–1183. Cited on page **343**.

Hintikka, J. 1953. Distributive normal forms in the calculus of predicates. *Acta Philosophica Fennica*, **6**, 71. Cited on page **48**.

Hintikka, J. 1955. Form and content in quantification theory. *Acta Philosophica Fennica*, **8**, 7–55. Cited on page **125**.

Hintikka, J. 1968. Language-games for quantifiers. Pages 46–73 of: N. Rescher (ed.), *Studies in Logical Lheory*, Oxford: Blackwell Publishers. Cited on page **125**.

Hintikka, J., and Rantala, V. 1976. A new approach to infinitary languages. *Annals of Mathematical Logic*, **10**(1), 95–115. Cited on page **276**.

Hodges, W. 1985. *Building Models by Games*. London Mathematical Society Student Texts, vol. 2. Cambridge: Cambridge University Press. Cited on page **126**.

Hodges, W. 1993. *Model Theory*. Encyclopedia of Mathematics and its Applications, vol. 42. Cambridge: Cambridge University Press. Cited on page **125**.

Huuskonen, T. 1995. Comparing notions of similarity for uncountable models. *Journal of Symbolic Logic*, **60**(4), 1153–1167. Cited on pages **276** and **278**.

Huuskonen, T., Hyttinen, T., and Rautila, M. 2004. On potential isomorphism and nonstructure. *Archive für mathematische Logic*, **43**(1), 85–120. Cited on page **277**.

Hyttinen, T. 1987. Games and infinitary languages. *Annales Academi Scientiarum Fennic Series A I Mathematica Dissertationes*, 32. Cited on pages **275**, **276**, and **277**.

Hyttinen, T. 1990. Model theory for infinite quantifier languages. *Fundamenta Mathematicae*, **134**(2), 125–142. Cited on pages **275** and **276**.

Hyttinen, T. 1992. On nondetermined Ehrenfeucht-Fraïssé games and unstable theories. *Zeitschrift für mathematische Logik und Grundlagen der Mathematik*, **38**(4), 399–408. Cited on page **277**.

Hyttinen, T., and Rautila, M. 2001. The canary tree revisited. *Journal of Symbolic Logic*, **66**(4), 1677–1694. Cited on page **277**.

Hyttinen, T., and Shelah, S. 1999. Constructing strongly equivalent nonisomorphic models for unsuperstable theories. Part C. *Journal of Symbolic Logic*, **64**(2), 634–642. Cited on page **277**.

Hyttinen, T., Shelah, S., and Tuuri, H. 1993. Remarks on strong nonstructure theorems. *Notre Dame J. Formal Logic*, **34**(2), 157–168. Cited on page **277**.

Hyttinen, T., Shelah, S., and Väänänen, J. 2002. More on the Ehrenfeucht-Fraïssé-game of length ω_1. *Fundamenta Mathematicae*, **175**(1), 79–96. Cited on page **277**.

Hyttinen, T., and Tuuri, H. 1991. Constructing strongly equivalent nonisomorphic models for unstable theories. *Annals of Pure and Applied Logic*, **52**(3), 203–248. Cited on pages **275** and **276**.

Hyttinen, T., and Väänänen, J. 1990. On Scott and Karp trees of uncountable models. *Journal of Symbolic Logic*, **55**(3), 897–908. Cited on pages **255**, **275**, **276**, and **277**.

Jech, T. 1997. *Set theory*. Second edn. Perspectives in Mathematical Logic. Berlin: Springer-Verlag. Cited on pages **11**, **24**, **28**, **59**, **61**, and **189**.

Juhász, I., and Weiss, W. 1978. On a problem of Sikorski. *Fundamenta Mathematicae*, **100**(3), 223–227. Cited on page **277**.

Karp, C. 1964. *Languages with Expressions of Infinite Length*. Amsterdam: North-Holland Publishing Co. Cited on page **171**.

Karp, C. 1965. Finite-quantifier equivalence. Pages 407–412 of: *Theory of Models (Proc. 1963 Internat. Sympos. Berkeley)*. Amsterdam: North-Holland. Cited on pages **71** and **171**.

Karttunen, M. 1979. Infinitary languages $N_{\infty\lambda}$ and generalized partial isomorphisms. Pages 153–168 of: *Essays on Mathematical and Philosophical Logic (Proc. Fourth Scandinavian Logic Sympos. and First Soviet-Finnish Logic Conf., Jyväskylä, 1976)*. Synthese Library, vol. 122. Dordrecht: Reidel. Cited on page **276**.

Karttunen, M. 1984. Model theory for infinitely deep languages. *Annales Academiae Scientiarum Fennicae Series A I Mathematica Dissertationes*, 96. Cited on pages **275** and **276**.

Keisler, H. J., and Morley, M. 1968. Elementary extensions of models of set theory. *Israel Journal for Mathematics*, **6**, 49–65. Cited on page **126**.

Keisler, H. J. 1965. Finite approximations of infinitely long formulas. Pages 158–169 of: *Theory of Models (Proc. 1963 Internat. Sympos. Berkeley)*. Amsterdam: North-Holland. Cited on page **222**.

Keisler, H. J. 1970. Logic with the quantifier "there exist uncountably many". *Annals of Mathematical Logic*, **1**, 1–93. Cited on pages **314** and **343**.

Keisler, H. J. 1971. *Model Theory for Infinitary Logic*. Amsterdam: North-Holland Publishing Co. Studies in Logic and the Foundations of Mathematics, Vol. 62. Cited on pages **171** and **222**.

Kolaitis, P., and Väänänen, J. 1995. Generalized quantifiers and pebble games on finite structures. *Annals of Pure and Applied Logic*, **74**(1), 23–75. Cited on page **343**.

Kolaitis, P., and Vardi, M. 1992. Infinitary logics and 0-1 laws. *Information and Computation*, **98**(2), 258–294. Selections from the 1990 IEEE Symposium on Logic in Computer Science. Cited on page **48**.

Krynicki, M., Mostowski M., and Szczerba, L. (Eds.). 1995. *Quantifiers*. Berlin: Kluwer Academic Publishers. Cited on page **343**.

Kueker, D. 1968. Definability, automorphisms and infinitary languages. Pages 152–165 of: *The Syntax and Semantics of Infinitary Languages*. Springer Lecture Notes in Math., Vol. 72. Berlin: Springer-Verlag. Cited on page **171**.

Kueker, D. 1972. Löwenheim-Skolem and interpolation theorems in infinitary languages. *Bulletin of the American Mathematical Society*, **78**, 211–215. Cited on pages **125** and **222**.

Kueker, D. 1975. Back-and-forth arguments and infinitary logics. Pages 17–71 of: *Infinitary Logic: in Memoriam Carol Karp*. Lecture Notes in Math., Vol. 492. Berlin: Springer-Verlag. Cited on pages **171** and **275**.

Kueker, D. 1977. Countable approximations and Löwenheim-Skolem theorems. *Annals of Mathematical Logic*, **11**(1), 57–103. Cited on pages **85**, **125**, and **222**.

Kuratowski, K. 1966. *Topology. Vol. I*. New edition, revised and augmented. Translated from the French by J. Jaworowski. New York: Academic Press. Cited on page **223**.

Kurepa, G. 1956. Ensembles ordonnés et leurs sous-ensembles bien ordonnés. *Les Comptes Rendus de l'Académie des sciences*, **242**, 2202–2203. Cited on pages **255**, **275**, and **278**.

Lindström, P. 1966. First order predicate logic with generalized quantifiers. *Theoria*, **32**, 186–195. Cited on page **342**.

Lindström, P. 1973. A characterization of elementary logic. Pages 189–191 of: *Modality, Morality and Other Problems of Sense and Nonsense*. Lund: CWK Gleerup Bokförlag. Cited on pages **112** and **126**.

Lopez-Escobar, E. G. K. 1965. An interpolation theorem for denumerably long formulas. *Fundamenta Mathematicae*, **57**, 253–272. Cited on pages **222** and **223**.

Lopez-Escobar, E. G. K. 1966a. An addition to: "On defining well-orderings". *Fundamenta Mathematicae*, **59**, 299–300. Cited on page **222**.

Lopez-Escobar, E. G. K. 1966b. On defining well-orderings. *Fundamenta Mathematicae*, **59**, 13–21. Cited on pages **183** and **222**.

Lorenzen, P. 1961. Ein dialogisches Konstruktivitätskriterium. Pages 193–200 of: *Infinitistic Methods (Proc. Sympos. Foundations of Math., Warsaw, 1959)*. Oxford: Pergamon. Cited on page **125**.

Łoś, J. 1955. Quelques remarques, théorèmes et problèmes sur les classes définissables d'algèbres. Pages 98–113 of: *Mathematical Interpretation of Formal Systems*. Amsterdam: North-Holland Publishing Co. Cited on page **126**.

Löwenheim, L. 1915. Über Möglichkeiten im Relativkalkul. *Mathematische Annalen*, **76**, 447–470. Cited on page **125**.

Luosto, K. 2000. Hierarchies of monadic generalized quantifiers. *Journal of Symbolic Logic*, **65**(3), 1241–1263. Cited on page **343**.

Makkai, M. 1969a. An application of a method of Smullyan to logics on admissible sets. *Bulletin de l'Acadámie Polonaise des Science, Série des Sciences Mathématiques, Astronomiques et Physiques*, **17**, 341–346. Cited on page **222**.

Makkai, M. 1969b. On the model theory of denumerably long formulas with finite strings of quantifiers. *Journal of Symbolic Logic*, **34**, 437–459. Cited on pages **222** and **223**.

Makkai, M. 1977. Admissible sets and infinitary logic. Pages 233–281 of: *Handbook of Mathematical Logic*, Studies in Logic and the Foundations of Math., Vol. 90. Amsterdam: North-Holland. Cited on pages **171**, **213**, and **222**.

Makowsky, J. A., and Shelah, S. 1981. The theorems of Beth and Craig in abstract model theory. II. Compact logics. *Archiv für mathematische Logik und Grundlagenforschung*, **21**(1–2), 13–35. Cited on page **343**.

Malitz, J. 1969. Universal classes in infinitary languages. *Duke Mathematical Journal*, **36**, 621–630. Cited on page **223**.

Mekler, A., and Oikkonen, J. 1993. Abelian p-groups with no invariants. *Journal of Pure and Applied Algebra*, **87**(1), 51–59. Cited on page **277**.

Mekler, A., and Shelah, S. 1986. Stationary logic and its friends. II. *Notre Dame Journal of Formal Logic*, **27**(1), 39–50. Cited on page **343**.

Mekler, A., and Shelah, S. 1993. The canary tree. *Canadian Mathematical Bulletin*, **36**(2), 209–215. Cited on page **277**.

Mekler, A., Shelah, S., and Väänänen, J. 1993. The Ehrenfeucht–Fraïssé game of length ω_1. *Transactions of the American Mathematical Society*, **339**(2), 567–580. Cited on pages **253** and **277**.

Mekler, A., and Väänänen, J. 1993. Trees and Π_1^1-subsets of $^{\omega_1}\omega_1$. *Journal of Symbolic Logic*, **58**(3), 1052–1070. Cited on pages **273**, **276**, and **277**.

Mildenberger, H. 1992. On the homogeneity property for certain quantifier logics. *Archive für mathematische Logic*, **31**(6), 445–455. Cited on page **343**.

Morley, M. 1968. Partitions and models. Pages 109–158 of: *Proceedings of the Summer School in Logic (Leeds, 1967)*. Berlin: Springer. Cited on pages **126** and **222**.

Morley, M. 1970. The number of countable models. *Journal of Symbolic Logic*, **35**, 14–18. Cited on page **152**.

Morley, M., and Vaught, R. 1962. Homogeneous universal models. *Mathematica Scandinavica*, **11**, 37–57. Cited on page **343**.

Mortimer, M. 1975. On languages with two variables. *Zeitschrift für mathematische Logik und Grundlagen der Mathematik*, **21**, 135–140. Cited on page **38**.

Moschovakis, Y. 1972. The game quantifier. *Proceedings of the American Mathematical Society*, **31**, 245–250. Cited on page **201**.

Mostowski, A. 1957. On a generalization of quantifiers. *Fundamenta Mathematicae*, **44**, 12–36. Cited on pages **291**, **314**, and **342**.

Mycielski, J. 1992. Games with perfect information. Pages 41–70 of: *Handbook of Game Theory with Economic Applications, Vol. I*. Handbooks in Econom., vol. 11. Amsterdam: North-Holland. Cited on page **28**.

Nadel, M., and Stavi, J. 1978. $L_{\infty\lambda}$-equivalence, isomorphism and potential isomorphism. *Transactions of the American Mathematical Society*, **236**, 51–74. Cited on page **277**.

Nešetřil, J., and Väänänen, J. 1996. Combinatorics and quantifiers. *Commentationes Mathematicae Universitatis Carolinae*, **37**(3), 433–443. Cited on page **343**.

Oikkonen, J. 1990. On Ehrenfeucht–Fraïssé equivalence of linear orderings. *Journal of Symbolic Logic*, **55**(1), 65–73. Cited on page **275**.

Oikkonen, J. 1997. Undefinability of κ-well-orderings in $L_{\infty\kappa}$. *Journal of Symbolic Logic*, **62**(3), 999–1020. Cited on page **276**.

Peters, S., and Westerståhl, D. 2008. *Quantifiers in Language and Logic*. Oxford: Oxford University Press. Cited on page **343**.

Rantala, V. 1981. Infinitely deep game sentences and interpolation. *Acta Philosophica Fennica*, **32**, 211–219. Cited on page **276**.

Rosenstein, Joseph G. 1982. *Linear Orderings*. Pure and Applied Mathematics, vol. 98. New York: Academic Press Inc. [Harcourt Brace Jovanovich Publishers]. Cited on page **71**.

Rotman, B., and Kneebone, G. T. 1966. *The Theory of Sets and Transfinite Numbers*. London: Oldbourne. Cited on page **11**.

Schwalbe, U., and Walker, P. 2001. Zermelo and the early history of game theory. *Games and Economomic Behaviour*, **34**(1), 123–137. With an appendix by Ernst Zermelo, translated from German by the authors. Cited on page **28**.

Scott, D. 1965. Logic with denumerably long formulas and finite strings of quantifiers. Pages 329–341 of: *Theory of Models (Proc. 1963 Internat. Sympos. Berkeley)*. Amsterdam: North-Holland. Cited on page **171**.

Scott, D., and Tarski, A. 1958. The sentential calculus with infinitely long expressions. *Colloqium Mathematicum*, **6**, 165–170. Cited on page **170**.

Shelah, S. 1971. Every two elementarily equivalent models have isomorphic ultrapowers. *Israel Journal for Mathematics*, **10**, 224–233. Cited on page **125**.

Shelah, S. 1975. Generalized quantifiers and compact logic. *Transactions of the American Mathematical Society*, **204**, 342–364. Cited on page **343**.

Shelah, S. 1985. Remarks in abstract model theory. *Annals of Pure and Applied Logic*, **29**(3), 255–288. Cited on page **343**.

Shelah, S. 1990. *Classification theory and the number of nonisomorphic models*. Second edn. Studies in Logic and the Foundations of Mathematics, vol. 92. Amsterdam: North-Holland Publishing Co. Cited on pages **241**, **275**, and **276**.

Shelah, S., and Väänänen, J. 2000. Stationary sets and infinitary logic. *Journal of Symbolic Logic*, **65**(3), 1311–1320. Cited on page **277**.

Shelah, S., and Väisänen, P. 2002. Almost free groups and Ehrenfeucht–Fraïssé games for successors of singular cardinals. *Annals of Pure and Applied Logic*, **118**(1–2), 147–173. Cited on page **277**.

Sierpiński, W. 1933. Sur un problème de la théorie des rélations. *Annali della Scuola Normale Superiore di Pisa, II. Ser.*, **2**, 285–287. Cited on page **223**.

Sikorski, R. 1950. Remarks on some topological spaces of high power. *Fundamenta Mathematicae*, **37**, 125–136. Cited on page **277**.

Skolem, T. 1923. Einige Bemerkungen zur axiomatischen Begündung der Mengenlehre. Pages 217–232 of: *5. Kongreß Skandinav. in Helsingfors vom 4. bis 7. Juli 1922 (Akademische Buchhandlung)*. Cited on page **125**.

Skolem, T. 1970. *Selected Works in Logic*. Edited by Jens Erik Fenstad. Oslo: Universitetsforlaget. Cited on page **125**.

Smullyan, R. 1963. A unifying principal in quantification theory. *Proceedings of the National Academy of Sciences U.S.A.*, **49**, 828–832. Cited on page **125**.

Smullyan, R. 1968. *First-Order Logic*. Ergebnisse der Mathematik und ihrer Grenzgebiete, Band 43. New York: Springer-Verlag New York, Inc. Cited on page **125**.

Spencer, J. 2001. *The Strange Logic of Random Graphs*. Algorithms and Combinatorics, vol. 22. Berlin: Springer-Verlag. Cited on page **48**.

Svenonius, L. 1965. On the denumerable models of theories with extra predicates. Pages 376–389 of: *Theory of Models (Proc. 1963 Internat. Sympos. Berkeley)*. Amsterdam: North-Holland. Cited on page **208**.

Tarski, A. 1958. Remarks on predicate logic with infinitely long expressions. *Colloqium Mathematicum*, **6**, 171–176. Cited on page **170**.

Todorčević, S. 1981a. Stationary sets, trees and continuums. *Publications de l'Institut Mathématique* , **29**(43), 249–262. Cited on page **278**.

Todorčević, S. 1981b. Trees, subtrees and order types. *Annals of Mathematical Logic*, **20**(3), 233–268. Cited on page **277**.

Todorčević, S., and Väänänen, J. 1999. Trees and Ehrenfeucht–Fraïssé games. *Annals of Pure and Applied Logic*, **100**(1–3), 69–97. Cited on pages **255**, **258**, and **276**.

Tuuri, H. 1990. *Infinitary languages and Ehrenfeucht–Fraïssé games*. PhD in Mathematics, University of Helsinki. Cited on page **276**.

Tuuri, H. 1992. Relative separation theorems for $L_{\kappa^+\kappa}$. *Notre Dame Journal of Formal Logic*, **33**(3), 383–401. Cited on page **276**.

Väänänen, J. 1977. Remarks on generalized quantifiers and second order logics. *Prace Nauk. Inst. Mat. Politech. Wrocław.*, 117–123. Cited on page **343**.

Väänänen, J. 1991. A Cantor–Bendixson theorem for the space $^{\omega_1}\omega_1$. *Fundamenta Mathematicae*, **137**(3), 187–199. Cited on page **277**.

Väänänen, J. 1995. Games and trees in infinitary logic: A survey. Pages 105–138 of: Krynicki, M., Mostowski, M., and Szczerba, L. (eds), *Quantifiers*. Quad. Mat. Kluwer Academic Publishers. Cited on pages **276** and **277**.

Väänänen (Ed.), J. 1999. *Generalized quantifiers and computation*. Lecture Notes in Computer Science, vol. 1754. Berlin: Springer-Verlag. Cited on page **343**.

Väänänen, J. 2007. *Dependence Logic*. London Mathematical Society Student Texts, vol. 70. Cambridge: Cambridge University Press. Cited on pages **205** and **208**.

Väänänen, J. 2008. How complicated can structures be? *Nieuw Archief voor Wiskunde*, **June**, 117–121. Cited on page **277**.

Väänänen, J., and Veličković, B. 2004. Games played on partial isomorphisms. *Archive für mathematische Logic*, **43**(1), 19–30. Cited on page **248**.

Väänänen, J., and Westerståhl, D. 2002. On the expressive power of monotone natural language quantifiers over finite models. *Journal of Philosophical Logic*, **31**(4), 327–358. Cited on page **343**.

Väisänen, P. 2003. Almost free groups and long Ehrenfeucht–Fraïssé games. *Annals of Pure and Applied Logic*, **123**(1–3), 101–134. Cited on page **277**.

van Benthem, J. 1984. Questions about quantifiers. *Journal of Symbolic Logic*, **49**(2), 443–466. Cited on page **343**.

Vaught, R. 1964. The completeness of logic with the added quantifier "there are uncountably many". *Fundamenta Mathematicae*, **54**, 303–304. Cited on page **343**.

Vaught, R. 1973. Descriptive set theory in $L_{\omega_1\omega}$. Pages 574–598. Lecture Notes in Math., Vol. 337 of: *Cambridge Summer School in Mathematical Logic (Cambridge, England, 1971)*. Berlin: Springer. Cited on pages **201**, **209**, **213**, and **277**.

Vaught, R. 1974. Model theory before 1945. Pages 153–172 of: *Proceedings of the Tarski Symposium (Proceedings of Symposia in Pure Mathematics, Vol. XXV, Univ. California, Berkeley, Calif., 1971)*. Providence, R.I.: Amer. Math. Soc. Cited on page **125**.

von Neumann, J., and Morgenstern, O. 1944. *Theory of Games and Economic Behavior*. Princeton, New Jersey: Princeton University Press. Cited on page **28**.

Walkoe, Jr., Wilbur John. 1970. Finite partially-ordered quantification. *Journal of Symbolic Logic*, **35**, 535–555. Cited on page **207**.

Index